BIOENGINEERING OF THE SKIN

Water and the Stratum Corneum

SECOND EDITION

DERMATOLOGY: CLINICAL & BASIC SCIENCE SERIES

Series Editor Howard I. Maibach, M.D.

Published Titles:

Protective Gloves for Occupational Use
Gunh Mellström, J.E. Walhberg, and Howard I. Maibach

Bioengineering of the Skin: Water and the Stratum Corneum, Second Edition
Peter Elsner, Enzo Berardesca, and Howard I. Maibach

Bioengineering of the Skin: Cutaneous Blood Flow and Erythema
Enzo Berardesca, Peter Elsner, and Howard I. Maibach

Bioengineering of the Skin: Methods and Instrumentation
Enzo Berardesca, Peter Elsner, Klaus P. Wilhelm, and Howard I. Maibach

Bioengineering of the Skin: Skin Surface, Imaging, and Analysis
Klaus P. Wilhelm, Peter Elsner, Enzo Berardesca, and Howard I. Maibach

Bioengineering of the Skin: Skin Biomechanics
Peter Elsner, Enzo Berardesca, Klaus-P. Wilhelm, and Howard I. Maibach

Skin Cancer: Mechanisms and Human Relevance
Hasan Mukhtar

Dermatologic Research Techniques
Howard I. Maibach

The Irritant Contact Dermatitis Syndrome
Pieter van der Valk, Pieter Coenrads, and Howard I. Maibach

Human Papillomavirus Infections in Dermatovenereology
Gerd Gross and Geo von Krogh

Contact Urticaria Syndrome
Smita Amin, Arto Lahti, and Howard I. Maibach

Skin Reactions to Drugs
Kirsti Kauppinen, Kristiina Alanko, Matti Hannuksela, and Howard I. Maibach

Dry Skin and Moisturizers: Chemistry and Function
Marie Lodén and Howard I. Maibach

Dermatologic Botany
Javier Avalos and Howard I. Maibach

Hand Eczema, Second Edition
Torkil Menné and Howard I. Maibach

Pesticide Dermatoses
Homero Penagos, Michael O'Malley, and Howard I. Maibach

Nickel and the Skin: Absorption, Immunology, Epidemiology, and Metallurgy
Jurij J. Hostýneck and Howard I. Maibach

The Epidermis in Wound Healing
David T. Rovee and Howard I. Maibach

DERMATOLOGY: CLINICAL & BASIC SCIENCE SERIES

BIOENGINEERING OF THE SKIN

Water and the Stratum Corneum

SECOND EDITION

Edited by
Joachim Fluhr
Peter Elsner
Enzo Berardesca
Howard I. Maibach

informa
healthcare

New York London

Informa Healthcare USA, Inc.
52 Vanderbilt Avenue
New York, NY 10017

10 9 8 7 6 5 4 3 2 1

International Standard Book Number-10: 0-8493-1443-7 (Hardcover)
International Standard Book Number-13: 978-0-8493-1443-8 (Hardcover)

Library of Congress Cataloging-in-Publication Data

Bioengineering of the skin : water and stratum corneum/ edited by Joachim Fluhr...[et al.]-- 2nd ed.
p. ; cm. -- (Dermatology)
Icludes bibliographical references and index. ISBN-10: 0-8493-1443-7 (alk. paper)
1. Skin--Measurement. 2. Bioengineering. 3. Skin absorption. 4. Dehydration (Physiology) I. Fluhr, Joachim. II. Dermatology (CRC Press)
[DNLM: 1. Skin--metabolism. 2. Biomedical Engineering--methods. 3. Body Water 4. Skin Diseases--diagnosis. 5. Water Loss, Insensible--physiology. WR 102 B6155 2004]

QP88.5.B557 2004
617'.05--dc22 2004045727

Visit the Informa Web site at
www.informa.com

and the Informa Healthcare Web site at
www.informahealthcare.com

Series Preface

Our goal in creating the Dermatology: Clinical and Basic Science Series is to present the insights of experts on emerging applied and experimental techniques and theoretical concepts that are, or will be, at the vanguard of dermatology. These books cover new and exciting multidisciplinary areas of cutaneous research, and we want them to be the books every physician will use to become acquainted with new methodologies in skin research. These books can be also given to graduate students and postdoctoral fellows when they are looking for guidance to start a new line of research.

The series consists of books that are edited by experts, with chapters written by the leaders in each particular field. The books are richly illustrated and contain comprehensive bibliographies. Each chapter provides substantial background material relevant to its subject. These books contain detailed tricks of the trade and information regarding where the methods presented can be safely applied. In addition, information on where to buy equipment and helpful Web sites for solving both practical and theoretical problems are included.

We are working with these goals in mind. As the books become available, the efforts of the publisher, book editors, and individual authors will contribute to the further development of dermatology research and clinical practice. The extent to which we achieve this goal will be determined by the utility of these books.

Howard I. Maibach, M.D.

Preface

The stratum corneum is the interface between the sometimes harsh environment and the internal milieu of the human body. Our understanding of the stratum corneum has grown over the last two decades from seeing it as an inert surrounding sheet to understanding it as a biological active compartment. The second edition of *Bioengineering of the Skin: Water and the Stratum Corneum* reflects ten years of progress not only of the noninvasive biophysical assessment of skin physiology parameter but also of the perfection of the available devices. Several new authors have added innovative aspects to this edition. The revised chapters reflect the progress in their respective fields. Bioengineering of the skin, or more precisely the biophysical assessment of skin physiology, is moving rapidly from a descriptive approach to a deeper understanding of biophysical and biochemical processes in the stratum corneum (e.g., on stratum corneum barrier function as well on stratum corneum hydration). The current research utilizing bioengineering methods offers reliable and reproducible approaches for product testing in the pharmaceutic and cosmetic industries and in basic research. Functional data can be acquired with low follow-up costs.

We are thankful to all the authors of both new and revised chapters, and we are grateful to Yvonne Herzberg and Erika Dery for assistance in the editorial process. We hope that this new edition of a successful book will enhance communication within the research community and will be useful for scientists starting in the skin bioengineering field.

Joachim W. Fluhr, Peter Elsner, Enzo Berardesca, Howard I. Maibach
Jena, Rome, San Francisco

Biographies

Joachim W. Fluhr, M.D., is senior dermatologist and head of the Skin Physiology Laboratory at the Friedrich-Schiller-University of Jena, Germany. Dr. Fluhr received his medical training at the University of Mainz, Germany, and the University of Strasbourg, France. He earned his M.D. degree in 1992. He served as a resident at the Department of Dermatology at the Municipal Hospital Karlsruhe. He was a visiting researcher at the Department of Dermatology, IRCCS Policlinico S. Matteo, Pavia. Dr. Fluhr did a postdoctoral research fellowship between 2000 and 2002 at the University of California, San Francisco with Peter Elias and Ken Feingold.

Dr. Fluhr is member of the board of the International Society of the Skin (ISBS) and is secretary and cofounder of the International Society of Skin Pharmacology and Physiology (ISP). He has published more than 60 scientific papers and book chapters. His current research interests are the regulatory functions of the stratum corneum pH, models in irritant contact dermatitis, and technical comparison of bioengineering instruments.

Peter Elsner, M.D., studied medicine at Julius-Maximilians-University, Wuerzburg, Germany, from 1974 to 1981, and was trained as a dermatologist and allergologist at the Department of Dermatology, Wuerzburg University, from 1983 to 1987. He received his doctoral degree in 1981 and his lectureship in dermatology in 1987. From 1988 to 1989, he was visiting research dermatologist at the Department of Dermatology, University of California, San Francisco, and from 1991 to 1997, he was consultant and associate professor, Department of Dermatology, University of Zurich, Switzerland. Since 1997, he has been professor and chairman, Department of Dermatology and Allergology, Friedrich-Schiller-University of Jena, Germany. Dr. Elsner has published more than 200 original papers and 20 books. He is a member of more than 30 scientific societies and has served as chairman of the International Society for Bioengineering and the Skin and as a member of the Scientific Committee for Cosmetics and Non-Food Products (SCCNFP) of the European Commission and the European Group on Efficacy Measurement of Cosmetics and Other Topical Products (EEMCO).

Enzo Berardesca, M.D., is director of clinical dermatology at the San Gallicano Dermatological Institute, Rome, Italy. Dr. Berardesca obtained his training at the University of Pavia and received his M.D. degree in 1979. He served as resident and dermatologist in the Department of Dermatology, IRCCS Policlinico S. Matteo, Pavia, from 1982 to 1987, and as assistant researcher in the Department of Dermatology, University of California School of Medicine in San Francisco in 1987. From 1988 to 2001 he was at the Department of Dermatology of the University of Pavia, acting as head of the Dermatoallergology Unit and of the Skin Bioengineering Lab.

Dr. Berardesca was chairman of the International Society for Bioengineering and the Skin from 1990 to 1996. He has organized several international meetings in Europe on skin bioengineering and irritant contact dermatitis. He is a member of the editorial board of *Skin Pharmacology, Skin Research and Technology, The American Journal of Clinical Dermatology,* and the *Journal of Cutaneous and Ocular Toxicology.* He is a member of the Society for Investigative Dermatology, the European Society for Dermatological Research, and the Italian Group for Research on Contact Dermatitis (GIRDCA), and is vice-chairman of the European Group for Efficacy Measurements on Cosmetics and Other Topical Products (EEMCO) group. His current major research interests are irritant dermatitis, barrier function, and noninvasive techniques to investigate skin physiology, with particular regard to racial differences in skin function, sensitive skin, and efficacy evaluation of topical products. He is the author of 5 books and more than 200 papers and book chapters.

Howard Maibach, M.D., is a professor of dermatology at the University of California, San Francisco and has been a long-term contributor to experimental research in dermatopharmacology and to clinical research on contact dermatitis, contact urticaria, and other skin conditions.

Contributors

Lora Bankova
Department of Dermatology
Friedrich-Schiller-University
Jena, Germany

Enzo Berardesca
San Gallicano Dermatological Institute
Rome, Italy

Marianne Brandt
proDERM GmbH
Schenefeld/Hamburg, Germany

K. De Paepe
Department of Toxicology, Dermato-Cosmetology and Pharmacognosy
Vrije Universiteit Brussel
Brussels, Belgium

Chris Edwards
Department of Dermatology
Royal Gwent Hospital
Newport, United Kingdom

Peter Elias
Dermatology Service
Veterans Affairs Medical Center
San Francisco, California

Peter Elsner
Department of Dermatology
Friedrich-Schiller-University
Jena, Germany

Manigé Fartasch
Department of Dermatology
University of Erlangen/Nürnberg
Erlangen, Germany

Joachim W. Fluhr
Department of Dermatology
Friedrich-Schiller-University
Jena, Germany

Silke M. Fuchs
Department of Dermatology
Friedrich-Schiller-University
Jena, Germany

Bernard Gabard
Iderma AG/SA
Muenchenstein, Switzerland

Francesca Giusti
Department of Dermatology
University of Modena and Reggio Emilia
Modena, Italy

Gary L. Grove
cyberDERM, inc.
Media, Pennsylvania

Steven B. Hoath
The Skin Sciences Institute
Cincinnati Children's Hospital Medical Center
University of Cincinnati College of Medicine
Cincinnati, Ohio

Evi Houben
Department of Toxicology, Dermato-Cosmetology and Pharmacognosy
Vrije Universiteit Brussel
Brussels, Belgium

Gabriel Khazaka
Courage + Khazaka electronic GmbH
Köln, Germany

M. Kucharekova
Department of Dermatology
University Medical Centre
Nijmegen, The Netherlands

Jean-Luc Lévêque
L'Oreal Research
Clichy, France

Magnus Lindberg
Department of Medicine, Occupational and Environmental Dermatology
Karolinska Institute
Stockholm, Sweden

Marie Lodén
ACO Hud AB
Upplands Väsby, Sweden

Howard I. Maibach
Department of Dermatology
University of California
School of Medicine
San Francisco, California

Ronald Marks
Department of Dermatology
University of Wales College of Medicine
Cardiff, United Kingdom

Bozena B. Michniak
University of Medicine and Dentistry of New Jersey
New Jersey Medical School
Newark, New Jersey

Ingrid Nicander
Karolinska Institutet
Division of Medical Engineering
Novum Huddinge, Sweden

Stig Ollmar
Karolinska Institutet
Division of Medical Engineering
Novum Huddinge, Sweden

Giovanni Pellacani
Department of Dermatology
University of Modena and Reggio Emilia
Modena, Italy

Jana Praessler
Department of Dermatology
Friedrich-Schiller-University
Jena, Germany

Grazia Primavera
San Gallicano Dermatological Institute
Rome, Italy

Vera Rogiers
Department of Toxicology, Dermato-Cosmetology and Pharmacognosy
Vrije Universiteit Brussel
Brussels, Belgium

Stefania Seidenari
Department of Dermatology
University of Modena and Reggio Emilia
Modena, Italy

Jørgen Serup
Department of Dermatology
Bispebjerg Hospital
Copenhagen, Denmark

Robert Sinaiko
Department of Dermatology
University of California
School of Medicine
San Francisco, California

R.A. Tupker
Department of Dermatology
St. Antonius Ziekenhuis
Nieuwegein, The Netherlands

P.G.M. van der Valk
Department of Dermatology
University Medical Centre
Nijmegen, The Netherlands

Clarissa D. Villarama
University of California San Francisco
San Francisco, California

Marty O. Visscher
The Skin Sciences Institute
Cincinnati Children's Hospital Medical Center
University of Cincinnati College of Medicine
Cincinnati, Ohio

Philip W. Wertz
Dows Institute
University of Iowa
Iowa City, Iowa

Randy Wickett
University of Cincinnati
College of Pharmacy
Cincinnati, Ohio

Klaus-P. Wilhelm
proDERM GmbH
Schenefeld/Hamburg, Germany

Gil Yosipovitch
Department of Dermatology
Wake Forest University Medical School
Winston-Salem, North Carolina

Charles Zerweck
KGL Inc. Skin Study Center
Broomall, Pennsylvania

Hongbo Zhai
Department of Dermatology
University of California
School of Medicine
San Francisco, California

Table of Contents

PART I Introduction

Chapter 1
Water–Lipid Interaction 3
Bozena B. Michniak and Philip W. Wertz

Chapter 2
Water–Keratin Interactions 15
Jean-Luc Lévêque

Chapter 3
Stratum Corneum pH and Ions — Distribution and Importance 27
Joachim W. Fluhr, Lora Bankova, and Peter Elias

Chapter 4
Occlusion 49
Jana Praessler and Joachim W. Fluhr

PART II Transepidermal Water Loss and Barrier Function

Chapter 5
Transepidermal Water Loss Measurements in Dermato-Cosmetic Sciences 63
V. Rogiers, E. Houben, and K. De Paepe

Chapter 6
Ultrasound and Water in the Stratum Corneum 77
Stefania Seidenari and Giovanni Pellacani

Chapter 7
Standardization of Measurements and Guidelines 83
Grazia Primavera, Joachim W. Fluhr, and Enzo Berardesca

Chapter 8
Transepidermal Water Loss and Its Relation to Barrier Function and Skin Irritation 97
P.G.M. van der Valk, M. Kucharekova, and R.A. Tupker

Chapter 9
Transepidermal Water Loss and Allergic Contact Dermatitis 105
Stefania Seidenari and Francesca Giusti

Chapter 10
Prediction of Irritancy 113
R.A. Tupker

Chapter 11
Transepidermal Water Loss and Racial Differences 129
Grazia Primavera and Enzo Berardesca

Chapter 12
Sensitive Skin and Transepidermal Water Loss 135
Clarissa D. Villarama and Howard I. Maibach

Chapter 13
Transepidermal Water Loss and Barrier Function of Aging Human Skin 143
Klaus-P. Wilhelm, Marianne Brandt, and Howard I. Maibach

Chapter 14
Atopic Dermatitis and Other Skin Diseases 159
Manigé Fartasch

Chapter 15
Transepidermal Water Loss and Dry Skin 171
Marie Lodén

Chapter 16
Bioengineering Correlates of the Sensitive Skin Syndrome: The Sensory Irritation Component 187
Robert Sinaiko and Howard I. Maibach

Chapter 17
Barrier Recovery 197
Silke M. Fuchs, Joachim W. Fluhr, and Howard I. Maibach

Chapter 18
Testing the Efficacy of Moisturizers 211
Bernard Gabard

Chapter 19
Dynamic Measurements: The Plastic Occlusion Stress Test, Moisture Accumulation Test, and Sorption–Desorption Test 237
Grazia Primavera and Enzo Berardesca

PART III Water Content

Chapter 20
Assessment of Stratum Corneum Hydration: Corneometer CM 825 249
Gabriel Khazaka

Chapter 21
Hardware and Measuring Principles: The NOVA Dermal Phase Meter 263
Randy Wickett

Chapter 22
Hardware and Measuring Principles: The Computerized DermaLab Transepidermal Water Loss Probe 275
Gary L. Grove and Charles Zerweck

Chapter 23
Electrical Assessment of Skin Hydration: Standardization of Measurements and Guidelines 287
Grazia Primavera, Joachim W. Fluhr, and Enzo Berardesca

Chapter 24
Hydration and Moisturizers 295
Marie Lodén

Chapter 25
Infant Skin Hydration 307
Marty O. Visscher and Steven B. Hoath

Chapter 26
Hydration and Atopic Dermatitis 323
Chris Edwards and Ronald Marks

Chapter 27
Within and beyond the Skin Barrier as Seen by Electrical Impedance 335
Stig Ollmar and Ingrid Nicander

Chapter 28
Hydration in Psoriasis and Eczema: The Dry Surface–High Evaporative
Water Loss Paradox 351
Jørgen Serup

Chapter 29
Hydration and Lipids 359
Philip W. Wertz and Bozena B. Michniak

Chapter 30
Xerotic Skin and Impairment of Barrier Function Associated with Itch 369
Gil Yosipovitch

Chapter 31
Testing and Efficacy of Barrier Creams 375
Hongbo Zhai and Howard I. Maibach

Chapter 32
Testing of Moisturizers 387
Marie Lodén and Magnus Lindberg

Index 407

Part I

Introduction

1 Water–Lipid Interaction

Bozena B. Michniak and Philip W. Wertz

CONTENTS

I. Introduction 3
II. Lipid Classes Found in the Epidermis 4
A. Fatty Acids 4
B. Phospholipids and Cholesterol 4
C. Glycolipids 5
D. Ceramides 6
E. Cholesterol Sulfate 6
F. Cholesteryl Esters 7
III. Stratum Corneum Lipids 7
A. Chemistry 7
B. Phase Behavior 8
IV. Stratum Corneum Lipid Liposomes 10
A. Experimental Applications 10
B. Commercial Applications 10
References 11

I. INTRODUCTION

The interactions of lipids and water are fundamental to all living organisms [1]. Water is ubiquitous and is often taken for granted, but its properties are unique. The water molecule is an electrical dipole, as the hydrogen atoms have a partial positive charge and the oxygen has a partial negative charge. This results in extensive hydrogen bonding among water molecules as well as potentially between water molecules and solutes. Each water molecule is able to hydrogen bond to four neighbors. Water hydrogen bonds with the hydrogens attached to oxygen or nitrogen molecules in the polar head groups of lipids; similarly, oxygens and nitrogens from the polar head groups can hydrogen bond with hydrogens from water.

Nonpolar lipids, defined as lipids that do not form a lipid–water phase, include triglycerides, cholesterol esters, wax esters, and so forth [2]. These lipids can be suspended in water, forming emulsions. In some cases, for example, fatty alcohols, a surface monolayer can be formed. In the case of a fatty alcohol, the hydroxyl group hydrogen bonds with water at the surface, and the aliphatic chain sticks into the atmosphere above the water. These lipids will not be discussed further in this chapter.

0-8493-1443-7/05/$0.00+$1.50

Polar lipids, which are capable of forming lamellar or hexagonal phases in the presence of excess water [2] will be the main topic of this chapter, with an emphasis on lipids found in epidermis [3,4]. This chapter will include some discussion of fatty acids, phospholipids and phospholipid–cholesterol systems, roles of glucosylceramides, and the ceramide–cholesterol–fatty acid mixture of the stratum corneum. Some comments on cholesterol sulfate and synthetic surfactants will be included.

II. LIPID CLASSES FOUND IN THE EPIDERMIS

A. Fatty Acids

Free fatty acids and their salts generally have very low solubility in water but, depending on conditions, are capable of forming a stable monolayer on top of an aqueous subphase [5], micelles [6], bilayered vesicles [7], or multilamellar tubules — often referred to as myelin figures [8]. When a fatty acid salt, or soap, is dispersed in water, a few molecules go into solution as individual molecules. Once a certain concentration of dissolved monomer is reached — the critical micellar concentration — micelles form. Micelles are globular aggregates of fatty acid molecules with the hydrocarbon chains on the inside and the carboxylate groups on the outside. This arrangement maximizes hydrogen bonding between the carboxylate group and water as well as avoids unfavorable interaction between the aliphatic chains and water. Micelles composed of fatty acid salts are, in part, stabilized by charge repulsion among the carboxylate groups. This is also true of some synthetic surface active agents; for example, sodium dodecyl sulfate. Micelles are capable of incorporating nonpolar lipids or hydrocarbons into their interiors. This is how soap solutions remove grease or oils from surfaces.

Free fatty acid–fatty acid salt mixtures are capable of forming bilayered vesicles when the fatty acid and its salt are present in approximately a 1:1 molar ratio. Carboxylic acids in general have a pKa of about 4; however, when the acids are incorporated into bilayers, they behave like much weaker acids, with apparent pKa values of about 7. Regardless of the mixture of fatty acids and fatty acid salts introduced into an aqueous medium, the degree of ionization will be determined by the pH of the system. The pH at which the micelle-to-lamellar phase transition occurs depends somewhat on the chain length of the fatty acid.

B. Phospholipids and Cholesterol

Most biological membranes, including those in the viable epidermis, are composed mainly of phospholipids and sterols, with cholesterol being the dominant sterol in the animal kingdom [9]. The most common phospholipids include sphingomyelin, phosphatidylcholine, and phosphatidylethanolamine. Phosphatidylserine and phosphatidylinositol are generally present in lower proportions. The phosphoglycerides generally contain a saturated fatty acid at the *sn*-1 position and an unsaturated fatty acid at the *sn*-2 position [10]. Because of the different polar head groups and fatty acid combinations that can be present in phosphoglycerides, there can be many possible molecular species. This diversity of structure probably reflects the numerous functional roles of membrane lipids. Sphingomyelin, in addition to being a phospholipid, is also a sphingolipid. In epidermis, sphingomyelin is one of the precursors of the ceramides of the stratum corneum. The phospholipids and cholesterol are not randomly distributed throughout the cellular membrane system. Each membranous organelle has its own characteristic lipid composition [11,12]. For example, nuclear membranes tend to be rich in sphingomyelin and cholesterol, whereas mitochondria contain the relatively unusual four-chain phospholipid, cardiolipin. Furthermore, in biological membranes there is generally an asymmetric distribution of lipids between the leaflets of the bilayer [13,14]. For example, in the plasma membrane of viable mammalian cells, phosphatidylserine and phosphatidylethanolamine are preferentially localized in the inner, or cytoplasmic, leaflet. This distribution is actively maintained through the action of lipid translocater proteins. During apoptosis, lipid translocation is disrupted and a scramblase facilitates a loss of plasma membrane lipid asymmetry [14].

When phospholipids with or without cholesterol are dispersed in an aqueous medium, vesicles spontaneously form [15]. Depending on the composition and method of preparation, vesicles prepared from phospholipids with or without cholesterol may be small (<50 nm diameter) unilamellar vesicles, large (ca. 50 to 2000 nm diameter) unilamellar vesicles, multilamellar vesicles, or plurilamellar vesicles. Multilamellar vesicles consist of concentric spheres of bilayers separated by thin layers of water. Pleurilamellar liposomes consist of one or more small unilamellar vesicles inside a large unilamellar vesicle. Various types of phospholipids–cholesterol liposomes have been used extensively to model biological membranes.

C. Glycolipids

Glycolipids consisting of monohexosylceramides, lactosylceramide, gangliosides, and sulfatides are present in minor amounts in essentially all viable mammalian cells. Galactosylceramides, or cerebrosides, are abundant lipids in tissue of the nervous system, and glucosylceramides are present in relatively high levels in the viable portion of the epidermis [3]. The glycolipids serve in a range of specific recognition functions. Liposomes containing glycolipids have been used to study such processes as cell adhesion [16] and signal transduction [17], and they may be useful for targeting liposomes to a specific tissue location [18-20].

The epidermis contains several structurally unusual acylglucosylceramides, the most abundant and most studied of which consists of 30- through 34-carbon ω-hydroxyacids amide-linked to sphingosine with a glucose moiety β-glycosidically attached to the primary hydroxyl group of the base and linoleate ester-linked to the ω-hydroxyl group [21,22]. Acylglucosylceramide, and an acylceramide produced by deglycosylation (see below), appear to be unique to keratinizing epithelia and can be considered differentiation markers [23-25]. It has been proposed that acylglucosylceramide is involved in assembly of the contents of lamellar granules [26]. Specifically, it has been suggested that acylglucosylceramide induces both the flattening and aggregation of lipid vesicles to produce the stacks of lamellae seen within the granules. With synthetic vesicles, acylglucosylceramide has been shown to cause flattening and aggregation. As noted, acylglucosylceramide is found in keratinizing epithelia, all of which produce lamellar granules, and is not found in other tissues that do not synthesize lamellar granules [25,27]. In a comparison of the epidermal lipids from different orders of vertebrates, it can be shown that mammals, birds, and reptiles produce lamellar granules and contain acylglucosylceramides [28,29]. Amphibians and fish neither produce lamellar granules nor contain acylglucosylceramides.

In addition to the possible role of lamellar granule–associated acylglucosylceramide in the assembly of lamellar granule contents, it has been suggested that about two-thirds of the acylglucosylceramide is associated with the bounding membrane of the granule [30]. This pool of acylglucosylceramide would be introduced to the cell periphery when the bounding membrane of the lamellar granule fuses into the cell plasma membrane. It has been suggested that the glucose and linoleate are removed at this time and that the resulting ω-hydroxyceramide becomes ester-linked to acidic amino acid side chains on the outer surface of the cornified envelope [30,31].

D. Ceramides

Ceramides are present in many tissues as minor membrane components, but they are major components of the intercellular membranes in the stratum corneum [3,32]. The basic building blocks for human ceramides include sphingosines and dihydrosphingosines, phytosphingosines, and 6-hydroxysphingosines as the basic component, in addition to normal fatty acids, α-hydroxyacids, and ω-hydroxyacids. Ceramide production in the plasma membrane through the enzymatic hydrolysis of sphingomyelin has been implicated as the initiating biochemical event leading to apoptosis in a variety of cell types [33-35].

An acylceramide is derived from acylglucosylceramide associated with the lamellar granule contents after extrusion into the intercellular space. For human epidermis, variants of the acylceramide have been reported that contain sphingosine [32], 6-hydroxysphingosine [36], or phytosphingosine [37] as the base component. On the basis of transmission electron microscopic images [38] and x-ray diffraction data [39,40] it appears that acylceramide plays a critical role in organization of the lipid in the intercellular spaces of the stratum corneum [41,42]. Several similar models have been proposed to account for this unique role [43-45]. This is discussed at greater length in the section dealing specifically with stratum corneum lipids.

E. Cholesterol Sulfate

Although cholesterol sulfate is a minor stratum corneum lipid component, it may have an influence on lipid organization [46], and it has been implicated in the regulation of the desquamation process. It has been demonstrated that hydrolysis of cholesterol sulfate is associated with cell shedding both from human skin *in vivo* [47] and from the surface of a mouse skin organ culture model [48]. Conversely, in the hereditary disease recessive X-linked ichthyosis, cholesterol sulfate accumulates to elevated levels in the stratum corneum, and desquamation does not proceed normally [49]. As a result, the skin surface can become scaly. Cholesterol sulfate has been shown to inhibit serine proteases that may be involved in degradation of desmosomal proteins [50]. This is probably the basis for its role in regulation of desquamation.

F. Cholesteryl Esters

The cholesteryl ester fraction from stratum corneum consists almost entirely of cholesteryl oleate [51]. It has been proposed that oleate from the *sn*-2 position of phosphoglycerides is transferred to cholesterol in the final stages of the keratinization process. Cholesterol esters are not membrane lipids. Cholesteryl oleate is a liquid at skin temperature, and it has been suggested that it phase separates into isolated liquid phase droplets within the stratum corneum. This could provide a mechanism for its exclusion from the lipid lamellae. Because oleic acid is a known permeability enhancer [52], keeping it out of the lipid lamellae would protect barrier function.

III. STRATUM CORNEUM LIPIDS

A. Chemistry

The lipids of the stratum corneum consist mainly of ceramides, cholesterol, and fatty acids, with small proportions of cholesterol sulfate and cholesterol esters [4].

The free fatty acids found in epidermis range from 16 to 28 carbons in length and are predominantly saturated [51]. In the stratum corneum, these fatty acids represent about 10% of the lipid mass, and the 22- and 24-carbon entities are the most abundant. With the exception of a small proportion of cholesterol sulfate, the fatty acids are the only ionizable lipids in either human or porcine stratum corneum. It has been suggested that ionized fatty acids may be important for the formation of lamellar phases in this compartment [30].

Structures of the ceramides identified in human stratum corneum are summarized in Figure 1.1. Every possible acid–base combination is represented, including three linoleate-containing acylceramides, shown at the top.

As noted above, the acylceramides derived from acylglucosylceramides are structurally highly unusual. The ω-hydroxyacids are 30 to 34 carbons in length, which is long enough to completely span a typical bilayer. Furthermore, the ester-linked linoleate on the ω-hydroxyl group makes it possible to insert this portion of the molecule into an adjacent bilayer, thus riveting the two bilayers together at a molecular level. Transmission electron microscopy using ruthenium tetroxide post-fixation [38] and small-angle x-ray diffraction [39,40] both reveal a predominant 13-nm lamellar spacing, which is thought to be a trilamellar lipid structure. In experiments in which lamellae were reconstituted from extracted stratum corneum lipids, the 13-nm periodicity could be reproduced, clearly demonstrating that this repeat unit does not require any protein. However, if acylceramide was removed from the lipid mixture, then essentially only a 5-nm periodicity was observed [39,40]. It has been proposed that acylceramide molecules in the outer lamellae of the trilaminar units insert their linoleate tails into the central lamella, thus riveting the three lamellae together at a molecular level [44,42].

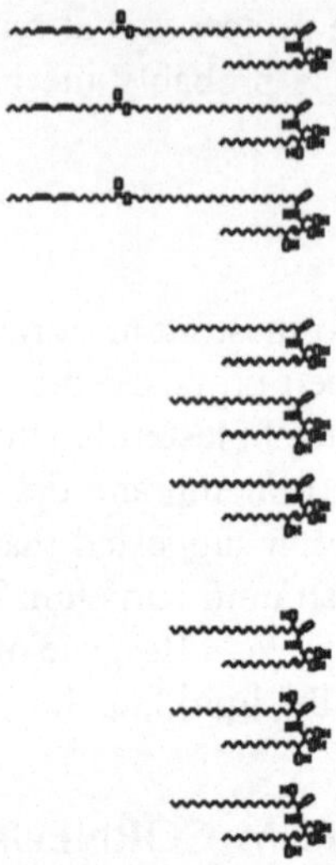

FIGURE 1.1 Representative structures of human epidermal ceramides. The top set of three structures represents the acylceramides containing, from top to bottom, sphingosine, phytosphingosine, and 6-hydroxysphingosine as the long-chain base component. The middle and lower sets of three structures show the ceramides in which the same long-chain bases bear amide-linked normal and α-hydroxyacids, respectively.

All of the ceramides as well as the free fatty acids in stratum corneum are cylindrical in shape, which makes them well suited for the formation of highly ordered membrane domains. This should be advantageous for the barrier function of the skin; however, gel phase or solid membranes could be brittle and incompatible with a pliable skin. The stratum corneum is saturated with cholesterol, and cholesterol crystals have been detected by x-ray diffraction [46]. Cholesterol is capable of making membranes either fluid or stiff, depending on the chemical nature of the membrane components and the relative proportions. It has been suggested that cholesterol in the stratum corneum serves to provide a degree of flexibility to what could otherwise be a rigid and brittle membrane system [30].

B. Phase Behavior

Most biological membranes, including those in the viable epidermis, exist under physiological conditions in a predominantly fluid liquid crystalline phase; however, there may be coexisting membrane rafts in a more ordered state [53]. If cooled, at a certain temperature the liquid crystalline phase can undergo a transition to a more rigid lamellar phase, called a solid or gel phase [2]. The temperature at which this happens is the phase transition temperature and is determined by the chemical properties of the component lipids.

Some membrane lipids, for example, dioleoylphosphatidylethanolamine, are capable of promoting the formation of a nonlamellar inverted hexagonal phase [54]. When incorporated into liposomes, lipids that favor the inverted hexagonal phase tend to be fusogenic, and it is thought that formation of this hexagonal phase is involved in fusion events in biological membrane systems [55]. This could be significant, for example, in the fusion of the lamellar granule-bounding membrane into the cell plasma membrane. Cubic phases would be transient intermediates in going from a lamellar phase to a hexagonal phase and back again to a lamellar phase [2,56]. Stable cubic phase membranes have been identified in a number of biological systems [56].

Obviously, at the phase transition temperature at which a gel phase would melt into a liquid crystalline phase, the two lamellar phases would coexist. However, it has also been recognized that through lateral segregation of lipids within a membrane, gel phase lipid domains can be formed within an otherwise liquid crystalline membrane [57,58]. Sphingolipid- and cholesterol-rich membrane rafts are thought to be involved in trans-membrane signaling, endocytosis, and exocytosis [59]. Some of these sphingolipid- and cholesterol-rich membrane rafts have proteins called caveolins associated with them [60]. These membrane regions form caveolae (little caves), which appear as small invaginations at the plasma membrane. The caveolae are thought to be involved in the trafficking of cholesterol into and out of cells, among other functions.

The classic view of the phase behavior of the intercellular lamellae of the stratum corneum was put forth by Forslind as the domain mosaic model [61]. In this model, it was proposed that islands of gel phase domains are imbedded in a continuous liquid crystalline domain. Molecules diffusing through these lamellae would do so more rapidly through the more fluid liquid crystalline domains compared with the relatively rigid and impermeable gel phase domains. The regions of least diffusional resistance would be the phase boundaries because these would be the regions with the greatest frequency of chain packing defects. There is now ample evidence from different laboratories, using different physical techniques, that has been interpreted as support for the occurrence of gel phase domains in the stratum corneum [52,62,63]. These gel phase domains could support barrier function, whereas the continuous liquid crystalline domain could provide the flexibility necessary for pliable skin.

In lipid reconstitution experiments, it has been shown that cholesterol sulfate is capable of reducing the proportion of cholesterol in the crystalline state and of causing a shift from a predominantly 13-nm lamellar phase toward an increased proportion of a 4.3-nm lamellar period [46]. These findings have implications for the stratum corneum in recessive X-linked ichthyosis.

More recently, a single gel phase model has been proposed for the stratum corneum lipids [64]. However, the details proposed for this alternative model are not supported by the available x-ray diffraction data [65]. It is possible that a single-phase liquid ordered phase [59,66] could satisfy both the functional needs of the barrier and could fit the physical data. A liquid ordered phase is favored by lipid mixtures rich in both sphingolipids and cholesterol, and the stratum corneum lipid composition meets both of these criteria. A liquid ordered phase has some of the properties of a gel phase, including low permeability, but has some properties more closely resembling a liquid crystalline phase. Although the aliphatic chains are not melted, there is more rapid lateral movement of lipid molecules in this type of lamellar phase.

IV. STRATUM CORNEUM LIPID LIPOSOMES

A. Experimental Applications

The first reported attempt at the preparation of liposomes from stratum corneum lipids was that of Gray and White in 1979 [67]. Because of uncertainty about the composition of stratum corneum lipids, the composition of plasma membranes from granular cells was used as an approximation. As a result, a mixture of ceramides, cholesterol, fatty acids, and granular cells was used. Although vesicles were demonstrated by both freeze-fracture electron microscopy and negative staining, the inclusion of glucosylceramides and the relative proportions of the other lipid components were not representative of the stratum corneum. This was an important step toward preparing stratum corneum lipid liposomes and was one of the first reports of liposomes without phospholipids.

In 1986, liposomes were prepared from porcine epidermal ceramides, cholesterol, palmitic acid, and cholesterol sulfate in the proportions found in the stratum corneum [68]. This was achieved by sonication at 80°C, and electron microscopic methods demonstrated the formation of small unilamellar vesicles. It was shown that vesicles could be formed if either fatty acids or cholesterol sulfate were omitted, but that at least one of these ionizable lipids was required for vesicle formation. These vesicles were subsequently used to investigate roles of acylceramides and acylglucosylceramides. It was demonstrated that acylceramide or acylglucosylceramide could induce aggregation and fusion of the vesicles [69], and the fusion process could also be facilitated by calcium ions [70]. It was further demonstrated that the ability of acylceramide and acylglucosylceramides to induce aggregation did not depend on the nature of the ester-linked fatty acid. The possibility of forming a hexagonal phase was also shown [71].

A number of investigations used liposomes to study the interactions of sodium dodecyl sulfate (SDS) with stratum corneum lipids [72-76]. At relatively low concentrations, SDS readily partitioned into the liposomes until it represented about 20% of the mass of the vesicle. SDS increased the liposomal permeability. At still-higher SDS concentrations, the liposomes were destroyed with the formation of mixed micelles.

B. Commercial Applications

Although there has been a great deal of interest in developing phospholipid-based liposomes for both cosmetic [77,78] and medical applications [79,80], there has been relatively little interest in commercial development of stratum corneum lipid liposomes. Stratum corneum lipid–based liposomes, or liposomes closely resembling stratum corneum lipids in composition, could have several advantages over their phospholipid counterparts, especially for topical use. For one thing, liposomes made of ceramides, cholesterol, and long-chain saturated fatty acids resembling the endogenous lipids should be more biocompatible, thereby minimizing risk of adverse skin reactions. A second potential advantage could be in stability to oxidation. The stratum corneum lipids are highly saturated, which confers long-term stability on exposure to air. Finally, the substitution of stratum corneum lipid liposomes for phospholipid-based liposomes for topical use could deny surface bacteria a source of phosphorous that could promote their growth.

Among the reasons that there has not been much effort in developing application for ceramide-based liposomes has been a lack of ceramides on an affordable commercial scale. There are now several commercial sources of ceramides [81], and these materials are now being used in a range of skin moisturizer products and at least one hair conditioner, although not yet in liposomal form.

REFERENCES

1. Attenborough, D., *Life on Earth*, Little Brown, Boston, 1980.
2. Larsson, K., *Lipids — Molecular Organization, Physical Functions and Technical Applications*, Oily Press, Dundee, 1994, chap. 3.
3. Yardley, H.J. and Summerly, R., Lipid composition and metabolism in normal and diseased epidermis, *Pharmacol. Therap.*, 13, 357, 1981.
4. Law, S. et al. Regional variation in content, composition and organization of porcine epithelial barrier lipids revealed by thin-layer chromatography and transmission electron microscopy, *Arch. Oral Biol.*, 40, 1085, 1995.
5. Peng, J.B., Barnes, G.T., and Gentle, I.R., The structures of Langmuir-Blodgett films of fatty acids and their salts, *Adv. Colloid Interface Sci.*, 91, 163, 2001.
6. Lindman, B. and Wennerstrom, H., Micelles. Amphiphile aggregation in aqueous solution, *Topics Curr. Chem.*, 87, 1, 1980.
7. Apel, C.L., Deamer, D.W., and Mautner, M.N., Self-assembled vesicles of monocarboxylic acids and alcohols: conditions for stability and for encapsulation of biopolymers, *Biochim. Biophys. Acta*, 1559, 1, 2002.
8. Blanchette-Mackie, E.J. and Amende, L.M., Electron microscopic visualization of fatty acids in tissues, *J. Electron Micros. Tech.*, 7, 205, 1987
9. Bloch, K., Sterol molecule: structure, biosynthesis and function, *Steroids*, 57, 378, 1992.
10. Yamashita, A., Sugiura, T., and Waku, K., Acyltransferases and transacylases involved in fatty acid remodeling of phospholipids and metabolism of bioactive lipids in mammalian cells, *J. Biochem.*, 122, 1, 1997.
11. Gray, G.M. and Yardley, H.J., Mitochondria and nuclei of pig and human epidermis: isolation and lipid composition, *J. Invest. Dermatol.*, 64, 423, 1975.
12. Dodd, B.J. and Gray, G.M., The lipid composition of rat-liver plasma membranes. *Biochim. Biophys. Acta*, 150, 397, 1968.
13. Frickenhaus, S. and Heinrich, R., Kinetic and thermodynamic aspects of lipid translocation in biological membranes, *Biophys. J.*, 76, 1293, 1999.
14. Bevers, E.M. et al. Lipid translocation across the plasma membrane of mammalian cells, *Biochim. Biophys. Acta*, 1439, 317, 1999.
15. Lasic, D.D. et al. Spontaneous vesiculation, *Adv. Colloid Interface Sci.*, 89-90, 337, 2001.
16. Handa, K., Kojima, N., and Hakomori, S., Analysis of glycolipid-dependent cell adhesion based on carbohydrate-carbohydrate interaction, *Methods Enzymol.*, 312, 447, 2000.
17. Hakomori, A. et al. Signal transduction through glycol(sphingo)lipids. Introduction and recent studies on glycol(sphingo)lipid-enriched microdomains, *Ann. N. Y. Acad. Sci.*, 845, 1, 1998.

18. Lian, T. and Ho, R.J., Cholera toxin B-mediated targeting of lipid vesicles containing ganglioside GM1 to mucosal epithelial cells, *Pharm. Res.*, 14, 1309, 1997.
19. Pagnan, G. et al. GD2-mediated melanoma cell targeting and cytotoxicity of liposome-entrapped fenretinide, *Int. J. Cancer*, 81, 268, 1999
20. Duffels, A. et al. Synthesis of high-mannose type neoglycolipids: active targeting of liposomes to macrophages in gene therapy, *Chemistry-A European Journal*, 6, 1416, 2000.
21. Wertz, P.W. and Downing, D.T. Acylglucosylceramides of pig epidermis: structure determination, *J. Lipid Res.*, 24, 753, 1983.
22. Abraham, W., Wertz, P.W., and Downing, D.T., Linoleate-rich acylglucosylceramides from pig epidermis: structure determination by proton magnetic resonance, *J. Lipid Res.*, 26, 761, 1985.
23. Brod, J. et al. Acylceramides and lanosterol–lipid markers of terminal differentiation in cultured human keratinocytes: modulating effect of retinoic acid, *In Vitro Cell. Dev. Biol.*, 27A, 163, 1991.
24. Madison, K.C. et al. Lipid composition of cultured murine keratinocytes, *J. Invest. Dermatol.*, 87, 253, 1986.
25. Wertz, P.W. et al. Lipids of epidermis and keratinized and non-keratinized oral epithelia, *Comp. Biochem. Physiol.*, 83B, 529, 1986.
26. Wertz, P.W. and Downing, D.T., Glycolipids in mammalian epidermis: structure and function in the water barrier, *Science*, 217, 1261, 1982.
27. Schoephoerster, R.T. et al. A survey of polar and nonpolar lipids of mouse organs, *Comp. Biochem. Physiol.*, 82B, 229, 1985.
28. Burken, R.R. et al. The occurrence of "molecular rivets" in epidermis from different orders of vertebrates, *Clin. Res.*, 34, 741A, 1986.
29. Wertz, P.W., *Biology of the Integument*, Vol. 2, Springer, Berlin, 1986, 815.
30. Wertz, P.W., Lipids and barrier function of the skin, *Acta Derm. Venereol.*, 208, 1, 2000.
31. Swartzendruber, D.C. et al. Evidence that the corneocyte has a chemically bound lipid envelope, *J. Invest. Dermatol.*, 88, 709, 1987.
32. Wertz, P.W. and Downing, D.T., Ceramides of pig epidermis: structure determination, *J. Lipid Res.*, 24, 759, 1983.
33. Monick, M.M. et al. Ceramide regulates lipopolysaccharide-induced phosphatidylinositol 3-kinase and Akt activity in human alveolar macrophages, *J. Immunol.*, 167, 5977, 2001.
34. Marchetti, C. et al. Induction of apoptosis by 1,4-benzothiazine analogs in mouse thymocytes, *J. Pharmacol. Exp. Ther.*, 300, 1053, 2002.
35. Maddens, S. et al. Kit signaling inhibits the sphingomyelin-ceramide pathway through PLC gamma 1: implication in stem cell factor radioprotective effect, *Blood*, 100, 1294, 2002.
36. Stewart, M.E. and Downing, D.T., A new 6-hydroxy-4-sphingenine-containing ceramide in human skin, *J. Lipid Res.*, 40, 1434, 1999.
37. Ponec, M. et al. New acylceramide in native and reconstructed epidermis, *J. Invest. Dermatol.*, 120, 581, 2003.
38. Madison, K.C. et al. Presence of intact intercellular lamellae in the upper layers of the stratum corneum, *J. Invest. Dermatol.*, 88, 714, 1987.
39. White, S.H., Mirejovsky, D., and King, G.I., Structure of lamellar lipid domains and corneocyte envelopes of murine stratum corneum. An X-ray diffraction study, *Biochemistry*, 27, 3725, 1988.

40. Bouwstra, J.A. et al. The structure of human stratum corneum as determined by small angle X-ray scattering, *J. Invest. Dermatol.*, 97, 1005, 1991.
41. McIntosh, T.J., Stewart, M.E., and Downing, D.T., X-ray diffraction analysis of isolated skin lipids: reconstitution of intercellular lipid domains, *Biochemistry*, 35, 3649, 1996.
42. Bouwstra, J.A. et al. Role of ceramide 1 in the molecular organization of the stratum corneum lipids, *J. Lipid Res.*, 39, 186, 1998.
43. Swartzendruber, D.C. et al. Molecular models of the intercellular lipid lamellae in mammalian stratum corneum, *J. Invest. Dermatol.*, 92, 251, 1989.
44. Wertz, P.W., *Formation and Structure of Human Hair*, Birkhauser, Basel, 1996, 227.
45. Bouwstra, J.A. et al. The lipid organization in the skin barrier, *Acta Derm. Venereol.*, 208, 23, 2000.
46. Bouwstra, J.A. et al. pH, cholesterol sulfate and fatty acids affect stratum corneum lipid organization, *J. Invest. Dermatol. Symp. Proc.*, 3, 69, 1989.
47. Long, S.A. et al. Human stratum corneum polar lipids and desquamation, *Arch. Dermatol. Res.*, 277, 284, 1985.
48. Ranasinghe, A.W. et al. Lipid composition of cohesive and desquamated corneocytes from mouse ear skin, *J. Invest. Dermatol.*, 86, 187, 1986.
49. Williams, M.L., A new look at the ichthyoses: disorders of lipid metabolism, *Pediat. Dermatol.*, 3, 476, 1986.
50. Sato, J. et al. Cholesterol sulfate inhibits proteases that are involved in desquamation of stratum corneum, *J. Invest. Dermatol.*, 111, 189, 1998.
51. Wertz, P.W. et al. The composition and morphology of epidermal cyst lipids, *J. Invest. Dermatol.*, 89, 419, 1987.
52. Golden, G.M., McKie, J.E., and Potts, R.O., Role of stratum corneum lipid fluidity in transdermal drug flux, *J. Pharm. Sci.*, 76, 25, 1987.
53. Brown, D.A. and London, E., Functions of lipid rafts in biological membranes, *Annu. Rev. Cell Dev. Biol.*, 14, 111, 1998.
54. Johnson, M. and Edwards, K., Phase behavior and aggregate structure in mixtures of dioleoylphosphatidylethanolamine and poly(ethylene glycol)-lipids, *Biophys. J.*, 80, 313, 2001.
55. Bailey, A.L. and Cullis, P.R., Modulation of membrane fusion by asymmetric transbilayer distribution of amino lipids, *Biochemistry*, 33, 12573, 1994.
56. Norlen, L., Skin barrier formation: the membrane folding model, *J. Invest. Dermatol.*, 117, 823, 2001.
57. Clerc, S.G. and Thompson, T.E., Permeability of dimyristoyl phosphatidyl choline/dipalmitoyl phosphatidylcholine bilayer membranes with coexisting gel and liquid-crystalline phases, *Biophys. J.*, 68, 2333, 1995.
58. Piknova, B., Marsh, D., and Thompson, T.E., Fluorescence-quenching study of percolation and compartmentalization in two-phase lipid bilayers, *Biophys. J.*, 71, 892, 1996.
59. Brown, D.A. and London, E., Structure and function of sphingolipid- and cholesterol-rich membrane rafts, *J. Biol. Chem.*, 275, 17221, 2000.
60. Razani, B. and Lisanti, M.P. Caveolins and caveolae: molecular and functional relationships, *Exp. Cell Res.*, 271, 36, 2001.
61. Forslind, B., A domain mosaic model of the skin barrier, *Acta Derm. Venereol.*, 74, 1, 1994.
62. Krill, S.L., Knutson, K., and Higuchi, W.I., The stratum corneum thermotropic phase behavior, *Biochim. Biophys. Acta*, 1112, 281, 1992.

63. Bouwstra, J.A. et al. Phase behavior of lipid mixtures based on human ceramides: coexistence of crystalline and liquid phases, *J. Lipid Res.*, 42, 1759, 2001.
64. Norlen, L., Skin barrier structure and function: the single gel phase model, *J. Invest. Dermatol.*, 117, 830, 2001.
65. Bouwstra, J.A., Pilgram, G.S., and Ponec, M., Does the single gel phase exist in stratum corneum? *J. Invest. Dermatol.*, 118, 897, 2002.
66. Ge, M. et al. Electron spin resonance characterization of liquid ordered phase of detergent-resistant membranes from RBL-2H3 cells, *Biophys. J.*, 77, 925, 1999.
67. Gray, G.M. and White, R.J., Epidermal lipid liposomes: a novel non-phospholipid membrane system, *Biochem. Soc. Trans.*, 7, 1129, 1979.
68. Wertz, P.W. et al. Preparation of liposomes from stratum corneum lipids, *J. Invest. Dermatol.*, 87, 582, 1986.
69. Abraham, W., Wertz, P.W., and Downing, D.T., Effect of epidermal acylglucosylceramides and acylceramides on the morphology of liposomes prepared from stratum corneum lipids, *Biochim. Biophys. Acta*, 939, 403, 1988.
70. Abraham, W. et al. Stratum corneum lipid liposomes: calcium-induced transformation into lamellar sheets, *J. Invest. Dermatol.*, 88, 212, 1987.
71. Abraham, W., Structures formed by epidermal lipids *in vitro*, *Semin. Dermatol.*, 11, 121, 1992.
72. Downing, D.T. et al. Partition of sodium dodecyl sulfate into stratum corneum lipid liposomes, *Arch. Dermatol. Res.*, 285, 151, 1993.
73. de la Maza, A. et al. Influence of the level of ceramides in the permeability of stratum corneum lipid liposomes caused by sodium dodecyl sulfate, *Chem. Phys. Lipids*, 94, 181, 1998.
74. Lopez, O. et al. Solubilization of liposomes by sodium dodecyl sulfate: new mechanism based on direct formation of mixed micelles, *Arch. Biochem. Biophys.*, 367, 153, 1999.
75. Lopez, O. et al. Influence of ceramides in the solubilization of stratum corneum lipid liposomes by C(12)-betaine/sodium dodecyl sulfate mixtures, *Int. J. Pharmacol.*, 187, 231, 1999.
76. Lopez, O. et al. Different stratum corneum lipid liposomes as models to evaluate the effect of sodium dodecyl sulfate, *Biochim. Biophys. Acta*, 1508, 196, 2000.
77. Meybeck, A., *Liposome Dermatics*, Springer, Berlin, 1992, 341.
78. Lautenschlager, H., *Handbook of Cosmetic Science and Technology*, Marcel Dekker, New York, 1991, 201.
79. Gregoriadis, G., *Liposome Dermatics,* Springer, Berlin, 1992, 346.
80. Maurer, N., Fenske, D.B., and Cullis, P.R., Developments in liposomal drug delivery systems, *Expert Opin. Biol. Ther.*, 1, 923, 2001.
81. Michniak, B.B. and Wertz, P.W., *Handbook of Cosmetic Science and Technology*, Marcel Dekker, New York, 2001, 361.

2 Water–Keratin Interactions

Jean-Luc Lévêque

CONTENTS

I. Introduction 15
II. SC and Hair: Similarities and Differences 17
III. Sorption Isotherms of Keratins and Other Related Techniques 18
IV. Tightly Bound Water 20
V. Bound Water 21
VI. Free Water 21
VII. Influence of Lipids and Hydrosoluble Materials 22
VIII. SC and Hair Swelling 23
IX. Conclusion 24
References 24

I. INTRODUCTION

Keratins are a family of proteins produced by certain epidermal cells in higher vertebrates. Their function is to form an interface between the living medium and the external environment. This function is essential, as it helps to maintain the physical and chemical equilibrium without which life could not exist. As an example, body hair contributes to temperature homeostasis, but it also protects the skin against ultraviolet irradiation. The stratum corneum (SC: the first 10 to 20 μm of the epidermis) controls exchanges between the "inside" and "outside" environments. In the animal kingdom, feathers not only make flight possible but also protect against temperature variations, whereas hooves, nails, and horns are necessary for travel, the search for food, and the maintenance of the species.

These proteins fulfill their role as an interface between the body and the environment thanks to their remarkable physical and chemical properties, which have been the subject of intensive research over the last 50 years. Although all keratins belong to the same family, they show a number of differences in their chemical structure and, above all, in their macromolecular organization. Despite these differences, all keratins contain the amino acid cystine (R-CH2-S-S-CH2-R), responsible for strong covalent bonds not only within a given polypeptide chain but also between chains, which makes these molecules poorly reactive and confers remarkable physical properties.

0-8493-1443-7/05/$0.00+$1.50

Initially, it was the enormous economic value of wool that served as the impetus for research by chemists in large private and university laboratories of the countries concerned (mainly Australia and the United Kingdom), with the aim of modifying this raw material. Later, keratins were used as model proteins in studies by biologists, chemists, biophysicists, and physicists.

As outlined at the beginning of this chapter, keratins have an important interface function, but this can be influenced by environmental conditions that modify their nature and, therefore, their properties. Among these environmental factors, water is one of the most important. Because of their highly intense dipolar nature, water molecules can form bonds with many side-chain endings of keratin. It can also cause internal reorganization of the hydrophobic groups. Water can thereby have a major effect on the physical properties of keratin (Figure 2.1).

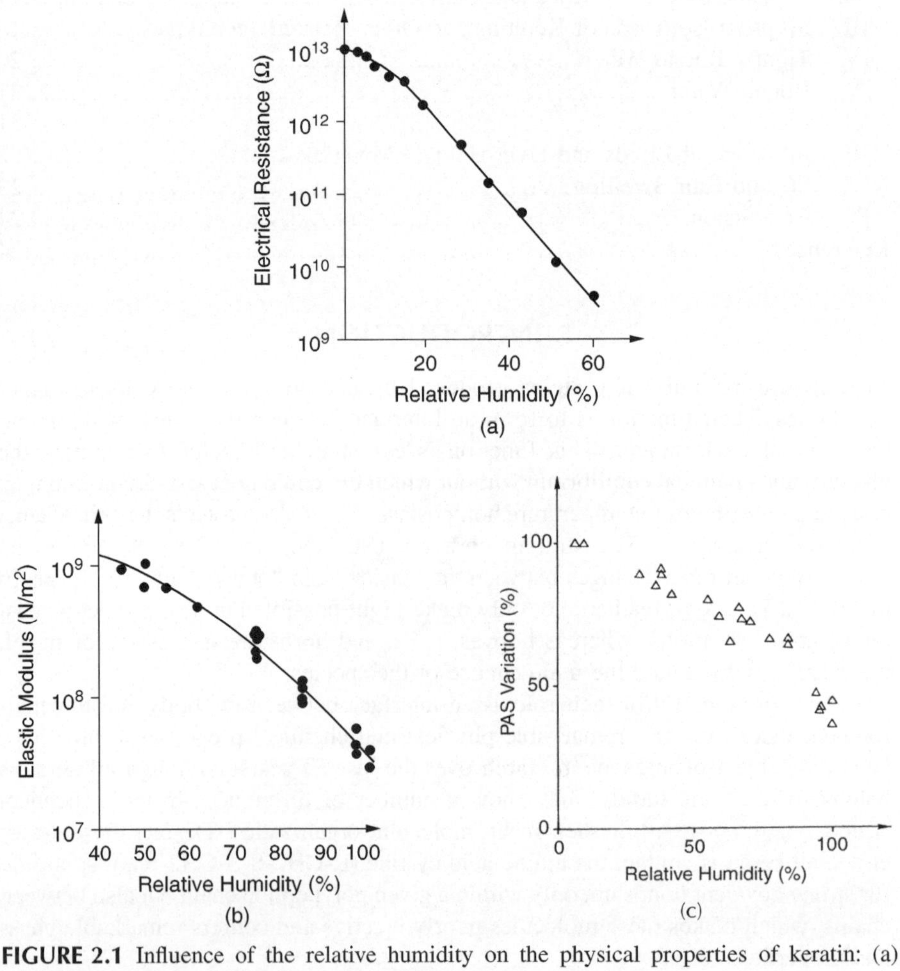

FIGURE 2.1 Influence of the relative humidity on the physical properties of keratin: (a) electrical resistance of natural hair, (b) elastic modulus of SC samples, and (c) photoacoustic properties of SC.

II. SC AND HAIR: SIMILARITIES AND DIFFERENCES

Like the eyelashes and nails (which will not be dealt with here), the SC and hair are part of the large family of keratinized tissues [1]. Both the SC and hair are composed of cells (disk-shaped corneocytes and fusiform cortical cells, respectively) that are connected by an intercellular cement. In hair, the cortical cells are protected by about 10 layers of cuticle cells. These layers play a role in limiting the exchanges between the cortex and the environment. They also limit the water swelling of the structure.

At the microscopic scale, the hair is composed of macrofibrils, which are themselves formed of microfibrils with a diameter of about 80 Å. These microfibrils probably correspond to the tonofilaments in the SC. Protofibrils (or intermediate filaments) with a diameter of about 20 Å have also been described. In hair, these protofibrils contain two α-helices, demonstrated by Pauling in 1951. Questions are now being raised as to the presence of these helices in the SC, which does not produce the X-ray diffraction pattern characteristic of α-keratin. According to Garson et al., the keratin in the SC would be an intermediate between β-keratin and that of feathers [2].

The constituents of these two "tissues" also present similarities and differences. Without going into too much detail, one of the major differences is the fact that lipids represent about 10 to 15% of the mass of the SC but only 2% of hair. The presence of these lipids explains, at least in part, the slow equilibrium kinetics with water.

In SC, lipids form bilayers that, using the transmission electronic microscope, appear to fall under two main types of organization: regular and irregular alternations of clear and dark bands [3]. A chemical model, based on the interdigitation of ceramide molecules, has been proposed to represent this type of organization [4]. X-ray diffraction allows us to measure the reticular distance of these bilayers. They are characterized by two reticular distances at 6.5 and 4.5 nm [2], the second figure being probably the third order of a 13.5-nm reticular distance [5]. It is worth noting that there is no swelling of these bilayers during hydration, as no change in these reticular distances can be detected. This would mean that water molecules are located in the zone between the polar heads of two successive bilayers. Concerning hair, the situation is quite different: Only one diffraction pattern can be distinguished, at 4.2 nm, and the pattern is not always present [6]. From a chemical and physical point of view, there are also important differences.

Using Raman spectroscopy, Gniadecka et al. discovered major differences among SC, hair, and nail in the behavior of the lipids and proteins [7]; a greater fluidity is found in SC. Using differential scanning calorimetry, Golden et al. suggested a correlation between SC lipid fluidity and water content [8]. These studies clearly demonstrate that there is an interaction between lipids and water [9], but a comprehensive model of this interaction still does not exist.

Table 2.1 shows the very large difference in the cystine content of the constituent proteins. The low concentration of cystine in the SC probably explains its low-elasticity module [10]; similarly, the covalent cystine bonds (which are mainly perpendicular to the axis of the hair) probably limit its swelling and, therefore, its

TABLE 2.1
Amino Acid Analyses of Low Sulfur Fractions (residues per 100 residues) According to Fraser

Residue	Hair	Nail	Stratum Corneum
Alanine	6.9	6.5	7.1
Arginine	7.2	6.7	4.5
Aspartic acid	9.3	8.9	9.8
Half cystine	7.6	7.4	2.9
Glutamic acid	16.6	15.2	15.8
Glycine	5.2	6.6	11.6
Histidine	0.7	0.9	1.6
Isoleucine	3.7	3.9	4.6
Leucine	10.2	9.9	9.3
Lysine	3.5	4.5	5.7
Methionine	0.4	0.8	0.5
Phenylalanine	2.0	2.2	3.4
Proline	3.8	4.1	2.9
Serine	9.0	8.7	7.2
Threonine	5.5	5.0	5.4
Tyrosine	2.5	2.8	2.0
Valine	6.1	5.9	5.7

Source: From Fraser, R.B.D., *Keratins*. C. Thomas, New York, 1972.

maximum water uptake. This is illustrated in Figure 2.2, which shows the absorption isotherms for water of the SC and hair. In the zone of moderate humidity, the pattern is similar for the two tissues, probably because the nature of the substrate controls the phenomenon, as we shall see later. Above 80% relative humidity, the open structure of the SC modifies this process. Because of the stratum corneum's strong dilation (the thickness of the SC can increase by more than 100%, though that of the hair is limited to 15%), the structural interaction of the polypeptide chains with the water molecules falls sharply, disappearing above a regain of 30% (see Section VI). Let us first look at the hydration phenomena from an experimental viewpoint and then go into what is known of the different types of water binding to keratin.

III. SORPTION ISOTHERMS OF KERATINS AND OTHER RELATED TECHNIQUES

To investigate the hydration of any substrate, the first step is to determine the sorption isotherm, which reflects the quantity of water that can bind at a given temperature and relative humidity. Such curves are shown for the SC and natural hair in Figure 2.2. As for many other substances, particularly those composed of protein, these curves have a sigmoid aspect and can be divided into three sections.

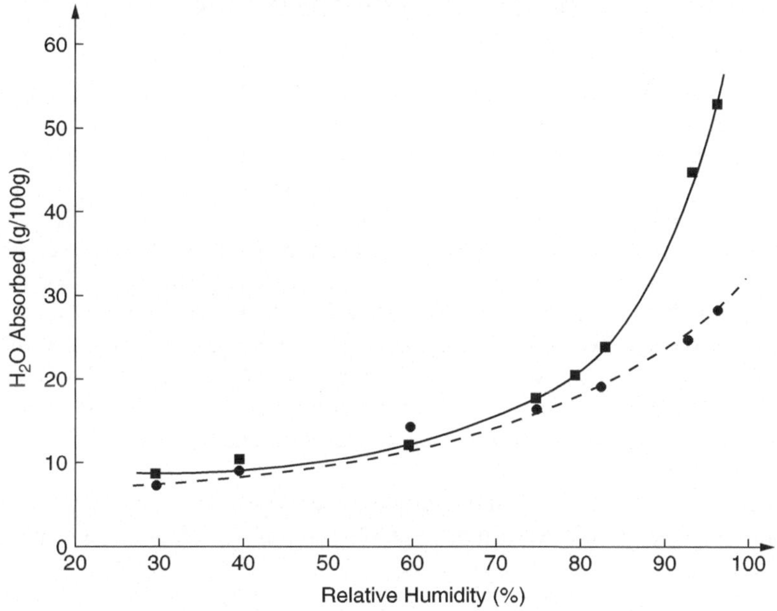

FIGURE 2.2 Sorption isotherms (23°C) for hair (~) and SC (U).

The first section corresponds to water contents between 0 and 9%, at which water uptake is very rapid. This section is the most difficult to study, because it requires deep dehydration of the studied specimen. Beyond about 70% relative humidity, the increase in water content is very rapid — particularly in the case of the SC, which can take up several times its own weight in water, whereas natural hair can only take up about 30% of its dry weight.

As mentioned previously, the two curves diverge widely in this zone of humidity. This type of curve closely reflects the interactions between the chemical groups of a given material and water molecules. The sorption isotherm provides information on the interactions with the various sites in terms of the number of molecules, as well as the condensation of several molecules either on a given site or on a water molecule that has already been bound. The problems relating to the binding energies can only be tackled by means of calorimetric measurements, with or without weight measurements [11].

Finally, spectroscopic methods such as infrared measurements and magnetic resonance spectroscopy deserve a mention, as they can be used to investigate protein–water interactions (see below). X-ray diffraction can also be used to detect the influence of water molecules on the fine structural elements such as microfilaments and protofilaments of the hair or SC [12].

We will now examine the different ways in which water molecules can bind to keratin, using some experimental findings.

IV. TIGHTLY BOUND WATER

When SC or hair specimens have been profoundly dehydrated (moderate heating in a vacuum), water uptake at a relative humidity between 0 and 20% apparently corresponds to the binding of water molecules in a single layer on a number of particular binding sites. Many experimental methods have been used to investigate this phenomenon, including calorimetric measurement [13], infrared absorption [14], mathematical analysis of isotherms for normal or modified samples [15], depolarizing thermocurrents, [16] and electrical conductance [17]. Binding energy with the protein, in the case of human SC, can be seen in Figure 2.4 [18]. The high interaction energy (58 kJ/mol) probably corresponds with a double hydrogen bond with the polar side-groups of two close keratin filaments. This stage of hydration corresponds to the first stage of primary hydration described by Hansen and Yellin [14]. In the case of wool, the drying energy is also high in this region (67 kJ/mol) according to Watt et al. [19], whereas Escoubés found a sorption energy value of 61 kJ/mol for human hair [20].

These values are close to the hydration energy at NH_3+ sites; in contrast, the hydration energy of carboxyl groups (COO–) is only 31 kJ/mol. Watt and Leeder

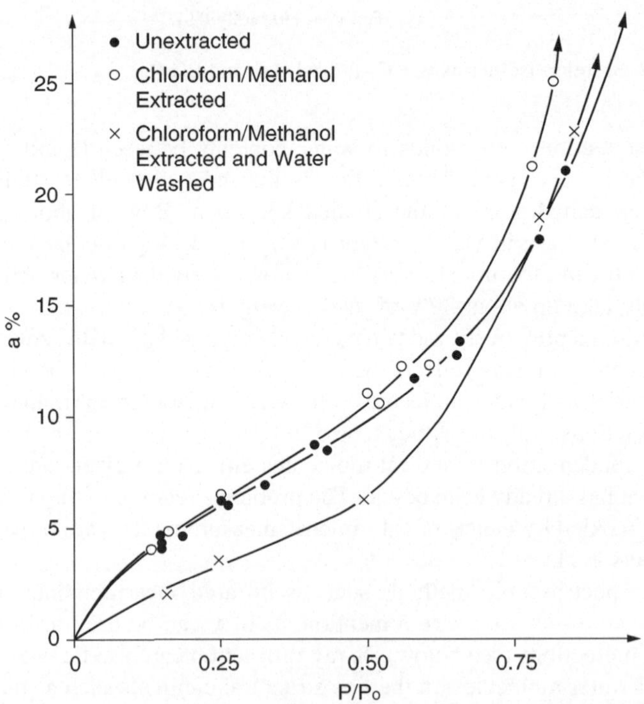

FIGURE 2.3 Sorption isotherms. Water content (a) of SC as expressed by the ratio of the amount of water over the dry weight sample without lipids. ● — ●, unextracted; ○ — ○, chloroform/methanol extracted; and × — ×, chloroform/methanol extracted and water washed.

consider that steric hindrance may explain the better accessibility of COO– groups at low humidity [21]. Thermocurrent studies also indicate preferential binding to carboxyl groups [16]. Precise mathematical analysis of the isotherms in the region of 0 to 20% relative humidity allows one to calculate the number of strongly bound water molecules: BET thermodynamic analysis and Harkins Jura analysis gave values of 5.5 and 7.8%, respectively [12].

V. BOUND WATER

The precise work by Watt and Leeder showed that the percentages of water molecules bound to carboxyl and amino groups accounted for only 46% of the sites occupied at 65% relative humidity; that is, 2.3 and 2.6 molecules at each site on average [21]. At the beginning of the sorption process, there is thus a condensation of water molecules on other hydrophilic sites (e.g., the peptide group or free OH) via hydrogen bonds. This explains the gradual decrease in binding energy illustrated in Figure 2.4. This process allows water molecules to insert between polypeptide chains and leads to a reduction in the interaction between the chains and, thus, facilitates the plasticization of the structure.

The extent to which this process takes place depends on the macromolecular conformation of the chains. In materials containing a large degree of crystalline matter, the quantity of water at equilibrium would tend to be lower than in amorphous substances. As an example, Zeine (70% of α-helices) contains 50% less water than lactoglobulin (10% of α-helices). Contrary to the case of highly bound water, the water molecules bound by hydrogen bonds can mutually exchange protons that migrate within the network of hydrogen bonds. This mechanism explains why electrical conductance and bound-water content are strongly related.

Experiments using differential thermal analysis by Bulgin and Vinson [22] and Walkley [13] gave a figure of about 34% of SC dry weight, which is in agreement with the values obtained by Hansen and Yellin (30 to 35%) [14], who used infrared measurements. These percentages correspond to the normal SC. They are in good accordance with recent results obtained by Gniadecka et al. [7], who conclude that about 20% of the SC total water content is in a free form (tetrahedron water structure). In the case of psoriatic plaques and the stratum disjunctum from patients with senile xerosis, the percentage of bound water appears to be lower [23].

VI. FREE WATER

The experiments cited above show that beyond this threshold value of about 34%, the water present in the horny layer gives rise to a fusion peak at about 0°C, indicating the presence of "freezable" water known as free water. It is noteworthy that this peak does not appear in hair keratin (which never reaches 34% of water content); in other words, hair does not contain free water or freezable water. This is supported by the magnetic resonance spectroscopy experiments by Clifford and Sheard [24] and Lynch and Marsden [25], who found that water molecules are restricted in their movements even at 60°C but do not freeze, whatever the water content. However, some experimental data have shown that above a water content of 18 to 19%, water molecules have peculiar properties. This is the case of the electrical properties of keratin. Above this water content value, the activation energy of electrical conductance falls from 125 to 58.5 kJ/mol [26]. Similarly, a thermal current peak (a dielectric polarization phenomenon) appears at 130°K, corresponding to a characteristic of the free-water fraction [27].

It thus appears that above a water content value of about 19%, water molecules in hair condense in aggregates that retain the same polarization properties as in the liquid state, but lose the rotational properties of free water, as indicated by magnetic resonance spectroscopy experiments. As a result, the term "free water" is totally inappropriate; "intermediate water" would be more suitable.

As stated above, in normal physiologic conditions, SC water mainly exists in a bound state. When it is totally hydrated, however, Transmission Electron Microscopy studies show that water exist as "pools" or "voids" that can disrupt the structure and even shift lamellar bilayers, creating amorphous intercellular zones [28]. These modifications are, however, reversible [29].

VII. INFLUENCE OF LIPIDS AND HYDROSOLUBLE MATERIALS

Figure 2.3 represents the effect of extracting lipids and water-soluble material (WSM) on the sorption isotherm, and Figure 2.4 shows the changes in enthalpy with water content in the same conditions. The use of statistical thermodynamic methods (BET) has shown that lipid extraction slightly but significantly increases the value of the monolayer (5 to 6%), whereas extraction of WSM reduces it (5 to 3.5%).

In Figure 2.3, it can be seen that water content increases by about 10 to 20%, indicating that some binding sites for water molecules are probably inaccessible when lipids are present. This phenomenon appears to be most marked in the range of high humidities. According to Hey et al. [12], the x-ray diagram for the SC, which gives a 10-Å line (corresponding to the average distance between the keratin chains), shifts by about 1.3 Å when the relative humidity reaches 65%. They logically concluded from this result that, in this relative humidity range, the water is located between the keratin microfibrils.

We have shown that more water penetrates the structure in the high relative humidity range when lipids are removed. This last result, combined with the x-ray results of Hey et al. [12], means that a certain number of lipids are likely to be located inside the corneocytes. This agrees with one of the conclusions reached by Goldsmith and Baden in 1970 [30]: WSM could be an important element in water binding, but the most important phenomenon observed during their extraction was the considerable limitation of the equilibrium kinetics between SC and the atmosphere. Thus, the role of WSM could be to facilitate the passage of water between the lipid bilayers (intercellular spaces) into corneocytes.

There has been far less work done on hair and wool. The amount of lipids in hair has been estimated at about 2% by weight [31], but their influence on hydration has not, to our knowledge, been described. However, permanent waving and bleaching appear to induce greater water uptake in conditions of high humidity.

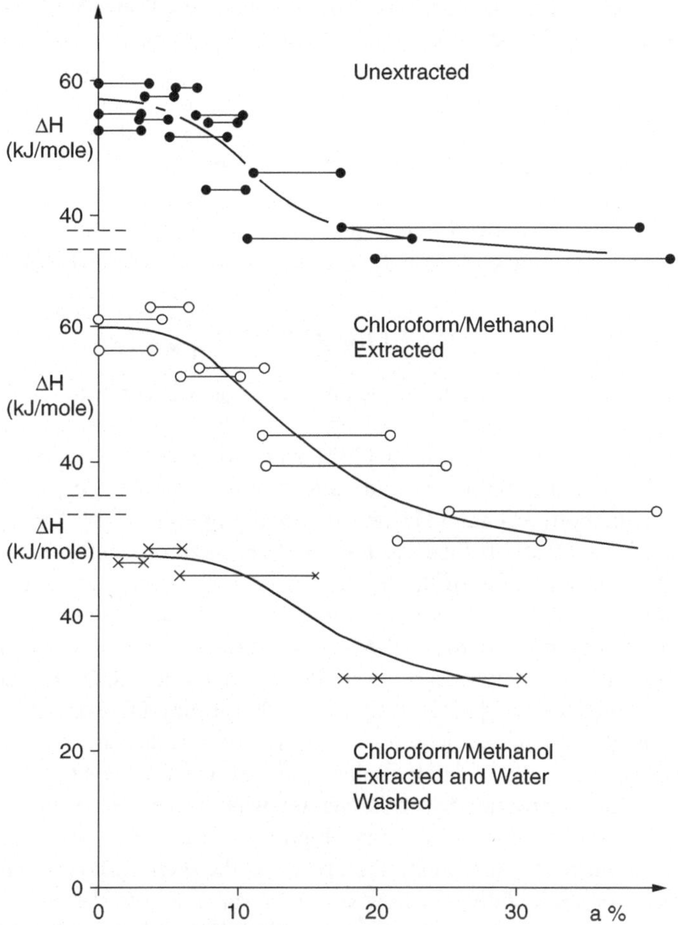

FIGURE 2.4 Enthalpimetric measurements. Water content (a) of SC as expressed by the ratio of the amount of water over the dry weight sample. s — •, unextracted; chloroform/methanol extracted; and chloroform/methanol extracted and water washed.

VIII. SC AND HAIR SWELLING

There are very few data concerning SC and hair swelling. In regard to SC, a recent study carried out using confocal microscopy concluded that there was 20 to 30% swelling in the lateral dimension versus 3 to 5% swelling in the longitudinal one [32]. Lipid extraction produces a higher lateral swelling, which would be linked, according the authors, to the intercellular lipid structure. It must be noticed that removing polar lipids washes away 12% of water-soluble materials (natural moisturizing factors) that, as mentioned earlier, play an important role in the water–SC interaction. The importance of these natural moisturizing factors can be inferred from the huge swelling percentage of simple corneocytes [33].

Both previous and recent studies on hair demonstrate a 12 to 15% lateral swelling in water versus a 1 to 2% swelling in the longitudinal dimension [34]. This swelling anisotropy in hair and the SC can be explained by the location of the water molecules in the keratin structure. X-ray diffraction shows that there is only very weak interaction between water molecules and the helices (unpublished results).

Water molecules are mainly located in the amorphous part of keratin, between the microfibrils, which are moved away. Removing the hair cuticle by means of a razor blade, for example, results in a much more marked swelling, which demonstrates the role of the cuticle in strongly limiting the water sorption by the hair cortex.

IX. CONCLUSION

Water has considerable influence on the physical properties of both hair and the SC. For example, there is a coefficient of about 100 between the elasticity modules measured in dry and humid conditions [35], whereas electrical conductance is multiplied by a factor of 10,000 between dry and hydrated samples [36]. Major changes are also seen in both optical [37] and thermal properties [38]. The equilibrium between keratin and water is thus essential for keratinized tissues to fulfill their role correctly. Our understanding of these interactions is far from complete. As we saw earlier in this chapter, WSMs play an important role in the hydration of the SC. A major advance was made when WSMs were shown to be filaggrin degradation products [39]. Unfortunately we still do not know exactly how these products (mainly amino acids) influence the degree of hydration, but it may involve a direct effect or the facilitation of water molecule access to the hydrophilic sites on keratin. Similarly, little is known on the role of intercellular lipids, except for their barrier function.

The information presented in this chapter was derived from data obtained *in vitro.* The situation *in vivo* is of course highly different; for example, there is a hydration gradient between the external layer and the internal layers. This gradient is responsible for the continuous flow of water vapor across the SC. In addition, sweat secretion — both continuous and discontinuous — plays an important role in the hydration of the SC.

Finally, surface lipids (essentially sebum) can form an emulsion with water and probably play a role in maintaining surface hydration (even though they do not form a barrier to water vapor). All these problems are difficult to approach *in vivo,* as almost all the methods we have for quantifying hydration are indirect. Fortunately, more specific methods are beginning to emerge: In particular, near-infrared spectroscopy, magnetic resonance imaging, and spectroscopy should prove invaluable in this setting.

REFERENCES

1. Fraser, R.B.D., *Keratins,* C. Thomas, New York, 1972.
2. Garson, J.C., Doucet, J., Lévêque, J.L., and Tsoucaris, G., Oriented structure in human stratum corneum revealed by X-ray diffraction, *J. Invest. Dermatol.*, 96, 43, 1991.

3. Landman, L., Epidermal permeability barrier. Transformation of lamellar granular-disks into cellular sheets by a membrane-fusion process, a freeze fracture study, *J. Invest. Dermatol.*, 87, 202, 1986.
4. Schwartzendruber, D.C., Wertz, P.W., Madison, K.C., and Downing, D.T., Evidence that the corneocyte has a chemically bound envelope, *J. Invest. Dermatol.*, 88(6), 709, 1987.
5. Bouwstra, J.A. et al., Structural investigation of human stratum corneum by small angle x-ray scattering, *J. Invest. Dermatol.*, 97, 1005, 1991.
6. Briki, F., Busson, B., Salicru, B., Esteve, F., and Doucet, J., Breast cancer diagnosis using hair, *Nature*, 400, 412, 1999.
7. Gniadecka, M., Nielsen, O.F., Christensen, D.H., and Wulf, H.C., Structure of water, proteins and lipids in intact human skin, hair and nail. *J. Invest. Dermatol.*, 110, 393, 1998.
8. Golden, G.M., Guzek, D.B., Harris, R.R., McKie, J.E., and Potts, R.O., Lipid themotropic transitions in human stratum corneum. *J. Invest. Dermatol.*, 86, 255, 1986.
9. Imokawa, G., Kuno, H., and Kawai, M., Stratum corneum lipids as a bound water modulator. *J. Invest. Dermatol.*, 96, 845, 1991.
10. Lévêque, J.L., Poelman, M.C., de Rigal, J., and Kligman, A.M., Are corneocytes elastic? *Dermatologica,* 176, 65, 1988.
11. Lévêque, J.L., Escoubez, M., and Rassneur, L., Water-keratin interaction in human stratum corneum, *Bioeng. Skin,* 3, 227, 1987.
12. Hey, J., Taylor, D.J., and Derbishyre, W., Water sorption by human callus, *Biochem. Biophys. Acta,* 540, 518, 1978.
13. Walkley, K., Bound water in stratum corneum measured by differential scanning calorimetry, *J. Invest. Dermatol.*, 59, 225, 1972.
14. Hansen, J.R. and Yellin, W., NMR and infrared spectroscopic studies of stratum corneum hydration, in *Water Structure and the Water–Polymer Interface*, Jellinek, N.J., Ed., Plenum, London, 1972.
15. Cuong, V.L. and Nhan, G.L., Multilayer adsorption of moisture in wool and its application in fabric steaming, *Text. Res. J.,* 62, 648, 1992.
16. Lévêque, J.L., Garson, J.C., and Boudouris, G., Water in keratin: study of the depolarization thermal current peak. II. *Biopolymers,* 16, 1725, 1977.
17. Lévêque, J.L., Garson, J.C., and Boudouris, G., L'eau dans les kératines: etude par mesure de Ia conductibilité électrique, *C.R. Acad. Sci. Paris,* 288, 1679, 1979.
18. Lévêque, J.L., Escoubés, M., Rassneur, L. Water-keratin interaction in human stratum corneum. *Bioeng. Skin.*, 3, 227, 1987.
19. Watt, I.C., Kennett, R.H., and James J.F.P., The dry weight of wool, *Text Res. J.,* 29, 975, 1959.
20. Escoubez, C., personal communication.
21. Watt, I.C. and Leeder, J.D., Role of carboxyl group in water absorption by keratin, *J. Appl. Chem.,* 18, 1, 1968.
22. Bulgin, J.J. and Vinson, R., The use of differential thermal analysis to study the bound water in stratum corneum membranes, *Biochem. Biophys. Acta,* 136, 551, 1967.
23. Takonouchi, M., Susuki, H., and Tagami, H., Hydration characteristics of pathologic stratum corneum — evaluation of bound water, *J. Invest. Dermatol.,* 87, 574, 1986.
24. Clifford, J. and Sheard, B., Nuclear magnetic resonance investigation of the state of water in human hair, *Biopolymers*, 4, 1057, 1966.

25. Lynch, L.J. and Marsden, K.H., An NMR study of keratin hydration, *J. Chem. Phys.,* 56, 5681, 1969.
26. Algie, J.E. and Gamble, R.A., Dielectric properties of wool and horn containing absorbed water. *Kolloid Z. Z. Polym.*, 251, 554, 1973.
27. Lévêque, J.L., Garson, J.C., Pissis, P., and Boudouris, G., Free water in hair keratin? A depolarization thermal current study. *Biopolymers*, 20, 2469, 1981.
28. Warner, R.R., Boissy, Y.I., and Spears, M.J. Water disrupts stratum corneum lipid lamellae, damage is similar to surfactants. *J. Invest. Dermatol.* 113, 960, 1999.
29. van Hal, D.A., Jeremiasse, E., Junginger, H.E., and Bouwstra, J.A. Structure of fully hydrated human stratum corneum: a freeze fracture electron microscopy study. *J. Invest. Dermatol.* 106, 89, 1996.
30. Goldsmith, L.A. and Baden, H.P., Uniquely oriented epidermal lipid, *Nature,* 225, 1052, 1970.
31. Hilterhaus, S. and Zahn W., Contribution to the chemistry of human hair: lipid chemical aspects of permanently waved hair, *Int. J. Cosmet. Sci.,* 11, 167, 1989.
32. Norlen, L., Emilson, A., and Forslind, B., Stratum corneum swelling. Biophysical and computer assisted quantitative assessments, *Arch. Dermatol. Res.,* 289, 506, 1997.
33. Laffon, E., Dilhaire, S., Lévêque, J.L., and Corcuff, P., An improved technique for optical interferometric imaging of isolated cells, *Cytometry*, 24, 93, 1996.
34. Zviak, C., Ed., *The Science of Hair Care.* Marcel Dekker, New York, 1986.
35. Lévêque, J.L. and Rassneur, L., Mechanical properties of stratum corneum: influence of water and lipids, in *Physical Nature of the Skin*, Marks, R.M. et al., Eds., MTP Press Limited, Lancaster, 1988.
36. Campbell, S.D., Kraning, K.K., Schibli, E.G., Momii, S.T., Hydration characteristics and chemical resistivity of stratum corneum using a noninvasive four-point microelectrode method, *J. Invest. Dermatol.*, 69, 290–295, 1977.
37. Solan, J.L. and Laden, K., Factors affecting the penetration of light through stratum corneum, *J. Soc. Cosmet. Chem.,* 28, 125, 1977.
38. Deffond, D., Lévêque, J.L., Scot, J.C., and Saint-Léger, D., A photoacoustic investigation of the influence of constituents of stratum corneum on ultraviolet absorbtion, *Photodermatology,* 2, 279, 1985.
39. Harding, C. and Scot, I., Histidine-rich proteins (filaggrins): structural and functional heterogeneity during epidermal differentiation, *J. Mol. Biol.,* 170, 651, 1983.

3 Stratum Corneum pH and Ions — Distribution and Importance

Joachim W. Fluhr, Lora Bankova, and Peter Elias

CONTENTS

I. pH of Stratum Corneum — Formation and Importance 28
  A. Introduction 28
  B. Methods of Measurement 28
  C. Mechanisms Leading to SC Acidification 29
    1. Histidine-to-Urocanic Acid Pathway 29
    2. Phospholipid-to-Free Fatty Acid Pathway 30
    3. Membrane Antiporters Regulate SC Acidification and Function 30
  D. Factors Affecting Human Surface pH 31
    1. Endogenous Factors, Unrelated to Pathological Features 31
    2. Gender 31
    3. Endogenous Factors Related to Clinical Pathological Situations 33
    4. Exogenous Factors 34
  E. Importance of pH for the Barrier 34
  F. pH Antimicrobial Function 35
II. Ions 36
  A. Calcium 36
    1. Distribution of the Calcium Ion within the Epidermis and the Cellular Compartments 36
    2. Ontogenesis of the Calcium Gradient 36
    3. Visualization of the Calcium Distribution in the Epidermis 36
    4. Barrier Disruption and Calcium (Figure 3.3) 37
    5. Electric Potential and Calcium 38
    6. Origin of the Calcium Gradient 38
    7. Calcium as a Second Messenger 39
    8. Importance of the Calcium Ion 39
    9. Pathologic Conditions and Calcium 40
  B. Magnesium 40
  C. Potassium 40
  D. Sodium 40
References 41

0-8493-1443-7/05/$0.00+$1.50

I. pH OF STRATUM CORNEUM — FORMATION AND IMPORTANCE

A. Introduction

The universal measure for acidity is pH, defined as the negative logarithm of the H^+ content in an aqueous solution. pH is intended to be a measure of the activity of hydrogen ions in solution.

With regard to skin bioengineering measurements, a specified definition of pH has been introduced: pH as measured by flat glass electrode at the skin surface with a hydrated skin electrode interfaced.

The usual pH range at the surface of nonoccluded human skin is between 4 and 6. The acidic character of the skin was first mentioned by Heuss [1], later Schade and Marchionini [2] introduced the term "acid mantle" to describe the acidic character of the skin.

At any rate, it has only recently been discovered that the epidermis also displays a characteristic rising gradient of the pH toward the lower stratum corneum (SC) layers [3–5]. At the granular layer level, the pH values of 6.8 are almost equal to the physiologic pH, characteristic of the body's internal environment [6]. Using sequential tape striping, an increase of 0.2 to 0.3 units was shown for each step [3,4,6]; such a sharp decline over a distance of only 10 to 15 μm requires a 100- to 1000-fold increase in protons (i.e., H^+ ion concentration). Furthermore, the acidic pH appears to be heterogeneously distributed within the lower SC; that is, within membrane domains both at and immediately above the SG (stratum granulosum)–SC interface, whereas the corneocyte cytosol at these levels is neutral [7,8]. A "dual pH gradient" was first mentioned in the 1950s [9]. These initial results were confirmed in a recent study by Denda et al. showing that thc SG–SC interface displays a more acidic character [10] (Figure 3.1).

B. Methods of Measurement

At present, the preferred method of measurement of the skin surface pH is the potentiometric assessment with a flat glass electrode [11]. Recently, more sophisticated,

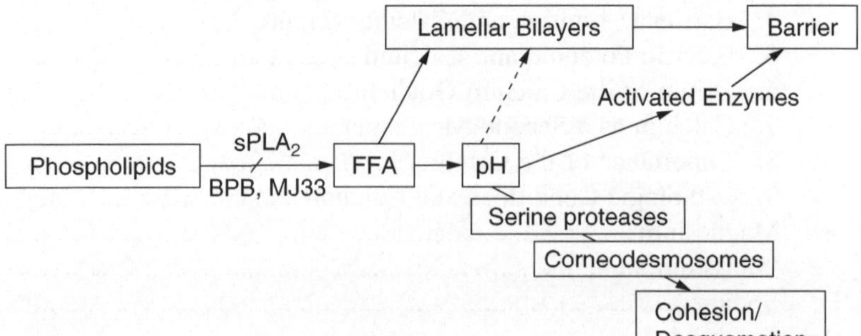

FIGURE 3.1 Structural and functional consequences of stratum corneum acidification.

high-resolution methods that focus on microdomains, such as fluorescence lifetime imaging microscopy [7,8] and electron spin resonance imaging [12], have been employed to assess SC pH.

C. Mechanisms Leading to SC Acidification

The "acid mantle" is the combined effect of the action of a number of endogenous and exogenous mechanisms. The endogenous mechanisms include enzymatic processes and membrane antiporters/pumps, whereas the exogenous mechanisms comprise free fatty acids from sweat and sebum and hydrolytic products of filaggrin, microbial metabolites, and lactic acid from eccrine glands.

As both appendage-free and germ-free skin possess an acidic pH (excluding an absolute requirement for exogenous mechanisms), endogenous mechanisms clearly suffice and would appear to predominate. Because epidermis retains a more neutral pH in early infancy [13] but is fully cornified, bulk proteolysis as a nonspecific companion to cornification can be excluded.

At present, three endogenous mechanisms have been identified that not only could influence SC pH but that also could regulate one or more key SC functions.

1. Histidine-to-Urocanic Acid Pathway

The histidine-to–urocanic acid pathway is largely responsible for SC hydration [14,15]. Early in cornification, filaggrin, the predominant, histidine-enriched, basic protein in F-type keratohyalin granules, disperses around keratin filaments within the stratum compactum [16]. Filaggrin undergoes proteolysis to free amino acids (histidine, glutamine, and arginine), urocanic acid, pyrrolidone carboxylic acid, and ornithine/citrulline/aspartic acid, which make up much of the osmotically active material that largely accounts for the ability of the SC to remain hydrated [17–19].

The succeeding enzymatic processes of nonoxidative histidine deamination have recently been identified as contributing to SC acidification [20–22]. The key enzyme in this process is the histidine-ammonia-lyase (histidase) It is found primarily in the liver and the skin [21,23–27]; in the skin, it is almost entirely located in the SC and functions optimally at a neutral pH [22,24–26].

The main product of the histidine deamination is the trans-urocanic acid (tUCA), which was shown to play an important role not only in skin hydration but also in several non-pH-dependent skin protective functions.

Ultraviolet protection: The tUCA acts as an endogenous sunscreen.

Immunosuppression: One of the metabolites in the breakdown of the tUCA (cis-UCA) is a potent immunosuppressive molecule that is hypothesized to favor the development of skin cancer [21,22].

Although the histidine-to-UCA pathway is a potent mechanism in SC acidification, there are clues indicating that it does not entirely explain the acidity of the skin surface:

The optimum pH, at which histidase exhibits its catalytic activity is 8.2 to 8.5 (i.e., as high as the characteristic pH in the outer SC [24–27]).

Another limiting step in the histidine cascade is the filaggrin-to-histidine proteolysis, which is humidity sensitive [17,28]; this pathway might become inoperative under certain conditions (e.g., fully hydrated SC).

The key metabolite in this pathway, tUCA, is extremely polar, and polar molecules tend to remain in the cytosol [22], away from the relatively hydrophobic membrane domains, where it is supposed to act.

If the UCA pathway were the sole source of SC acidification, patients with histidinemia and his/his knockout mice should display gross cutaneous abnormalities [27,29,30]: Such abnormalities were not observed.

Filaggrin is accumulated in the SC before birth, and UCA levels peak in SC immediately after birth, following initial exposure of neonatal skin to a dry environment [19], indicating that this pathway is blocked during fetal development.

2. Phospholipid-to-Free Fatty Acid Pathway

A second pathway contributing to SC acidification is the secretory phospholipase A_2 pathway. A product of this pathway is a pool of free fatty acids, which were shown to influence not only the bulk SC pH but also two further SC functions [31].

The processing of phospholipids by $sPLA_2$ generates nonessential free fatty acids, one of the three key families of SC barrier lipids. Applications of either bromphenacyl bromide or 1-hexadecyl-3(trifluoroethyl)-sn-glycero-2-phosphomethanol, which are inhibitors of $sPLA_2$, exert profound effects on barrier homeostasis and epidermal lipid processing [32,33]. These inhibitors not only perturb permeability barrier homeostasis but also elevate SC bulk pH, with additional functional alterations in SC integrity/cohesion [31]. Pertinently, coapplications of an acidic buffer with the $sPLA_2$ inhibitors corrected the integrity/cohesion abnormality but not the critical requirement for free fatty acids as structural components of the barrier. The phospholipid-to–free fatty acids pathway contributes directly to SC acidification, affecting both SC barrier function and integrity/cohesion (Figure 3.1; adapted from Fluhr and Elias [34]).

3. Membrane Antiporters Regulate SC Acidification and Function

Non-energy-requiring, integral membrane transporters that extrude protons in exchange for sodium (NHEs) are ubiquitous in mammalian tissues [35–39]. NHEs comprise a family of at least six 80- to 90-kDa proteins [37,38]; however, the only one expressed in keratinocytes is NHE1 [39–41], which is upregulated under ambient conditions present in the outer epidermis: high calcium, osmolarity, or acidity [35,36] (i.e., it could acidify extracellular membrane domains selectively near the SG–SC interface or in membrane domains in the lower SC).

D. Factors Affecting Human Surface pH

Skin surface pH can be influenced by a number of endogenous and exogenous factors:

1. Endogenous factors, unrelated to pathological features (e.g., developmental changes in pH)
2. Endogenous factors, related to clinical pathological situations (e.g., increased pH in atopic dermatitis)
3. Exogenous factors (e.g., influence of detergents on pH)

1. Endogenous Factors, Unrelated to Pathological Features

a. Racial Differences

Controversial data exist concerning the influence of race on skin pH. The skin surface pH in African Americans displayed significantly lower values in the upper SC, whereas deeper sections of the SC assessed after tape stripping did not reveal any difference in the pH gradient [4]. Similar results were presented in an earlier study of Warrier et al. [42]. In contrast, an early publication from Draize showed that the average pH on multiple areas was higher in black (pH 5.21) males than in white males (pH 4.85) [43].

b. Topographic Variation and Gender Differences

It is generally considered that skin surface pH is relatively constant in different body areas except for the intertriginous areas (axilla, inguinal and submammary folds, and finger webs) and sebaceous gland-rich areas (breast skin and back). This phenomenon might be related to a lowered filaggrin–histidine–urocanic acid breakdown caused by high humidity conditions or by sebaceous and eccrine gland distribution.

The standard test site for pH measurements (the volar forearm) was measured in a number of studies, displaying values ranging from pH 5.06 ± 0.11 (male) and 5.26 ± 0.12 (female) [44] to 4.84 ± 0.09 (male) and 5.74 ± 0.14 (female) [45] and to 5.07 ± 0.62 for both sexes [46]. Yosipovitch et al. presented the following values for the intertriginous areas and sebaceous gland-rich areas: axilla: 5.84 ± 0.1 (male) and 5.99 ± 0.11 (female) [44]; inguinal area: 6.22 ± 0.13 both sexes. Breast skin showed a markedly higher pH (5.54 ± 0.1 in males and 5.83 ± 0.17 in females) [44], and back pH was shown to be 5.22 ± 0.08 [45]. A difference between left and right arm could not be detected in two independent studies [47,48]. In addition, studies on the vulvar surface pH have shown equally elevated values 5.99 ± 0.45 [49].

2. Gender

The information contained in the literature considering gender differences in pH values is contradictory. Blank showed higher values on the volar forearm for females than for males [50], and these data have been confirmed by more recent studies showing pH levels of 5.26 ± 0.12 (female) vs. 5.06 ± 0.11 (male) [44], 5.74 ± 0.14

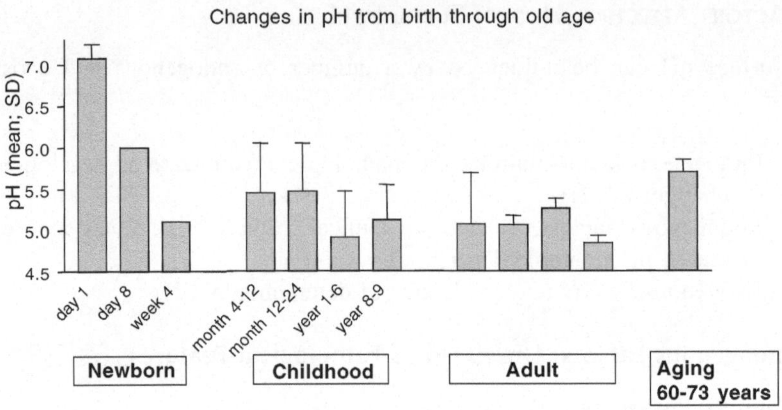

FIGURE 3.2 Changes in pH during different periods of life.

(female) vs. 4.84 ± 0.09 (male) [45] and 5.80 (females) compared to 5.54 (males) on the forearm [47]. Other studies could not detect gender-related pH differences on the forehead and cheek [52,53].

d. Developmental Changes in pH

The effects of age-related pH changes can be divided into four phases (Figure 3.2; adapted from Fluhr and Elias [34]):

1. Newborn period with the formation of an acid mantle within the first weeks after birth
2. Childhood period with elevated SC pH values
3. Adult period with lowest SC pH values over lifetime
4. Aging period with an increase in SC pH values

The "acid mantle" is not completely developed at birth — the human newborn SC displays a near-neutral surface pH (6.5) [54]. In the first postnatal month, however, a rapid decline in the pH is observed, reaching values around 5.4 at 7 to 14 days after birth [13,54–56], followed by a more gradual change in the next 6 to 12 months until the steady-state characteristic of the adults is reached. Healthy young children display pH values lower than those measured in adults. On the forehead, pH values of 4.67 ± 0.39 were reported [57], values on the back were 4.80 ± 0.44, on the abdomen they were 5.04 ± 0.43, and on the cheek they were 5.43 ± 0.42, whereas the volar forearm shows higher values (e.g., 4.86 ± 0.45 [57] and 4.91 ± 0.56 [46]). In diapered regions the surface pH remains higher [54]. Later in infancy, as in a study group of children between 8 and 35 months, values became normal after 6 months, with values at the volar forearm around 5.5 ± 0.6 [13]. The forearm of children at the age of 3.5 years displays a surface pH on the volar forearm of 4.9 ± 0.6 [46]. At the age of 8 to 9 years, a pH of 5.12 ± 0.43 was reported on healthy skin for the volar forearm [58] (Figure 3.2; adapted from Fluhr and Elias [34]).

Aged persons show elevated SC pH values. For example, Zlotogorski showed that subjects overs 80 years of age demonstrate a mean SC pH of 5.1 on the forehead and 5.4 on the cheek — about half a unit higher than the pH in younger adults [52]. Thune et al. also showed a positive correlation between age and pH (i.e., higher pH at higher age) [59]. These results were partially confirmed over the ankle by Wilhelm et al., who again showed higher values for aged persons [53].

e. *Circadian Rhythm*

Recent studies also report a circadian rhythm of skin surface pH in forearm, shin, axillae, and facial regions [60–62]. Yosipovitch et al. reported SC pH maxima between 2:00 and 4:00 PM, with minimal values around 8:00 PM [62]. Le Fur et al. found the lowest values to be around 4:00 AM, with a plateau occurring during the daytime [60]. The axillae showed a significant decrease during a single evening series of measurements (5.49 ± 0.23) when compared with a single morning measurement (5.87 ± 0.23) [61]. Again, these data indicate the influence of hormones underlying a circadian cycle like glucocorticosteroids. Yosipovitch et al. showed a variation over 24 hours of 0.4 pH units for both men and women [62].

3. Endogenous Factors Related to Clinical Pathological Situations

In general, dermatites are associated with higher skin surface pH. This increase in pH may be attributed to increased transepidermal water loss, which could lead to decreased urocanic acid generation and, therefore, increased pH. An additional explanation could be the increased release of interstitial fluid (neutral pH) resulting from impaired barrier function and the consequent increase in SC pH.

a. *Eczematous Diseases*

Eczematous skin diseases are associated with an elevated skin surface pH. Eczematous skin of the volar forearm showed an elevated surface pH (compared with healthy skin) in a large group of atopic dermatitis children (aged 8 to 9 years) [58]. Seidenari and Giusti showed elevated pH values in both involved and uninvolved skin of children with atopic dermatitis compared with the skin of nondiseased children [57], at the same time, involved skin was characterized by higher values than was uninvolved skin [57]. Beare et al. reported an alkaline pH in children affected by seborrheic dermatitis compared with healthy children in both affected and unaffected sites [63].

b. *Ichthyosis*

Ichthyotic patients were reported to have an elevated surface pH in different areas both in involved and uninvolved skin compared with healthy subjects [64]. The only exception was the axillary regions, which displayed lower pH in ichthyotic patients compared with normal patients, probably because of the reduced sweating of these patients. The higher pH in ichthyosis vulgaris was paralleled by an impaired alkali resistance test [64]. More recently, Ohmann and Vahlquist confirmed the elevated surface pH in ichthyosis vulgaris compared with X-linked recessive ichthyosis patients and healthy volunteers. However, the pH gradient showed markedly lower pH values in X-linked recessive ichthyosis and slightly lower values in ichthyosis vulgaris [3]. One possible explanation for theses findings in ichthyosis might be the increased activity of serum protease enzymes that could occur with elevated pH. In contrast, a low pH would decrease the protease activity and consequently lead to SC retention, as occurs in X-linked recessive ichthyosis.

c. Other Diseases

In addition, other disease states that are associated with accelerated aging show altered skin pH values. For example, patients with non-insulin-dependent diabetes mellitus showed higher pH values in intertriginous regions compared with normal control subjects [44]. Increased pH, in term, might be one factor that increases susceptibility of diabetic skin to bacterial and yeast colonization or infections. Yosipovitch et al. also showed an elevated skin surface pH in patients undergoing hemodialysis compared with healthy control subjects, even though hemodialyzed patients displayed an overall decrease in blood pH [45].

4. Exogenous Factors

a. Detergents and Cleansing Products

Washing procedures have an alkalinizing effect on the skin surface pH, independent of the pH of the washing solution. This effect can be observed after a single washing, and it usually normalizes within 8 hours [65]. Even short-term washing with tap water results in an increase in surface pH in infants [66]. However, repetitive washing with sodium lauryl sulfate showed a more pronounced alkalization effect than the single washing procedure did [59,67,68]. Furthermore, the more alkaline surface pH persists for more than 12 hours in a repeated washing model [69]. A crossover study with acidic and neutral cleanser showed an increase of propionibacteria in alkaline cleanser–treated skin, whereas acidic cleansers reduced the propionibacteria colonization [70]. Thus, the use of a neutral or acidic cleanser is less harmful to the surface pH than is the use of alkaline soap. However, it remains unclear whether the increase in SC pH is the result of a direct effect of the soaps on SC or whether soaps add base equivalents to SC (e.g., OH).

b. SC Hydration

Overall skin/SC hydration appears to play a key role in skin surface pH. For example, a dramatic increase in surface pH can be observed after both 3 days (e.g., from pH 4.9 to 7.1 [71]) and 5 days (i.e., from 4.38 to 7.05 [72]) of occlusion. The pH increase was paralleled in both studies by an increase of bacterial colonization. A negative correlation has been shown between the sweat inhibition and skin surface pH values (i.e., a good sweat inhibition results in low pH values) [73]. An older article from 1942, in contrast, reported that the influence of sweating resulted in a more acidic skin surface pH [74]. It also has been shown that diaper dermatitis is correlated with high pH [75]. Children wearing diapers showed higher pH values early after birth [54].

E. Importance of pH for the Barrier

Changes in pH could influence one or more key SC functions. Barrier repair after acute perturbations proceeds normally when the pH is acidic, whereas recovery is delayed at a neutral pH independent of the ionic environment of the bathing solution [76]. Moreover, lamellar body secretion is unimpeded at neutral pH; instead, the postsecretory processing of newly secreted polar lipids into mature lamellar bilayers is impeded at a neutral pH. Furthermore, exposure to a neutral pH correlates with inhibition of β-Glucocerebrosidase activity in the lower SC, as shown by *in situ* zymography [76]. In addition, neonatal SC exhibits a neutral pH with normal basal barrier function [54–56], but a delay in barrier recovery after acute perturbations, in conjunction with decreased lipid processing (preliminary observations from this laboratory). Together, these results support the hypothesis that changes in extracellular pH influence barrier homeostasis by modulating the expression or activity of lipid processing enzymes (Figure 3.1). Altered surface pH values after tape stripping regain normal values in humans within 3 to 4 days [77]. Pretreatment with acidic alpha-hydroxy acids (pH 3.4 to 4.4, compared to an alkaline vehicle) revealed significant higher resistance against a barrier challenge with sodium lauryl sulfate [78]. Ultrastructural findings support the positive effect of glycolic acid on the degradation of desmosomes [79]. These studies again show the importance of an acidic SC pH for barrier homeostasis and for SC integrity and cohesion.

Permeability barrier homeostasis requires the secretion of both lipid precursors and colocalized hydrolytic enzymes from epidermal lamellar bodies and the subsequent, postsecretory processing of polar lipids into their more nonpolar products [80]. Whereas changes in extracellular ions regulate lamellar body secretion [81], the epidermal hydrolytic enzymes generate the lipid products required to form the mature membrane structures that mediate the barrier: β-glucocerebrosidase, acidic sphingomyelinase, secretory phospholipase A_2, and steroid sulfatase. Two of these key enzymes — secretory phospholipase A_2 [82,83] and steroid sulfatase [84] — exhibit neutral pH optima; epidermal β-glucocerebrosidase and acidic sphingomyelinase display acidic pH optima [85,86]. Blockade of these enzymes results in abnormal SC structure and function in human diseases, such as Gaucher disease [86–88] and X-linked recessive ichthyosis [88]; in murine transgenic knockout models of these disorders [85,86]; and in inhibitor-based models [86]. Thus, it appears that the activity of these lipid-processing enzymes, which is critical for barrier integrity, is dependent on the SC-localized pH for the enzymes' appropriate activities.

F. pH Antimicrobial Function

The microflora of the skin comprise transient, temporary-resident, and permanent-resident species [67,70]. The predominant resident flora include several species of coagulase-negative staphylococci, propionibacteria, and yeasts, whose relative proportions depend on habitat (e.g., coagulase-negative staphylococci prevail on dry, lipid-depleted surfaces, whereas propionibacteria predominate on sebaceous gland–enriched skin sites [70,89]). Although the density and composition of the skin flora is dependent on a variety of factors, pH is preeminent [13,72,75,90,91]. Whereas propionibacteria grow well at pH 6.0 to 6.5, growth slows at pH 5.5 [70,92]. In contrast, *Staphylococcus aureus* grows best at pH 7.5 but continues to proliferate slowly at pH 5.0 to 6.0 [70,89,92]. Thus, the acidic pH of the SC restricts colonization by pathogenic flora and encourages persistence of normal microbial flora. Pertinently, both intertriginous and inflamed skin display an increased skin pH [44,59,63,75,93], as does newborn SC, which only gradually develops an acidic pH over the first 1 to 3 months of life [54–56,94]; hence the increased propensity for colonization by normal flora and the reduced resistance to pathogens that characterize these settings. However, the developmental delay in acidification in neonates often is complicated by increased hydration and urea or fecal contamination, which can further increase SC pH, with a subsequent increased risk of infection and diaper dermatitis [54,72,75,93]. Thus, SC pH is closely linked to microbial colonization and possible pathogenesis.

II. IONS

Both localization and distribution of ions play a crucial role in barrier homeostasis. Calcium and magnesium were shown to be important regulators in barrier homeostasis, whereas the role of potassium, sodium, chloride, and phosphate ions has still not been investigated in depth.

A. Calcium

1. Distribution of the Calcium Ion within the Epidermis and the Cellular Compartments

The calcium distribution in the epidermis demonstrates a characteristic pattern, with little calcium within the basal and spinous layers and a gradual increase toward the outer SG [95,96]. However, calcium largely disappears from the SC above the SG–SC interface.

Calcium also shows a characteristic distribution regarding cell compartments — in the basal cell layer, calcium is found in intracellular compartments: the cytosol, mitochondria, and nuclear chromatin. In the granular layer, calcium is mainly located in the intercellular space — the only intracellular calcium is within the mitochondria and lamellar bodies (limiting membranes and discs). The most apical granular cells demonstrate dense extracellular deposits and high free intracellular Ca in the cytosol [97]. Although some corneocytes in the lower cornified layer demonstrate small amounts of calcium, most of them are devoid of the ion.

2. Ontogenesis of the Calcium Gradient

During fetal epidermal development the epidermal calcium gradient develops late in gestation, parallel to the emergence of a competent permeability barrier [98]. The calcium gradient is not present before barrier formation, as shown by ion precipitation or by proton-induced X-ray emission analysis of fetal mouse and rat skin [98].

3. Visualization of the Calcium Distribution in the Epidermis

The calcium gradient was first demonstrated using ultrastructural cytochemistry [97,99] and then later by several quantitative methods in adult human and murine

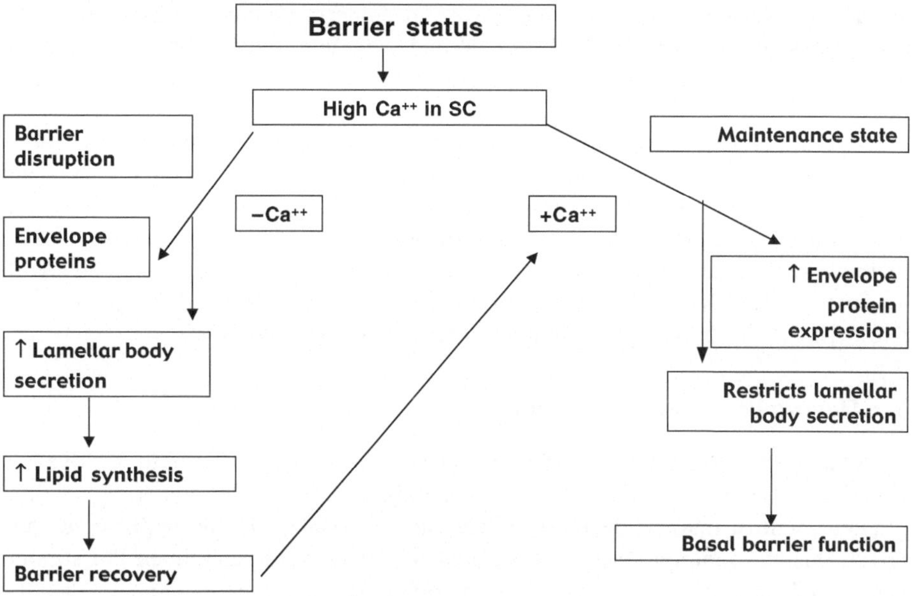

FIGURE 3.3 The role of calcium in barrier recovery.

epidermis [99] and by using ion capture cytochemistry and PIXE [96], all of which methods provided similar results.

4. Barrier Disruption and Calcium (Figure 3.3)

Acute barrier disruption by organic solvent treatment, detergents, or tape stripping leads to the loss of the calcium gradient because of the passive loss of calcium from the upper epidermis [95,96]. However, numerous calcium precipitates are present within the extracellular spaces of the lower SC [95]. Between 1 and 24 hours after acetone treatment, calcium-containing precipitates gradually begin to reappear both in the cytosol of upper SG cells and in the extracellular domains [95,96], resulting in a parallel restoration of the epidermal calcium gradient. In the case of tape stripping, calcium persists in small foci where the SC is not removed (in these sites the calcium distribution does not change, and accelerated secretion of lamellar bodies is not observed [95]). Furthermore, immersion in calcium-containing solutions replenishes and maintains the calcium content in the SG, blocks the expected increase in lamellar body secretion and lipid synthesis, and leads to retardation of barrier repair [95]. Conversely, removal of calcium as a result of acute barrier disruption leads to an acceleration of both lamellar body secretion and lipid synthesis [95].

Prolonged occlusion (24 to 48 hours) with a vapor-impermeable wrap blocks the restoration of the calcium gradient after acute barrier disruption [100], as the barrier fails to normalize under these conditions. The artificial correction of the barrier function by occlusion blocks the passive return of calcium ions to the epidermis.

The prolonged occlusion of tape-stripped epidermis induces transitional cells and delays the restoration of the epidermal calcium gradient. When the SC is restored, the transitional cells disappear in parallel with normalization of the epidermal calcium gradient.

Chronic barrier disruption leads to an increased calcium loss through the cornified layer. In contrast to acute perturbation, chronic defects produce a constant increase in the flux of water through the epidermis. Thus, an increase in calcium content is observed in different animal models with chronic barrier disruption (e.g., in the essential fatty acid–deficient mouse and in animals treated topically with lovastatin [a HMGCoA reductase inhibitor], both of which are models with a defective permeability barrier and a hyperproliferative epidermis) [100].

5. Electric Potential and Calcium

Lee et al. demonstrated, using calcium capture cytochemistry, that both positive and negative iontophoresis on hairless mouse skin induce the loss of epidermal calcium gradient, with a marked decrease of the calcium content in the upper epidermis [101]. Positive iontophoresis is associated with increased calcium in the stratum basale and dermis, whereas negative iontophoresis increases calcium in the SC. Denda et al. also showed that an external electric potential — both positive and negative currents affect the ionic gradients in the epidermis and, hence, influence the skin barrier homeostasis [102].

6. Origin of the Calcium Gradient

The calcium gradient was shown to be formed and maintained as a result of the action of passive processes within the epidermis. Although the epidermis possesses a wide array of energy-dependent Ca pumps [103], indirect clues as well as direct evidence that permeability barrier status regulates the formation of the epidermal calcium gradient and passive diffusion were presented and seem to be sufficient to explain the gradient's origin.

Indirect studies have shown that

1. The calcium gradient disappears after acute barrier disruption and returns simultaneously with the barrier recovery [95,96,104].
2. The calcium gradient and a competent barrier appear simultaneously during fetal development [98].
3. The restoration of the barrier function leads to the reestablishment of the calcium gradient [105].
4. Sustained extremes in environmental humidity cause calcium gradient changes in parallel with the humidity-induced modulations in barrier function [106].

Direct evidence was presented in a recent study by Elias et al. [107]:

1. The calcium gradient returns rapidly under vapor-permeable dressings lipid synthesis; induced delay in barrier recovery results in a delayed reappearance of the calcium gradient.
2. Cold exposure also blocks barrier recovery and retards return of the calcium gradient.
3. Artificial barrier restoration results in the return of a calcium gradient in cold-exposed skin.

7. Calcium as a Second Messenger

Changes in the free intracellular calcium level act as an important second messenger in cells, translating external environmental signals into internal signals and generating cellular responses [108].

8. Importance of the Calcium Ion

a. *Signaling the Lamellar Body Secretory Response*

The key event in the homeostatic response to barrier disruption is lamellar body secretion [109,110]. Fluctuations in calcium in the outer epidermis were shown to regulate lamellar body secretion, leading to a restoration of the permeability barrier homeostasis [110]. The abrupt reduction in calcium level in the outer epidermis triggers lamellar body secretion [101,109], whereas the high levels of calcium present in unperturbed epidermis restrict organelle secretion to rates that, although low, are sufficient to maintain barrier homeostasis under basal conditions [100].

Further, the association of calcium with lamellar body disc membranes and contents indicates that divalent cations may contribute to both lamellar body secretion and to the formation of intercorneocyte membrane bilayers [97].

b. *Regulator of Keratinocyte Differentiation [111,112]*

In vitro data indicate that lowered extracellular calcium concentrations promote continued proliferation of keratinocytes, whereas higher calcium concentrations lead to expression of terminal differentiation markers, stratification, and cultures that are differentiating [108]. In cultured keratinocytes, most of the differentiation-specific proteins are regulated both at a transcriptional and, in some cases, at a posttranscriptional level (e.g., transglutaminase 1 and profilaggrin) by external calcium levels [112–114]. The increased levels of calcium present normally in the outer epidermis correlate spatially with the sites in which epidermis generates late, differentiation-specific proteins, such as involucrin, filaggrin, and loricrin. However, whether modulations in epidermal calcium would dominate over the variety of other endogenous factors that coregulate keratinocyte differentiation is not known [115]. In a recent study, Elias et al. demonstrated that barrier-induced fluctuations in epidermal calcium coordinately, but divergently, regulate the two major differentiation programs that generate a normal SC — the lipid-enriched extracellular matrix and the corneocyte [106]. The mRNA levels of epidermal differentiation-specific proteins change in parallel with alterations in the levels of calcium in the outer epidermis [106], and these changes occur both when calcium levels are changed indirectly (i.e., with barrier disruption) or when they are changed directly (i.e., with exposure to a low/high calcium bath, independent of barrier status).

9. Pathologic Conditions and Calcium

An abnormal ion profile is seen in aging, atopic dermatitis, and psoriasis [116,117]. The psoriatic epidermis displays an altered calcium gradient and an increased concentration of calcium in the lower epidermis [118]. In a recent study, it was also shown that the reestablishment of the calcium gradient in chronic plaque psoriasis after occlusive therapy could be an important factor in the therapeutic effect of occlusion in this pathologic condition [119].

B. Magnesium

The localization of magnesium also plays an important role in barrier homeostasis [10,120]. The magnesium ion is localized, like calcium, in the upper epidermis and shows a gradient similar to that of calcium. Using gel blotting it was shown that magnesium is at a higher concentration in the nucleus than in the cytosol following tape stripping. After acute barrier perturbation, the gradation of magnesium disappears [10]. The application of some magnesium salts and a mixture of calcium and magnesium salts accelerates barrier recovery [120].

A number of studies indicate a role of magnesium in the differentiation of the keratinocytes and barrier homeostasis as it is required for the activity of rab-geranylgeranyl transferase, which is an essential enzyme in the terminal differentiation of the keratinocyte [121–123].

C. Potassium

Warner et al. reported the lowest potassium concentration between the upper epidermis and the SC in human skin [124]. A gradient peaking similar to calcium at the SG–SC interface was found for potassium by Mauro et al., using microbeam proton-induced X-ray emission [104]. This study also shows that epidermal concentration of potassium decreases after barrier abrogation, although to a lesser extent than does calcium. Later, using another technique (blotting to gel containing chemical indicators), Denda et al. demonstrated a peak of the potassium concentration in the spinous layer and the lowest values in the granulous layer [10]. The compartmental distribution of potassium within the cell did not show a clear pattern [10]. A close interaction between calcium and potassium was indicated [125], as both ions play an important role in keratinocyte differentiation and barrier homeostasis.

D. Sodium

Sodium does not show a gradation in the epidermis, and a barrier insult does not affect its distribution [10,126]. Regarding compartmental distribution, sodium was shown to be absent in the nucleus [10]. A correlation between sodium and epidermal homeostasis could not be demonstrated [10].

Potassium and chloride demonstrate gradients at baseline that remain unchanged after barrier disruption.

REFERENCES

1. Heuss, E., Die Reaktion des Scheisses beim gesunden Menschen, *Monatsh. Prakt. Dermatol.*, 14, 343, 1892.
2. Schade, H.M., A., Zur physikalischen Chemie der Hautoberfläche, *Arch. Dermatol. Syphil.*, 154, 690, 1928.
3. Ohman, H. and Vahlquist, A., The pH gradient over the stratum corneum differs in X-linked recessive and autosomal dominant ichthyosis: a clue to the molecular origin of the "acid skin mantle?" *J. Invest. Dermatol.*, 111(4), 674, 1998.
4. Berardesca, E. et al., Differences in stratum corneum pH gradient when comparing white Caucasian and black African-American skin, *Br. J. Dermatol.*, 139(5), 855, 1998.
5. Turner, N.G., Cullander, C., and Guy, R.H., Determination of the pH gradient across the stratum corneum, *J. Invest. Dermatol. Symp. Proc.*, 3(2), 110, 1998.
6. Ohman, H. and Vahlquist, A., *In vivo* studies concerning a pH gradient in human stratum corneum and upper epidermis, *Acta Derm. Venereol.*, 74(5), 375, 1994.
7. Behne, M. et al., Neonatal development of the stratum corneum pH gradient: localization and mechanisms leading to emergence of optimal barrier function, *J. Invest. Dermatol.*, 120(6), 998, 2003.
8. Behne, M. et al., NHE1 regulates the stratum corneum permeability barrier homeostasis, *J. Biol. Chem.*, 277(49), 47399, 2002.
9. Szakall, A., Über die Eigenschaften, Herkunft und physiologischen Funktionen der H-Ionenkonzentration bestimmenden Wirkstoffe in der verhornten Epidermis, *Arch. Klin. Exp. Dermatol.*, 201, 331, 1955.
10. Denda, M., Hosoi, J., and Asida, Y., Visual imaging of ion distribution in human epidermis, *Biochem. Biophys. Res. Commun.*, 272(1), 134, 2000.
11. Dikstein, S. and Zlotogorski, A., Measurement of skin pH, *Acta Derm. Venereol. Suppl.*, 185, 18, 1994.
12. Kroll, C. et al., Influence of drug treatment on the microacidity in rat and human skin — an *in vitro* electron spin resonance imaging study, *Pharm. Res.*, 18(4), 525, 2001.
13. Giusti, F. et al., Skin barrier, hydration, and pH of the skin of infants under 2 years of age, *Pediatr. Dermatol.*, 18(2), 93, 2001.
14. Rawlings, A.V. et al., Stratum corneum moisturization at the molecular level, *J. Invest. Dermatol.*, 103(5), 731, 1994.
15. Harding, C. et al., Review article: dry skin, moisturization, and corneodesmolysis, *Int. J. Cosmetic Sci.*, 22, 21, 2000.
16. Harding, C.R. and Scott, I.R., Histidine-rich proteins (filaggrins): structural and functional heterogeneity during epidermal differentiation, *J. Mol. Biol.*, 170(3), 651, 1983.
17. Scott, I.R., Harding, C.R., and Barrett, J.G., Histidine-rich protein of the keratohyalin granules. Source of the free amino acids, urocanic acid and pyrrolidone carboxylic acid in the stratum corneum, *Biochim. Biophys. Acta*, 719(1), 110, 1982.

18. Warner, R.R., Bush, R.D., and Ruebusch, N.A., Corneocytes undergo systematic changes in element concentrations across the human inner stratum corneum, *J. Invest. Dermatol.*, 104(4), 530, 1995.
19. Scott, I.R. and Harding, C.R., Filaggrin breakdown to water binding compounds during development of the rat stratum corneum is controlled by the water activity of the environment, *Dev. Biol.*, 115(1), 84, 1986.
20. Krien, P.M. and Kermici, M. Evidence for the existence of a self-regulated enzymatic process within the human stratum corneum — an unexpected role for urocanic acid, *J. Invest. Dermatol.*, 115(3), 414, 2000.
21. Baden, H.P. and Pathak, M.A., The metabolism and function of urocanic acid in skin, *J. Invest. Dermatol.*, 48(1), 11, 1967.
22. Schwarz, W. et al., Distribution of urocanic acid in human stratum corneum, *Photodermatology*, 3(4), 239, 1986.
23. Kuroda, Y. et al., A new sensitive method for assay of histidase in human skin and detection of heterozygotes for histidinemia, *Clin. Chim. Acta*, 96(1–2), 139, 1979.
24. Sano, H. et al., Isolation of a rat histidase cDNA sequence and expression in *Escherichia coli* — evidence of extrahepatic/epidermal distribution, *Eur. J. Biochem.*, 250(1), 212, 1997.
25. Baden, H.P. and Gavioli, L., Histidase activity in rat liver and epidermis, *J. Invest. Dermatol.*, 63(6), 479, 1974.
26. Bhargava, M.M. and Feigelson, M., Studies on the mechanisms of histidase development in rat skin and liver. I. Basis for tissue specific developmental changes in catalytic activity, *Dev. Biol.*, 48(2), 212, 1976.
27. Taylor, R.G., Levy, H.L., and McInnes, R.R., Histidase and histidinemia. Clinical and molecular considerations, *Mol. Biol. Med.*, 8(1), 101, 1991.
28. Harding, C.R., Ellis, K., and Scott, I.R., Alterations in the processing of human filaggrin following skin occlusion *in vitro* and *in vivo*, *J. Invest. Dermatol.*, 100, 579, 1993.
29. Taylor, R.G. et al., Identification of the mutation in murine histidinemia (his) and genetic mapping of the murine histidase locus (Hal) on chromosome 10, *Genomics*, 16(1), 231, 1993.
30. Kacser, H., Bulfield, G., and Wallace, M.E., Histidinaemic mutant in the mouse, *Nature*, 244(5411), 77, 1973.
31. Fluhr, J.W. et al., Generation of free fatty acids from phospholipids regulates stratum corneum acidification and integrity, *J. Invest. Dermatol.*, 117(1), 44, 2001.
32. Mao-Qiang, M. et al., Secretory phospholipase A2 activity is required for permeability barrier homeostasis, *J. Invest. Dermatol.*, 106(1), 57, 1996.
33. Mao-Qiang, M. et al., Extracellular processing of phospholipids is required for permeability barrier homeostasis, *J. Lipid Res.*, 36(9), 1925, 1995.
34. Fluhr, J.W. and Elias, P.M., Stratum corneum pH: formation and function of the "acid mantle," *Exogenous Dermatol.*, 1(4), 163, 2002.
35. Demaurex, N. and Grinstein, S., Na+/H+ antiport: modulation by ATP and role in cell volume regulation, *J. Exp. Biol.*, 196, 389, 1994.
36. Mahnensmith, R.L. and Aronson, P.S., The plasma membrane sodium-hydrogen exchanger and its role in physiological and pathophysiological processes, *Circ. Res.*, 56(6), 773, 1985.
37. Orlowski, J. and Grinstein, S., Na+/H+ exchangers of mammalian cells, *J. Biol. Chem.*, 272, 22373, 1997.

38. Orlowski, J., Kandasamy, R.A., and Shull, G.E., Molecular cloning of putative members of the Na/H exchanger gene family. cDNA cloning, deduced amino acid sequence, and mRNA tissue expression of the rat Na/H exchanger NHE-1 and two structurally related proteins, *J. Biol. Chem.*, 267(13), 9331, 1992.
39. Sarangarajan, R. et al., Molecular and functional characterization of sodium-hydrogen exchanger in skin as well as cultured keratinocytes and melanocytes, *Biochim. Biophys. Acta*, 1511, 181, 2001.
40. van Hooijdonk, C. et al., Demonstration of an Na+/H+ exchanger in mouse keratinocytes measured by the novel pH-sensitive fluorochrome SNARF-calcein, *Cell Proliferation*, 30(8–9), 351, 1997.
41. Behne, M. et al., Functional role of the sodium/hydrogen antiporter, NHE1, in the epidermis: pharmacologic and NHE1 null-allele mouse studies, *J. Invest. Dermatol.*, 114(4), 797, 2000.
42. Warrier, A.G. et al., A comparison of black and white skin using noninvasive methods. *J. Soc. Cosmetic Chem.*, 47, 229, 1996.
43. Draize, J.H., The determination of the pH of the skin of man and common laboratory animals, *J. Invest. Dermatol.*, 5, 77, 1942.
44. Yosipovitch, G. et al., Skin surface pH in intertriginous areas in NIDDM patients. Possible correlation to candidal intertrigo, *Diabetes Care*, 16(4), 560, 1993.
45. Yosipovitch, G. et al., Skin surface pH, moisture, and pruritus in haemodialysis patients. *Nephrol. Dial Transplant.*, 8(10), 1129, 1993.
46. Fluhr, J.W., Pfisterer, S., and Gloor, M., Direct comparison of skin physiology in children and adults with bioengineering methods. *Pediatr. Dermatol.*, 17(6), 436, 2000.
47. Ehlers, C. et al., Females have lower skin surface pH than men. A study on the surface of gender, forearm site variation, right/left difference and time of the day on the skin surface pH, *Skin Res. Technol.*, 7(2), 90, 2001.
48. Treffel, P. et al., Hydration, transepidermal water loss, pH and skin surface parameters: correlations and variations between dominant and non-dominant forearms, *Br. J. Dermatol.*, 130(3), 325, 1994.
49. Elsner, P. and Maibach, H.I., The effect of prolonged drying on transepidermal water loss, capacitance and pH of human vulvar and forearm skin, *Acta Derm. Venereol.*, 70(2), 105, 1990.
50. Blank, I.A., Measurement of pH of the skin surface. II. pH of the exposed surfaces of adults with no apparent skin lesions, *J. Invest. Dermatol.*, 2, 75, 1939.
51. Ehlers, C. et al., Comparison of two pH meters used for skin surface pH measurement: the pH meter "pH900" from Courage & Khazaka versus the pH meter "1140" from Mettler Toledo, *Skin Res. Technol.*, 7(2), 84, 2001.
52. Zlotogorski, A., Distribution of skin surface pH on the forehead and cheek of adults, *Arch. Dermatol. Res.*, 279(6), 398, 1987.
53. Wilhelm, K.P., Cua, A.B., and Maibach, H.I., Skin aging. Effect on transepidermal water loss, stratum corneum hydration, skin surface pH, and casual sebum content, *Arch. Dermatol.*, 127(12), 1806, 1991.
54. Visscher, M.O. et al., Changes in diapered and nondiapered infant skin over the first month of life, *Pediatr. Dermatol.*, 17(1), 45, 2000.
55. Behrendt, H. and Green, M., Skin pH pattern in the newborn infant, *J. Dis. Child*, 95, 35, 1958.
56. Fox, C., Nelson, D., and Wareham, J., The timing of skin acidification in very low birth weight infants, *J. Perinatol.*, 18(4), 272, 1998.

57. Seidenari, S. and Giusti, G., Objective assessment of the skin of children affected by atopic dermatitis: a study of pH, capacitance and TEWL in eczematous and clinically uninvolved skin, *Acta Derm. Venereol.*, 75(6), 429, 1995.
58. Eberlein-Konig, B. et al., Skin surface pH, stratum corneum hydration, trans-epidermal water loss and skin roughness related to atopic eczema and skin dryness in a population of primary school children, *Acta Derm. Venereol.*, 80(3), 188, 2000.
59. Thune, P. et al., The water barrier function of the skin in relation to the water content of stratum corneum, pH and skin lipids. The effect of alkaline soap and syndet on dry skin in elderly, non-atopic patients, *Acta Derm. Venereol.*, 68(4), 277, 1988.
60. Le Fur, I. et al., Analysis of circadian and ultradian rhythms of skin surface properties of face and forearm of healthy women, *J. Invest. Dermatol.*, 117(3), 718, 2001.
61. Burry, J., Coulson, H.F., and Roberts, G., Circadian rhythms in axillary skin surface pH. *Int. J. Cosmetic Sci.*, 23, 207, 2001.
62. Yosipovitch, G. et al., Time-dependent variations of the skin barrier function in humans: transepidermal water loss, stratum corneum hydration, skin surface pH, and skin temperature, *J. Invest. Dermatol.*, 110(1), 20, 1998.
63. Beare, J.T. et al., The pH of the skin surface of children with seborrheic dermatitis compared with unaffected children, *Br. J. Dermatol.*, 70, 233, 1958.
64. Tippelt, H., Zur Hautoberflächen-pH-Mesung und Alkaliresistenzprobe bei Hautgesunden und Ichthyosis vulgaris-Kranken, *Dermatologica*, 139(3), 201, 1969.
65. Post, A., Gloor, M., and Gehring, W., Über den Einfluss der Hautwaschung auf den Haut-pH-Wert, *Dermatol. Monatsschr.*, 178, 216, 1992.
66. Gfatter, R., Hackl, P., and Braun, F., Effects of soap and detergents on skin surface pH, stratum corneum hydration and fat content in infants, *Dermatology*, 195(3), 258, 1997.
67. Korting, H.C. et al., Influence of repeated washings with soap and synthetic detergents on pH and resident flora of the skin of forehead and forearm. Results of a cross-over trial in health probationers, *Acta Derm. Venereol.*, 67(1), 41, 1987.
68. Grunewald, A.M. et al., Damage to the skin by repetitive washing, *Contact Dermatitis*, 32(4), 225, 1995.
69. Grunewald, A.M. et al., Efficacy of barrier creams, *Curr. Probl. Dermatol.*, 23, 187, 1995.
70. Korting, H.C. et al., Differences in the skin surface pH and bacterial microflora due to the long-term application of synthetic detergent preparations of pH 5.5 and pH 7.0. Results of a crossover trial in healthy volunteers, *Acta Derm. Venereol.*, 70(5), 429, 1990.
71. Hartmann, A.A., Effect of occlusion on resident flora, skin-moisture and skin-pH, *Arch. Dermatol. Res.*, 275(4), 251, 1983.
72. Aly, R. et al., Effect of prolonged occlusion on the microbial flora, pH, carbon dioxide and transepidermal water loss on human skin, *J. Invest. Dermatol.*, 1978. 71(6), 378.
73. Hölzle, E., Antiperspirants. *Dermatopharmacology of Topical Preparations*, Treffel, P., ed., Springer, Berlin, 2000, p. 401.
74. Bernstein, E.T. and F. Hermann, The acidity on the surface of the skin, *N.Y. State J. Med.*, 42, 436, 1942.
75. Berg, R.W., Milligan, M.C., and Sarbaugh, F.C., Association of skin wetness and pH with diaper dermatitis, *Pediatr. Dermatol.*, 11(1), 18, 1994.

76. Mauro, T. et al., Barrier recovery is impeded at neutral pH, independent of ionic effects: implications for extracellular lipid processing, *Arch. Dermatol. Res.*, 290(4), 215, 1998.
77. Wilhelm, D., Elsner, P., and Maibach, H.I. Standardized trauma (tape stripping) in human vulvar and forearm skin. Effects on transepidermal water loss, capacitance and pH, *Acta Derm. Venereol.*, 71(2), 123, 1991.
78. Berardesca, E. et al., Alpha hydroxyacids modulate stratum corneum barrier function, *Br. J. Dermatol.*, 137(6), 934, 1997.
79. Fartasch, M., Teal, J., and Menon, G.K., Mode of action of glycolic acid on human stratum corneum: ultrastructural and functional evaluation of the epidermal barrier, *Arch. Dermatol. Res.*, 289(7), 404, 1997.
80. Elias, P.M., McNutt, N.S., and Friend, D.S., Membrane alterations during cornification of mammalian squamous epithelia: a freeze-fracture, tracer, and thin-section study, *Anat. Rec.*, 189(4), 577, 1977.
81. Elias, P.M. and Feingold, K.R., Coordinate regulation of epidermal differentiation and barrier homeostasis, *Skin Pharmacol. Appl. Skin Physiol.*, 14(Suppl 1), 28, 2001.
82. Suzuki, N. et al., Structures, enzymatic properties, and expression of novel human and mouse secretory phospholipase A(2)s, *J. Biol. Chem.*, 275(8), 5785, 2000.
83. Tischfield, J.A., A reassessment of the low molecular weight phospholipase A2 gene family in mammals, *J. Biol. Chem.*, 272(28), 17247, 1997.
84. Hobkirk, R., Steroid sulfotransferase and steroid sulfate sulfatase: characterization and biological roles, *Can. J. Biochem. Cell Biol.*, 63, 1127, 1985.
85. Holleran, W.M. et al., Processing of epidermal glucosylceramides is required for optimal mammalian cutaneous permeability barrier function, *J. Clin. Invest.*, 91(4), 1656, 1993.
86. Schmuth, M. et al., Permeability barrier disorder in Niemann-Pick disease: sphingomyelin-ceramide processing required for normal barrier homeostasis, *J. Invest. Dermatol.*, 115(3), 459, 2000.
87. Holleran, W.M. et al., beta-Glucocerebrosidase activity in murine epidermis: characterization and localization in relation to differentiation, *J. Lipid Res.*, 33(8), 1201, 1992.
88. Zettersten, E. et al., Recessive x-linked ichthyosis: role of cholesterol-sulfate accumulation in the barrier abnormality, *J. Invest. Dermatol.*, 111(5), 784, 1998.
89. Leyden, J.J., Stewart, R., and Kligman, A.M., Updated in vivo methods for evaluating topical antimicrobial agents on human skin, *J. Invest. Dermatol.*, 72(4), 165, 1979.
90. Jensen, P.J. et al., Plasminogen activator inhibitor type 2: an intracellular keratinocyte differentiation product that is incorporated into the cornified envelope, *Exp. Cell Res.*, 217(1), 65, 1995.
91. Chikakane, K. and Takahashi, H., Measurement of skin pH and its significance in cutaneous diseases, *Clin. Dermatol.*, 13(4), 299, 1995.
92. Korting, H.C. et al., Influence of topical erythromycin preparations for acne vulgaris on skin surface pH, *Clin. Invest.*, 71(8), 644, 1993.
93. Brook, I., Microbiology of secondarily infected diaper dermatitis, *Int. J. Dermatol.*, 31(10), 700, 1992.
94. Beare, J. et al., The effect of age on the pH of the skin surface in the first week of life, *Br. J. Dermatol.*, 72, 62, 1960.
95. Menon, G.K. et al., Localization of calcium in murine epidermis following disruption and repair of the permeability barrier, *Cell Tissue Res.*, 270(3), 503, 1992.

96. Mao-Qiang, M. et al., Calcium and potassium inhibit barrier recovery after disruption, independent of the type of insult in hairless mice, *Exp. Dermatol.*, 6(1), 36, 1997.
97. Menon, G.K., Grayson, S., and Elias, P.M., Ionic calcium reservoirs in mammalian epidermis: ultrastructural localization by ion-capture cytochemistry, *J. Invest. Dermatol.*, 84(6), 508, 1985.
98. Elias, P.M. et al., Formation of the epidermal calcium gradient coincides with key milestones of barrier ontogenesis in the rodent, *J. Invest. Dermatol.*, 110(4), 399, 1998.
99. Forslind, B., Quantitative X-ray microanalysis of skin. Particle probe evaluation of the skin barrier function, *Acta Derm. Venereol. Suppl. (Stockh.)*, 134, 1, 1987.
100. Menon, G.K., Elias, P.M., and Feingold, K.R., Integrity of the permeability barrier is crucial for maintenance of the epidermal calcium gradient, *Br. J. Dermatol.*, 130(2), 139, 1994.
101. Lee, S.H. et al., Iontophoresis itself on hairless mouse skin induces the loss of the epidermal calcium gradient without skin barrier impairment, *J. Invest. Dermatol.*, 111(1), 39, 1998.
102. Denda, M. and Kumazawa, N., Negative electric potential induces alteration of ion gradient and lamellar body secretion in the epidermis, and accelerates skin barrier recovery after barrier disruption, *J. Invest. Dermatol.*, 118(1), 65, 2002.
103. Mauro, T.M. et al., Ion channels are linked to differentiation in keratinocytes, *J. Membr. Biol.*, 132(3), 201, 1993.
104. Mauro, T. et al., Acute barrier perturbation abolishes the Ca2+ and K+ gradients in murine epidermis: quantitative measurement using PIXE, *J. Invest. Dermatol.*, 111(6), 1198, 1998.
105. Vicanova, J. et al., Normalization of epidermal calcium distribution profile in reconstructed human epidermis is related to improvement of terminal differentiation and stratum corneum barrier formation, *J. Invest. Dermatol.*, 111(1), 97, 1998.
106. Elias, P.M. et al., Modulations in epidermal calcium regulate the expression of differentiation-specific markers, *J. Invest. Dermatol.*, 119(5), 1128, 2002.
107. Elias, P. et al., Origin of the epidermal calcium gradient: regulation by barrier status and role of active vs passive mechanisms, *J. Invest. Dermatol.*, 119(6), 1269, 2002.
108. Goldsmith, L.A., Physiology, biochemistry, and molecular biology of the skin. 1991, New York ; Oxford: Oxford University Press.
109. Menon, G.K., Feingold, K.R., and Elias, P.M., Lamellar body secretory response to barrier disruption, *J. Invest. Dermatol.*, 98(3), 279, 1992.
110. Elias, P.M. and Menon, G.K., Structural and lipid biochemical correlates of the epidermal permeability barrier, *Adv. Lipid Res.*, 24, 1, 1991.
111. Bikle, D.D. et al., Changes in calcium responsiveness and handling during keratinocyte differentiation. Potential role of the calcium receptor, *J. Clin. Invest.*, 97(4), 1085, 1996.
112. Yuspa, S.H. et al., Expression of murine epidermal differentiation markers is tightly regulated by restricted extracellular calcium concentrations in vitro, *J. Cell Biol.*, 109(3), 1207, 1989.
113. Hennings, H. et al., Calcium regulation of growth and differentiation of mouse epidermal cells in culture, *Cell*, 19(1), 245, 1980.
114. Pillai, S. et al., Calcium regulation of growth and differentiation of normal human keratinocytes: modulation of differentiation competence by stages of growth and extracellular calcium, *J. Cell Physiol.*, 143(2), 294, 1990.

115. Eckert, R.L., Crish, J.F., and Robinson, N.A., The epidermal keratinocyte as a model for the study of gene regulation and cell differentiation, *Physiol. Rev.*, 77(2), 397, 1997.
116. Forslind, B. et al., Elemental analysis mirrors epidermal differentiation, *Acta Derm. Venereol.*, 79(1), 12, 1999.
117. Pallon, J. et al., Pixe analysis of pathological skin with special reference to psoriasis and atopic dry skin, *Cell Mol. Biol. (Noisy-le-grand)*, 42(1), 111, 1996.
118. Menon, G.K. and Elias, P.M., Ultrastructural localization of calcium in psoriatic and normal human epidermis, *Arch Dermatol.*, 127(1), 57, 1991.
119. Hwang, S.M. et al., Basis of occlusive therapy in psoriasis: correcting defects in permeability barrier and calcium gradient, *Int. J. Dermatol.*, 40(3), 223, 2001.
120. Denda, M. et al., Some magnesium salts and a mixture of magnesium and calcium salts accelerate skin barrier recovery, *Arch Dermatol. Res.*, 291(10), 560, 1999.
121. Song, H.J. et al., The genes encoding geranylgeranyl transferase alpha-subunit and transglutaminase 1 are very closely linked but not functionally related in terminally differentiating keratinocytes, *Biochem. Biophys. Res. Commun.*, 235(1), 10, 1997.
122. Seabra, M.C. et al., Rab geranylgeranyl transferase. A multisubunit enzyme that prenylates GTP-binding proteins terminating in Cys-X-Cys or Cys-Cys, *J. Biol. Chem.*, 267(20), 14497, 1992.
123. Novick, P. and Brennwald, P., Friends and family: the role of the Rab GTPases in vesicular traffic, *Cell*, 75(4), 597, 1993.
124. Warner, R.R., Myers, M.C., and Taylor, D.A., Electron probe analysis of human skin: determination of the water concentration profile, *J. Invest. Dermatol.*, 90(2), 218, 1988.
125. Hennings, H., Holbrook, K.A., and Yuspa, S.H., Potassium mediation of calcium-induced terminal differentiation of epidermal cells in culture, *J. Invest. Dermatol.*, 81(1 Suppl), 50, 1983.
126. von Zglinicki, T. et al., Water and ion distribution profiles in human skin, *Acta Derm. Venereol.*, 73(5), 340, 1993.

4 Occlusion

Jana Praessler and Joachim W. Fluhr

CONTENTS

I. Definition 49
II. Epidermal Effects 49
A. Role of Water 49
B. Effects on Intraepidermal Processes 50
III. Influence of the Occlusion Time 51
IV. Occlusive Materials and Wound Healing 52
V. Absorption 53
VI. Therapeutic Use 53
VII. Measurement Parameters and Techniques 54
VIII. Conclusions 55
References 56

I. DEFINITION

According to the definition from Bucks [1], occlusion is the complete impairment of passive transepidermal water loss at the application site. Thus, the main influence on the skin under occlusion seems to be the retention of water [2,3] that is detectable by the decrease of transepidermal water loss and the increase of water holding capacity [4]. Furthermore, examination of the effects of occlusion focuses on the influence of subsequent effects like temperature, gas exchange, stimulation of enzymes, growth and migration of cells, and drug and substance penetration. Changing of the epidermal ambiance (e.g., by growth of bacterial flora) has to be considered as well.

II. EPIDERMAL EFFECTS

A. Role of Water

Water content obviously plays a crucial role in the homeostasis of the stratum corneum (SC). Denda et al. [5] found that dry environment leads to epidermal hyperplasia, mast cell degeneration, and cytokine secretion. They also showed that these effects can be prevented by occlusion. Another group [6] showed that occlusion also works therapeutically on dry skin, leading to clinical improvement. The authors

0-8493-1443-7/05/$0.00+$1.50

measured an increased tissue mass, despite a stable amount of cell layers and thickness; an increase of water-extractable material and keratin; and a decrease in the ether-extractable materials. These phenomena were summarized as an increase in the SC capacity to absorb water.

In contrast, prolonged water contact may cause irritant contact dermatitis. It is well known that, for example, prolonged and repetitive use of occlusive rubber gloves supports the development of occupational contact dermatitis [7–9]. Irritation could experimentally be prevented by additional use of cotton gloves worn between the skin and the occlusive rubber gloves [10]. If water-absorbing material like cotton prevents irritation under occlusion, it is further proof that irritation after occlusion is mainly provoked by retention of water. Warner et al. [11] proposed a pathway with the primary disruption of the intercellular lipid structure being by water and with an assumed breakdown of corneodesmosomes [12], leading to increased skin permeability [13]. These observations could correlate with findings of Lindberg et al. [14], who found widening of the spinous intercellular space after occlusion.

Under occlusion, water leads to thickening of the SC (threefold after 4 hours, fourfold after 24 hours) resulting from corneocyte swelling and the storage of water and water-soluble substances in intercellular cisternae [13,15]. These observations present a dynamic dimension of the brick and mortar model that is able to adapt to external stress factors like hyperhydration.

Not only the content but the flux of water through the SC seems to be of importance. In hairless mice, it was shown that barrier recovery after disruption was prolonged under total occlusion but normal under vapor-permeable membranes [16]. The parameters of regulation are the stimulation of DNA transcription [17] and the subsequent lipid synthesis [16]. They are retarded by total occlusion and water retention, respectively, after acute barrier disruption.

In several studies, no erythema occurred after occlusive water exposure [13]. In one study, a blanching effect occurred 1 hour after the end of occlusion by rubber gloves (clinical assessment) [9]. However, missing erythema does not necessarily mean an absence of damage [13].

B. Effects on Intraepidermal Processes

Occlusion influences most of the components of the epidermal homeostasis. On the cellular level, occlusion inhibits mitotic activity in nonhealthy skin (e.g., in psoriasis [18] or after acute barrier disruption by tape stripping [19]), but not in healthy epidermis [20]. This effect occurs independent of the type of the occlusive film used [21]. In healthy skin, keratinocytes show newly formed perinuclear and cytoplasmic vacuoles after occlusion [21]. Levine et al. [22] pointed out that occlusion exerts different effects depending on the character of wounds: In deep lesions (removal of the entire epidermis and papillary dermis), epidermal cell proliferation is increased under occlusion, whereas it is decreased in superficial wounds (removal of the SC) [22]. Proksch et al. [17] also showed an inhibition of DNA synthesis under complete occlusion after an acute barrier insult induced by tape stripping. This observation seems to be specific for the condition after treatment with tape stripping, because delayed barrier repair was also shown in human skin under occlusion after barrier disruption by tape stripping but not after barrier perturbation by treatment with sodium lauryl sulfate [23]. Van de Kerkhof et al. [24] could not confirm the inhibition of DNA synthesis in human skin using hydrocolloidal dressings after acute barrier disruption by tape stripping [24].

Infiltration of mononuclear cells from dermal layers has been demonstrated [14]. Langerhans cells progressively move toward the central layers in the epidermis [25] and show intracellular alterations, indicating an activation [26]. These processes happen without immediately visible erythema or dermatitis, respectively [26,27]. Only after prolonged water contact (e.g., under occlusion) can irritation and inflammation be observed clinically [28].

At the anatomical site of the sweat glands, occlusion leads to the induction of miliaria [21,29]. Occlusive tapes applied for several days lead to anhydrosis at the occlusive site that requires at least 3 weeks for recovery. The degree of anhidrosis correlates with the occlusive property of the tape [30]. The downregulation of sweat gland activity needs to be further investigated.

The change of epidermal activity is noticeable because of the altered amount of enzymes that are responsible for barrier homeostasis and proliferation. Occlusion in psoriasis, for example, leads to recurring expression of Calmodulin, basal keratin, and ki-67 in basal keratinocytes [18], whereas the activity of urokinase decreases [31]. These findings confirm a decrease of keratinocyte migration and proliferation in psoriasis after occlusion [19,31,32]. In healthy skin, after superficial barrier disruption, proliferation markers (like ornithine decarboxylase activity, keratin 16 expression) are expressed in a comparable amount under both occlusive and non-occlusive conditions [24].

The changes and the influence of the concentration of O_2 on wound healing under occlusion obviously varies under different conditions. It seems that the tension of oxygen has to be adjusted to the depth of the wound. Higher pO_2 is required for the growth of keratinocytes [33], whereas a lower oxygen tension promotes the proliferation of fibroblasts, the production of angiogenesis factors, and the synthesis of collagen [34–36]. The tension of oxygen can be adjusted by using dressings with appropriate oxygen permeability [21]. An initial deficiency of oxygen correlates with an acidic tissue pH and subsequently promotes granulocyte, macrophage, and lymphocyte immigration as well as an activation of the complement system. Moreover, a low pH acts bacteriostatically [37,38].

Occlusion induces several processes in the SC and the viable epidermis, leading to morphological changes. Thus, occlusion has to be considered not only as supportive treatment but as a pharmacologic-like intervention [21].

III. INFLUENCE OF THE OCCLUSION TIME

Different investigators introduced test settings for prolonged occlusion in human skin. They used multiple measurement parameters, such as transepidermal water loss, water-holding capacity, prehydration state, capacitance-based SC hydration measurement, microbial flora, surface pH, and CO_2 [37,39,40]. They showed damage to the barrier function after only 1 day of occlusion and maximal values of transepidermal water loss (TEWL), water-holding capacity, prehydration state, and SC hydration after 2 days [39]. There was also an increase in the amount of microbial flora, pH, and CO_2 [37,40]. Under long-term occlusion (4 or 8 days), there was no further increase but even a decrease of these values [37,39]. Some sort of hardening effect has to be discussed. Under these experimental conditions, water content and CO_2, however, showed a higher steady state [37] that correlated with the morphological observation of lacking skin dryness [39].

In short-term occlusion, changes in SC hydration are already detectable after 30 minutes. Two different groups showed, by using either dielectric probe technique [3] or TEWL and capacitance [41], an increase of SC water content up to 50% after an occlusion of 30 minutes could be achieved, dropping again to basal values 1 h after stopping the occlusion. The second group even demonstrated, by using echography, a dermal hydration lasting 2 hours after occlusive treatment [41].

IV. OCCLUSIVE MATERIALS AND WOUND HEALING

Not only solid materials like membranes and dressings have occlusive effects. All topically applied products exert an occlusion to a greater (creams) or lesser (lotions, powders) extent [4,42]. They stay on the surface or penetrate into the intercellular spaces of the SC, forming a water-impermeable, occlusive membrane [43].

The effect of occlusion is different even for solid materials in both quantity and quality. Berardesca et al. [44] found that occlusive dressings may vary in skin surface water loss. Materials with similar skin surface water loss values showed different values for their water holding capacity. Semipermeable membranes are of high therapeutic value because of the optimized water-vapor gradient [16,45] that allows a moist ambiance as well as a transepidermal water flux and an oxygen contact. In psoriasis, for example, the optimal occlusive dressing has a water-vapor permeability of 500 g/m^2 per day (~20 g/m^2 per hour), almost corresponding to the amount of water evaporization in normal skin [18]. Dressings with different permeability activities were significantly less effective. This result supports the data indicating that occlusive dressings have to be specific to the character of wounds.

Hydrocolloidal dressings consist of two components [46]: a carrier film (mostly polyurethane foil) that is permeable for gas but not for water or bacteria, and an absorbing core of hydrocolloidal particles, like gelatin, that builds up an osmotic gradient, storing secreted fluids and transforming itself into a gel-like structure. Hydrocolloidal dressings adapt to wounds with loss of tissue substance, thus covering just the wound ground and not affecting the adjacent healthy skin by maceration. They accelerate all processes of wound healing [46], such as angioneogenesis, proliferation of fibroblasts, and epithelialization. They also act bacteriostatically, reducing the formation of scars and the risk of sensibilization to topically applied compounds. Hydrocolloidal dressings are proven to be more effective than saline gauze compresses in long-term wound treatment [47].

V. ABSORPTION

Absorption under occlusion is already enhanced by blocking evaporation. Increased absorption has been shown for several substances with different chemical structures [1,4,48]. As there is an increase in the water content of the SC under occlusion and a swelling of the corneocytes as well as an enlargement of the intercellular water cisternae [13], an enhanced absorption of water-soluble substances can be observed. Furthermore, the intercellular lamellar bilayers can be disturbed under occlusive conditions. However, several investigations did not support these assumptions. It was actually shown that absorption of hydrophilic or amphiphilic molecules is not necessarily increased under occlusion, in contrast to absorption of lipophilic molecules [1,49,50]. The more lipophilic the drug is, the more absorption takes place via the intercellular route. Hydrophilic compounds are likely to penetrate into the skin via the intracellular keratin matrix [50] or via the appendages [51]. It has been shown that absorption is increased in follicular-rich areas and in "active" follicles with hair growth and sebum production [52]. Whether absorption on the follicular way is increased because of occlusion remains to be explored further.

The penetration of water itself is increased under occlusion [53]. Moreover, water can act as an occlusive compound (e.g., as a bath) and can thus enhance the uptake of substances (e.g., psoralen) [4]. In addition to occlusion, other factors have an effect on the penetration rate of topically applied drugs, such as temperature, dosing conditions, and the vehicle. The vehicle or parts of it, like lipids, influence the release of drugs from the formulation and may also interact with the SC in a druglike manner [54]. Kim et al. [55] investigated the drug release of hydrocortisone from a liposome formulation both with and without occlusion. They found no differences concerning the depth of the penetration, but after occlusion there was a markedly higher concentration of hydrocortisone in the SC than in the viable epidermis [55]. This tissue partitioning within the epidermis has to be considered when it comes to measurement techniques investigating absorption. In contrast to this result, Gloor et al. [56] demonstrated a quick and high bioavailability of steroidal substances in liposomal formulation after occlusion, with effects even in the deeper dermis.

VI. THERAPEUTIC USE

Occlusion is in widespread use in the management of wound care. The occlusive dressings produce a warm and moist ambiance. An ideal temperature of 37°C leads to activation of phagocytes, which is followed by a reduction of wound detritus [57]. The film structure of the dressing material allows the keratinocytes to grow from the borders of the wound along the occlusive material and to swim on the surface of the wound. Keratinization in nonoccluded wounds comes mainly from the stem cells of the hair follicles and is situated on the ground of the wound [57]. It was also reported by Collawn [58] that after CO_2 laser treatment, occlusion enhances cell migration from the hair follicles, diminishing eschar formation and leading to a more rapid healing. In occluded, epithelium-covered wounds, the recovery of the dermal collagen matrix is enhanced by occlusion [57].

The treatment of wounds has to be adjusted to the depth of the wound [59] and to the phase of wound healing [57]. Strongly secreting wounds such as leg ulcers, for example, require an absorbing activity (hydrocolloids) and a more frequent change of dressings. Otherwise, a moist wound chamber could lead to a delay of healing and to an enlargement of the wound by maceration of the surrounding healthy skin. In wounds with low secretion, the frequency of dressing changes can be diminished, which prevents microbial colonization of the wound. It is reported that bacterial proliferation under occlusive dressing increases [37,40,59,60], and the number of organisms is strictly related to the degree of hydration [61]. A qualitative change in the microbial pattern, especially with the increase of coagulase-negative staphylococci, lipophilic bacteria, and diphtheroids has been reported [40,61]. Despite these facts, the infection rates are significantly lower under occlusive dressings [59]. Occlusion even prevents the dissemination of methicillin-resistant *Staphylococcus aureus* [59].

In psoriasis, occlusion has been used since the nineteenth century [62]. It leads to a reduction of scale and thickness [63], but it has no influence on erythema [9,10,63]. Changes can be seen after long-term occlusion of 2 weeks, but no additional effects are seen after 10 weeks [63]. To improve the therapeutic affectivity of occlusion, this treatment can be combined with antipsoriatic medication such as dithranol [18] or topical corticosteroids [64]. The effect of occlusion consists of the normalization of the expression of proliferation markers [18,31,65], the reduction of the acanthosis [65], the decrease of the mitotic rate [66], and a subsequent decrease in proliferation [63], along with the formation of a granular cell layer [67] and, as a consequence, an improvement of the barrier function.

VII. MEASUREMENT PARAMETERS AND TECHNIQUES

For the assessment of the effects of occlusion in the SC and epidermis, several noninvasive biophysical methods are used to evaluate the flux and the content of water, such as transepidermal water loss, capacitance, water holding capacity, hygroscopicity, and the dielectric probe technique [2,44,68–70]. Although these methods focus on similar parameters — the changes in skin conditions resulting from the altered influence of water — they are complementary. The measurements show dynamic variation over time [21,37]. Fluhr et al. [39] showed an increase of both TEWL and hygroscopicity after long-term occlusion, but with no correlation between the two parameters. Using B-scanning echography, Seidenari et al. [41] also found that hydration is different in the single skin compartments after short-term occlusion. The increases in dermal hydration were detectable longer than differences of TEWL and capacitance [41]. Ryatt et al. demonstrated, by using laser Doppler velocimetry [71], vasodilatory effects on dermal capillaries resulting from enhanced absorption of hexyl nicotinate after short-term occlusion [2]. The authors also found a correlation between laser Doppler velocimetry and the values of SC hydration.

Because the tension of CO_2 can give an indirect measure of the integrity of the SC barrier function [21], it is determined via the carbon dioxide emission rate. The CO_2 is generated by keratinocytes with an increase in impaired SC and can be detected on the skin surface. This sensitive method correlates with temperature, skin thickness, and sweating [21,72,73] — parameters that are strongly influenced by occlusion.

Changes in the quantity of proliferation markers are demonstrated via immunohistochemistry. There are several publications in psoriatic skin that show an inhibition of disease-specific increased proliferation markers after occlusion [18,21,31,65]. Grubauer et al. [16] investigated the regulation of epidermal lipid synthesis biochemically and histochemically via lipid staining, presenting an inhibition of normal lipid synthesis under occlusion after acute barrier perturbation. Autoradiography was also used to demonstrate the localization of DNA synthesis within the epidermal layers [17].

Warner et al. [13] developed a method to investigate the effects of water in the epidermis. Sections of frozen epidermal material were analyzed via transmission electron microscopy, detecting both intercellular dilations, with separation of lipid bilayers and being filled with water, and newly formed amorphous lipids [13]. Under prolonged skin occlusion with water, a swelling of keratinocytes can be observed, with a further widening of the intercellular spaces forming cisternae. These cisternae contain water, material from disruption processes, and external applied fluids [13].

VIII. CONCLUSIONS

Investigating and using the effects of occlusion, one has to consider different influences, such as the basal condition (intact or impaired SC), the kind of experimentally induced barrier disruption (tape stripping, SLS treatment), the depth of the wound, the duration of occlusion, the time point of measurement after occlusion, and the type of tested skin (animal or human). Figure 4.1 summarizes the factors influenced by occlusion. Statements about occlusion have to be made while taking into account the specific settings.

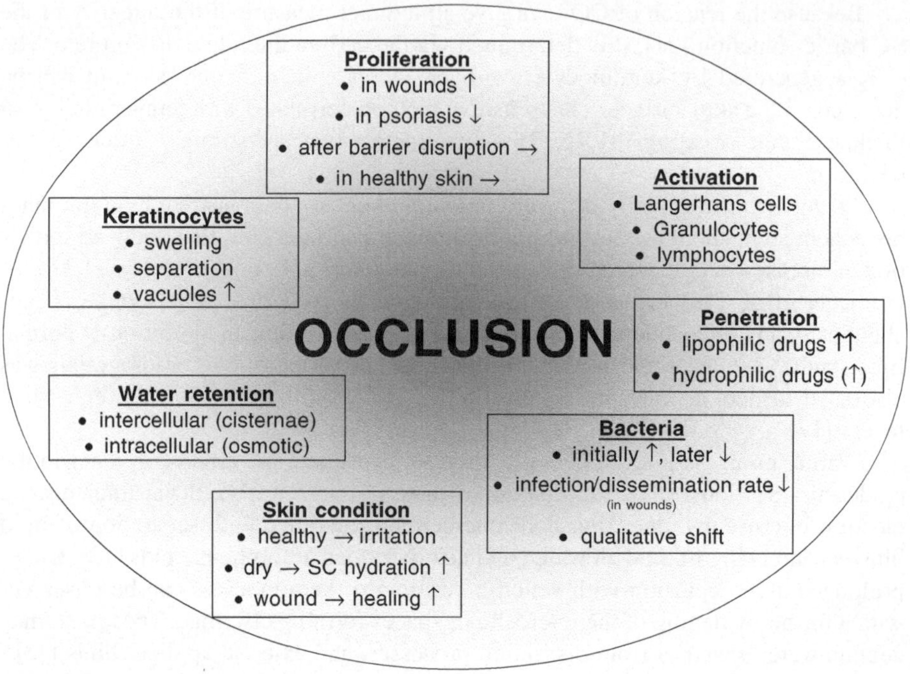

FIGURE 4.1 Summary of the factors influenced by occlusion.

REFERENCES

1. Bucks, D.A. et al., Bioavailability of topically administered steroids: a "mass balance" technique, *J. Invest. Dermatol.*, 91(1), 29, 1988.
2. Ryatt, K.S. et al., Methodology to measure the transient effect of occlusion on skin penetration and stratum corneum hydration in vivo, *Br. J. Dermatol.*, 119(3), 307, 1988.
3. Treffel, P. and Gabard, B., Skin hydration, in *Dermatopharmacology of Topical Preparations*, Treffel, P. (Ed.), Springer, Berlin, 2000, p. 319.
4. Schaefer, H. and Redelmeier, T.E., *Skin Barrier: Principles of Percutaneous Absorption*, Karger, Basel, 1996, p. 167–168.
5. Denda, M., Influence of dry environment on epidermal function, *J. Dermatol. Sci.*, 24(Suppl 1), S22, 2000.
6. Anderson, R.L. et al., The effect of *in vivo* occlusion on human stratum corneum hydration — dehydration *in vitro*, *J. Invest. Dermatol.*, 61(6), 375, 1973.
7. Van der Valk, P.G. and Maibach, H.I., Post-application occlusion substantially increases the irritant response of the skin to repeated short-term sodium lauryl sulfate (SLS) exposure, *Contact Dermatitis*, 21(5), 335, 1989.
8. Mathias, C.G., Prevention of occupational contact dermatitis, *J. Am. Acad. Dermatol.*, 23(4 Pt 1), 742, 1990.
9. Ramsing, D.W. and Agner, T., Effect of glove occlusion on human skin. I. Short-term experimental exposure, *Contact Dermatitis*, 34(1), 1, 1996.

10. Ramsing, D.W. and Agner, T., Effect of glove occlusion on human skin. II. Long-term experimental exposure, *Contact Dermatitis*, 34(4), 258, 1996.
11. Warner, R.R. et al., Water disrupts stratum corneum lipid lamellae: damage is similar to surfactants, *J. Invest. Dermatol.*, 113(6), 960, 1999.
12. Menon, G.K. and Elias, P.M., Morphologic basis for a pore-pathway in mammalian stratum corneum, *Skin Pharmacol.*, 10(5–6), 235, 1997.
13. Warner, R.R., Stone, K.J., and Boissy, Y.L., Hydration disrupts human stratum corneum ultrastructure, *J. Invest. Dermatol.*, 120(2), 275, 2003.
14. Lindberg, M., Johannesson, A., and Forslind, B., The effect of occlusive treatment on human skin: an electron microscopic study on epidermal morphology as affected by occlusion and dansyl chloride, *Acta Derm. Venereol.*, 62(1), 1, 1982.
15. Van Hal, D.A. et al., Structure of fully hydrated human stratum corneum: a freeze-fracture electron microscopy study, *J. Invest. Dermatol.*, 106(1), 89, 1996.
16. Grubauer, G., Elias, P.M., and Feingold, K.R., Transepidermal water loss: the signal for recovery of barrier structure and function, *J. Lipid Res.*, 30(3), 323, 1989.
17. Proksch, E. et al., Barrier function regulates epidermal DNA synthesis, *J. Clin. Invest.*, 87(5), 1668, 1991.
18. Wollina, U., Geyer, A., and Gebhardt, M., Topical therapy of psoriasis vulgaris with self-adhesive polyurethane dressings, *Dermatol. Wochenschr.*, 178, 412, 1992.
19. Fisher, L.B. and Maibach, H.I., Physical occlusion controlling epidermal mitosis, *J. Invest. Dermatol.*, 59(1), 106, 1972.
20. Fisher, L.B. and Maibach, H.I., The effect of occlusive and semipermeable dressings on the mitotic activity of normal and wounded human epidermis, *Br. J. Dermatol.*, 86(6), 593, 1972.
21. Berardesca, E. and Maibach, H.I., Skin occlusion: treatment or drug-like device? *Skin Pharmacol.*, 1(3), 207, 1988.
22. Levine, R., Agren, M.S., and Mertz, P.M., Effect of occlusion on cell proliferation during epidermal healing, *J. Cutan. Med. Surg.*, 2(4), 193, 1998.
23. Welzel, J., Wilhelm, K.P., and Wolff, H.H., Skin permeability barrier and occlusion: no delay of repair in irritated human skin, *Contact Dermatitis*, 35(3), 163, 1996.
24. van de Kerkhof, P.C. et al., Epidermal regeneration and occlusion, *Acta Derm. Venereol.*, 75(1), 6, 1995.
25. Sjoborg, S. et al., A new method for the visualization of the epidermal Langerhans cell and its application on normal and allergic skin, *Acta Derm. Venereol. Suppl. (Stockh.)*, 58(79), 23, 1978.
26. Lindberg, M. and Forslind, B., The effects of occlusion of the skin on the Langerhans' cell and the epidermal mononuclear cells, *Acta Derm. Venereol.*, 61(3), 201, 1981.
27. Mikulowska, A., Reactive changes in the Langerhans' cells of human skin caused by occlusion with water and sodium lauryl sulphate, *Acta Derm. Venereol.*, 70(6), 468, 1990.
28. Kligman, A.M., Hydration injury to human skin, in *Bioengineering of the Skin: Water and the Stratum Corneum*, Elsner, P., Berardesca, E., and Maibach, H.I., Eds., CRC Press, Boca Raton, Florida, 1994, p. 251–255.
29. Sulzberger, M.B., Griffin, T.B., and Wiley, H.S., Miliaria and anhidrosis. I. Experimental production in volunteers, *Dermatologica*, 135(5), 414, 1967.
30. Orentreich, N., Berger, R.A., and Auerbach, R., Anhidrotic effects of adhesive tapes and occlusive film, *Arch. Dermatol.*, 94(6), 709, 1966.
31. Wollina, U. et al., Immunhistologische Lokalistion von Plasminogenaktivatoren in gesunder und pathologischer Epidermis, *Dermatol. Monatsschr*, 176, 11, 1991.

32. Morioka, S. et al., Migrating keratinocytes express urokinase-type plasminogen activator, *J. Invest. Dermatol.*, 88(4), 418, 1987.
33. Pentland, A.P. and Marcelo, C.L., Modulation of proliferation in epidermal keratinocyte cultures by lowered oxygen tension, *Exp. Cell Res.*, 145(1), 31, 1983.
34. Knighton, D.R. et al., Oxygen tension regulates the expression of angiogenesis factor by macrophages, *Science*, 221(4617), 1283, 1983.
35. Knighton, D.R., Silver, I.A., and Hunt, T.K., Regulation of wound-healing angiogenesis-effect of oxygen gradients and inspired oxygen concentration, *Surgery*, 90(2), 262, 1981.
36. Alvarez, O.M., Mertz, P.M., and Eaglstein, W.H., The effect of occlusive dressings on collagen synthesis and re-epithelialization in superficial wounds, *J. Surg. Res.*, 35(2), 142, 1983.
37. Faergemann, J. et al., Skin occlusion: effect on *Pityrosporum orbiculare*, skin PCO2, pH, transepidermal water loss, and water content, *Arch. Dermatol. Res.*, 275(6), 383, 1983.
38. Bibel, D.J. et al., Microbial adherence to vulvar epithelial cells, *J. Med. Microbiol.*, 23(1), 75, 1987.
39. Fluhr, J.W. et al., Effects of prolonged occlusion on stratum corneum barrier function and water holding capacity, *Skin Pharmacol. Appl. Skin Physiol.*, 12(4), 193, 1999.
40. Aly, R. et al., Effect of prolonged occlusion on the microbial flora, pH, carbon dioxide and transepidermal water loss on human skin, *J. Invest. Dermatol.*, 71(6), 378, 1978.
41. Seidenari, S., Belletti, B., and Pellacani, G., Time course of skin changes induced by short-term occlusion by TEWL, capacitance and B-scanning echography, *Skin Res. Technol.*, 2, 52, 1996.
42. Lehmann, L. et al., Stability and occlusivity of dermatological ointments, *Z. Hautkrankheiten*, 8(72), 585, 1997.
43. Halkier-Sorensen, L., Efficacy of skin care products and different mixtures of lipids on barrier function, in *Dermatopharmacology of Topical Preparations: A Product Development-Oriented Approach*, Treffel, P. (Ed.), Springer, Berlin, 2000, p. 339.
44. Berardesca, E. et al., Effect of occlusive dressings on the stratum corneum water holding capacity, *Am. J. Med. Sci.*, 304(1), 25, 1992.
45. Visscher, M. et al., Effect of semipermeable membranes on skin barrier repair following tape stripping, *Arch. Dermatol. Res.*, 293(10), 491, 2001.
46. Fritsch, P. and Schuler, G., Hydrokolloidale Verbände, *Hautarzt*, 9(43), 597, 1992.
47. Alm, A. et al., Care of pressure sores: a controlled study of the use of a hydrocolloid dressing compared with wet saline gauze compresses, *Acta Derm. Venereol. Suppl. (Stockh.)*, 149, 1, 1989.
48. Scheuplein, R.J. and Blank, I.H., Permeability of the skin, *Physiol. Rev.*, 51(4), 702, 1971.
49. Behl, C.R. et al., Hydration and percutaneous absorption: I. Influence of hydration on alkanol permeation through hairless mouse skin, *J. Invest. Dermatol.*, 75(4), 346, 1980.
50. Treffel, P. et al., Effect of occlusion on in vitro percutaneous absorption of two compounds with different physicochemical properties, *Skin Pharmacol.*, 5(2), 108, 1992.
51. Illel, B. et al., Follicles play an important role in percutaneous absorption, *J. Pharm. Sci.*, 80(5), 424, 1991.

52. Lademann, J. et al., Investigation of follicular penetration of topically applied substances, *Skin Pharmacol. Appl. Skin Physiol.*, 14(Suppl 1), 17, 2001.
53. Blank, I.H. et al., The diffusion of water across the stratum corneum as a function of its water content, *J. Invest. Dermatol.*, 82(2), 188, 1984.
54. Wiechers, J.W., The barrier function of the skin in relation to percutaneous absorption of drugs, *Pharm. Weekbl. Sci.*, 11(6), 185, 1989.
55. Kim, M.K. et al., Delivery of hydrocortisone from liposomal suspensions to the hairless mouse skin following topical application under non-occlusive and occlusive conditions, *J. Microencapsul.*, 15(1), 21, 1998.
56. Gloor, M. and Gehring, W., Einwirkung von Liposomen auf die Haut, *Z. Körperpflegemittel-, Parfümerie-, Riechstoff- Aerosol-Industrie*, 4, 137, 1991.
57. Rippmann, K. and Seidel, H., Die feuchte Wundbehandlung, *Haut*, 1, 18, 1992.
58. Collawn, S.S., Occlusion following laser resurfacing promotes reepithelialization and wound healing, *Plast. Reconstr. Surg.*, 105(6), 2180, 2000.
59. Kannon, G.A. and Garrett, A.B., Moist wound healing with occlusive dressings. A clinical review, *Dermatol. Surg.*, 21(7), 583, 1995.
60. Rajka, G. et al., The effect of short-term occlusion on the cutaneous flora in atopic dermatitis and psoriasis, *Acta Derm. Venereol.*, 61(2), 150, 1981.
61. Marples, R.R., The effect of hydration on the bacterial flora of the skin, in *Skin Bacteria and Their Role in Infection*, McGraw-Hill, New York, 1965, p. 33–41.
62. Hebra, F. and Kaposi, M., *Lehrbuch der Hautkrankheiten*, Band I, Verlag von Ferdinand Enke, Erlangen, 1874, p. 367.
63. Gottlieb, A.B. et al., Occlusive hydrocolloid dressings decrease keratinocyte population growth fraction and clinical scale and skin thickness in active psoriatic plaques, *J. Dermatol. Sci.*, 1(2), 93, 1990.
64. Faber, W.R. et al., Effectiveness and usability of short-time treatment of psoriasis vulgaris with two different corticosteroids under occlusion with a hydrocolloid dressing (Contreet), *JEADV*, 5, 157, 1995.
65. Petzold, D.G., Braun-Falco, O., and Wenig, K.-H., Effects of plastic-foil-occlusion on psoriasis lesions, *Arch. Klin. Exp. Dermatol.*, 238, 160, 1970.
66. Fry, L. and McMinn, R.M., Observations on mitosis in psoriatic epidermis, *Br. J. Dermatol.*, 82(1), 19, 1970.
67. Fry, L., Almeyda, J., and McMinn, R.M., Effect of plastic occlusive dressings on psoriatic epidermis, *Br. J. Dermatol.*, 82(5), 458, 1970.
68. Rogiers, V., EEMCO guidance for the assessment of transepidermal water loss in cosmetic sciences, *Skin Pharmacol. Appl. Skin Physiol.*, 14(2), 117, 2001.
69. Berardesca, E., EEMCO guidance for the assessment of stratum corneum hydration: electrical methods, *Skin Res. Technol.*, 3, 126, 1997.
70. Fluhr, J.W. et al., Comperative study of five instruments measuring stratum corneum hydration (Corneometer CM 820 and CM 825, Skicon 200, Nova DPM 9003, DermaLab). Part I. *In vitro*, *Skin Res. Technol.*, 5, 161, 1999.
71. Bircher, A. et al., Guidelines for measurement of cutaneous blood flow by laser Doppler flowmetry. A report from the Standardization Group of the European Society of Contact Dermatitis, *Contact Dermatitis*, 30(2), 65, 1994.
72. Thiele, F.A. and Malten, K.E., Insensible water loss. The inter-subject variation related to skin temperature, forearm circumference and sweat gland activity, *Trans. St Johns Hosp. Dermatol. Soc.*, 58(2), 199, 1972.
73. Thiele, F.A. and van Kempen, L.H., A micro method for measuring the carbon dioxide release by small skin areas, *Br. J. Dermatol.*, 86(5), 463, 1972.

Part II

Transepidermal Water Loss and Barrier Function

5 Transepidermal Water Loss Measurements in Dermato-Cosmetic Sciences

V. Rogiers, E. Houben, and K. De Paepe

CONTENTS

I. Introduction 64
II. TEWL Measurement Techniques 64
III. Variables Affecting TEWL Measurements 66
A. Environmental Conditions–Linked Variables 66
1. Probe Temperature 66
2. Air Circulation 66
3. Environmental Temperature and Relative Humidity 66
4. Light Sources 66
5. Ionizing Radiation 66
B. Person-Linked Variables 66
1. Age and Gender 66
2. Ethnic Differences 67
3. Anatomical Sites 67
4. Skin Surface Temperature and Sweating 68
5. Skin Damage and Diseases 68
6. Circadian Rhythm 68
7. Stress 68
III. Applications of TEWL Measurements 69
A. Claim Support for Cosmetic Products 69
1. Skin Mildness 69
2. Reduction of Irritative Skin Reactions 70
3. Modulation of the Barrier Function 70
4. Protection of the Skin in Occupational Settings 70
5. Increase in Skin Hydration, Improvement of Moisturizing Properties of Skin Care Products 70
6. Positive Effect on the Shaving Process 70
7. Protective Effects against Ultraviolet Damage 70

0-8493-1443-7/05/$0.00+$1.50

B. Development of Innovative Cosmetic Ingredients and Finished Products 71
C. Use of Noninvasive Methodology in Safety Testing of Cosmetics on Human Skin 71
D. Improvement of Topical Therapeutic Treatment by the Use of Noninvasive Methodology 71
IV. Conclusion 72
References 72

I. INTRODUCTION

The stratum corneum (SC) plays an important role in the clinical appearance of the skin as a result of its water-holding capacity and lipidic content. In addition, it acts as a barrier to protect the body from percutaneous absorption of a wide variety of xenobiotics, from desiccation, and from insults by a number of environmental conditions [1].

A lot of research has been devoted to the elucidation of the structure and function of this barrier, and it has become clear now that the SC consists of protein-rich stacked keratinocytes with lipid-rich intercellular bilayers, which are greatly responsible for these unique properties [2,3].

From the deeper, highly hydrated layers of the epidermis and dermis, a passive flux of water takes place toward the more superficial SC layers, which have a relatively low water content [4]. This is the so-called transepidermal water loss (TEWL), which can be measured by noninvasive, biophysical methods under conditions in which no water originating from the sweat glands is added because of sweating [5].

Part of the diffusing water is evaporating at the surface of the SC, depending on the climatological conditions, and part is retained in the SC by the presence of natural moisturizing factors, which play a crucial role in skin hydration. Of major importance is the integrity of the barrier function. When the barrier is damaged, water can easily escape, and TEWL increases.

A generally accepted relationship exists between the TEWL and the degree of barrier disruption [6,7]. As a consequence, TEWL measurement seems to be a very useful tool to evaluate the effects on the barrier function of a number of products and conditions.

II. TEWL MEASUREMENT TECHNIQUES

Most techniques are based on the open-chamber method, equipped with a set of two thermistors and two hygrosensors placed at different but constant distances from the skin surface and measuring the water gradient evaporating through the SC [8].

Well-known instruments available on the market and based on this principle are the Evaporimeter®, the Tewameter®, and the DermaLab® system. Although only a limited number of comparison studies among different types of instruments have been described in the literature [9–13], it is generally believed that these commercially available instruments are quite reliable and that results obtained from different instruments correlate well [14–16].

It must, however, be mentioned that differences in calibration procedure exist between the different instruments and that variation may occur, even between apparatus of the same type and from the same firm [12]. It is, therefore, more correct to express results of studies as relative values rather than absolute numbers.

Recently, a portable device using a closed chamber system for measuring TEWL has been described [17,18]. Its measuring principle is based on a computed TEWL estimation from the progressive increase in relative humidity inside a closed chamber resulting from continuous water evaporation from the skin surface, on which the chamber is placed. The device has been compared with the conventional open-chamber methodology (with DermaLab) by conducting *in vivo* measurements on both healthy and diseased skin [17]. The sensitivity of the device was found to be similar to that of the DermaLab. Although all absolute values measured were much lower than those obtained for the open chamber method, they correlated well with each other.

The data of Nuutinen et al. [18] support these findings. The authors describe a highly significant correlation coefficient of 0.99, a reproducibility of 4% at an evaporation rate of 40 mg/m^2 per hour, and a reproducibility of measurements on healthy volunteers situated between 8 and 10%. Very recently, in our own laboratory, a Vapometer device, which is also based on the closed chamber system [19], was tested. Its properties were compared with those of a Tewameter, and general conclusions can be summarized as follows:

For all measurements done, the absolute values (expressed in milligrams per square meter per hour) obtained with a Vapometer were significantly lower than those values found with a Tewameter. There was, however, a good correlation between both apparatus. This was the case for measurements carried out during tape stripping, sodiumlaurylsulfate (SLS) damage of the skin, and topical cream application.

Probe temperature and humidity were easier to control for a Tewameter than for a Vapometer, but the reproducibility of both techniques seemed to be quite similar.

Interindividual spreading of repeated measurements on the nasolabial skin of a group of young women was quite comparable for both apparatus and was situated for the Vapometer between 15 and 20% (variation coefficient).

The sensitivity of the Vapometer® was somewhat lower than that of the Tewameter, as diurnal changes on face skin, clearly detected by the latter apparatus, could not be found back with the former [19].

In our hands, the closed-chamber technique was found to be useful and reliable, particularly in those cases in which a small and easily portable device is highly needed (e.g., in the nasolabial zone, where effects from the airflow of the nose are to be expected [19]).

III. VARIABLES AFFECTING TEWL MEASUREMENTS

A number of detailed review articles have appeared on this topic [8,9,20–23]. Here, only a summary of the most important person- and environmental conditions–linked variables is given.

A. Environmental Conditions–Linked Variables

1. Probe Temperature

The importance of this variable was in particular demonstrated for the Tewameter [9]: The probe must be warmed up to skin temperature. A device now exists (Tewameter TM210) for which this can be done automatically.

2. Air Circulation

For an open-chamber system, a box to cover the measuring site seems appropriate [9,22]. For a closed-chamber system, this is not necessary [19].

3. Environmental Temperature and Relative Humidity

The temperature of the measuring room must be standardized and kept below 22°C (sweating threshold); the relative humidity should be set close to but lower than 50% [9].

4. Light Sources

Light sources that produce warmth should be avoided because they can affect the ambient air and skin temperature [20]. Recently, it was found that conventional phototherapy in nonhemolytic hyperbilirubinemia-term infants [24] and in preterm babies [25–27] increased TEWL significantly. The latter finding has important implications for the amounts of maintenance fluids given to preterm infants [28].

5. Ionizing Radiation

Radiation therapy following breast-conserving surgery for breast cancer is accompanied by an increase in TEWL caused by radiation dermatitis. The barrier impairment was found to be comparable to changes observed with ultraviolet radiation exposure [29].

B. Person-Linked Variables

1. Age and Gender

Neither age nor gender seems to have a major effect on TEWL measurements [9,20,22]. However, in the case of an immature barrier, as is present in most premature infants, a high TEWL is observed [28]. In particular, in the first 2 weeks of life, this may lead to poor temperature control and difficulty in fluid balance. The TEWL can be reduced by manipulating the ambient humidity or by covering the skin [28]. Normally, a remarkable rate of barrier maturation is seen during the first few days of postnatal life [30].

2. Ethnic Differences

As mentioned before [8], some differences have been reported between Black and White human skin [31,32] and between White and Hispanic skin [33]. More recently, TEWL measurements were carried out in the context of an iontophoresis study with saline. Significant differences in TEWL were found among four ethnic groups. The rank order was reported to be White > Asian > Hispanic > Black [34].

Aramaki et al. [35] compared Japanese and German women, both living in Germany, during a SLS forearm test and came to the conclusion that no significant differences of the barrier function could be observed by TEWL measurements done in both groups.

3. Anatomical Sites

TEWL measurements are greatly affected by anatomical sites [8], and it is believed that the ranking order is palm > sole > forehead = postauricular skin = nail = dorsum of hand > forearm = upper arm = thigh = chest = abdomen = back [22].

Regional variations in skin barrier function resulting from iontophoresis with saline were studied, and no differences in TEWL values between the upper arm and chest were found, although some limited changes were observed at the abdomen [36], pointing to a quite similar barrier function in the upper arm, chest, and abdomen.

The effect of anatomical location on TEWL measurements after acute barrier disruption by acetone treatment or tape stripping was studied by Fluhr et al. [37]. They concluded that the lipid-rich skin areas studied, such as the forehead, were the most vulnerable body sites to barrier disruption.

Some studies on regional variations on specific body sites (e.g., face and forearm) have been carried out [38–40]. On women's faces, a symmetrical distribution around the facial median line was reported, with some minor variations within the forehead and chin areas [38]. The cheeks exhibited a distinct gradient, with the highest values seen in the paranasal zones and the lowest values found on the cheek bones. For a given individual, the mean values on both cheeks were nearly identical and the patterns within the two cheeks were superimposable. The authors concluded that TEWL values differed considerably between different facial areas but that they followed a characteristic distribution [38]. Similar studies on men's faces are not yet available in the current literature.

If absolute TEWL values are needed or precise changes on specific face skin areas are to be evaluated, it probably is necessary to perform these studies on facial skin.

For forearm skin (both genders), however, TEWL values were reported to be 10% significantly lower at the forearm midpoint than at the extremities [39]. TEWL on the dominant forearm might be significantly higher than the one on the nondominant forearm [40,41], but this has not been found to be true by all investigators [9,42].

4. Skin Surface Temperature and Sweating

Both factors are important and can be appropriately controlled by a 30-minute preconditioning of the test persons, before the measurements, in a temperature- and relative humidity–controlled room. In addition, physical activity should be kept to a minimum [8]. In our laboratory, consumption of stimulating drinks containing caffeine is also prohibited before and during TEWL studies.

5. Skin Damage and Diseases

There are several skin diseases in which the lipid composition of the intercellular matrix of the stratum corneum is different from that of healthy skin. For example, atopic dermatitis patients have a reduced ceramide content, and patients with lamellar ichthyosis show a lower free fatty acid concentration and an altered ceramide profile [44]. In both types of patients, elevated TEWL values have been reported, accompanied by defects in lateral lipid packing and lamellar organization in the stratum corneum [44]. Other skin pathologies, including psoriasis and contact dermatitis, are characterized by increased TEWL levels [45,46]. In addition, skin damage, such as is present on the hands of hospital workers, leads to higher TEWL levels and higher clinical Dermatology Life Quality Index scores, as seen recently by our group when we evaluated the hand skin of nurses working in an operating room in comparison with that of a control group of female administrative employees [47].

In numerous articles on skin irritation with detergents [48–51], organic solvents [52], skin stripping [53,54], and other agents, changes in TEWL have been clearly shown [8]. Recently, radiation therapy was added to this list [55]. Indeed, it was shown that during radiation therapy for breast cancer, TEWL levels increased progressively because of the development of acute radiation dermatitis.

6. Circadian Rhythm

TEWL has been reported to be higher in the evening and night than in the morning [56]. In addition, cutaneous barrier repair, measured by TEWL, after tape stripping of volar forearm skin in human volunteers seems to be time-dependent, but independent of changes in skin temperature and cortisol level, as the lowest values were seen between 20 and 23 hours and the highest TEWL measurements were done around 3 AM [54].

Circadian rhythmicity was also clearly observed by Le Fur et al., in the face as well as in the forearm skin of healthy women [57]. Even ultradian rhythms with periods of 8 and 12 hours could be detected for face and forearm skin, respectively [57].

7. Stress

Only recently have the effects of psychological stress on human skin have been studied. Three different stressors (a psychologic interview, sleep deprivation, and exercise) were imposed on healthy women. Several dermatologic measures were examined, including TEWL and recovery of barrier function after tape stripping [58].

It was seen that interview stress and sleep deprivation caused a significant delay in the recovery of skin barrier function, whereas exercise stress did not. It was suggested that acute psychosocial and sleep deprivation stress disrupted skin barrier function homeostasis in women and that this disruption might be related to stress-induced changes in cytokine secretion [58]. In another study, the effect of stress of marital dissolution on skin barrier recovery (TEWL level) was investigated [59]. It was concluded that no correlation existed between the degree of stress and barrier strength (number of tape strippings required to disrupt skin barrier). However, individuals with high stress recovered slower than the individuals with low stress [59].

Mechanical stress is an important physiological parameter. The effect of controlled strain on human skin *in vivo* was studied by submitting human skin to extension from 0 to 20% as imposed by a Densiscore device. However, an extension force up to 20% elongation did not significantly alter the barrier function (TEWL) [60]. On the contrary, when TEWL was measured in obese individuals, significantly increased values were observed but were not correlated with a high body mass index [61].

III. APPLICATIONS OF TEWL MEASUREMENTS

TEWL measurements are, among other bioengineering techniques, often used in cosmetology for research and regulatory testing. In addition, in pharmaceutical, clinical, and toxicological studies, and even in fundamental research, a wide field of applications has been developed over the last 20 years.

A. Claim Support for Cosmetic Products

By implementing the Sixth Amendment (Council Directive 93/35/EEC) in the European Union Cosmetic Legislation, the manufacturer or primary importer within the community is obliged to prove the effects claimed for a given product, meaning that this information must be made available, on request, to the competent authorities. Proof of claim is part of a technical dossier that must be compiled for each cosmetic product sold on the European market. Several methodologies to be used on human subjects have been proposed to support cosmetic claims [62,63], and one of these consists of the use of bioengineering methods, among which measurement of barrier function is a powerful one. Some claim-support examples, for which TEWL measurements have been used, are cited here.

1. Skin Mildness

Cleansing products, sometimes containing aggressive anionic tensides, may cause skin irritation and inflammation or induce a rough and dry aspect of the stratum corneum. These products affect barrier function, and as a consequence, TEWL measurements can be used to objectively compare the properties of different tensides or mixtures on human skin. "Skin compatibility" and "skin mildness" of cleansing products are claims that are made frequently, and TEWL measurement techniques can be applied to prove these. Measurements are usually performed before and after a patch test. This topic has been discussed in extenso by Schrader and others [64–66].

2. Reduction of Irritative Skin Reactions

The strategy usually used to test irritative skin reactions consists of skin damage resulting from exposure to SLS, tape stripping, acetone treatment, and so forth, followed by topical application of the product that is tested for its "repair" qualities. A variety of examples of this strategy can be found in the literature [8,52,67]. Illustrative examples from our own laboratory are studies on the "healing" properties of starch and starch-containing bath products added to bath water [68], the beneficial effects of a moisturizing cream on irritant and allergic contact dermatitis [49], and the effects of body lotions enriched with skin identical lipids [69].

3. Modulation of the Barrier Function

This can be achieved by the deposition of lipids on or in the skin by cosmetic products [70,71]. From our own laboratory, we can mention some studies on the use of body lotions enriched with skin identical lipids [69,72].

4. Protection of the Skin in Occupational Settings

The claim to protect the skin in occupational settings is often made for protective creams with ingredients such as petrolatum, silicones, and so forth [48,73,74].

5. Increase in Skin Hydration, Improvement of Moisturizing Properties of Skin Care Products

When the moisturizing properties of a cosmetic are measured (e.g., by capacitance measurements), it is useful to also determine the effect of the product on the TEWL. In case the latter effect increases, attention should be paid to the composition of the test product because it affects skin barrier integrity, which will eventually lead to skin damage and dehydration [75].

6. Positive Effect on the Shaving Process

Shaving additives are intended to improve shaving closeness, to provide a fresh feeling, and to decrease or prevent irritation in the beard area. These products have been around for quite some time, but their application as additives to be used with electrical shavers is new. Only limited information is available about these products because of patent restrictions [76].

7. Protective Effects against Ultraviolet Damage

Ultraviolet light, and in particular ultraviolet-B radiation, is at the origin of erythema and sunburn. This is accompanied by an increase in TEWL levels. Sunscreens can decrease sun exposure and diminish the concomitant side reactions. As a consequence, they can also minimize skin barrier damage and changes in TEWL [77].

B. Development of Innovative Cosmetic Ingredients and Finished Products

TEWL measurements can help in the development of novel cosmetic ingredients and products that act on the barrier function [78,79]. In particular, interest in barrier lipids has been shown. These lipids may affect repair after SLS damage [80–83], after tape stripping [71], in the case of aged skin [84], or when atopic dermatitis [85] is involved.

C. Use of Noninvasive Methodology in Safety Testing of Cosmetics on Human Skin

Skin compatibility testing has become an important tool in a well-executed cosmetic safety assessment program. As a first step in safety assessment, existing data in the scientific literature are reviewed and information from the manufacturers of the different ingredients or mixtures is carefully checked. All existing *in vivo* information on ingredients, as well as on similar products, is collected. In certain cases, *in vivo* animal tests are performed; however, with the implementation of the Seventh Amendment (Council Directive 2003/15/EC), this step will become impossible in the near future. After a battery of *in vitro* tests has shown that ingredients or products are nonirritating, human clinical testing can be considered. In routine clinical practice, patch tests are often performed and then evaluated subjectively and graded using a clinical rating scale. By using noninvasive bioengineering techniques, such as TEWL methodology, readings can be done objectively [86]. During a European Centre for the Validation of Alternative Methods Workshop, this topic was discussed in detail [87].

D. Improvement of Topical Therapeutic Treatment by the Use of Noninvasive Methodology

As discussed for cosmetic claim substantiation, TEWL measurements can also be applied for improving the performance of pharmaceutical ingredients, products, or treatments. Numerous examples have been published, and only a few illustrative examples are mentioned here, including

Topical corticosteroid therapy for acute radiation dermatitis could be improved by the use of TEWL measurements in a prospective randomized, double-blind study in which patients, undergoing fractionated radiation therapy for breast cancer, were treated with different topical products [55].

TEWL loss in jaundiced preterm infants receiving phototherapy could be decreased by the application of clear topical ointment [24].

Appropriate dimethicone skin protection products could be identified and applied to treat surfactant users to prevent irritant contact dermatitis [48].

Atopic dermatitis lesions could be healed much quicker by adding rice starch to bath water. The application of TEWL measurements could prove this finding in a scientific way. For years this application has been known of in "popular medicine," but it has never been proven to be correct [68].

IV. CONCLUSION

If all the variables discussed above are taken into account during TEWL measurements, based on open- or closed-chamber technology, reliable and objective measurements can be obtained that can be used in a wide variety of applications in the field of cosmetology as well as in dermatology.

REFERENCES

1. Tagami, H., Hashimoto-Kumasaka, K., and Terui, T., The stratum corneum as a protective biological membrane of the skin, in *Skin: Interface of a Living System. Perspective for Skin Care System in the Future*, Tagami, H., Parish, J.H., and Ozawa, T., Eds., Elsevier, Amsterdam, 1998, pp. 23–37.
2. Schaefer, H. and Redelmeier, T.E., *Skin Barrier: Principles of Percutaneous Absorption*. Karger, Basel, 1996.
3. Elias, P.M., Stratum corneum architecture, metabolic activity and interactivity with subjacent cell layers, *Exp. Dermatol.*, 5, 191, 1996.
4. Wilson D.R. and Maibach, H., Transepidermal water loss. A review, in *Cutaneous Investigation in Health and Disease: Noninvasive Methods and Instrumentation*, Lévêque, J.L., Ed., Marcel Dekker, New York, 1989, pp. 113–133.
5. Batt, M.D. and Fairhurst, E., Hydration of the stratum corneum, *Int. J. Cosmet. Sci.*, 8, 253, 1986.
6. Grubauer, G., Feingold, K.R., and Elias, P.M., Relationship of epidermal lipogenesis to cutaneous barrier function, *J. Lipid Res.*, 28, 746, 1987.
7. Jass, H.E. and Elias, P.M., The living stratum corneum: implication for cosmetic formulation, *Cosmet. Toiletries*, 106, 47, 1991.
8. Rogiers, V. and the EEMCO Group., EEMCO guidance for the assessment of transepidermal water loss measurements in cosmetic sciences, *Skin Pharmacol. Appl. Skin Physiol.*, 14, 117, 2001.
9. Rogiers, V., Transepidermal water loss measurements in patch test assessment: the need for standardisation, in *Irritant Dermatitis: New Clinical and Experimental Aspects*, Elsner, P. and Maibach, H.I., Eds., Karger, Basel, 1995, Vol. 23, pp. 152–158.
10. Schott, R.C., Oliver, G.J.A., Dugard, P.H., and Singh, H.J., A comparison of techniques for the measurement of transepidermal water loss, *Arch. Dermatol. Res.*, 274, 57, 1982.
11. Schulz, A., Elsner, P., and Burg, G., Quantification of irritant contact dermatitis *in vivo*: comparison of the Dermatest® system with the Evaporimeter®, *Contact Dermatitis*, 24, 235, 1991.
12. Pinnagoda, J. et al., Comparability and reproducibility of results of water loss measurements: a study of four evaporimeters, *Contact Dermatitis*, 20, 241, 1989.
13. Grove, G.L. et al., Comparative metrology of the Evaporimeter and the DermaLab TEWL Probe, *Skin Res. Technol.*, 5, 1, 1999.
14. Barel, A.O. and Clarys, P., Study of the stratum corneum barrier function by transepidermal water loss measurements: comparison between two commercial instruments, *Skin Pharmacol.*, 8, 186, 1995.
15. Kalia, Y.N., Pirot, F., and Guy, R.H. Homogeneous transport in a heterogeneous membrane: water diffusion across human stratum corneum *in vivo*, *Biophys. J.*, 71, 2692, 1996.

16. Rodrigues, L. et al., Transepidermal water loss kinetic modelling approach for the parametrization of skin water dynamics, *Skin Res. Technol.*, 5, 72, 1999.
17. Tagami, H., Kobayashi, H., and Kikuchi, K., A portable device using a closed chamber system for measuring transepidermal water loss: comparison with the conventional method, *Skin Res. Technol.*, 8, 7, 2002.
18. Nuutinen, J. et al., A closed unventilated chamber for the measurement of transepidermal water loss, *Skin Res. Technol.*, 9, 85, 2003.
19. Houben, E., De Paepe, K., and Rogiers, V., The closed chamber Vapometer® used for the measurement of transepidermal water loss: comparison with an open chamber system, submitted.
20. Lévêque, J.L., Measurement of transepidermal water loss, in *Cutaneous Investigation in Health and Disease: Noninvasive Methods and Instrumentation*, Lévêque, J.L., Ed., Marcel Dekker, New York, 1989, pp. 135–153.
21. Nilsson, G.E., Measurement of water exchange through skin, *Med. Biol. Eng. Comput.*, 15, 209, 1977.
22. Pinnagoda, J. et al., Guidelines for transepidermal water loss (TEWL) measurement, *Contact Dermatitis*, 22, 164, 1990.
23. Berardesca, E. and Maibach, H.I., Monitoring the water-holding capacity in visually non-irritated skin by plastic occlusion stress test, *Clin. Exp. Dermatol.*, 15, 107, 1990.
24. Wananukul, D. and Praisuwanna, P., Transepidermal water loss during conventional phototherapy in non-hemolytic hyperbilirubinemia term infants, *J. Med. Assoc. Thai.*, 84, S46, 2001.
25. Wananukul, S. and Praisuwanna, P., Clear topical ointment decreases transepidermal water loss in jaundiced preterm infants receiving phototherapy, *J. Med. Assoc. Thai.*, 85, 102, 2002.
26. Grunhagen, D.J., Transepidermal water loss during halogen spotlight phototherapy in preterm infants, *Pediatr. Res.*, 51, 402, 2002.
27. Maayan-Metzger, A. et al., Transepidermal water loss and skin hydration in preterm infants during phototherapy, *Am. J. Perinatol.*, 18, 393, 2001.
28. Rutter, N., Clinical consequences of an immature barrier, *Semin. Neonatal.*, 5, 281, 2000.
29. Schmuth, M. et al., Permeability barrier function of skin exposed to ionizing radiation, *Arch. Dermatol.*, 137, 1079, 2001.
30. Okah, F.A. et al., Surface electrical capacitance as a non-invasive bedside measure of epidermal barrier maturation in the newborn infant, *Pediatrics*, 96, 688, 1995.
31. Wilson, D., Berardesca, E., and Maibach, H.I., *In vivo* transepidermal water loss and skin surface hydration in assessment of moisturization and soap effects, *Int. J. Cosmet. Sci.*, 10, 201, 1988.
32. Berardesca, E. and Maibach, H.I., Radical differences in sodium-lauryl-induced cutaneous irritation: comparison of black and white subjects, *Contact Dermatitis*, 18, 65, 1988.
33. Berardesca, E. and Maibach, H.I., Sodium-lauryl-induced cutaneous irritation: comparison of black and white subjects, *Contact Dermatitis*, 19, 136, 1988.
34. Singh, J. et al., Effect of saline iontophoresis on skin barrier function and cutaneous irritation in four ethnic groups, *Food Chem. Toxicol.*, 38, 717, 2000.
35. Aramaki, J. et al., Differences of skin irritation between Japanese and European women, *Br. J. Dermatol.*, 146, 1052, 2002.
36. Singh, J. et al., Regional variations in skin barrier function and cutaneous irritation due to iontophoresis in human subjects, *Food Chem. Toxicol.*, 39, 1079, 2001.

37. Fluhr, J.W. et al., Impact of anatomical location on barrier recovery, surface pH and stratum corneum hydration after acute barrier disruption, *Br. J. Dermatol.*, 146, 770, 2002.
38. Lopez, S. et al., Transepidermal water loss, temperature and sebum levels on women's facial skin follow characteristic patterns, *Skin Res. Technol.*, 6, 31, 2000.
39. Chilcott, R.P. and Farrar, R., Biophysical measurements of human forearm skin *in vivo*: effects of site, gender, chirality and time, *Skin Res. Technol.*, 6, 64, 2000.
40. Schnetz, E. et al., Intra- and inter-individual variations in transepidermal water loss on the face: facial locations for bioengineering studies, *Contact Dermatitis*, 40, 243, 1999.
41. Treffel, P. et al., Hydration, transepidermal water loss, pH and skin surface parameters: correlations and variations between dominant and non-dominant forearms, *Br. J. Dermatol.*, 130, 325, 1994.
42. Rodrigues, L. and Pereira, L.M., Basal transepidermal water loss: right/left forearm difference and motoric dominance, *Skin Res. Technol.*, 4, 135, 1998.
43. Oestmann, E. et al., Skin barrier function in healthy volunteers as assessed by transepidermal water loss and vascular response to hexyl nicotinate: intra- and inter-individual variability, *Br. J. Dermatol.*, 128, 130, 1993.
44. Pilgram, G.S. et al., Aberrant lipid organization in stratum corneum of patients with atopic dermatitis and lamellar ichthyosis, *J. Invest. Dermatol.*, 117, 710, 2001.
45. Lavrijsen, A.P.M. et al., Barrier function parameters in various keratinisation disorders: transepidermal water loss and vascular response to hexyl nicotinate, *Br. J. Dermatol.*, 124, 547, 1993.
46. Hachem, J.P. et al., Combination therapy improves the recovery of the skin barrier function: an experimental model using a contact allergy patch test combined with TEWL measurements, *Dermatology*, 202, 314, 2001.
47. Hachem, J.P. et al., Evaluation of biophysical and clinical parameters of skin barrier function among hospital workers, *Contact Dermatitis*, 46, 220, 2002.
48. Zhai, H. et al., A bioengineering study on the efficacy of a skin protectant lotion in preventing SLS-induced dermatitis, *Skin Res. Technol.*, 6, 77, 2000.
49. De Paepe, K. et al., Beneficial effects of a skin tolerance-tested moisturizing cream on the barrier function in experimentally-elicited irritant and allergic contact dermatitis, *Contact Dermatitis*, 2001, 44, 337, 2001.
50. Di Nardo, A. et al., SLS induced irritant contact dermatitis: a correlation study between ceramides and *in vivo* parameters of irritation, *Contact Dermatitis*, 36, 86, 1996.
51. Aramaki, J. et al., Which bioengineering assay is appropriate for irritant patch testing with sodium lauryl sulphate? *Contact Dermatitis*, 45, 286, 2001.
52. Schnetz, E. et al., Multicentre study for the development of an *in vivo* model to evaluate the influence of topical formulations on irritation, *Contact Dermatitis*, 42, 336, 2000.
53. Visscher, M. et al., Effect of semipermeable membranes on skin barrier repair following tape stripping, *Arch. Dermatol. Res.*, 293, 491, 2001.
54. Denda, M. and Tsuchiya, T., Barrier recovery rate varies time-dependently in human skin, *Br. J. Dermatol.*, 142, 881, 2000.
55. Schmuth, M. et al., Topical corticosteroid therapy for acute radiation dermatitis: a prospective, randomized, double-blind study, *Br. J. Dermatol.*, 146, 983, 2002.
56. Yosipovitch, G. et al., Time-dependent variations of the skin barrier function in humans: transepidermal water loss, stratum corneum hydration, skin surface pH, and skin temperature, *J. Invest. Dermatol.*, 110, 20, 1988.

57. Le Fur, I. et al., Analysis of circadian and ultradian rhythms of skin surface properties of face and forearm of healthy women, *J. Invest. Dermatol.*, 117, 718, 2001.
58. Altemus, M. et al., Stress-induced changes in barrier function in healthy women, *J. Invest. Dermatol.*, 117, 309, 2001.
59. Muizzuddin, N. et al., Impact of stress of marital dissolution on skin barrier recovery: tape stripping and measurement of transepidermal water loss, *Skin Res. Technol.*, 9, 34, 2003.
60. Lévêque, J.L. et al., Structure and function of human stratum corneum under deformation, *Dermatology*, 205, 353, 2002.
61. Loffler, H., Aramaki, J.U., and Effendy, I., The influence of body mass index on skin susceptibility to sodium lauryl sulphate, *Skin Res. Technol.*, 8, 19, 2002.
62. Masson, P., The contribution of the European Cosmetics Directive towards international harmonisation: impact on the evaluation of safety and efficacy, in *Cosmetics: Controlled Efficacy Studies and Regulation*, Elsner, P., Merk, H.F., and Maibach, H.I., Eds., Springer, Berlin, 1999, pp. 20–35.
63. COLIPA (The European Toiletry and Perfumery Association), Guidance for the evaluation of the efficacy of a cosmetic product, 1998.
64. Schrader, K., On the problem of *in vivo* cleansing of the human skin, in *Cosmetics: Controlled Efficacy Studies and Regulation*, Elsner, P., Merk, H.F., and Maibach, H.I., Eds., Springer, Berlin, 1999, pp. 92–106.
65. Lambers, H. and Pronk, H., Biophysical methods for stratum corneum characterisation, in *Cosmetic Lipids and the Skin Barrier*, Förster, T. Ed., Marcel Dekker, Basel, 2002, pp. 185–225.
66. Harvell, J. and Maibach, H.I., Bioengineering methods to evaluate *in vivo* skin irritation: application to *in vitro* validation, in *Alternative Methods in Toxicology, Irritiation, Phototoxicity, Sensitisation*, Rougier A, Goldberg, A.M. and Maibach, H.I., Eds., New York, Liebert, 1994, Vol. 10, pp. 9–11.
67. Loden, M. et al., Instrumental and dermatologist evaluation of the effects of glycerine and urea on dry skin in atopic dermatitis, *Skin Res. Technol.*, 7, 209, 2001.
68. De Paepe, K. et al., Effect of rice starch as a bath additive on the barrier function of healthy but SLS-damaged skin and skin of atopic patients, *Acta Derm. Venereol.*, 82, 184, 2002.
69. De Paepe, K. et al., Body lotions enriched with skin identical lipids: a TEWL study of aged skin and SLS-induced scaly skin, *Eur. Cosmetics*, 7, 38, 1999.
70. Gabard, B. and Bieli, E., Measurement of lipid deposition on the skin of the forearm: comparison of different bath oils, *J. Soc. Cosmetic Chem.*, 42, 299, 1991.
71. Mao-Qiang, M. et al., A natural lipid mixture improves barrier function and hydration in human and murine skin, *J. Soc. Cosmet. Chem.*, 47, 157, 1997.
72. De Paepe, K. et al., Ceramides/cholesterol/FFA-containing body lotions: effects on the TEWL of aged and SLS-damaged skin, in *Conference Proceedings. Active Ingredients Conference*, Verlag für chemische Industrie, H. Ziolkowsky GmbH, Paris, 1996, pp. 97–111.
73. Schluter-Wigger, W. and Elsner, P., Efficacy of 4 commercially available protective creams in the repetitive irritation test, *Contact Dermatitis*, 34, 278, 1996.
74. Korstanje, C., Barrier creams, in *Handbook of Cosmetic Science and Technology*, Barel, A.O., Paye, M., and Maibach, H.I., Eds. Marcel Dekker, Basel, 2001, pp. 557–566.

75. De Paepe, K. et al., De vergelijkende studie van dermato-cosmetische producten voor de droge tot zeer droge huid: objectieve evaluatie aan de hand van efficiënti-emetingen, *Tijdschr. Geneeskunde*, 55, 268, 1999.
76. Wesdorp, C.A., Scheren met en zonder additieven, in *Proceedings of 18th Badecos Symposium*, Rogiers, V., Ed., Vrije Universiteit Brussel, Brussels, 1998, pp. 14–22.
77. Haratake, A. et al., UVB induced alterations in permeability barrier function: roles for epidermal hyperproliferation and thymocyte-mediated response, *J. Invest. Dermatol.*, 108, 769, 1997.
78. Philippe, M. et al., Synthesis of 2-N-oleylamino-octadecane-1,3-diol: a new ceramide highly effective for the treatment of skin and hair, *Int. J. Cosmet. Sci.*, 17, 133, 1995.
79. Lintner, K. et al., The effect of a synthetic ceramide-2 on transepidermal water loss after stripping or sodium lauryl sulphate treatment: an *in vivo* study, *Int. J. Cosmet. Sci.*, 19, 1915, 1997.
80. Imokawa, G. et al., Water-retaining function in the stratum corneum and its recovery properties by synthetic pseudoceramides, *J. Soc. Cosmetic Chem.*, 40, 273, 1989.
81. Berardesca, E. et al., in *Surfactant Damaged Skin: Which Treatment? The Environmental Threat to the Skin*, Marks, D. and Plewig, G., Eds., Martin Dunitz, London, 1991, pp. 283–285.
82. Farin, F. et al., Human skin-identical ceramides, *Cosmetic Toiletries*, 3, 126, 1995.
83. De Paepe, K., Roseeuw, D., and Rogiers, V., Les céramides dans les préparations dermo-cosmétiques, *J. Méd. Esth. Chir. Derm.*, 26, 179, 1999.
84. Zettersten, E. et al., Optimal ratios of topical SC lipids improve barrier recovery in chronobiologically aged skin, *J. Am. Acad. Dermatol.*, 37, 403, 1997.
85. Chamlin, S.L. et al., Ceramide-dominant barrier repair lipids alleviate childhood atopic dermatitis: changes in barrier function provide a sensitive indicator of disease activity, *J. Am. Acad. Dermatol.*, 47, 198, 2002.
86. Rogiers, V. et al., The potential use of non-invasive methods in safety assessment of cosmetic products. The report and recommendations of an ECVAM/EEMCO Workshop (ECVAM Workshop 36). ATLA 27, 515, 1999.
87. Ennen, J., Use of non-invasive methods in safety testing of cosmetic products, in *Intensive Course in Dermato-Cosmetic Sciences*, Part II, Rogiers, V. (Ed.), Vrije Universiteit Brussel, Brussels, 2002, pp. 125–138.

6 Ultrasound and Water in the Stratum Corneum

Stefania Seidenari and Giovanni Pellacani

CONTENTS

I. Introduction 77
II. Ultrasound and Water Content in Irritant Reactions 79
References 80

I. INTRODUCTION

Because the penetration depth of ultrasound waves is inversely related to their frequency, high-frequency ultrasound (20 to 150 MHz) can be employed for the study of the skin, which, thanks to its superficial location, can be explored achieving a high resolution and magnification.[1–16]

The generation and detection of ultrasound is based on the pulse-echo principle:[17,18] Employing 20 MHz B-scanning, each echo signal is converted into a pixel, which is depicted by one false color out of 256 colors composing a fictional scale, where 0 corresponds to absence of echogenicity and 255 to maximum in echogenicity. Thus, a cross-sectional image of the skin is represented on the monitor. The positioning of each pixel from the surface to the depth is established according to the interval between echo transmission and echo return (assuming a constant velocity throughout the tissue of 1580 m/s).

According to tissue structure, water content, and different physiological and pathological modifications of the skin, different physical phenomena such as reflection, refraction, scattering, and attenuation occur during the propagation of the ultrasound waves in biological tissues, and this determines the formation of the ultrasonic image and should be considered for the interpretation of the scans.

Looking at a 20-MHz ultrasound image of normal skin, a hyperreflective band-like structure (the so-called entry echo) visually coinciding with the epidermis is observable at the skin surface; it is probably generated by the impedance jump between the coupling medium and the epidermis. Below the hyperreflecting epidermal band, we can see the dermis. The main source of the echogenicity of the dermis is represented by the regularly aligned bundles of collagen fibers which appear as moderately or highly reflective structures. Their different thickness and arrangement is the cause of the different echostructure of the lower part of the corium, which

0-8493-1443-7/05/$0.00+$1.50

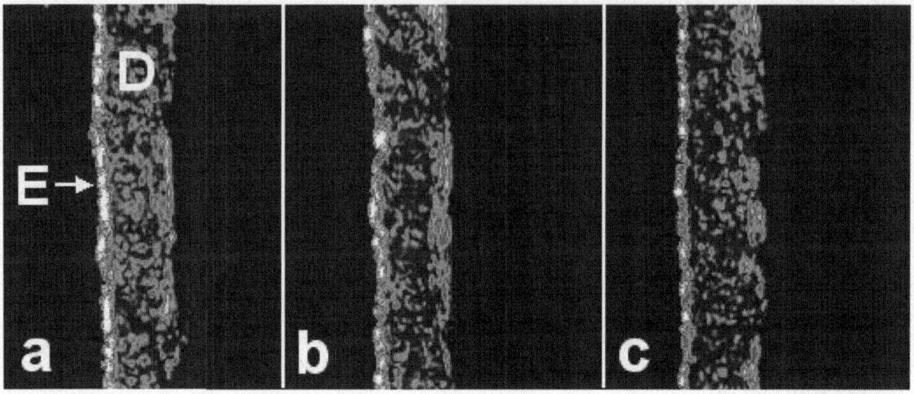

FIGURE 6.1 Sonographic images of the skin of the volar forearm. (a) Baseline, (b) after a 30-minute occlusion with saline solution, (c) after a 30-minute occlusion with a moisturizer. The entry echo is attenuated after the application of the substances. E, epidermis; D, dermis.

usually seems more echogenic compared with the upper part. In contrast, echogenicity of the dermis is inversely related to its water content: All inflammatory processes, accompanied by edema, appear as echo-poor areas within the reflective connective tissue.[1–7,9–11]

The introduction of dedicated software for the elaboration of B-scan images has transformed the ultrasound technique from a morphological to a quantitative assessment method, enabling the objective evaluation of the images, the quantification of visual data, and the expression of these data as numbers, which can be used for statistical evaluation.[3] By ascribing fictional values to the echoes' amplitudes, the selection of amplitude bands of interest and the segmentation of the image (i.e., the enhancement of areas of interest), the calculation of the extension of areas formed by pixels sharing similar amplitude values is possible. This method can be employed for assessment of the water content in the skin, as its variations influence the physical properties and the echogenicity of the epidermis and the dermis. Absorption of water by the stratum corneum induces an attenuation of the entry echo (Figure 6.1).[14,16] Performing 30-minute patch tests with saline solution on the volar forearm of healthy volunteers, we observed an increase in transepidermal water loss (TEWL) and hydration values, as measured by capacitance, immediately after patch test removal and 20 min later.[14] At the same time, image analysis on echographic pictures showed an attenuation of the superficial hyperreflecting band corresponding to the epidermis. The decrease in epidermal echogenicity was inversely related to hydration values (Figure 6.2).

In order to study the effects of skin care products on the sonographic aspect of the skin, we carried out two patch tests on the volar forearm of healthy subjects, one with 60 mg of petrolatum and the other with an oil-in-water emulsion.[15,16] Both tests were applied for 1 hour, and measurements were performed at the beginning of the experiment, immediately after removal of the test chamber, and at 15, 60, and 180 minutes later. On application of the two different formulations, we observed significant increases, compared with baseline, in capacitance values at all times of

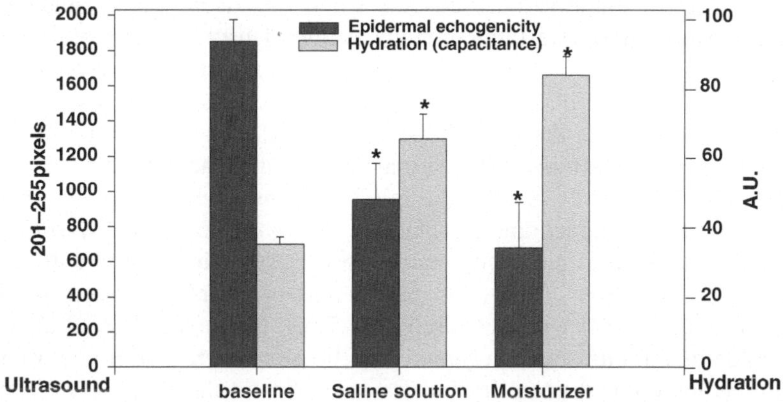

FIGURE 6.2 Variations in the extension of areas reflecting within the 201–255 interval (in pixels) in the epidermis and in capacitance values after application of saline solution and of a moisturizer. Echographic (in black) and capacitance values (in gray). * $P < .05$: significant in respect to baseline values.

assessment for the moisturizer and from 15 to 180 minutes for petrolatum. The increase in TEWL values was detected immediately after patch test removal and 15 minutes later. Ultrasound showed that the two products induced an attenuation of the superficial hyperreflecting band corresponding to the epidermis, and image analysis on echographic pictures revealed that 201–255 pixel areas had significantly diminished at all assessment times after application of the tested products. Thus, by sonography, increased hydration of the stratum corneum corresponds to an attenuation of the epidermis echo accompanied by a reduction in superficial 201–255 pixel areas, whose extension is inversely related to capacitance values. Whereas only a slight decrease in superficial echogenicity was induced by the water-poor petrolatum, a marked reduction in epidermal reflectivity was noticed with the oil-in-water emulsion.

II. ULTRASOUND AND WATER CONTENT IN IRRITANT REACTIONS

Irritant substances induce a toxic damage to the epidermis while at the same time altering its water content. The introduction of B-scanning methods has enabled the visual assessment of skin responses to irritant substances through the representation of the dynamics of the reaction, which can be followed up for hours or days without any interference with the natural evolution of the skin response. Using this method, it is possible to assess the degree of both the inflammatory and the epidermal component of irritant responses, including subclinical ones.

For the assessment of epidermal damage and variations in water content, a 201–255 amplitude interval, evaluating the superficial hyperreflecting band corresponding to the epidermis, was identified based on the results of experimental studies.[2,5–7,10,11] Variations in the epidermal component of skin reflectivity are highly specific for single irritant substances and vary according to modifications of water dynamics in the stratum corneum. Sodium lauryl sulfate application onto the skin induces a characteristic decrease of the superficial hyperreflecting band corresponding to the epidermis, which in strong reactions disappears completely. Moreover, a decrease in dermal echogenicity, which is subepidermal in the first phase and spreads to the underlying dermal tissue as the reaction grows in intensity, is observable. The degree of attenuation of epidermal echogenicity is related to the concentration of the irritant substance, as demonstrated by patch testing with sodium lauryl sulfate at different concentrations. At sodium lauryl sulfate patch test sites, the extension of 201–255 pixel areas is inversely related to TEWL, whereas it is associated with dehydration.[2,7] Thus, the morphological alterations of the epidermis in ultrasound scans seem to be more correlated to the water flux through the epidermis than to the real water content of the stratum corneum.

REFERENCES

1. Seidenari, S. and Di Nardo, A., B-scanning evaluation of allergic reactions with binary transformation and image analysis, *Acta Derm. Venereol.,* 175, 3, 1992.
2. Seidenari, S. and Di Nardo, A., B-scanning evaluation of irritant reactions with binary transformation and image analysis, *Acta Derm. Venereol.*, 175, 9, 1992.
3. Seidenari, S. and Di Nardo, A., Cutaneous reactivity to allergens at 24-h increases from the antecubital fossa to the wrist: an echographic evaluation by means of a new image analysis system, *Contact Dermatitis,* 26, 171, 1992.
4. Seidenari, S., Reactivity to nickel sulfate at sodium lauryl sulfate pre-treated skin sites is higher in atopics: an echographic evaluation by means of image analysis performed on 20 MHz B-scan recordings, *Acta Derm. Venereol.*, 74, 245, 1994.
5. Seidenari, S., Echographic evaluation with image analysis of irritant reactions induced by nonanoic acid and hydrochloric acid, *Contact Dermatitis,* 31, 146, 1994.
6. Seidenari, S., Zanella, C., and Pepe, P., Echographic evaluation of sodium lauryl sulfate induced irritation in mice, *Contact Dermatitis*, 30, 41, 1994.
7. Seidenari, S. and Belletti, B., Instrumental evaluation of subclinical irritation induced by sodium lauryl sulfate, *Contact Dermatitis*, 30, 175, 1994.
8. Seidenari, S. et al., Echographic evaluation with image analysis of normal skin: variations according to age and sex, *Skin Pharmacol.*, 7, 201, 1994.
9. Di Nardo, A., Schiavi, M.E., and Seidenari, S., Instrumental evaluation of subclinical responses at 0.05% nickel sulfate patch tests, *Skin Res. Technol.,* 1, 21, 1995.
10. Seidenari, S., Pepe, P., and Di Nardo, A., Sodium hydroxide-induced irritant dermatitis as assessed by computerized elaboration of 20 MHz B-scan images and by TEWL measurement: a method for investigating skin barrier function, *Acta Derm. Venereol.,* 75, 97, 1995.
11. Schiavi, M.E., Belletti, B., and Seidenari, S., Ultrasound description and quantification of irritant reactions induced by dithranol at different concentrations. A comparison with visual assessment and colorimetric measurements, *Contact Dermatitis*, 34, 272, 1996.
12. Pellacani, G. and Seidenari, S., Variations in facial skin thickness and echogenicity with site and age, *Acta Derm. Venereol.*, 79, 366, 1999.
13. Seidenari, S. et al. Thickness and echogenicity of the skin in children as assessed by 20 MHz ultrasound, *Dermatology,* 201, 218, 2000.

14. Seidenari, S., Belletti, B., and Pellacani, G. Time course of skin changes induced by short-term occlusion with water: evaluation by TEWL, capacitance and B-scan echography, *Skin Res. Technol.*, 2, 52, 1996.
15. Pellacani, G. et al., Epidermal echogenicity as a hydration parameter: the effect of moisturizers on the skin as evaluated by 20 MHz B-scanning, *J. Soc. Cosmetic Chem.*, 48, 243, 1997.
16. Pellacani, G., Belletti, B., and Seidenari, S., Evaluation of the short-term effects of skin care products: a comparison between capacitance values and echographic parameters of epidermal hydration, *Curr. Probl. Dermatol.*, 26, 177, 1998.
17. Altmeyer, P. et al., General phenomena of ultrasound in dermatology, in *Ultrasound in Dermatology,* Altmeyer, P., el-Gammal, S., Hoffmann, K., Eds., Springer, Berlin, 1992, p. 55.
18. Bamber, J.C. and Tristam, M., Diagnostic ultrasound, in *The Physics of Medical Imaging*, Webb, S., Ed., Adam Hilger, Bristol, 1988, p. 319.

7 Standardization of Measurements and Guidelines

Grazia Primavera, Joachim W. Fluhr, and Enzo Berardesca

CONTENTS

I. Introduction 84
II. Sources of Error and Associated Variables 84
A. Instrument-Related Variables 85
1. Instrumental Variability, Start-Up, and Use 85
2. Zeroing 85
3. Measuring 85
4. Zero Drift 86
5. Use of the Probe Protection Covers 86
6. Calibration 87
7. Accuracy 87
B. Environment-Related Variables 87
1. Air Convections 87
2. Ambient Air Temperature 88
3. Ambient Air Humidity 88
4. Light Sources 88
5. Skin Cleansing 88
C. Individual-Related Variables 89
1. Age, Sex, and Race 89
2. Anatomical Sites 89
3. Sweating 90
4. Skin Surface Temperature 90
5. Skin Damage and Diseases 90
6. Circadian Rhythm 90
7. Intra- and Interindividual Variation 90
III. Conclusions 91
References 92

0-8493-1443-7/05/$0.00+$1.50

I. INTRODUCTION

Measurement of transepidermal water loss (TEWL) is widely used to characterize the water barrier function of skin (both in physiological and pathological conditions), to perform predictive irritancy tests, and to evaluate the efficacy of therapeutic treatments on diseased skin.

TEWL assessment can be performed using different techniques [1,2] (closed-chamber method, ventilated-chamber method, and open-chamber method). Some of these techniques only have a historical value, and their use is not widespread today. Earlier methods placed a precisely weighed amount of hygroscopic salt in an unventilated chamber attached to the skin surface for a certain period. TEWL was then determined by weighing the salt before and after application. Other techniques either measured changes in thermal conductivity or infrared radiation absorption in a continuous flow of dry nitrogen passing over a defined skin surface, or worked in a closed, unventilated chamber. The latter technique tended to occlude the skin, and in addition, both methods interfered with the microclimate near the surface of the skin, thus affecting TEWL to varying extents. A variation on the open-chamber method, the water vapor evaporation gradient estimation method, however, has proven to be more useful and forms the basis of the commercially available instruments of today.

To obtain reliable TEWL values with all these devices, it is necessary to have a standardization of procedures. The measurement of TEWL is valid only within the boundary layer of diffusion surrounding the human body [3], which has a depth of about 10 to 30 μm under normal conditions [4]. Thus, it is apparent that any environment or instrument-related variables that influence the depth of this boundary layer would affect the gradient and, hence, the measured TEWL value. Furthermore, the extreme sensitivity of the instrument also indicates that any disturbances in the microclimate, whether the result of environmental, instrumental, or individual related variables, are immediately and instantly displayed as a fluctuation, indicating an error in the measured TEWL value [5]. In this chapter, therefore, attention is focused on the standardization of measurements by determining the variables associated with the method of measuring TEWL, assessing the extent to which they influence the measured value, and elaborating the techniques by which they may be overcome.

II. SOURCES OF ERROR AND ASSOCIATED VARIABLES

Three categories of factors and sources of variation exist in measuring TEWL, including instrumental, environmental, and individual (person-linked factors) variables. A detailed account of these influencing variables is given in this section.

A. Instrument-Related Variables

1. Instrumental Variability, Start-Up, and Use

Commercially available instruments to measure the TEWL in a direct way are the Evaporimeter® and the Tewameter®, based on the open-chamber evaporation gradient method. Therefore, most scientific literature on TEWL refers to these apparatus [6]. A third new apparatus, the DermaLab system, is quite similar to the Evaporimeter and the Tewameter. The Vapometer® and the H4300 are both based on the closed-chamber evaporation gradient methods and appeared on the market only recently [7]. Only limited literature is yet available for the last two instruments, and their performances should be confirmed [7–9].

Tewameter systematically provides higher values than the Evaporimeter because of calibration procedure differences and because variation between different instruments of the same type may occur [10,11]. This leads to the conclusion that calibration is critical.

A high reproducibility of results was found for individual instruments (i.e., a low intrainstrumental variability was found [10]). However, large differences were found among the same types of instruments — a high interinstrumental variability [10]; it depends on the age of the instrument. The newer instruments respond faster (i.e., stabilization time is shorter in some devices, e.g., the TM 300 [Tewameter] as a result of the pretreating of the probe), and the older versions appear to measure much lower TEWL values. This difference may be attributed to the sensors of the probe, which are reported to undergo a slight aging (12). Thus, the calibration of the instrument should be checked frequently.

To enable successful and reliable interlaboratory comparison of results overcoming the effect of the interinstrumental variability, an additional calibration procedure incorporating a calibration for an actual *in vitro* measured water loss can be adopted (13).

The instruments based on the open chamber evaporation gradient method should be turned on at least 15 minutes before measurements are performed, and if the instrument is being used intermittently during the day, it should not be switched off between measurements [12,14].

2. Zeroing

After the warm-up period, the instrument should be "zeroed" only if necessary [13]. Regularly calibrated and well-maintained instruments will not require this zeroing daily if the offset knob is not used for zeroing between measurements.

3. Measuring

a. *Surface Plane*

It is advisable to measure on a horizontal plane to avoid a chimney effect. Furthermore, the contact pressure of the probe on the skin surface should be held low but constant [3,12,15]. To obtain a constant probe pressure, a built-in spring system should be provided; however, that is not the case in any of the instruments available at the moment. In some instruments, screens and grids can be used. These elevate the probe and, thus, also the sensors. This will directly affect TEWL values, and one should be careful in comparing data gathered with or without these devices [4].

b. Contact Time

The time the probe is applied to the skin should be as short as possible. Stabilization of the TEWL value is usually reached by 30 to 45 seconds after starting the measuring [16,17]. The skin area measured is approximately 1 cm^2, and the smallest amount measured corresponds to 0.00000006 g/cm^2 per second (2.16 $g/m^2 h$) [5]. Such precision also means that disturbances in the microclimate are immediately detected as a fluctuation in TEWL [5]. The prescribed stabilization period is 30 to 45 seconds [18,19] for baseline TEWL measurements only. If measurements are made on a diseased or damaged SC barrier, high water evaporation rates are expected. Furthermore, high ambient relative humidities require a longer stabilization period. This prescribed period is not a hard-and-fast rule, as it may also vary from instrument to instrument [19].

4. Zero Drift

Displacement of the water evaporation (WE) zero level in between measurements is attributed to the abrupt humidity changes as well as to the temperature changes of the probe, as a result of the measurement itself [13].

a. Humidity Changes

After each measurement, some condensation vapor remains in the probe, causing a moisture gradient to persist [12]. Therefore, the instrument will continue for some time to indicate a "nonzero" WE value, which will of course disappear when the water vapor in the probe is evaporated [12]. An equilibrium time should be taken into consideration before the next measurement is started. The process can be accelerated by moving the probe carefully. A well-maintained instrument will return to zero within about 2 to 4 minutes postmeasurement [19].

b. Temperature Changes

The temperature-dependent variability of the sensors of the probe and of the amplifiers in the probe handle are important when measuring TEWL [5,12]. During a measurement, the temperature of the probe increases as a result of heat absorbed from the measured skin surface and also from the operator's hand. The skin temperature of the test persons should be measured on the test site, and the measuring probe should then be warmed up to this particular temperature. This can be done on a part of the skin surface on which no measurements are carried out. The measuring probe itself should not be touched before and during measurements and can be handled with the electrical wire, with a coating, by wearing gloves, or by using a clump to hold the probe.

5. Use of the Probe Protection Covers

Although it is recommended to apply, whenever possible, the protection covers supplied with the instruments [9], it is important to recognize the variables that they introduce [20]. The measurement principle of the instrument is based on the estimation of the evaporated water from the vapor pressure gradient within the boundary layer and its depth depending on the site, air speed, and convections — forced or free [3,4]. In the absence of convection currents or draughts, a mean depth of about 10 to 30 μm may be assumed for this boundary layer [4]. The use of the protection cover with the screen and the grid elevates the probe, and therefore the sensors, above the water vapor boundary layer surrounding the skin because of the added height (6 to 7 mm) of the stainless steel screen. This will influence absolute TEWL measurements [3,4,12,13]. Therefore, measurements made with the screen and without the screen cannot be directly compared [12,13]. With the screen, the TEWL values will be lower than without it, and the difference becomes greater as the TEWL rate increases [13,21]. Use of these protection covers should thus be stated clearly in reports and publications.

6. Calibration

As a general rule, calibration or checking the calibration of the instruments according to the manufacturer's specifications should be performed at regular intervals [12]. However, only a limited number of comparative studies between different types of instruments have been described in the literature [15,22–24]. The message to be taken from this is that the Tewameter systematically provides higher values than the Evaporimeter because of calibration procedure differences and that variation between different instruments of the same type may occur (10). This leads to the conclusion that calibration is critical. It is, therefore, more correct to talk of relative values than of real absolute values.

7. Accuracy

Few publications testing for accuracy for the different instruments are available [4,26]. In fact, for the Evaporimeter, an underestimation of the water evaporation rate is reported [4]. At evaporation rates of about 20 g/m^2 per hour, the underestimation is only about 10%. When the evaporation rate exceeds 80 g/m^2 per hour, however, the underestimation may be about 50% in still air [4]. It was assumed that the presence of the probe may restrict the flux of water vapor, particularly at the high evaporation rates. This was also reported by Scott et al. [25], who stated that this underestimation is at evaporation rates above 75 g/m^2 per hour.

B. Environment-Related Variables

1. Air Convections

Air convections are the main source of disturbance resulting in rapid fluctuations of the measurements [5]. They are commonly produced by disturbances in the room, such as people moving around, opening and closing doors, breathing across the measurement zone, air conditioners, and so forth. As these disturbances are difficult to avoid, the use of a covering box was proposed to shield as much as possible the probe from undesirable air turbulence [12,15]. A shielding box should have an open top, covered with a cotton cloth so as not to build up occlusivity. However, such shields might increase the relative humidity in the space around the measured skin area. Therefore, temperature and relative humidity in this box should be recorded.

2. Ambient Air Temperature

The most important effect of the temperature of the air is that it influences the skin temperature both directly (by convection) and indirectly (by central thermoregulatory effects) [27,28]. A distinction must be made between the temperature of the measuring room and the temperature (climate) at which the volunteers live. Therefore, an adaptation time of 15 to 30 minutes is mandatory.

It was found that fluctuations in the temperature of the measuring room affected stratum corneum (SC) hydration and TEWL [15]. It is advised that the room temperature be controlled below 22°C. It must, however, be mentioned that a room temperature of 18°C, as suggested in some articles, seems impossible because test persons complain of the cold and usually refuse to continue the study.

Because ambient air temperature affects TEWL measurement, seasonal variations should be avoided. Even if one works in a temperature-controlled room, it has been observed that the TEWL baseline is not stable. No significant difference between summer and winter baseline TEWL could be observed by Agner and Serup [29]; in aged people, however, it has been shown that SC lipid levels of different body sites were depleted in winter in comparison with spring and summer. These changes contributed to the increased susceptibility of aged skin to perturbation of barrier function and xerosis, particularly during the winter months [30]. Sweating in the summer and feeling cold in the winter seem to be obvious problems. As a consequence, it is evident that geographical variations also may affect TEWL measurements.

3. Ambient Air Humidity

The relationship of the TEWL to ambient air humidity is not linear [3,12]. The message is that ambient relative humidity is a complex but important variable in TEWL measurements. It is generally advised that you work in a temperature-controlled room with additional control over the relative humidity. The latter should be set close to but lower than 50%. The remarks made earlier for seasonal and geographical variations also apply here.

4. Light Sources

Any light source close to the test site affecting the ambient air temperature, the probe temperature, or the temperature of the skin surface of the test persons should be avoided [2].

5. Skin Cleansing

Cleansing of the skin with surface-active agents and solvents could increase TEWL [31] because of damage to the barrier function of the skin [15,32]. Increased TEWL values could also result from the removal of occlusive substances from the skin surface, such as cream [33], and possibly also from the removal of sebum. It is also obvious that exposure of the skin to water-containing products could result in elevated water loss from the surface (e.g., when increased TEWL values are obtained for 10 minutes after a rapid cleansing of the skin with a mild product [15]). A rapid superficial cleansing of the skin with water-free ether does not appear to increase the water loss values [33]. It is also important to keep in mind that agents in cleansers may deposit on the skin surface and reduce TEWL.

C. Individual-Related Variables

1. Age, Sex, and Race

Age, sex, and race have no marked effect on TEWL [2,12,15,34]. However, during certain periods of life, significant differences may occur. For example, impaired epidermal barrier properties could be demonstrated in immature infants of less than 30 weeks gestation. However, a remarkable rate of barrier maturation was seen during the first few days of postnatal life [35]. In fact, based on the almost identical values for the parameters of TEWL, SC, hydration, and pH value, the skin physiology of children differs very little in SC hydration and barrier function from that of adults [36].

From adulthood through senescence, the age dependence of the cutaneous permeability barrier function, and hence baseline TEWL, have been controversially discussed. Wilhelm and Maibach [36] indicate that there is evidence that baseline TEWL is reduced in aged individuals as compared with values from mid-adulthood. Furthermore, with increasing age, significant decreased levels of all major barrier lipids have been observed, contributing to an increased susceptibility of aged skin to perturbation of the barrier function and xerosis [37–39].

Some differences have been reported between Black and White human skin [40–42]. This is also the case for White and Hispanic subjects [43].

2. Anatomical Sites

The regional variation in TEWL is related to the varying skin structure, particularly the different lipid fractions between individual locations [44], the different thickness between anatomical sites [45], and the regional distribution of eccrine sweat glands, which are concentrated on the palms and soles, face, and upper trunk.

TEWL values of different anatomical sites found in the literature from 1977 to 1988 indicate the following ranking: palm > sole > forehead = postauricular skin = nail = dorsum of hand > forearm = upper arm = thigh = chest = abdomen = back [12]. In fact, skin at different body sites shows distinct patterns of barrier recovery that are likely to be related to structural and physiological differences [46]. Lipid-rich skin areas (e.g., the forehead) are the most vulnerable to barrier disruption [46]. Measurements are often carried out on the volar forearms. TEWL on the dominant forearm might be significantly higher than on the nondominant forearm [47,48], but this has not been found to be the case by all investigators [15,49]. Recently, it has been suggested that for the evaluation of cosmetic formulations, facial skin would be more suitable for measuring than the volar forearm [50].

3. Sweating

Physical, thermal, and emotional sweating are important variables that need to be controlled for [10,51]. If the ambient air temperature is below 20°C and the skin temperature is below 30°C, thermal sweat gland activity is unlikely, provided that the skin is not exposed to forced convection and no excessive body heat is produced (as a result of physical exercise) [52–54]. Therefore, a premeasurement 15- to 30-minute rest in a temperature-controlled room of 20° to 22°C is, in most studies, taken into consideration. In addition, physical activity is kept to a minimum. It must, however, be mentioned that it is impossible to control insensible perspiration.

4. Skin Surface Temperature

The skin temperature of the investigator's hand might affect the probe temperature [15]. Therefore, the application of a protection cover of the probe was advised. However, the measuring probe itself is also significantly warmed up by contact with the skin of the test person [12,15]. Skin temperature is an important factor, and preconditioning of the test persons in a temperature-controlled room is required. This is of particular importance for those instruments that do not provide the temperature of the probe on the display. One can also heat the probe to a given temperature [55] or normalize the TEWL values obtained to 30°C [56].

5. Skin Damage and Diseases

Skin diseases in which the barrier function is significantly altered, including burns, psoriasis, some ichthyotic disorders, contact dermatitis, and atopic dermatitis, are characterized by increased TEWL values [2,16,57–59]. Most of the changes in the barrier function are caused by chemical contacts [60,61], surfactant damage [62,63], or diseased states (dermatitis; e.g., psoriasis [64] and eczema [59]), resulting in an increased water evaporation rate within the range of 20 to 60 g/m^2 per hour. However, more severe damage to the barrier such as burns [13] gives rise to much higher evaporation rates (above 100 g/m^2 per hour) that should be considered in the interpretation of the results.

6. Circadian Rhythm

Diurnal variation of TEWL has been described [65]. Fluctuations of the TEWL may be mainly temperature dependent. TEWL has been reported to undergo a circadian rhythm and to be higher in the evening and during the night than in the morning [66,67]. More recently, Le Fur et al. have found a bimodal rhythm for TEWL, with two peaks located at 8:00 and 16:00 hours [67].

7. Intra- and Interindividual Variation

For most skin sites, important interindividual variations occur, which are usually also dependent on the measuring device [3,12,15,48,68,69]. It should be taken into account that some skin sites, including some parts of the forehead, the palm of the hand, and the wrist, should be avoided because of their very high interindividual variability. The intraindividual variation per site usually is smaller [15,69]. A wide interindividual range of variation is particularly seen when individuals undergo certain treatments. A typical example are TEWL measurements after sodium lauryl sulfate damage of the skin [70]. It seems also that obese individuals show significantly increased TEWL after an irritant patch test with sodium lauryl sulfate [71].

III. CONCLUSIONS

TEWL measurements produce reliable results on the conditions that all variables are controlled and standardized conditions are rigorously applied. The European Expert Group Measurement of Cosmetics recommendations with respect to the measurement conditions are as follows [13]:

> If possible, measurements should be carried out in a temperature- and relative humidity-controlled room. Usually it is suggested to keep the temperature between 20° and 22°C ± 1°C and the relative humidity lower than 60%. However, depending on the purpose and design of the study conditions may be different.
>
> The skin temperature of the test persons should be measured on the test site and the measuring probe should then be warmed up to this particular temperature. This can be done on a part of the skin surface where no measurements are carried out. When test products need to be applied or skin damage must be provoked, relevant controls have to be included, e.g., corresponding spots on left and right forearms. Only TEWL values from the same anatomical sites are expected to be comparable.
>
> Measurement should be carried out in a room with limited air circulation. A shielding box with an open top can be used if doubt exists whether undesirable air turbulence is present or not.
>
> TEWL measurements of a single experiment should, whenever possible, be completed within one season. Measurements during hot summer and freezing winter days should be avoided, with the exception of course when the aim of the study asks for this kind of environmental conditions.
>
> Direct and close light sources should be avoided.
>
> The measuring probe itself should not be touched before and during measurements and can be handled with the electrical wire, a coating or by wearing gloves.
>
> The measuring surface should be placed in a horizontal plane and the probe should be applied perpendicularly to this surface with a constant but light pressure. Measurements within one experiment preferably should be performed by the same operator.
>
> TEWL measurements should be as short as possible in order to avoid occlusion. Before each measurement the zero-value should be displayed.

If skin cleansing is carried out before measurements take place its effect on TEWL should be investigated.

Long-term or repeated measurements are preferably done at comparable time periods (e.g., same hour per day, same number of hours after skin cleaning, etc.).

REFERENCES

1. Wilson, D.R. and Maibach, H., Transepidermal water loss: a review, in *Cutaneous Investigation in Health and Disease. Noninvasive Methods and Instrumentation*, Lévêque, J.L., Ed., Marcel Dekker, New York, 1989, pp. 113–133.
2. Lévêque, J.L., Measurement of transepidermal water loss, in *Cutaneous Investigation in Health and Disease. Noninvasive Methods and Instrumentation*, Lévêque, J.L., Ed., Marcel Dekker, New York, 1989, pp. 135–153
3. Nilsson, G.E., Measurement of water exchange through skin, *Med. Biol. Eng. Comput.*, 15, 209, 1977.
4. Wheldon, A.E. and Monteith, J.L., Performance of a skin Evaporimeter, *Med. Biol. Eng. Comput.*, 18, 201, 1980.
5. *A Guide to Water Evaporation Rate Measurement*, ServoMed, Vallingby, Stockholm, Sweden,1981.
6. Schultz, A., Elsner, P., and Burg, G., Quantification of irritant contact dermatitis *in vivo*: comparison of the Dermatest system with the Evaporimeter, *Contact Dermatitis*, 24, 235, 1991.
7. Grove, G. et al., Comparative metrology of the Evaporimeter and the DermaLab TEWL Probe, *Skin Res. Technol.*, 5, 1, 1999.
8. Grove, G. et al., Computerised evaporimetry using the DermaLab TEWL Probe, *Skin Res. Technol.*, 5, 9, 1999.
9. Pinnagoda, J. et al., Comparability and reproducibility of results of water loss measurements: a study of four evaporimeters, *Contact Dermatitis*, 20, 241, 1989.
10. Holbrook, K.A. and Odland, G.F., Regional differences in the thickness (cell layers) of the human stratum corneum: an ultrastructural analysis, *J. Invest. Dermatol.*, 62, 415, 1974.
11. ServoMed Evaporimeters, *Operation Handbook*, ServoMed, Vallingby, Stockholm, Sweden, 1981.
12. Pinnagoda, J. et al., Guidelines for transepidermal water loss (TEWL) measurement, *Contact Dermatitis*, 22, 164, 1990.
13. Rogiers, V., EEMCO guidance for the assessment of transepidermal water loss in cosmetic sciences, *Skin Pharmacol. Appl. Skin Physiol.*, 14(2), 117, 2001.
14. Rogiers, V., Transepidermal water loss measurements in patch test assessment: the need for standardisation, in *Irritant Dermatitis. New Clinical and Experimental Aspects*, Elsner, P. and Maibach, H.I., Eds., Karger, Basel, 1995, Vol. 23, pp. 152.
15. Potts, R.O., Stratum corneum hydration: experimental techniques and interpretations of results, *J. Soc. Cosmet. Chem.*, 37, 9, 1986.
16. Batt, M.D. and Fairhurst, E., Hydration of the stratum corneum, *Int. J. Cosmetic Sci.*, 8, 253, 1986.
17. Blichmann, C.W. and Serup, J., Reproducibility and variability of transepidermal water loss measurements, *Acta Derm. Venereol.*, 67, 206, 1987.
18. Pinnagoda, J. et al., Comparability and reproducibility of the results of water loss measurements: a study of 4 evaporimeters, *Contact Dermatitis*, 20, 241, 1989.

19. Nilsson, G.E., personal communication, 1987.
20. Agner, T. and Serup, J., Transepidermal water loss and air convection, *Contact Dermatitis*, 22, 120, 1990.
21. Barel, A.O. and Clarys, P., Study of the stratum corneum barrier function by transepidermal water loss measurements: comprehension between two commercial instruments: Evaporimeter and TEWAmeter, *Skin Pharmacol.*, 8, 186, 1995.
22. Schott, R.C. et al., A comparison of techniques for the measurement of transepidermal water loss, *Arch. Dermatol. Res.*, 274, 57, 1982.
23. Schulz, A., Elsner, P., and Burg, G., Quantification of irritant contact dermatitis *in vivo*: comparison of the Dermatest® system with the Evaporimeter, *Contact Dermatitis*, 24, 235, 1991.
24. Petro, A.J. and Komor, J.A., Correction to absolute values of evaporation rates measured by the Servo-med® Evaporimeter, *Bioeng. Skin*, 3, 271, 1987.
25. Scott, R.C. et al., A comparison of techniques for the measurement of transepidermal water loss, *Arch Dermatol. Res.*, 274, 57, 1982.
26. Grice, K.A., Transepidermal water loss, in *The Physiology and Pathophysiology of the Skin*, Jarret, A., Ed., Academic Press, London, 1980, Vol. 6, p. 2121.
27. Rothman, S., The role of the skin in thermoregulation: factors influencing skin surface temperature, in *Physiology and Biochemistry of the Skin*, University of Chicago Press, Chicago, IL, 1954, p. 258.
28. Agner, T. and Serup, J., Seasonal variation in skin resistance to irritants, *Br. J. Dermatol.*, 121, 323, 1989.
29. Rogers, J. et al., Stratum corneum lipids: the effect of ageing and the seasons, *Arch. Dermatol. Res.*, 288, 765, 1996.
30. Rodrigues, L. et al., About claims substantiation for topical formulations: an objective approach to skin care products biological efficacy, *J. Appl. Cosmetol.*, 14, 93, 1996.
31. Fartasch, M., Human barrier formation and reaction to irritation., *Curr. Probl. Dermatol.*, 23, 95, 1995.
32. Loden, M., The increase in skin hydration after application of emollients with different amounts of lipids, *Acta Derm. Venereol. (Stockh.)*, 72, 327, 1992.
33. Giusti, F. et al., Skin barrier, hydration, and pH of the skin of infants under 2 years of age, *Pediatr. Dermatol.*, 18(2), 93, 2001.
34. Okah, F.A. et al., Surface electrical capacitance as a non-invasive bedside measure of epidermal barrier maturation in the newborn infant, *Pediatrics*, 96, 688, 1995.
35. Fluhr, J.W., Pfisterer, S., and Gloor, M., Direct comparison of skin physiology in children and adults with bioengineering methods, *Pediatr. Dermatol.*, 17(6), 436, 2000.
36. Wilhelm, K.P. and Maibach, H.I., Transepidermal water loss and barrier function of aging human skin, in *Bioengineering of the Skin, Water and the Stratum Corneum*, Elsner, P., Berardesca, E., and Maibach, H.I., Eds., CRC Press, Boca Raton, 1994, pp. 133–145.
37. De Paepe, K. et al., Ceramide/cholesterol//FFA-containing body lotions: effects on the TEWL of aged and SDS-damaged skin, in *Conference Proceedings*, Active Ingredients Conference, Verlag für chemische Industrie, H. Ziolkowsky GmbH, Paris, 1996, pp. 97–111.
38. De Paepe, K. et al., Ceramides/cholesterol/FFA-containing cosmetics: the effect on barrier function, *SOFW J.*, 122, 199, 1996.

39. Wilson, D., Berardesca, E., and Maibach, H.I., *In vivo* transepidermal water loss and skin surface hydration in assessment of moisturization and soap effects, *Int. J. Cosmetic Sci.*, 10, 201, 1988.
40. Berardesca, E. and Maibach, H.I., Racial differences in sodium-lauryl induced cutaneous irritation. Comparison of black and white subjects, *Contact Dermatitis*, 18, 65, 1988.
41. Berardesca, E. et al., Differences in stratum corneum pH gradient when comparing White Caucasian and Black African-American skin, *Br. J. Dermatol.*, 139(5), 855, 1998.
42. Berardesca, E. and Maibach, H.I., Sodium-lauryl-sulphate-induced cutaneous irritation. Comparison of black and white subjects, *Contact Dermatitis*, 19, 136, 1988.
43. Yoshikawa, N. et al., Regional analysis of ceramides within the stratum corneum in relation to seasonal required for normal barrier homeostasis, *J. Invest. Dermatol.*, 115, 459, 2000.
44. Plewig, G. and Marples, R.R., Regional differences of cell sizes in the human stratum corneum, *J. Invest. Dermatol.* 54, 13, 1970.
45. Fluhr, J.W. et al., Impact of anatomical location on barrier recovery, surface pH and stratum corneum hydration after acute barrier disruption, *Br. J. Dermatol.*, 146(5), 770, 2002.
46. Treffel, P. et al., Hydration, transepidermal water loss, pH and skin surface parameters: correlations and variations between dominant and non-dominant forearms, *Br. J. Dermatol.*, 130, 325, 1994.
47. Rodrigues, L. and Pereira, L.M., Basal transepidermal water loss: right/left forearm difference and motoric dominance, *Skin Res. Technol.*, 4, 135, 1998.
48. Oestmann, E. et al., Skin barrier function in healthy volunteers as assessed by transepidermal water loss and vascular response to hexyl nicotinate: intra- and inter-individual variability, *Br. J. Dermatol.*, 128, 130, 1993.
49. Schnetz, E. et al., Intra- and inter-individual variations in transepidermal water loss on the face; facial locations for bioengineering studies, *Contact Dermatitis*, 40, 243, 1999.
50. Pinnagoda, J. et al., Transepidermal water loss with and without sweat gland inactivation. *Contact Dermatitis*, 21(1), 16, 1989.
51. Shahidullah, M. et al., Transepidermal water loss in patients with dermatitis, *Br. J. Dermatol*, 81, 722, 1969.
52. Baker, H. and Kligman, A.M., Measurement of transepidermal water loss by electrical hygrometry. Instrumentation and responses of physical and chemical insults, *Arch. Dermatol.*, 96, 441, 1967
53. Pinnagoda, J. et al., Transepidermal water loss: with and without sweat gland inactivation, *Contact Dermatitis*, 21, 16, 1989.
54. Piérard, G.E. et al., Surfactant induced dermatitis. A comparison of corneosurfametry with predictive testing on human and reconstructed skin, *J. Acad. Dermatol.*, 33, 462, 1995.
55. Mathias, C.G.T., Wilson, D., and Maibach, H.I., Transepidermal water loss as a function of skin temperature, *J. Invest. Dermatol.*, 77, 219, 1981.
56. Di Nardo, A. et al., Ceramide and cholesterol composition of the skin of patients with atopic dermatitis, *Acta Derm. Venereol.*, 78, 27, 1998.
57. Chao, K.N., Eisley, J.G., and Yang, W.J., Heat and water losses from burnt skin, *Med. Biol. Eng. Comput.*, 15, 598, 1977.
58. Proksch, E., Brasch, J., and Sterry, W., Integrity of the permeability barrier regulates epidermal Langerhans cell density, *Br. J. Dermatol.*, 134, 630, 1996.

59. Tagami, H. et al., Environmental effects on the functions of the stratum corneum, *J. Invest. Dermatol. Symp. Proc.*, 6(1), 87, 2001.
60. Chilcott, R.P., Brown, R.F., and Rice, P., Non-invasive quantification of skin injury resulting from exposure to sulfur mustard and Lewisite vapours, *Burns*, 26(3), 245, 2000.
61. Tupker, R.A. et al., The influence of repeated exposure to surfactants on the human skin as determined by transepidermal water loss and visual scoring, *Contact Dermatitis*, 20, 108, 1989.
62. Fartasch, M., Ultrastructure of the epidermal barrier after irritation, *Microsc. Res. Tech.*, 1, 37(3), 193, 1997.
63. Grice, K.A., Transepidermal water loss in pathologic skin, in *The Physiology and Pathophysiology of the Skin*, Jarret, A., Ed., Academic Press, London, 1980, p. 2147.
64. Chilcott, R.P. and Farrar, R., Biophysical measurements of human forearm skin *in vivo*: effects of site, gender, chirality and time, *Skin Res. Technol.*, 6(2), 64, 2000.
65. Yosipovitch, G. et al., Time-dependent variations of the skin barrier function in humans: transepidermal water loss, stratum corneum hydration, skin surface pH, and skin temperature, *J. Invest. Dermatol.*, 110, 20, 1988.
66. Reinberg, A., Le Fur, I., and Tschochler, E., Problem related to circadian rhythms in human skin and their validation, *J. Invest. Dermatol.*, 111, 708, 1998.
67. Le Fur, I. et al., Analysis of circadian and ultradian rhythms of skin surface properties of face and forearm of healthy women, *J. Invest. Dermatol.*, 117(3), 718, 2001.
68. Pinnagoda, J. et al., The intra- and inter-individual variability and reliability of transepidermal water loss measurements, *Contact Dermatitis*, 21(4), 255, 1989.
69. Goffin, V., Piérard-Franchimont, C., and Piérard, G.E., Passive sustainable hydration of the stratum corneum following surfactant challenge, *Clin. Exp. Dermatol.*, 24, 308, 1999.

8 Transepidermal Water Loss and Its Relation to Barrier Function and Skin Irritation

P.G.M. van der Valk, M. Kucharekova, and R.A. Tupker

CONTENTS

I. Definition of Transepidermal Water Loss 97
II. Stratum Corneum Constituents and Barrier Function 98
III. Measuring TEWL 99
IV. TEWL as a Measure for Barrier Function 100
V. TEWL as a Tool to Study the Irritant Potential of Chemicals 100
VI. TEWL as a Tool to Monitor Disease Activity 101
VII. TEWL as a Predictor for Susceptibility for Skin Irritation and Irritant Contact Dermatitis 101
References 102

I. DEFINITION OF TRANSEPIDERMAL WATER LOSS

An important function of the skin is protection against the loss of water. Cutaneous loss of water comprises sweat loss, desorption, and diffusional water loss. Transepidermal water loss (TEWL) is the constitutive steady-state water vapor loss from the skin, theoretically excluding desorption and sweat gland activity. It is considered to be the result of passive diffusion through the skin — the diffusional flux obeying the physicochemical laws of passive transport through a membrane. The horny layer is the most important rate-limiting step for the transport of water to the exterior [1].

However, the horny layer is not an ideal membrane: it is perforated by skin appendages, and its properties as a barrier are dependent both on terminal differentiation and on environmental influences. The desquamating horny layer is nonliving and constantly renewing, and therefore the formation of a good-quality horny layer may be hampered by skin diseases affecting epidermal differentiation, such as atopic dermatitis. External insults may also influence epidermal differentiation by damaging

0-8493-1443-7/05/$0.00+$1.50

keratinocytes, and they may also have direct effects on corneocytes as well as on the intercellular domain, resulting in loss of barrier capacity [2].

II. STRATUM CORNEUM CONSTITUENTS AND BARRIER FUNCTION

The stratum corneum is composed of nonviable cells, the corneocytes, which are ordered in a columnar pattern and are embedded in a lipid matrix [3]. The corneocytes are the protein-rich end products of keratinocyte differentiation, filled with a tight network of keratin fibers. The corneocytes also contain natural moisturizing factors (free amino acids and other small water-soluble molecules), which are for a great part responsible for the water-holding capacity of the stratum corneum [4]. During terminal differentiation, the plasma membrane is replaced with a tough, insoluble proteinaceous envelope. This layer, which is called the cornified envelope, consists of cross-linked proteins. First, the proteins involucrin, envoplakin, and periplakin are sequentially cross-linked by epidermal transglutaminases, which promote the formation of disulfide and (-glutamyl)lysine isopeptide bonds. Proteins such as loricrin, elafin, S100, and small proline-rich region proteins are incorporated later to form a mature envelope [5].

In the stratum granulosum, the lamellar intercellular domain is formed by the extrusion of the densely packed lipid content of organelles, called lamellar bodies. The lamellar bodies are small organelles formed in the midepidermis. In the stratum granulosum, they migrate to the periphery, where the contents, phospholipids, sphingomyelin, cholesterol sulfate, and glucosylceramides, are extruded as relatively polar discs. Extracellularly, the contents are mainly processed to ceramides, cholesterol, and fatty acids. Initially, the lamellae are segregated into focal accumulations. Later, the preformed lipids form multilayer lipid sheets with alternating strata of hydrophilic polar "heads" of the lipid molecules, aligned toward an aqueous phase, and lipophilic hydrocarbon chains aligned toward the inner side of a lipid bilayer. The lamellar sheets are covalently bound to the cornified envelope by ω-hydroxyceramide, forming a lipid envelope that stabilizes the intercellular domain. In the outer desquamating layers of the stratum corneum, the well-organized structure is lost. Moreover, the lipids, derived from the lamellar bodies, are mixed at the surface of the stratum corneum with sebum components such as squalene, wax esters, triglycerides, and free fatty acids [6,7].

The unique organization of the hydrophilic cells within the lipid, hydrophobic environment makes this 10- to 20-μm layer extremely efficient as a barrier. Its water permeability is comparable with a plastic membrane and is 1000 times more efficient than other membranes of living organisms [8]. The lamellar structure of the intercellular domain shows a well-balanced range of lipids. If the lipid composition is disturbed qualitatively or quantitatively by environmental influences or by disease, barrier function will be hampered. In particular, the long-chained ceramides are essential for the barrier function. It should, however, be emphasized that barrier function is the result both of the structured lipid domain and of the ordering and function of the corneocytes. The intercellular domain is most important for the water barrier. The role of sebum in barrier function is probably very limited [9].

We want to emphasize that stratum corneum barrier formation is a dynamic process. The skin is continuously exposed to environmental influences, which may influence barrier integrity. Initially, harm to the skin may lead to decrease of barrier function but also, subsequently, either to recovery and hardening or to disease, depending on the size and nature of the assault and on the relevant constitutional factors of the exposed individual [10]. Increased lipid synthesis, lamellar body secretion, and lipid processing are aimed at barrier repair. If these adaptive phenomena fail, chronic barrier dysfunction may be the result. Moreover, the stratum corneum is the end product of epidermal differentiation. This means that, for example, nutritional deficiencies, neuroendocrine responses, or skin diseases that influence epidermal growth and differentiation may change the structure and function of the stratum corneum and may, therefore, induce barrier abnormalities.

Standardized TEWL measurements as an indicator for water barrier function should, therefore, be interpreted with great care, because many factors related to (skin) diseases, barrier disruption, recovery, and adaptation are involved in the final study results [11–13].

III. MEASURING TEWL

TEWL is, by definition, diffusional water loss through the epidermis. Therefore, sweat gland activity should be eliminated as much as possible by measuring under neutral environmental conditions. To avoid thermal sweating and emotional sweating, subjects should be put at ease for 15 to 30 minutes in a room with a constant temperature of 20° to 22°C and a relative humidity of about 40% [11]. In case there is still sweating despite following these recommendations, sweating can be blocked by a 1-hour occlusive application of scopolamine hydrobromide [14]. To measure pure diffusional water loss, one should avoid desorption (loss of excessive unbound water); for example, by measuring soon (less than 1 hour) after removing patches in irritancy testing.

Many environment-related variables influence water diffusion through the horny layer. The most important variables are ambient air temperature, which influences skin temperature and, consequently, the horny layer barrier function; air humidity, which influences the water vapor gradient over the skin and the barrier function characteristics of the horny layer; and seasonal variation. The latter is a very important variable, as the skin is often dehydrated by low indoor relative humidity in the winter in many parts of the world. Low relative humidity has a strong effect on skin hydration and, consequently, on barrier function [15]. Abrupt decreases in environmental humidity may be especially harmful to the barrier [16].

Whole-body water loss can be measured gravimetrically. The water loss of the skin is estimated by determining weight reduction of a person on a weighing scale. However, if information is needed from limited areas for the researchers to be informed about local skin barrier characteristics, this approach will not be sufficient. For measurements accurately registering small local differences, techniques using ventilated or unventilated chambers or instruments making use of gradient estimation can be used.

In unventilated chambers, a hygroscopic material is put in the chamber and the water loss is measured gravimetrically, or TEWL is computed by the progressive increase in relative humidity [17]. In ventilated chambers, a carrier gas with known water content is passed through a chamber placed with its open end on the skin. The water loss is proportional to the increase of water content in the carrier gas. Ventilated or unventilated chamber techniques influence the microclimate of the measuring area. The technique of gradient estimation, however, influences the microclimate to a much smaller extent, giving more information on the actual *in vivo* loss (without instrumentation) of the measuring area [18]. The method is widely used because of its simplicity and its ease of application. Its principle is based on the presence of a boundary layer of diffusion of 7 to 10 mm surrounding the human body. In this layer, water loss can be calculated by computing partial pressure gradient and evaporation rate from relative humidity and temperature data measured perpendicularly at two distances from the skin. The humidity transducers and thermistors are placed in an open cylinder fixed to a probe, separating a small area from any disturbing drafts.

To avoid instrument-related disturbance of this microclimate among others, direct light, heat, draft, and pressure of the probe should be avoided. Pinnagoda et al. [11] and more recently Rogiers [12] defined guidelines for accurate and reproducible measurements.

IV. TEWL AS A MEASURE FOR BARRIER FUNCTION

If the barrier function of the skin is hampered either by environmental influences or by skin disease, TEWL may be increased. Therefore, TEWL measurements can be used to study the integrity of the skin barrier [19]. However, many variables are involved in water loss of the skin. If TEWL measurements are used as an indicator for the water barrier, the above-mentioned environmental factors should either be eliminated or should be standardized. Special care must be employed to extrapolate water barrier function to barrier function against other chemicals, because it has been shown that the barrier to TEWL may not correlate to penetration of chemicals *in vitro* [20].

However, TEWL is by definition an indicator for water loss *in vivo*, and disturbances in the water barrier function may pose a risk factor for skin irritation in an occupational setting [21].

V. TEWL AS A TOOL TO STUDY THE IRRITANT POTENTIAL OF CHEMICALS

TEWL measurements are widely used to determine the irritant potential of chemicals [22–24]. The clinical response to experimental exposure to surfactants correlates with TEWL. However, visible damage inflicted to the skin by other chemicals may not correlate with loss of water barrier capacity. Irritants may damage the viable layers of the skin without affecting the horny layer, which is the main barrier for water loss [2]. Moreover, TEWL measurements should preferably not be performed just once. To get a full picture of the effects on the barrier, evaporative water loss should be monitored during the eliciting phase and the healing phase. In this way, both primary effects on the horny layer caused by interaction of the chemical with horny layer components and secondary effects caused by effects on the deeper layers of the skin and subsequent disturbances in epidermal differentiation can be studied.

TEWL is a parameter for functional damage. It may not correlate with the penetration of other chemicals and may not correlate with morphological changes and subjective irritation (burning, stinging). Other clinical and instrumental parameters, such as laser Doppler flowmetry, chromometry, skin hydration measurement, and electrical impedance measurement, should be monitored to assess the full irritant potential of a chemical (Figure 8.1) [25,26].

VI. TEWL AS A TOOL TO MONITOR DISEASE ACTIVITY

In ichthyosis, psoriasis, and contact dermatitis, but in particular in atopic dermatitis, TEWL may be increased [28–31]. Barrier function in atopic dermatitis may be deficient in both diseased and clinically normal skin and may reflect disease activity.

We used TEWL to study barrier recovery with and without treatment by tape stripping and sodium dodecyl sulfate(SDS)-induced skin irritation [27]. John et al. [32] demonstrated an increased TEWL within the first year of training in a cohort of junior hairdressers, which indicates (subclinical) skin irritation. TEWL being a functional parameter may, however, not always correlate with other parameters of disease activity. It is reasonable that an increased thickness of the horny layer may compensate for any loss of horny layer barrier quality or that the skin disease does not influence the stratum corneum barrier.

VII. TEWL AS A PREDICTOR FOR SUSCEPTIBILITY FOR SKIN IRRITATION AND IRRITANT CONTACT DERMATITIS

If TEWL reflects barrier capacity, a high TEWL value may be indicative of increased penetration of irritants. This increased penetration may cause skin irritation and irritant contact dermatitis, and therefore TEWL may be a predictor for susceptibility to irritant skin reactions. Indeed, it has been found that a high preexposure TEWL may predispose to more severe irritant reactions [33]. Skin irritation may also lead to hardening of the skin (adaptation) and repair [34]. Consequently, TEWL point measurements will reflect the barrier function of the horny layer as a product of epidermal differentiation at that specific moment. Because epidermal differentiation is a dynamic process, influenced by endogenous and exogenous factors, the barrier function of its end product, the horny layer, may not reflect its constitutive susceptibility to prevent penetration. This may be a reason why preemployment measurements of barrier function as measured by TEWL do not predict the development of irritant contact dermatitis [35,36].

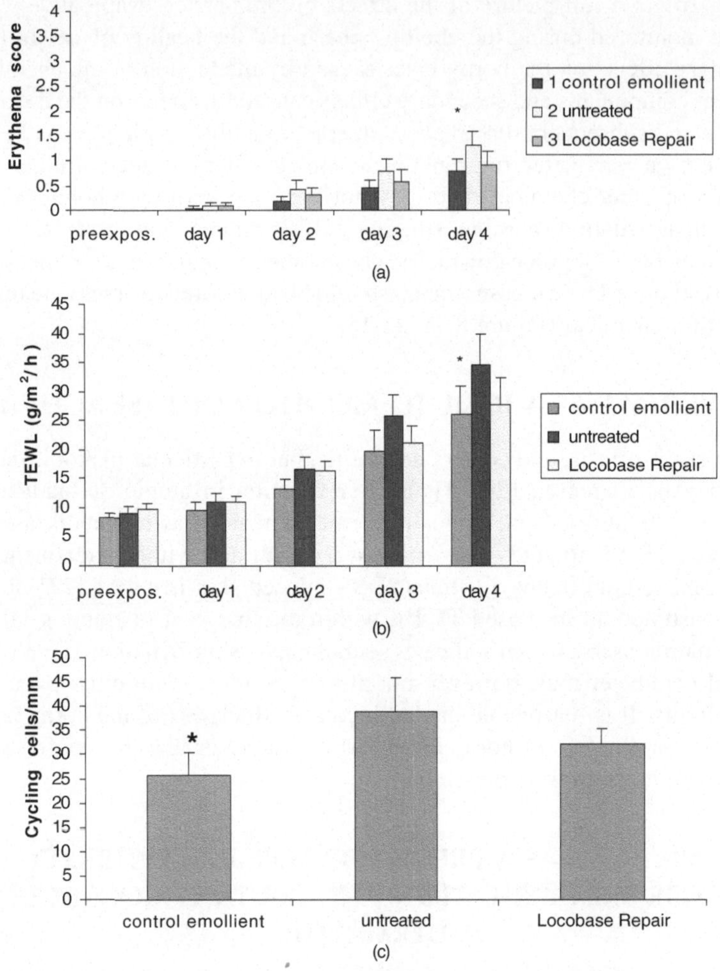

FIGURE 8.1 Multiple parameter assessment of skin irritation. (a) Erythema score in model of repetitive application of SDS (0.2%); *significant difference between the site treated with control emollient and the untreated site ($P < .03$; [19]). (b) Transepidermal water loss measurements in a model of repetitive application of SDS (0.2%); *significant difference between the site treated with control emollient and the untreated site ($P < .05$; [19]). (c) Cycling cells (Ki-67) 96 hours after repetitive application of SDS (0.2%); *significant difference between the site treated with control emollient and the untreated site ($P < .03$; [19]).

REFERENCES

1. Blank, I.H., Cutaneous barriers, *J. Invest. Dermatol.*, 45, 249, 1965.
2. van der Valk, P.G.M. et al., Eczematous (irritant and allergic) reactions of the skin and barrier function as determined by water vapor loss, *Clin. Exp. Dermatol.*, 10, 185, 1985.

3. Mackenzie, I.C., The cellular architecture of the stratum corneum, in *Stratum Corneum*, Marks, R., Plewig, G., Eds., Springer, Berlin, 1983, pp. 146–152.
4. Marty, J.P., NMF and cosmetology of cutaneous hydration, *Ann. Dermatol. Venereol.*, 129, 131, 2002.
5. Marshall, D., et al., Differentially expressed late constituents of the epidermal cornified envelope, *Proc. Natl. Acad. Sci.*, 98, 13031, 2001.
6. Denda, M., Role of lipids in skin barrier function, in *Cosmetic Lipids and the Skin Barrier, Cosmetic Science and Technology Series*, Förster, T., Ed., Marcel Dekker, 2002, Vol. 24, pp. 97–120.
7. Möller, H., The chemistry of natural and synthetic skin barrier lipids, in *Cosmetic Lipids and the Skin Barrier, Cosmetic Science and Technology Series*, Förster, T., Ed., Marcel Dekker, Düsseldorf, 2002, Vol. 24, pp. 1–35.
8. Potts, R.O. and Francoeur, M.L., The influence of stratum corneum morphology on water permeability, *J. Invest. Dermatol.*, 96, 495, 1991.
9. Abrams, K. et al., Effects of organic solvents on *in vitro* human skin water barrier function, *J. Invest. Dermatol.*, 101, 609, 1993.
10. Malten, K.E., Thoughts on irritant contact dermatitis, *Contact Dermatitis*, 7, 238, 1981.
11. Pinnagoda, J. et al., Guidelines for transepidermal water loss (TEWL) measurement. A report from the Standardization Group of the European Society of Contact Dermatitis, *Contact Dermatitis*, 22, 164, 1990.
12. Rogiers, V., EEMCO guidance for the assessment of transepidermal water loss in cosmetic sciences, *Skin Pharmacol. Appl. Skin Physiol.*, 14, 117, 2001.
13. Altemus, M. et al., Stress-induced changes in skin barrier function in healthy women, *J. Invest. Dermatol.*, 117, 309, 2001.
14. Pinnagoda, J. et al., Transepidermal water loss with and without sweat gland inactivation, *Contact Dermatitis*, 21, 16, 1989.
15. Pinnagoda, J., Standardization of measurements, in *Bioengineering of the Skin: Water and the Stratum Corneum*, Elsner, P., Berardesca, E., and Maibach, H.I., Eds., CRC Press, Boca Raton, Florida, 1994, pp. 59–65.
16. Sato, J. et al., Abrupt decreases in environmental humidity induce abnormalities in permeability barrier homeostasis, *J. Invest. Dermatol.*, 119, 900, 2002.
17. Tagami, H., Kobayashi, H., and Kikuchi, K., A portable device using a closed chamber system for measuring transepidermal water loss: comparison with the conventional method, *Skin Res. Technol.*, 5, 9, 1999.
18. Nilsson, G.E., Measurement of water exchange through skin, *Med. Biol. Eng. Comput.*, 15, 209, 1977.
19. Halkier-Sorensen, L. and Thestrup-Pedersen, K., The relationship between skin surface temperature, transepidermal water loss and electrical capacitance among workers in the fish processing industry: comparison with other occupations. A field study, *Contact Dermatitis*, 24, 345, 1991.
20. Chilcott, R.P. et al., Transepidermal water loss does not correlate with skin barrier function *in vitro*, *J. Invest. Dermatol.*, 118, 871.
21. Smit, H.A. et al., Variability in transepidermal water loss of the skin: evaluation of a method to assess susceptibility to contact dermatitis in epidemiological studies, *Int. Arch. Occup. Environ. Health*, 62, 509, 1990.
22. Wigger-Alberti, W. et al., Effects of various grit-containing cleansers on skin barrier function, *Contact Dermatitis*, 41, 136, 1999.
23. Tupker, R.A. et al., Irritancy ranking of anionic detergents using one-time occlusive, repeated occlusive and repeated open tests, *Contact Dermatitis*, 40, 316, 1999.

24. English, J.S., Ratcliffe, J., and Williams, H.C., Irritancy of industrial hand cleansers tested by repeated open application on human skin, *Contact Dermatitis*, 40, 84, 1999.
25. Ollmar, S. et al., Electrical impedance compared with other non-invasive bioengineering techniques and visual scoring for detection of irritation in human skin, *Br. J. Dermatol.*, 130, 29, 1994.
26. Fluhr, J.W. et al., Testing for irritation with a multifactorial approach: comparison of eight non-invasive measuring techniques on five different irritation types, *Br. J. Dermatol.*, 145, 696, 2001.
27. Kucharekova, M. et al., Effect of a lipid-rich emollient containing ceramide 3 in experimentally induced skin barrier dysfunction, *Contact Dermatitis*, 46, 331, 2002.
28. Frost, P. et al., Ichthyosiform dermatosis. III. Studies on transepidermal water loss, *Arch. Dermatol.*, 98, 230, 1968.
29. Berardesca, E. and Maibach, H.I., Noninvasive bioengineering assessment of psoriasis, *Int. J. Dermatol.*, 28, 157, 1989.
30. Lodén, M. et al., Friction, capacitance and transepidermal water loss (TEWL) in dry atopic and normal skin, *Br. J. Dermatol.*, 126, 137, 1992.
31. Shahidullah, M. et al., Transepidermal water loss in patients with dermatitis, *Br. J. Dermatol.*, 81, 722, 1969.
32. John, S.M., Uter, W., and Schwanitz, H.J., Relevance of multiparametric skin bioengineering in a prospectively-followed cohort of junior hairdressers, *Contact Dermatitis*, 43, 161, 2000.
33. Tupker, R.A. et al., Baseline transepidermal water loss (TEWL) as a prediction of susceptibility to sodium lauryl sulphate, *Contact Dermatitis*, 20, 265, 1989.
34. McOsker, D.E. and Beck, L.W., Characteristics of accommodated (hardened) skin, *J. Invest. Dermatol.*, 48, 372, 1967.
35. Goh, C.L. and Gan, S.L., The incidence of cutting fluid dermatitis among metalworkers in a metal fabrication factory: a prospective study, *Contact Dermatitis*, 31, 111, 1994.
36. Smit, H.A. et al., Susceptibility to and incidence of hand dermatitis in a cohort of apprentice hairdressers and nurses, *Scand. J. Work Environ. Health.*, 20, 113, 1994.

9 Transepidermal Water Loss and Allergic Contact Dermatitis

Stefania Seidenari and Francesca Giusti

CONTENTS

I. Introduction 105
II. TEWL Values in Patients with ACD 105
A. Eczematous Skin 105
B. Baseline Barrier Function at Healthy Skin Sites 106
C. Skin Hyperreactivity 106
D. ACD and Irritation 107
III. TEWL and Patch Tests 108
IV. TEWL and Topical Agents for ACD 109
References 110

I. INTRODUCTION

At a dermal level, the acute phase of eczema is characterized by an inflammatory process taking place in the deeper layers of the skin, expressed by vasodilatation and fluid extravasation with edema. In the epidermis, vacuolation and keratinocyte degeneration, with vesiculation and exudation, can be observed. These morphological features parallel epidermal dysfunction in allergic contact dermatitis (ACD). Objective monitoring of barrier impairment in ACD is of considerable interest in both clinical and experimental settings and can be performed by transepidermal water loss (TEWL) measurements.

II. TEWL VALUES IN PATIENTS WITH ACD

A. Eczematous Skin

At eczematous skin sites, an increase in TEWL values can be measured in both the acute and the chronic phase of contact dermatitis. Whereas barrier impairment in irritant contact dermatitis has been thoroughly studied, in spontaneous allergic contact eczema, variations in the degree of barrier impairment according to the intensity

0-8493-1443-7/05/$0.00+$1.50

of the eczematous response, the course, or the extent of the dermatitis have not been systematically investigated to date. This oversight may also be attributed to the fact that the relationship between constitutional or irritant-mediated barrier impairment and contact sensitization is complex, and in many patients, contact allergy and irritation may coexist. In most studies, when referring to nonoccupational eczema or monitoring the dermatitis in an occupational environment, no mention is made in regard to the influence on the barrier of a possible concomitant contact allergy.

B. Baseline Barrier Function at Healthy Skin Sites

At uninvolved skin sites, the baseline barrier function varies according to the activity of the eczema. When inflammatory lesions are present, barrier impairment may also be evident at surrounding healthy skin sites.[1,2] Baseline TEWL and capacitance values measured at six different skin sites on the forearms in nickel-allergic patients with circumscribed areas of eczema in a chronic phase were significantly higher with respect to those of healthy subjects.[3]

C. Skin Hyperreactivity

Functional alterations at uninvolved skin sites may provide a basis for a reduced threshold to irritant stimuli in subjects with active eczema.[4] Systemic mediators released by inflamed skin in the acute phase may induce and maintain this hyperirritability, which is only observed when eczematous lesions are in an active phase.[2,5]

In a study performed on 20 healthy volunteers and on 34 subjects with localized eczema in a chronic phase, comprising 14 patients with atopic dermatitis (AD) and 20 individuals with ACD, cutaneous reactions to a 30-minute application of 0.5% sodium lauryl sulfate (SLS) on six different areas of the forearms were investigated by measuring TEWL, capacitance, and skin echogenicity at different times after SLS exposure.[3,6] Post-SLS application measurements demonstrated a higher increase in TEWL in atopic or contact dermatitis patients than in healthy subjects, although this method was unable to detect any variation between the AD and the ACD group (Figure 9.1).[3] On the contrary, image processing of echographic images revealed a significant hyporeflectivity of the epidermis, expressing barrier function damage, 24 hours after SLS application only in atopic subjects, indicating a specific susceptibility of atopic skin to surfactants that is not present in ACD patients.[6]

After healing has taken place, barrier function reverts to normal. No significant differences in baseline TEWL values were observed between patients with localized or healed eczema and healthy controls.[7] Moreover, ACD patients with inactive disease showed normal baseline barrier function at uninvolved skin sites, unlike subjects with atopic dermatitis.[8] In experimental conditions, susceptibility to irritants on uninvolved skin in ACD patients without current manifestations appears normal. In 18 subjects with a history of ACD, postexposure (SLS and other irritants) TEWL values on the volar aspect of the forearms were lower than in 20 patients with past AD.[9] Investigating the response to 0.5% SLS patch tests applied for 24 hours on uninvolved skin in 30 hand eczema patients and 30 healthy volunteers by means of visual scoring and bioengineering methods, no difference was found in postexposure

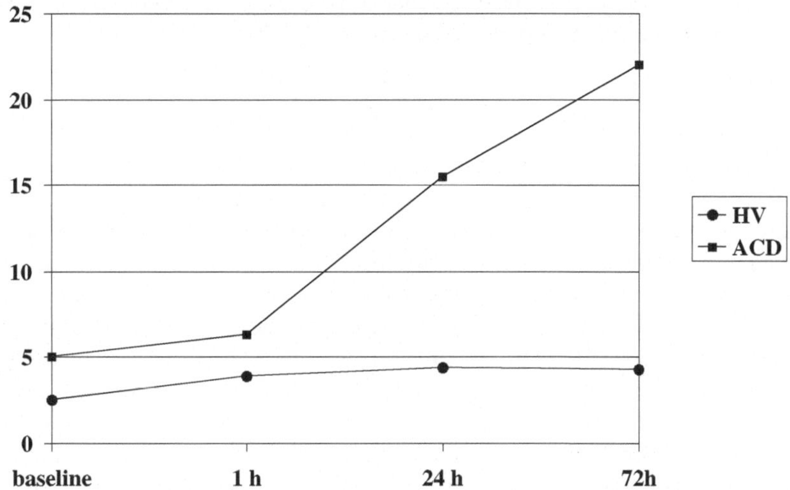

FIGURE 9.1 Post–sodium lauryl sulfate exposure transepidermal water loss values (g/m^2 per hour) on forearms in 14 allergic contact dermatitis (ACD) patients and 20 healthy volunteers (HV).

TEWL values between patients with chronic or healed eczema and controls, whereas patients with an acute disease showed an increased cutaneous reactivity.[10]

D. ACD and Irritation

It is well known that an irritant contact dermatitis may precede both the induction and the elicitation phase of ACD. SLS pretreatment of nickel patch test sites in nickel-sensitive individuals induces a more intense allergic reaction with respect to the application of nickel alone.[6,11] In addition to summation of immune and nonimmune mechanisms, this reaction may also be attributed both to increased allergen penetration resulting from barrier disruption and to the lowering of the threshold value for the elicitation of the allergic response. Moreover, irritants can increase the epidermal density of Langerhans' cells, and this may further enhance the allergic response, as shown by Proksch et al.,[12] who performed acetone pretreatment of allergic patch test areas in contact-sensitive patients. Studying the influence of SLS and nickel in combination on the patch test reactions in previously sensitized patients, Agner et al.[13] demonstrated a synergistic effect by visual scoring and colorimetry, whereas the response was found to be additive when assessed by TEWL measurements. The summing of irritants and subthreshold doses of allergens may cause enhanced damage if contact sensitization and atopy coexist: pretreatment of nickel patch test areas with SLS induced an earlier inflammatory response and a more pronounced skin damage in nickel-sensitive atopic subjects compared with nickel-sensitive nonatopics, followed by a more marked allergic reaction, as shown by TEWL and ultrasound measurements.[6] Recently, it has been suggested that a key role of irritants on ACD is to provide a "danger" signal, required for both the induction of contact sensitization and its elicitation.[14] On the basis of the above information, it is clear that the maintenance of the barrier function in sensitized patients is very important to prevent relapses of ACD.

III. TEWL AND PATCH TESTS

Patch testing is a widely used procedure in both clinical and experimental settings. Patch tests are routinely employed to diagnose ACD and are evaluated according to internationally agreed-on rules, using a clinical rating scale based on the degree of erythema and induration and on the presence or absence of vesicles. However, in experimental studies investigating the sensitizing properties of chemicals or the efficacy of drugs, visual assessment of patch test responses lacks objectivity and reproducibility and is, therefore, unsuitable. A number of instrumental techniques have been applied to patch test evaluation to quantify allergic responses precisely, producing continuous data grading of disease intensity suitable for statistical analysis.[15–24] Each method measures only one aspect of the inflammatory reaction: the evaporimeter evaluates the barrier disruption,[15,18] the laser Doppler flowmeter and the colorimeter quantify the component of the allergic response resulting from the increase in blood flow,[17,18,22–25] and the ultrasound measures the edema that is secondary to vasodilatation, increase in blood vessel permeability, and extravasation of water within the tissue.[16,18–21,24]

With regard to evaporimetry, van der Valk et al.[7] assessed TEWL in nickel-allergic patients undergoing 48-hour patch testing with nickel sulfate, SLS, dimethyl sulfoxide, and phenol. An increased TEWL was observed only in strong reactions to nickel. Investigating patch test responses to nickel and SLS in 12 nickel-sensitized patients by means of evaporimetry, Serup and Staberg[15] found that after 48 hours, only strong reactions resulted in increased TEWL. These data were confirmed by us, evaluating skin responses (+ or ++) to nickel sulfate at different concentrations in 12 nickel-sensitive women by means of evaporimetry and other instrumental methods: no significant concentration- or time-dependent modifications of TEWL were detectable (Table 9.1).[18] On the contrary, both the colorimetric and the echographic procedure provide numerical data that increase proportionally to the intensity of the patch test response.[19–21,24]

In a study performed in 18 patients with alopecia areata, aiming at evaluating their sensitization status before and 4 months after diphenylcyclopropenone contact immunotherapy, TEWL values on diphenylcyclopropenone patch test areas proved significantly lower after 4 months with respect to the beginning of immunotherapy, but they correlated to visual scores worse than laser Doppler flowmetry ones.[25]

Recently, iontophoresis has been suggested as a more comfortable alternative to patch testing: Delayed cutaneous responses induced via iontophoresis were demonstrated to be specific and to correlate well with positive patch test reactions.[26] A significant increase in TEWL was not observed in the responses induced by nickel iontophoresis in 16 nickel-allergic volunteers. In most cases, the positive reactions were weak, according to International Contact Dermatitis Research Group criteria, explaining the unchanged TEWL.

In conclusion, evaporimetry does not represent a suitable method for measuring the intensity of the skin response in an experimental setting simulating an acute

TABLE 9.1
Transepidermal Water Loss Values (g/m² per hour) at Patch Test Sites with Nickel Sulfate at Different Concentrations

Concentration (%)	Baseline	24 Hours	48 Hours	72 Hours
0.5	7.17	6.92	5.92	6.83
1	7.08	6.58	5.25	6.67
2	7.08	7.42	5.58	6.25
3	7.42	8.08	6.00	5.50
5	7.67	7.50	5.42	6.75

contact dermatitis by means of patch tests. The impairment of the barrier in acute contact eczema is probably secondary to the inflammatory process and is evident only for strong reactions. In the studies reported above, measurements were performed up to a maximum of 72 hours after the application of the contact sensitizers, and no mention was made of barrier status during recovery.

IV. TEWL AND TOPICAL AGENTS FOR ACD

To evaluate the efficacy of topical antiinflammatory agents, several models have been developed.[27,28] Inhibition of the elicitation of contact dermatitis was performed employing pre- or posttreatment of patch test areas in contact-allergic individuals by means of an antiinflammatory drug.[29–34] To assess the efficacy of these drugs, quantification of the inflammatory component of the allergic reaction is by far the most suitable method. This was performed employing colorimetry, laser Doppler flowmetry, and ultrasound. In spite of the fact that barrier damage in ACD does not appear to reflect the intensity of the inflammatory response, enhancement of barrier recovery or prevention of barrier damage may represent a useful tool for studying the influence of topical agents on ACD. Queille-Roussel et al. investigated the effects of four topical antiinflammatory drugs on experimentally induced nickel contact dermatitis. Lesions were treated twice daily with two steroids, two nonsteroidal antiinflammatory drugs, and a blank base for 4–5 days without occlusion. To quantify drug activity objectively, skin color, skin blood flow, and TEWL were measured both before the first application and again 6 hours after the last application of the drugs. Significant improvement of the spontaneous clinical evolution was only achieved by the most potent corticosteroid cream in comparison with the other creams and the untreated site. Colorimetric parameters and skin blood flow were highly correlated to visual grading, whereas TEWL appeared to be less related to clinical improvement.[31]

After performing nickel patch testing in patients with previously proven allergy to nickel, to create a standardized ACD, Zhai et al.[35] treated the patch test areas with a topical agent or with the placebo. Evaluation of the cutaneous reactions was based on visual scoring and TEWL, skin blood flow, and skin color. These authors found that TEWL monitoring helps evidence an improvement in barrier damage at treated sites compared with untreated ones, when evaluations are continued for 8–10 days.

Employing a similar model, Hachem et al.[36] evaluated the effects of topical corticosteroids on barrier function in ACD caused by nickel in 20 female volunteers. Corticosteroids improved TEWL values in the early inflammatory phase, whereas no significant influence on TEWL was observed later, probably because of the secondary effects of corticosteroids on the proliferation and differentiation of keratinocytes.[36] However, combining a topical corticosteroid with a moisturizing cream prevented delay in the healing of skin barrier disruption in ACD-induced inflammation, as evaluated by a significant decrease in TEWL values and an increase in capacitance ones.[37]

REFERENCES

1. Seidenari, S., Evaluation of barrier function and skin reactivity in occupational dermatoses, in *Handbook of Occupational Dermatology,* Kanerva, L. et al., Eds., Springer, Berlin, 2000, chap. 7.
2. De Fine Olivarius, F., Agner, T., and Menné, T., Skin barrier function and dermal inflammation. An experimental study of transepidermal water loss after dermal tuberculin injection compared with SLS patch testing, *Br. J. Dermatol.,* 129, 554, 1993.
3. Seidenari, S., Skin sensitivity, interindividual factors: atopy, in *The Irritant Contact Dermatitis Syndrome*, Van der Valk, P.G. and Maibach, H.I., Eds., CRC Press, Boca Raton, Florida, 1996, p. 266.
4. Hamami, I. and Marks, R., Abnormalities in clinically normal skin — a possible explanation of the "angry back syndrome," *Clin. Exp. Dermatol.*, 13, 328, 1998.
5. Pasche-Koo, F. and Hauser, C., How to understand the angry back syndrome, *Dermatology,* 184, 237, 1992.
6. Seidenari, S., Reactivity to nickel sulfate at sodium lauryl sulfate pretreated skin sites is higher in atopics: an echographic evaluation by means of image analysis performed on 20 MHz B-scan recordings, *Acta Derm. Venereol. (Stockh.),* 74, 245, 1994.
7. van der Valk, P.G.M. et al., Eczematous (irritant and allergic) reactions of the skin and barrier functions as determined by water vapour loss, *Clin. Exp. Dermatol.*, 10, 185, 1985.
8. Seidenari, S., unpublished data, 2002.
9. Tupker, R.A. et al., Susceptibility to irritants: role of barrier function, skin dryness and history of atopic dermatitis, *Br. J. Dermatol.,* 123, 199, 1990.
10. Agner, T., Skin susceptibility in uninvolved skin of hand eczema patients and healthy controls, *Br. J. Dermatol.,* 125, 140, 1991.
11. Seidenari, S. et al., Comparison of 2 different methods for enhancing the reaction to nickel sulfate patch tests in negative reactors, *Contact Dermatitis*, 35, 308, 1966.
12. Proksch, E. and Brasch, J., Influence of epidermal permeability barrier disruption and Langerhans' cell density on allergic contact dermatitis, *Acta Derm. Venereol. (Stockh.),* 77, 102, 1997.
13. Agner, T. et al., Combined effect of irritants and allergens, *Contact Dermatitis,* 47, 21, 2001.

14. McFadden, J.P. and Basketter, D.A., Contact allergy, irritancy and "danger," *Contact Dermatitis,* 42, 123, 2000.
15. Serup, J. and Staberg, B., Differentiation of allergic and irritant reactions by transepidermal water loss, *Contact Dermatitis,* 16, 129, 1987.
16. Serup, J. and Staberg, B., Ultrasound for assessment of allergic and irritant patch test reactions, *Contact Dermatitis,* 17, 80, 1987.
17. Gawkrodger, D.J., McDonagh, A.J.G., and Wright, A.L., Quantification of allergic and irritant patch test reactions using laser-Doppler flowmetry and erythema index, *Contact Dermatitis,* 24, 172, 1991.
18. Seidenari, S., Di Nardo, A., and Giannetti, A., Valutazione di reazioni a patch test eseguiti con solfato di nichel e laurilsolfato di Na (SLS) mediante sonografia in B scan, colorimetria, evaporimetria e laser Doppler flussimetria, *G. Ital. Dermatol. Venereol.,* 127, 15, 1997.
19. Seidenari, S. and Di Nardo, A., B-scanning evaluation of allergic reactions with binary transformation and image analysis, *Acta Derm. Venereol. (Stockh.),* 175, 3, 1992.
20. Seidenari, S. and Di Nardo, A., Cutaneous reactivity to allergens at 24-h increases from the antecubital fossa to the wrist: an echographic evaluation by means of a new image analysis system, *Contact Dermatitis,* 26, 171, 1992.
21. Seidenari, S., Echographic evaluation of subclinical allergic patch test reactions, *Contact Dermatitis,* 29, 156, 1993.
22. Quinn, A.G. et al., Quantification of contact allergic inflammation: a comparison of existing methods with a scanning laser Doppler velocimeter, *Acta Derm. Venereol. (Stockh.),* 73, 21, 1993.
23. Jemec, G.B.E. and Johansen, J.D., Erythema index of clinical patch test reactions, *Skin Res. Technol.,* 1, 26, 1995.
24. Seidenari, S. and Belletti, B., The quantification of patch test responses: a comparison between echographic and colorimetric methods, *Acta. Derm. Venereol. (Stockh.),* 78, 364, 1998.
25. Seo, K.I. and Eun, H.C., Loss of contact sensitization evaluated by laser Doppler blood flowmetry and transepidermal water loss measurement, *Contact Dermatitis,* 34, 233, 1996.
26. Brasch, J., Huttemann, M., and Proksch, E., Iontophoresis of nickel elicits a delayed cutaneous response in sensitized individuals that is similar to an allergic patch test reaction, *Contact Dermatitis,* 42, 36, 2000.
27. Wendt, H. and Frosch, P., *Clinico-Pharmacological Models for the Assay of Topical Corticoids,* Karger, Basel, 1982, 72.
28. Queille-Roussel, C., Le test de vasoconstriction en peau saine: techniques et applications, *Ann. Dermatol. Venereol.,* 115, 491, 1988.
29. Kaidbey, K.H. and Kligman, A.M., Assay of topical corticosteroids. Efficacy of suppression of experimental Rhus dermatitis in humans, *Arch. Dermatol.,* 112, 808, 1976.
30. Burrows, W.M. and Stoughton, R.B., Inhibition of induction of human contact sensitization by topical glucocorticosteroids, *Arch. Dermatol.,* 112, 175, 1976.
31. Queille-Roussel, C. et al., Objective assessment of topical anti-inflammatory drug activity on experimentally induced nickel contact dermatitis: comparison between visual scoring, colorimetry, laser Doppler velocimetry and transepidermal water loss, *Skin Pharmacol.,* 3, 248, 1990.

32. Seidenari, S., Di Nardo, A., and Giannetti, A., Assessment of topical corticosteroid activity on experimentally induced contact dermatitis: echographic evaluation with binary transformation and image analysis, *Skin Pharmacol.*, 6, 85, 1993.
33. Di Nardo, A. et al., Inhibition of elicitation of contact dermatitis in humans by mometasone furoate: evaluation by means of 20-MHz B scanning associated with image analysis, *Dermatology*, 195, 137, 1997.
34. Seidenari, S. et al., Parallel intraindividual evaluation of the vasoconstrictory action and the anti-allergic activity of topical corticosteroids, *Exp. Dermatol.*, 6, 75, 1997.
35. Zhai, H. et al., *In vivo* nickel allergic contact dermatitis: human model for topical therapeutics, *Contact Dermatitis,* 40, 205, 1999.
36. Hachem, J.P. et al., Efficacy of topical corticosteroids in nickel-induced contact allergy, *Clin. Exp. Dermatol.,* 27, 47, 2002.
37. Hachem, J.P. et al., Combination therapy improves the recovery of the skin barrier function: an experimental model using a contact allergy patch test combined with TEWL measurements, *Dermatology*, 202, 314, 2001.

10 Prediction of Irritancy

R.A. Tupker

CONTENTS

I. Introduction 113
II. Extrinsic Factors 114
A. Type of Compound 114
B. Concentration of the Compound 115
C. Duration of Exposure in the One-Time Occlusive Patch Test 115
D. Other Exposure Methods 116
III. Intrinsic Factors 117
A. Preexposure Barrier Function 117
B. Concomitant (Atopic) Dermatitis on Another Body Region 118
C. The Influence of Mucosal Atopy on Skin Susceptibility 119
D. Genetic Factors 119
E. Prior Exposure to Irritants and Other Personal Habits 120
IV. Predictive Testing in the Occupational Setting 120
V. Comments 121
References 125

I. INTRODUCTION

Chronic irritant contact dermatitis (ICD) is the most frequently encountered skin disease in many occupations in which workers are to some extent exposed to irritants. This disease represents a burden of suffering from irritation or itch and may lead to absence from work. It is, therefore, of great importance to find ways to lower the incidence of chronic ICD. In this process it is crucial to have insight into the factors that can predict irritation. Whether or not chronic ICD will develop depends on the balance between the sum of all extrinsic, harmful influences (irritants in the work and home environment, low humidity, high temperature, etc.) imposed on the skin on the one hand, and intrinsic, constitutional factors regulating the reparative capacity of the skin on the other hand.[1] Predictive irritancy testing as a method to understand and predict chronic ICD should be based on the etiological concepts of chronic ICD. Therefore, when considering predictive irritancy testing, it is helpful to make the same division into extrinsic and intrinsic factors, yielding two main work categories: predictive irritancy testing of various substances aimed to select the least irritant

0-8493-1443-7/05/$0.00+$1.50

substance, and predictive irritancy testing with one or more standard irritants aimed at selecting a population that is at risk of developing chronic ICD.

In finding new chemicals and finished products with low irritant potency, the industry making these products has to rely on human skin models, because the animal models have been abandoned on ethical grounds and because these methods had been proved to be poorly predictive in man.[2] To save testing on humans as much as possible, *in vitro* tests have been developed.[3–5] Exposure of human skin recombinants to sodium lauryl sulphate (SLS) resulted in a dose-dependent release of lactate dehydrogenase and inflammatory mediators.[3] Cytotoxicity in the *in vitro* neutral red uptake assay using human keratinocyte cell lines showed a good overall correlation with *in vivo* irritation potential in humans, with the exception of compounds with a high molecular weight.[4,5] It can be expected that *in vitro* testing will be used more extensively in future for screening chemicals with low irritancy. However, this tendency toward *in vitro* testing will not abrogate the need for *in vivo* tests on humans, which will remain the golden standard of predictive irritancy testing, at least for the near future. In this chapter, several factors are described that can influence the outcome of *in vivo* irritancy testing in humans, classified as extrinsic and intrinsic factors. The subject is not treated exhaustively; for a more complete survey on factors influencing irritant reactions the reader is referred to the guidelines on SLS exposure tests.[6] Predictive testing in the occupational setting is dealt with separately.

II. EXTRINSIC FACTORS

A. Type of Compound

Various irritants have different modes of action, and one irritant may have more than one way in which it causes skin irritation. In most cases, however, the action mechanism is not completely understood. Detergents may cause the so-called "soap-effect" (shining, wrinkling impression), scaling (or "roughness"), erythema, and fissures.[7] The putative action spectrum of detergents is broad (reviewed in Tupker et al.[6]). The epidermis is now considered a highly active site of lipid synthesis that is under the direct control of alterations in barrier status.[8] Barrier injury, regardless of the type (detergents, solvents, tape stripping, etc.) evokes a recovery response that leads to normalization of barrier function within hours to days. Perturbation of the ionic gradient in the epidermis, various cytokines (such as interleukin-1 and tumor necrosis factor (TNF)-α), and growth factors are involved in this process.[8]

Detergents and solvents exert a damaging influence on the stratum corneum, and therefore transepidermal water loss (TEWL) evaluation — which measures barrier function (see Chapter 9) — is the most appropriate noninvasive method to evaluate their effects. Using TEWL measurement, our group has demonstrated differences between detergents with respect to their irritant potency.[9] However, it is conceivable that TEWL measurement cannot detect responses from skin reactions by those types of substances that do not have a damaging effect on the stratum corneum. A recent study has illustrated this point by showing that different noninvasive measuring techniques and visual scoring were needed to evaluate the irritation reaction by five different types of damaging influences (SLS, tape stripping, tretinoin, ultraviolet light, and dithranol).[10]

B. Concentration of the Compound

Elevating the concentration of a compound may elicit qualitatively different reactions; for example, SLS and croton oil in low concentrations cause erythema, whereas high concentrations induce bullae.[11] Quantitative differences have been demonstrated in the form of clear dose–response relationships for SLS by means of TEWL measurements.[12–14] The biological effect of a substance depends on the amount of molecules (molarity) rather than on the total mass of the molecules (concentration).[15] Therefore, when the purpose of predictive testing is comparison of irritant potency among various compounds, these compounds should be tested on the basis of equal molarity; otherwise, a false impression may be obtained regarding differences in irritancy. This theoretical consideration has practical limitations in case mixtures of more than one compound are to be tested (such as in commercially available products).

C. Duration of Exposure in the One-Time Occlusive Patch Test

For many years closed-patch testing has been the favored test method, both for practical and traditional reasons. The 24-hour patch test is used in most studies, and applied on the forearm, the patch may be considered the gold standard for one-time exposure tests.[6]

Recently, SLS has been applied for 48 hours on the back and forearm to combine irritancy testing with routine allergic patch testing.[14] Visual scores on the back and forearm showed a good correlation, and the scores were more pronounced on the forearm than on the back. Visual scores on the forearm correlated well with TEWL values on the forearm, whereas visual scores on the back and TEWL values on the forearm had a low correlation.[14] In another study by the same group, the interrelationship between the concentration of SLS and exposure time (3, 6, 12, 24, and 48 hours) was determined.[13] There was a clear distinction between the concentrations used only at 12-, 24-, and 48-hour exposures.[13] The factor concentration had a higher effect on the outcome of the irritant response than did the factor exposure duration. To achieve the same irritant response as found with a doubled SLS dose, the application time must be tripled.[13] Recently, a simple 4-hour patch test has been designed to examine crude differences in effects of SLS on the skin, evaluated by a simple visual scoring system.[16] In an attempt to use this method to detect differences in susceptibility between summer and winter season, no significant results were found, although there was a trend toward a higher susceptibility in winter.[16] These differences were noted only for extremely high SLS concentrations (20 and 100%).

D. Other Exposure Methods

Although the one-time 24-hour occlusive patch test is a favored test method because of its easiness and rapidity to perform, the disadvantage of this technique is the fact that it mimics only an acute irritant reaction. The more common situation in real life is the development of chronic ICD resulting from multiple repeated exposures. To address this issue, the following test protocols have been used: repeated occlusive tests, repeated open tests, and immersion tests.[6]

Kligman and Wooding[17] developed a test method in which substances are brought into contact with the skin under occlusion for 24 hours a day for 20 consecutive days to be able to test substances with a very low irritant potency. Another way to test compounds of expected low irritancy is the chamber scarification technique, in which the skin is scratched before closed-patch tests are performed repeatedly, daily, for 3 days.[18,19] Using this method, in which the stratum corneum was bypassed, interindividual variation has been found, demonstrating that a subject's susceptibility toward irritants is also dependent on targets underneath the stratum corneum. Frosch and Kligman[18] have used an exposure model in which a substance is applied repeatedly during 5 consecutive days; the first day for 24 hours and the following days for 6 hours a day. To reflect better the conditions in daily practice, models have been developed in which detergents were applied in a multiple, repeated short-time way; namely, twice daily for 45 minutes under occlusion for 1 week[20,21] or for 3 weeks.[9,22,23] Using this test method, it was possible to rank detergents according to their irritant potency, with SLS being by far the most irritant detergent.[9,23]

A further step toward a better reflection of daily life conditions may be the repeated open-exposure technique. In this method, solutions are applied once daily on the forearm skin, using a glass or plastic ring, during 20 or 45 minutes for 1 week[24,25] or for more weeks.[23,26,27] Clear differences in irritancy resulting from the use of various detergents have been demonstrated using this method.[23,25]

By means of a repetitive open-irritation test, it was possible to test the efficacy of different barrier creams.[27] Using a repeated open-application test of a thin film over a surface of 3 cm^2 on the forearms, it has been proven that a children's hand cleanser was more irritating than industrial hand cleansers.[28]

Twice-daily, 30-minute immersions of various detergents induced different degrees of erythema and scaling, and SLS was shown to be the most irritating.[29] In a "use test," one hand was exposed to a 0.05% solution of a commercial dishwashing liquid, and the other hand was protected by a glove.[30] In 13 out of the 18 volunteers tested, clinical signs emerged during the first 2 weeks, accompanied by increased TEWL values.[30] The wash test represents another open test model, in which the act of washing is mimicked in a repetitive way.[31–33] Repetitive washing with grit-containing cleansers four times daily for 1 week lead to differences in TEWL values, skin redness, and hydration.[33]

A few studies have compared the patch test with the wash test with respect to the irritancy potential ranking order of a series of detergents.[29,31,32] In one study,[31] discrepancies were found in the ranking order obtained by the two techniques, whereas in the other study[32] similar rankings were found. The wash test, however, appeared to have a greater power of discrimination.[32] Comparing the one-time patch test with the immersion test, there was a relatively good correlation between these models, with the exception of soap, which was irritating in the patch test but not in the immersion test.[29] More recently, the same finding has been observed in our study investigating four different detergents, using one-time occlusive, repeated-occlusive, and repeated open tests.[23] Only in the one-time occlusive test did soap induce more erythema than the other detergents, whereas in the other models, SLS and sodium cocoylisethionate had higher visual scores than soap. By means of TEWL measurement, however, the mean ranking order of the detergents was similar in all models.[23] The fact that soap clinically was the most irritating substance in the one-time patch test could be explained by the earlier observation that the pH of a soap solution decreased after prolonged contact with the skin, as is the case in a 24-hour occlusive condition.[34] Soap with a low pH (7.5) induced erythema and pruritus, which did not occur at higher pH (9.5).[34] In another study on the effects of occlusion, it was found that postexposure occlusion by a plastic wrap caused more severe irritation compared with unoccluded exposures to SLS.[24] The degree of augmentation of the irritant response by occlusion differs between various types of substances. The irritating effect of polar compounds is less influenced by occlusion than is that of nonpolar compounds.[35]

Hence, it appears that the type of exposure may influence the outcome of the ranking order in irritancy testing. The central question is the following: Which exposure method offers the best prediction of real-life exposure? That supposes another question: What are the real-life conditions encountered? In most in-use situations the uncovered skin is exposed to irritants several times daily. In other situations, already-exposed skin is covered by "protective" gloves or other impermeable materials. In the first situation mentioned, the repeated open test seems the best way to simulate this scenario. In the second situation, however, the one-time or repeated occlusive test may be chosen.

III. INTRINSIC FACTORS

A. Preexposure Barrier Function

Using TEWL after 4 days of repeated exposure to SLS as a measure of susceptibility, a strong correlation with preexposure TEWL has been found ($R = 0.71$).[20] A similar correlation coefficient has been noted between TEWL values before and after a single 24-hour SLS exposure.[36] Subsequent studies have confirmed the strong correlation between pre- and postexposure TEWL in the one-time SLS exposure model.[37,38] In persons with a high preexposure TEWL, the barrier function is apparently impaired from the start. This allows the irritant to penetrate and damage the skin more easily, which renders the skin more susceptible — particularly in the initial phase of skin irritation. In other studies using one-time occlusive tests, these results could not be confirmed; no correlation[39] or low correlation[40] between pre- and postexposure TEWL values have been found. In long-term repeated-exposure models, lower correlations between pre- and postexposure TEWL also have been noted ($R = 0.31$).[22,26] After multiple repeated exposures, the skin will adapt to a certain degree.[8] Therefore, skin reactivity after multiple exposures is probably particularly determined by the ability to develop adaptation, and the influence of preexposure barrier function diminishes.[22] It must be stressed that the reported high values have been found for SLS only. Extrapolation to other types of irritants may yield only poor correlation values because those irritants might penetrate and attack the stratum corneum barrier in a different way.[10]

B. Concomitant (Atopic) Dermatitis on Another Body Region

It has been proven that existing dermatitis, irrespective of the type of dermatitis, on another location of the body may enhance reactivity to various irritants.[11] We have confirmed the validity of this phenomenon for atopic dermatitis (AD).[21] AD patients with more severe dermatitis exhibited higher levels of reactivity.[21] In a study by Agner,[37] only those patients with active hand eczema had an increased susceptibility to SLS on the upper arm, as opposed to chronic and healed eczema patients and normal controls. The cause of this propensity to hyperreact to chemicals in the presence of dermatitis elsewhere is yet unknown. Shahidullah[41] has observed barrier function impairment on the uninvolved skin sites related to the severity of dermatitis. Furthermore, TEWL of uninvolved sites of the forearm in patients with acute ICD of the hands was higher than that in patients with healed ICD of the hands, who in turn had higher TEWL values than did controls.[42] TEWL values on a distant site like the thigh were similar among persons with acute and healed ICD on the hands, but were higher compared with controls.[42] A correlation has been found between TEWL values measured on the forearm and those on the thigh, which was speculated to be a generalized barrier disturbance in ICD, although the presence of an atopic predisposition could not be ruled out.[42] Because the type of dermatitis present did not appear to influence the reactivity, the enhanced susceptibility noted was regarded as secondary to dermatitis activity per se.[43] However, van der Valk et al.[44] have found a significantly elevated TEWL before and after SLS exposure only in the subgroup of patients with a manifest AD and not in those with manifest ICD.

In a study on patients with a history of AD and patients with a history of contact dermatitis, higher preexposure and postexposure TEWL values were noted only in patients with a history of AD.[22] It was concluded that the lower preexposure barrier function in AD was at least in part responsible for the enhanced susceptibility in AD, in addition to other constitutional factors, such as elevated reactivity of cellular targets underneath the stratum corneum, and adaptation capacity.[22] In other studies, patients with active and those with inactive AD reacted more strongly to SLS as compared with nonatopic controls.[21,45] Also, in the open test model, subjects with a history of AD were more susceptible.[46]

The old statement about the pathogenesis of chronic ICD, namely, that its development depends on the balance between the sum of all extrinsic influences on the one hand and on intrinsic, constitutional factors regulating the reparative capacity of the skin on the other, is elaborated to a broader, hypothetical view.[8] In this view, the impaired barrier function in AD skin is considered to be of essential importance in its pathophysiology, in that it sustains the disease process. The abnormal T-cell is recruited within the skin after prior release of signaling mediators from the epidermis as a consequence of a barrier-initiated cascade of cytokines and growth factors.[8] This barrier-initiated cascade of cytokines is seen as a normal response, always accompanying the repair process after barrier perturbation. In normal skin, chronic exposure of the epidermis to damaging insults will result in a compensatory hyperplastic state of the epidermis, also called "hardening." However, in AD, skin exposure to these damaging insults induces recruitment and stimulation of inflammatory cell types in the epidermis, so that previously uninvolved skin sites will turn into involved sites. In healthy persons without a history of skin disease, it has been demonstrated that there were interindividual variations in TEWL time course after a one-time patch test, which were inversely correlated with the responses after repeated occlusive exposures ($R = 0.61$).[47] TEWL increases observed in the days after patch removal was associated with a higher TEWL value in the repeated test model, indicating the importance of adaptation.[47]

C. The Influence of Mucosal Atopy on Skin Susceptibility

The effect of mucosal atopy investigated in the experimental skin irritancy model is not as clearcut as the effect of its dermal counterpart, AD. Larger responses on graded SLS solutions were observed in AD patients and in patients with mucosal atopy without a history of AD compared with nonatopic controls.[45] In each atopic group, the effective SLS concentration causing irritation in 50% or more of the subjects (ED_{50}) was significantly lower compared with the controls. The ED_{50} values of active AD, inactive AD, mucosal atopy, and controls were 0.0625%, 0.31%, 0.19%, and 0.6%, respectively.[45] This was explained by the hypothesis that hyperreactivity of inflammatory cells, whether in the skin or the respiratory mucosa, is an intrinsic feature of the atopic diathesis. In a small study on a use test with a dishwashing liquid, persons with a history of mucosal atopy were not among the hyperreactors.[30] In a more recent study, AD patients had an impaired barrier function compared with mucosal atopy and control subjects, whereas the latter groups had a similar barrier function.[48] However, as has been stated before in this chapter, an individual's susceptibility is dependent not only on his or her barrier function but also on other factors, such as reactivity of cellular targets underneath the stratum corneum, and reparative capacity.

D. Genetic Factors

Recently, the influence of a genetic marker on irritant susceptibility has been investigated.[49] Visual irritant thresholds were determined using graded concentrations of SLS and benzalkonium chloride in a large group of nonatopics. Transition polymorphism has been identified in the TNFA gene. Individuals carrying a haplotype that includes the A allele are high secretors of TNF-α, which is a key mediator in the pathogenesis of ICD.[8] In the low-irritant-threshold groups of both SLS and benzalkonium chloride, a significantly increased number of persons having the A allele was found.[49] This study offers the first description of a non-barrier-related marker of susceptibility in nonatopics.

E. Prior Exposure to Irritants and Other Personal Habits

In guinea pig skin, eight applications of detergents induced an increasing irritation.[50] After 30 applications, however, there was a complete accomodation.[50] The same phenomenon has been noted by our group in multiple repeated exposures on humans.[22] Clinical changes were accompanied or preceded by a downgrade curve in the TEWL time course after an initial TEWL increase. A similar TEWL time course was observed on skin sites showing no or only mild clinical signs.[22] TEWL decrease after initial increase was also noted in a 21-day repeated-use test.[30] In another study, it has been shown that there was hyporeactivity toward a challenge with a one-time SLS patch (at sites already repeatedly exposed to SLS) after 6 and 9 weeks, but not after 3 weeks.[51] The hyporeactivity found in these studies signifies a functional adaptation, or accommodation ("hardening"), of the skin. Hardening of the skin may be an important factor in predictive testing on subjects with a different professional background. For this reason, the skin of a cleaner may be less susceptible than the skin of an office worker. The specificity of hardening is not exactly known until now. If the specificity of hardening is high, one must be cautious of this phenomenon in predictive irritancy tests comparing different chemicals.

Protective effects of moisturizers have been shown both during and after the irritation phase of repeated open- and occlusive-exposure methods. Therefore, one should be aware of the use of these products in experimental irritancy testing and in field studies. For a detailed review on the topic of moisturizers, the reader is referred to Chapter 32.

IV. PREDICTIVE TESTING IN THE OCCUPATIONAL SETTING

In the last decade, several field studies have been conducted in which the incidence or prevalence of hand dermatitis was investigated in a cohort of trainees in high-risk occupations. In these prospective studies, the aim was to examine the effect of one or more potential risk factors on the incidence of hand dermatitis. Table 10.1 shows the key features of these investigations. Important risk factors were history of AD,[57,61,63] history of hand dermatitis,[57] history of metal dermatitis,[57] the degree of chemical and mechanical exposure,[59,61] and insufficient recovery time during work.[61] In some studies, the factor history of AD has not been identified as a risk factor, probably because of selection bias.[56,58,59,65] Apparently, only those persons having no history of dermatitis had chosen these high-risk occupations. Dry skin and mucosal atopy were risk factors in a study on hairdressers and nurses.[59] It is possible that these high-risk persons belong to the group that may be regarded as having an "atopic skin diathesis" — a term coined by Lammintausta et al.[52] Atopic skin diathesis was denoted as the presence of dry skin and a low itch threshold for two of the three irritants (sweat, dust, and rough materials) in combination with white dermographism and facial pallor/infraorbital darkening. In their cross-sectional study on "wet" hospital work, Lammintausta et al.[52] found a strong influence of a history of AD, but a very weak influence of mucosal atopy. However, past or present mucosal atopy in combination with atopic skin diathesis was proven to increase the risk of hand dermatitis.[52] Diepgen has constructed an "atopy score"[53] based on the criteria for AD by Hanifin and Rayka[54] to better assess the likelihood of developing dermatitis in high-risk environments. In some field studies, however, the value of this atopy score was less contributive than was history of AD.[62,63] In a large-scale investigation on car industry workers, the following independent factors could predict the occurrence of hand dermatitis to a large extent: history of AD, history of hand dermatitis, wet work more than 3 hours per day, and dyshidrosis.[55] For hairdressers, predicting factors were atopic skin diathesis, wet work more than 4 hours per day, and exposure to permanent wave fluids.[55]

In contrast to the findings in experimental skin irritancy studies, prospective field studies could not demonstrate the importance of preexposure barrier function as a risk factor for hand dermatitis,[58,59,60,63] with the exception of the study by Coenraads and Pinnagoda.[56] In this relatively small study, the incidence of hand dermatitis was much lower than in the field studies conducted afterward. A possible reason for the elevated preexposure TEWL in the persons who turned out to be cases later on is the fact that these persons might have been already exposed during earlier occupations. In other studies, increased TEWL values were observed preceding the occurrence of hand dermatitis.[58,63]

The use of provocation tests as a tool for the prediction of risk for hand dermatitis has been investigated recently, as preexposure barrier function has been proven to be of no value for this purpose.[60,65] Tests with SLS or dimethyl sulfoxide were not able to predict hand dermatitis, in contrast to test with NaOH.[60] Parameters were calculated as delta TEWL values, defined as TEWL after provocation divided by TEWL before provocation.[60] The combination of the provocation values of dimethyl sulfoxide and NaOH with skin moisture had a high sensitivity (94%) but a low specificity (24%).[60] In the small study by Smith et al.,[65] nine of 24 apprentice hairdressers who were followed up developed hand dermatitis. In these cases, the preexposure irritation threshold tended to be lower than in noncases. Unfortunately, a 4-hour one-time SLS test was performed (and assessed visually) in this investigation, in which 16 out of the initial group of 42 persons were lost to follow-up for "administrative reasons."[65]

V. COMMENTS

What did we learn from the literature on prediction of irritancy? In regard to the use of predictive irritancy testing of various substances to select the least irritant compound, substantial progress has been made in the last decades. *In vitro* testing has replaced animal tests as a screening method that enables us to make a rough division between safe and not safe. The most reliable test methods are still the human models in which compounds are brought into contact with the skin in an occlusive (either one-time 24 hours or repeated) or open (repeated) way. The model of choice (occlusive vs. open) depends on the presumed circumstances under which these compounds are used in daily practice. Recently, a modification of the one-time 24-hour patch test was introduced by Basketter and colleagues[16] to examine crude responses. Indeed, this modification may be used as an intermediate method between the prescreening *in vitro* models on the one hand and the more delicate human test

TABLE 10.1
Prospective Cohort Studies on Factors Influencing the Development of Occupational Hand Dermatitis

Authors and References	No. of Persons	Type of Occupation	Observation Duration (months)	Incidence/ Prevalence of HD	Impact of Risk Factors					Other Remarks
					AD	Dry	Mucosal Atopy	Prework TEWL	Provocation Tests	
Coenraads and Pinnagoda[56]	54	Metal workers	3	7.4%				Higher in cases		
Nilsson and Bäck[57]	1857	Hospital	20	41%	+					Past HD and past metal dermatitis were more important
Goh and Gan[58]	54	Metal workers	6	20% (56% at week 12)						Cases had higher TEWL increases than did noncases; measured at 25° to 28°C
Smit et al.[59]	74	Hairdressers	24	32.8/100 person-years		+	+	–		
	111	Nurses	24	14.5/100 person-years		+	+	–		
Berndt et al.[60]	205	Metal workers	30	23%				–	TEWL after SLS, ; TEWL after NaOH, +; TEWL after DMSO	Combination of TEWL change after NaOH and DMSO and skin moisture had sensitivity of 94% and specificity of 24%

Berndt et al.[61,62]	201	Metal workers	30	23%	+				Most important risk factors: past AD, chemical and mechanical exposure, insufficient recovery time
John et al.[63]	92 (of which 66 eligible)	Hairdressers	36	77%	+		–		
Funke et al.[64]	2042	Car industry	36	Metal workers 15.3%; white collars: 6.9%					Different predictive values in various occupations in car industry; in first months, particularly high rate of first occurrence of HD
Smith et al.[65]	42 (follow-up in only 24)	Hairdressers	6	37.5%				Cases tend to have lower SLS threshold	16 of the initial group were lost for follow-up for administrative reasons, which lowers the power of the study

Note: HD, hand dermatitis; AD, history of atopic dermatitis; dry, dry skin; TEWL, transepidermal water loss, SLS, sodium lauryl sulfate; DMSO, dimethyl sulfoxide.

models on the other. These latter models should be considered to be more reliable until properly conducted comparative testing with the 4-hour modified test has proven equal predictivity.

A way to refine the human test methods is to assess the test responses by means of TEWL measurements. In that case, the guidelines on this topic should be adhered to in order to achieve trustful results.[6] It was stated in the literature in several instances that these guidelines were followed, but in fact authors made deviations on important aspects, such as performing measurements at 30 minutes after patch removal. These divergences will result in false outcomes, as the measured value is "contaminated" by postocclusive hydration. Therefore, when authors decide not to follow a particular aspect of the guidelines, I would suggest that they clarify this by stating that they do not adhere to the guidelines on this aspect and also that they give the reason for this deviation.

TEWL measurement should never be a goal per se, and it should always be accompanied by a good visual scoring system.[6] However, visual scoring has its limitations, such as with dark-skinned persons, in whom detection of erythema is more difficult.

Field studies on workers in high-risk occupations have shown that prework barrier function is not a valid predictor of the risk of hand dermatitis, in contrast to preexposure barrier function in experimental irritancy models. In these laboratory models, irritants are exposed for a relatively short time period on the skin.[20,36] In daily practice, however, the result of repeated exposures to damaging influences of various kinds is a complex interrelationship of different (partly unknown) factors, such as barrier function, individual reactivity of targets underneath the stratum corneum, history of AD, recovery time between exposures, severity of chemical and mechanical insults, ability to develop adaptation and seasonal influences, and so forth. Determining who is at risk probably demands a combination of anamnestic data and well-performed irritancy tests. Predictive irritancy testing by means of one or more standard irritants may be useful in the occupational setting. One probably needs for several irritants to be tested, as in reality there is exposure to different kinds of irritating factors.[60] It is a challenge to determine which testing method is most appropriate for a particular occupation. Once it has been decided which method is most appropriate, officially agreed guidelines for that method should be developed.

In view of the complex interrelationship of the above-mentioned factors in daily practice, multiple repeated-exposure models should be used as tools for predictive irritancy testing in field studies because only in this way can essential factors, such as the ability to develop adaptation, be tested. The disadvantage of such a repeated model, however, is its lower practicability in the field. A solution for this problem may be to perform TEWL measurements before and after some weeks of relevant exposure to the types of irritants involved in the occupation concerned, as field studies have demonstrated TEWL increases in workers who later turned out to have hand dermatitis.[58,63]

The near future will be an exciting period, with much progress to be expected. When risk groups are identified, special programs offering medical advice and reduction of exposure will be more effective in preventing the development of chronic ICD and, subsequently, the change or loss of jobs.[64]

REFERENCES

1. Hagerman, G., Ueber das "traumiterative" (toxische) Ekzem, *Dermatologica*, 115, 525, 1957.
2. Phillips, L. et al., A comparison of rabbit and human skin responses to certain irritants, *Toxicol. Appl. Pharmacol.*, 21, 369, 1972.
3. Ponec, M. and Kempenaar, J., Use of human skin recombinants as an *in vitro* model for testing the irritation potential of cutaneous irritants, *Skin Pharmacol.*, 8, 49,1995.
4. Lee, J.K., Kim, J.I., and Kim, P.Y., *In vitro* cytotoxicity tests on cultured human skin fibroblasts to predict skin irritation potential of surfactants, 14, 345, 2000.
5. Wilhelm, K.-P., Böttjer, B., and Sigers, C.-P., Quantitative assessment of primary skin irritants in vitro in a cytotoxicity model: comparison with *in vivo* human irritation tests, *Br. J. Dermatol.*, 145, 709, 2001.
6. Tupker, R.A. et al., Guidelines on sodium lauryl sulphate (SLS) exposure tests. A report from the standardization group of the European Society of Contact Dermatitis, *Contact Dermatitis*, 37, 53, 1997.
7. Frosch, P.J. and Kligman, A.M., The soap chamber test, *J. Am. Acad. Dermatol.*, 1, 35, 1979.
8. Elias, P.M., Wood, L.C., and Feingold, K.R., Epidermal pathogensis of inflammatory dermatoses, *Am. J. Contact Dermatitis*, 10, 119, 1999.
9. Tupker, R.A. et al., The influence of repeated exposure to surfactants on the human skin as determined by transepidermal water loss and visual scoring, *Contact Dermatitis*, 20, 108, 1989.
10. Fluhr, J.W. et al., Testing for irritation with a multifactorial approach: comparison of eight non-invasive measuring techniques on five different irritation types, *Br. J. Dermatol.*, 145, 696, 2001.
11. Björnberg, A., Skin Reactions to Primary Irritants in Patients with Hand Eczema. An Investigation with Matched Controls. Ph.D. Thesis. University of Gotenburg, 1968.
12. Agner, T. and Serup, J., Sodium lauryl sulphate for irritant patch testing — a dose-response study using bioengineering methods for determination of skin irritation, *J. Invest. Dermatol.*, 95, 543, 1990.
13. Aramaki, J. et al., Irritant patch testing with sodium lauryl sulphate: interrelation between concentration and exposure time, *Br. J. Dermatol.*, 145, 704, 2001.
14. Loffler, H. et al., Evaluation of skin susceptibility to irritancy by routine patch testing with sodium lauryl sulfate, *Eur. J. Dermatol.*, 11, 416, 2001.
15. Mathias, C.G.T., Clinical and experimental aspects of cutaneous irritation, in *Dermatotoxicology*, Marzulli, F.N., and Maibach, H.I., Hemisphere Publishing Corporation, Washington, DC, 1983, pp. 167–183.
16. Basketter, D.A. et al., Individual, ethnic and seasonal variability in irritant susceptibility of skin: the implications for a predictive human patch test, *Contact Dermatitis*, 35, 208, 1996.
17. Kligman, A.M. and Wooding, W.M., A method for the measurement and evaluation of irritants on human skin, *J. Invest. Dermatol.*, 49, 78, 1967.
18. Frosch, P.J. and Kligman, A.M., The chamber-scarification test for irritancy, *Contact Dermatitis*, 2, 314, 1976.
19. Andersen, K.E., Reproducibility of the chamber scarification test, *Contact Dermatitis*, 34, 181, 1996.
20. Tupker, R.A. et al., Baseline transepidermal water loss (TEWL) as a prediction of susceptibility to sodium lauryl sulphate, *Contact Dermatitis*, 20, 265, 1989.

21. Tupker, R.A. et al. Irritant susceptibility and wheal and flare reactions to bio-active agents in atopic dermatitis: I. Influence of disease severity. *Br. J. Dermatol.*, 133, 358, 1995.
22. Tupker, R. A. et al., Susceptibility to irritants: role of barrier function, skin dryness and history of atopic dermatitis, *Br. J. Dermatol.*, 123, 199, 1990.
23. Tupker, R.A. et al., Irritancy ranking of anionic detergents using one-time occlusive, repeated occlusive and repeated open tests. *Contact Dermatitis*, 40, 316, 1999.
24. van der Valk, P.G.M. and Maibach, H.I., Post-application occlusion substantially increases the irritant response to the skin to repeated short-term sodium lauryl sulfate (SLS) exposure. *Contact Dermatitis*, 21, 335, 1989.
25. Tupker, R.A. et al., Irritancy of sodium laurate and other anionic detergents using an open exposure model, *Skin Res. Technol.*, 3, 133, 1997.
26. Wilhelm, K.-P., Saunders, J.S., and Maibach, H.I., Increased stratum corneum turnover induced by subclinical irritant dermatitis, *Br. J. Dermatol.*, 122, 793, 1990.
27. Frosch, P.J., Kurte, A., and Pilz, B., Efficacy of skin barrier creams. III. The repetitive irritation test (RIT) in humans. *Contact Dermatitis*, 29, 113, 1993.
28. English, J.S., Ratcliffe, J., and Williams, H.C., Irritancy of industrial hand cleansers tested by repeated open application on human skin, *Contact Dermatitis*, 40, 84, 1999.
29. Smeenk, G., The influence of detergents on the skin (a clinical and biochemical study), *Arch. Klin. Exp. Dermatol.*, 235, 180, 1969.
30. Klein, G., Grubauer, G., and Fritsch, P., The influence of daily dish-washing with synthetic detergent on human skin, *Br. J. Dermatol.*, 127, 131, 1992.
31. Imokawa, G. and Mishima, Y., Cumulative effect of surfactants on cutaneous horny layers: adsorption onto human keratin layers *in vivo*, *Contact Dermatitis*, 5, 357, 1979.
32. Frosch, P.J., Irritancy of soaps and detergent bars, in *Principles of Cosmetics for the Dermatologist*, Frost, P. and Horwitz, S.N., Eds., CV Mosby, St. Louis, Missouri, 1982, pp. 5–12.
33. Wigger-Alberti, W. et al., Effects of various grit-containing cleansers on skin barrier function, *Contact Dermatitis*, 41, 136, 1999.
34. Blank, J.H. and Gould, J., Penetration of anionic surfactants into skin. III. Penetration from buffered sodium laurate solutions, *J. Invest. Dermatol.*, 37, 485, 1961.
35. Treffel, P. et al., Effect of occlusion on in vitro percutaneous absorption of two compounds with different physicochemical properties, *Skin Pharmacol.*, 5, 108, 1992.
36. Pinnagoda, J. et al., Prediction of susceptibility to an irritant response by transepidermal water loss, *Contact Dermatitis*, 20, 341, 1989.
37. Agner, T., Skin susceptibility in uninvolved skin of hand eczema patients and healthy controls, *Br. J. Dermatol.*, 125, 140, 1991.
38. Agner, T., Basal transepidermal water loss, skin thickness, skin blood flow and skin colour in relation to sodium lauryl sulphate-induced irritation in normal skin, *Contact Dermatitis*, 25, 108, 1991.
39. Freeman, S. and Maibach, H.I., Study of irritant contact dermatitis produced by repeat patch test with sodium lauryl sulphate and assessed by visual methods, transepidermal water loss, and laser Doppler velocimetry, *J. Am. Acad. Dermatol.*, 19, 496, 1988.
40. Berardesca, E. and Maibach, H.I., Racial differences in sodium lauryl sulphate induced skin irritation: black and white, *Contact Dermatitis*, 18, 65, 1988.
41. Shahidullah, M. et al., Transepidermal water loss in patients with dermatitis, *Br. J. Dermatol.*, 81, 722, 1969.

42. Effendy, I., Loeffler, H., and Maibach, H.I., Baseline transepidermal water loss in patients with acute and healed irritant contact dermatitis, *Contact Dermatitis*, 33, 371, 1995.
43. Gloor, M. et al., Irritationseffekt von Tensiden bei Patienten mit Gewerbeekzemen im Friseur-und Krankenpflegeberuf, *Dermatosen*, 33, 86, 1985.
44. van der Valk, P.G.M., Nater, J.P., and Bleumink, E., Vulnerability of the skin to surfactants in different groups of eczema patients and controls as measured by water vapour loss, *Clin. Exp. Dermatol.*, 10, 98, 1985.
45. Nassif, A. et al., Abnormal skin irritancy in atopic dermatitis and in atopy without dermatitis, *Arch. Dermatol.*, 130, 1402, 1994.
46. Lammintausta, K., Maibach, H.I., and Wilson, D., Susceptibility to cumulative and acute irritant dermatitis, *Contact Dermatitis*, 19, 84, 1988.
47. Tupker, R.A., Pinnagoda, J., and Nater, J.P., The transient and cumulative effect of sodium lauryl sulphate on the epidermal barrier assessed by transepidermal water loss: inter-individual variation, *Acta Derm. Venereol. (Stockh.)*, 70, 1, 1990.
48. Conti, A., Di Nardo, A., and Seidenari, S., No alteration of biophysical parameters in the skin of subjects with respiratory atopy, *Dermatology*, 192, 317, 1996.
49. Allen, M.H. et al., Association of TNFA gene polymorphism at position -308 with susceptibility to irritant contact dermatitis, *Immunogenetics*, 51, 201, 2000.
50. McOsker, D.E. and Beck, L.W., Characteristics of accommodated (hardened) skin, *J. Invest. Dermatol.*, 48, 372, 1967.
51. Widmer, J., Elsner, P., and Burg, G., Skin irritant reactivity following experimental cumulative irritant contact dermatitis, *Contact Dermatitis*, 30, 35, 1994.
52. Lammintausta, K. and Kalimo, K., Atopy and hand dermatitis in hospital wet work, *Contact Dermatitis*, 7, 301, 1981.
53. Diepgen, T.L., *Die atopische Hautdiathese. Epidemiologie, Klinik und berufsdermatologische Bedeutung*, Gentner, Stuttgart, 1991.
54. Hanifin, J.M. and Rayka, G., Diagnostic features of atopic dermatitis, *Acta Derm. Venereol. (Stockh.)*, 92, 44, 1980.
55. Diepgen, T.L., personal communication, 2002.
56. Coenraads, P.J. and Pinnagoda, J., Dermatitis and water vapour loss in metal workers. *Contact Dermatitis*, 13, 347, 1985.
57. Nilsson, E., and Bäck, O., The importance of anamnestic information of atopy, metal dermatitis and earlier hand eczema for the development of hand dermatitis in women in wet hospital work, *Acta Derm. Venereol. (Stockh.)*, 66, 45, 1986.
58. Goh, C.L. and Gan, S.L., Efficacies of barrier creams and afterwork emollient cream against cutting fluid dermatitis in metal workers: a prospective study. *Contact Dermatitis*, 31, 176, 1994.
59. Smit, H.A. et al., Individual susceptibility and the incidence of hand dermatitis in a cohort of hairdressers and nurses, *Scand. J. Work Environ. Health.*, 20, 113, 1994.
60. Berndt, U. et al., Is occupational irritant contact dermatitis predictable by cutaneous bioengineering methods? Results of the Swiss Metalworkers' Eczema Study (PROMETES), *Dermatology*, 198, 351, 1999.
61. Berndt, U. et al., Hand eczema in metalworker trainees — an analysis of risk factors, *Contact Dermatitis*, 43, 327, 2000.
62. Berndt, U. et al., Role of the atopy score and of single atopic features as risk factors for the development of hand eczema in trainee metal workers, *Br. J. Dermatol.*, 140, 922, 1999.

63. John, S.M., Uter, W., and Schwanitz, H.J., Relevance of multiparametric skin bioengineering in a prospectively-followed cohort of junior hairdressers, *Contact Dermatitis*, 43, 161, 2000.
64. Funke, U., Fartasch, M., and Diepgen, T.L., Incidence of work-related hand eczema during apprenticeship: first results of a prospective cohort study in the car industry, *Contact Dermatitis*, 44, 166, 2001.
65. Smith, H.R. et al., Skin irritation thresholds in hairdressers: implications for the development of hand dermatitis, *Br. J. Dermatol.*, 146, 849, 2002.

11 Transepidermal Water Loss and Racial Differences

Grazia Primavera and Enzo Berardesca

CONTENTS

I. Introduction 129
II. Barrier Function 129
III. Biophysical Parameters 130
IV. Irritation 132
V. Conclusion 132
References 132

I. INTRODUCTION

Even though it is well established that all humans belong to the same species, many physical differences exist among human populations. The use of bioengineering techniques is useful for investigating these differences, which could be the result of genetic, socioeconomic, or environmental factors [1].

II. BARRIER FUNCTION

Stratum corneum is equally thick in different races [2–5]. However, Weigand et al. [6] demonstrated that the stratum corneum in Blacks contains more cell layers and requires more cellophane tape strips to be removed than the stratum corneum of Whites, whereas Kompaore and Tsuruta [7] showed that Asian skin was significantly more sensitive to stripping than Black skin. Weigand also found great variance in values obtained from Black subjects, whereas data from White subjects were more homogeneous. No correlation was found between the degree of pigmentation and the number of cell layers. These data could be explained by the greater intercellular cohesion in Blacks, which results in an increased number of cell layers and an increased resistance to stripping. This mechanism may involve lipids [8] because the lipid content of the stratum corneum ranges from 8.5 to 14%, with higher values in Blacks [5,9]. This result was confirmed by Weigand et al. [6], who showed that delipidized specimens of stratum corneum were equal in weight in the two races.

0-8493-1443-7/05/$0.00+$1.50

Johnson and Corah [10] found that the mean electrical resistance of an adult Black skin is doubled in adult White skin, indicating an increased cohesion of the stratum corneum [10]. In fact, La Ruche and Cesarini [5] found that, in comparison with White skin, the Black skin stratum corneum is equal in thickness but is more compact: About 20 cell layers are observed in Blacks, vs. 16 layers in Whites.

Corcuff et al. [11] investigated the corneocyte surface area and spontaneous desquamation and found no differences between Black, White, or Asian skin. However, an increased desquamation (up to 2.5 times) was found in Blacks. The authors concluded that the differences may be related to a different composition of the intercellular lipids of the stratum corneum. Sugino et al. [12] found significant differences in the amount of ceramides in the stratum corneum, with the lowest levels in Blacks followed by Whites, Hispanics, and Asians. In this experiment, ceramide levels were inversely correlated with transepidermal water loss (TEWL) and directly correlated with water content. Meguro et al. [13] confirmed these correlations. These data may partially explain the controversial findings in the literature on the mechanisms of skin sensitivity.

Changes in skin permeability and barrier function have been reported: Kompaore et al. [7,14] evaluated TEWL and lag time after application of a vasoactive compound (methyl nicotinate) before and after removal of the stratum corneum by tape stripping. Before tape stripping, TEWL was 1.3 times greater in Blacks and Asians compared with Whites. No difference was found between Blacks and Asians, whereas after stripping they found a significantly higher TEWL in Blacks and Asians than in Whites. In particular, after stripping, Asians showed the highest TEWL (1.7 times greater than Whites). The authors conclude that, similar to previous studies [15,16], skin permeability measured by TEWL is higher in Blacks than in Whites. They also conclude that Asian skin has the highest permeability among the groups studied. However, these findings have not yet been confirmed by other groups. In fact, Sugino et al. [12] also included Asians in their study but found that baseline TEWL was, in decreasing order, Blacks > Whites ≥ Hispanics ≥ Asians. Another study [17] of Asian skin compared TEWL in Asians and Whites and found no statistically significant differences at baseline or after stripping; however, no vasoactive substance was applied.

Reed et al. [18] found differences in the recovery of the barrier between subjects with skin type II/III compared to those of skin type V/VI, but no differences were found between Whites in general and Asians. Darker skin recovered faster after barrier damage induced by tape stripping.

III. BIOPHYSICAL PARAMETERS

TEWL, skin conductance, and skin mechanical properties have been measured under basal conditions in Whites, Hispanics, and Blacks to assess whether skin color (melanin content) could induce changes in skin biophysical properties [19]. Differences that appear in skin conductance are more evident in biomechanical features such as skin extensibility, skin elastic modulus, and skin recovery. These features differ in dorsal and ventral sites according to race and highlight the influence of solar irradiation on skin and the role of melanin in maintaining the skin unaltered.

Wilson et al. [15] demonstrated higher *in vitro* TEWL values in Black compared to White skin taken from cadavers. They also found differences in Black and White skin physiology; in fact, TEWL increased with skin temperature. In their own study, they concluded that Black skin would have to have a greater TEWL increase to achieve the same temperature. Because TEWL depends on passive water vapor loss that is theoretically directly related to the ambient relative humidity and temperature [20], the increased TEWL in Black skin could be associated with an increase in temperature, as it is well established that a difference in Black and White temperature exists.

Most studies, using the forearm, back, and inner thigh [12–16,21,22], show a greater TEWL in Blacks compared with Whites; however, Warrier et al. [23], studying a larger sample size, have demonstrated that TEWL is lower in Blacks than Whites when measuring on the cheeks and legs. No racial differences in TEWL exist either on the volar or dorsal forearms. However, water content is increased in Hispanics on the volar forearm and is decreased in Whites (compared only to Blacks) on the dorsal forearm. These findings partially confirm previous observations [16,24]. Skin lipids may play a role in modulating the relation between stratum corneum water content and TEWL, resulting in higher conductance values in Blacks and Hispanics.

Racial differences in skin conductance are difficult to interpret in terms of stratum corneum water content because other physical factors, such as the skin surface or the presence of hair, can modify the quality of the skin–electrode contact. In all races, significant differences exist between the volar and dorsal forearms [19]. These results are in apparent contrast with TEWL recordings. Indeed, increased stratum corneum water content correlates with a higher TEWL [25]. These data may be explained on the basis of the different intercellular cohesion or lipid composition. Greater cell cohesion with a normal TEWL could result in increased skin water content.

Racial variability should be considered in terms of different skin responses to topical and environmental agents. Race provides a useful tool for investigating and comparing the effects of lifetime sun exposure and ambient relative humidity. Evolution has provided, over thousands of years, a genetic advantage to survive for those races living in a specific area with specific climatic conditions. Surviving in a harmful environment requires an optimal adaptation of outermost layers of our body, the skin, on a structural, biochemical, and molecular level. It is evident that melanin protection decreases sun damage: Differences between sun-exposed and sun-protected areas are not detectable in races with dark skin.

However, transepidermal water loss studies are characterized by a large interindividual variability and are biased by environmental effects and eccrine sweating. To bypass these influences, an *in vitro* technique for measuring TEWL was used to compare TEWL in two racial groups (Blacks and Whites) [15]. Black skin had a significantly higher mean TEWL than White skin. In both groups, however, a significant correlation between skin temperature and increased TEWL was found. The data confirm differences between races found in *in vivo* studies [16,24]. TEWL measurements with regard to Asian skin may be deemed inconclusive, as baseline measurements have found Asian skin to have TEWL values that are equal to Black skin and greater than White skin [14], that are less than those of other ethnic groups [12], and that are no different than those of other ethnic groups [17].

IV. IRRITATION

Irritation, as measured by TEWL [16,24], revealed a different pattern of reaction in Whites after chemical exposure to sodium lauryl sulfate (SLS). Blacks and Hispanics developed stronger irritant reactions after exposure. We applied 0.5% and 2.0% sodium lauryl sulfate to untreated, preoccluded, and predelipidized Black and White skin and quantified the resulting level of irritation using water content, TEWL, and laser Doppler velocimetry of the stratum corneum [16]. There was a statistical difference in irritation measuring TEWL only after 0.5% SLS application to the preoccluded area between the two groups. In fact, Blacks had 2.7 times higher TEWL levels than Whites, indicating that Blacks in the preoccluded state are more susceptible to irritation than are Whites. In another study, we compared differences in irritation between Hispanic and White skin [16]. We found higher values of TEWL for Hispanics compared with Whites after SLS-induced irritation. However, these values were not statistically significant. The reaction of Hispanic skin to SLS resembles that of Black skin when it is irritated with the same substance. Therefore, these data oppose the traditional clinical view, based on observing erythema, that darker skins are less reactive to irritants than are Whites.

V. CONCLUSION

Ethnic (racial) differences in skin physiology have been minimally investigated. The current experimental human model for skin is largely based on the physical and biochemical properties that are known for White skin. Thus, anatomical or physiological properties in skin of different races that may alter a disease process or a treatment of that disease are not being accounted for. Therefore, we still cannot answer the question, How resistant is Black skin compared to White? There exists reasonable evidence to support by means of objective measurements the theory that Black skin has a higher TEWL compared with White skin. Although some deductions have been made about Asian and Hispanic skin, the results are contradictory, and further evaluation of Asian and Hispanic skin needs to be done. Perhaps more specificity about the origin of their heritage should also be included, as "Asian" and "Hispanic" encompass a broad spectrum of people. Even now we remain optimistic that further knowledge will lead to refined claim support and more appropriate formulation for race-based skin care.

REFERENCES

1. Shriver, M.D., Ethnic variation as a key to the biology of human disease, *Ann. Intern. Med.*, 127, 401, 1997.
2. Freeman, R.G. et al., Sunlight as a factor influencing the thickness of epidermis, *J. Invest. Dermatol.*, 39, 295, 1962.

3. Thomson, M.L., Relative efficiency of pigment and horny layer thickness in protecting the skin of European and Africans against solar ultraviolet radiation, *J. Physiol. (Lond.)*, 127, 236, 1955.
4. Lock-Andersen, J. et al., Epidermal thickness, skin pigmentation and constitutive photosensitivity, *Photodermatol. Photoimmunol. Photomed.*, 13(4), 153, 1997.
5. La Ruche, G. and Cesarini, J.P., Histology and physiology of Black skin, *Ann. Dermatol. Venereol.* 119(8), 567, 1992.
6. Weigand, D.A., Haygood, C., and Gaylor, J.R., Cell layers and density of Negro and Caucasians stratum corneum, *J. Invest. Dermatol.*, 62, 563, 1974.
7. Kompaore, F. and Tsuruta, H., *In vivo* differences between Asian, Black and White in the stratum corneum barrier function, *Int. Arch. Occup. Environ. Health*, 65(1 suppl), S223, 1993.
8. Coderch, L. et al., Ceramides and skin function, *Am. J. Clin. Dermatol.*, 4(2), 107, 2003.
9. Rienertson, R.P. and Wheatley, V.R., Studies on the chemical composition of human epidermal lipids, *J. Invest. Dermatol.*, 32, 49, 1959.
10. Johnson, L.C. and Corah, N.L., Racial differences in skin resistance, *Science*, 139, 766, 1963.
11. Corcuff, P. et al., Racial differences in corneocytes, *Acta Derm. Venereol. (Stockh.)*, 71, 146, 1991.
12. Sugino, K., Imokawa, G., and Maibach, H., Ethnic difference of stratum corneum lipid in relation to stratum corneum function, *J. Invest. Dermatol.*, 100, 597, 1993.
13. Meguro, S. et al., Relationship between covalently bound ceramides and transepidermal water loss (TEWL), *Arch. Dermatol. Res.*, 292(9), 463, 2000.
14. Kompaore, F., Marty, J.P., and Dupont, C., *In vivo* evaluation of the stratum corneum barrier function in Blacks, Caucasians and Asians with two noninvasive methods, *Skin Pharmacol.*, 6, 200, 1993.
15. Wilson, D., Berardesca, E., and Maibach, H.I., *In vitro* transepidermal water loss: differences between Black and White human skin, *Br. J. Dermatol.*, 199, 647, 1988.
16. Berardesca, E. and Maibach, H.I., Racial differences in sodium lauryl sulphate induced cutaneous irritation: Black and White, *Contact Dermatitis*, 18, 136, 1988.
17. Yosipovitch, G. and Theng, C.T.S., Asian skin: its architecture, function, and differences from Caucasian skin, *Cosmetic Toiletries*, 117(9), 57, 2002.
18. Reed, J.T., Ghadially, R., and Elias, P.M., Effect of race, gender and skin type on epidermal permeability barrier function, *J. Invest. Dermatol.*, 102, 537, 1994.
19. Berardesca, E. et al., *In vivo* biophysical characterization of skin physiological differences in races, *Dermatologica*, 182, 89, 1991.
20. Baker, H., The skin as a barrier, in *Textbook of Dermatology*, Rook, A., Ed., Blackwell Scientific, Oxford, 1986, p. 355.
21. Reed, J.T., Ghadially, R., and Elias, P.M., Skin type, but neither race nor gender, influence epidermal permeability function, *Arch. Dermatol.*, 131(10), 1134, 1995.
22. Berardesca, E. et al., Differences in stratum corneum pH gradient when comparing White Caucasian and Black African-American skin, *Br. J. Dermatol.*, 139, 855, 1998.
23. Warrier, A.G. et al., A comparison of Black and White skin using noninvasive methods, *J. Soc. Cosmetic Chem.*, 47, 229, 1996.
24. Berardesca, E. and Maibach, H.I., Racial differences in sodium lauryl sulphate induced cutaneous irritation: Black and White, *Contact Dermatitis*, 18, 65, 1988.
25. Rietschel, R.L., A method to evaluate skin moisturizers *in vivo*, *J. Invest. Dermatol.*, 70, 152, 1978.

12 Sensitive Skin and Transepidermal Water Loss

Clarissa D. Villarama and Howard I. Maibach

CONTENTS

I. Epidemiology of Sensitive Skin 135
II. What Is Subjective/Sensory Irritation? 135
III. Pathomechanism 136
IV. Correlation of Self-Assessed Effects and Subclinical Change 136
V. Special Attributes of Skin of Stingers 138
VI. Barrier Function of Stingers 138
VII. Correlation of TEWL and Sensory Irritation 138
References 140

I. EPIDEMIOLOGY OF SENSITIVE SKIN

Sensitive skin is defined as having a reduced tolerance to frequent or prolonged use of cosmetics and toiletries, with symptoms ranging from visible signs of irritation such as erythema and scaling through more subjective neurosensory forms of discomfort such as stinging, burning, itching, tightness, and smarting [1].

A recent epidemiological study revealed that 10% of women and 5.8% of men described themselves as having very sensitive skin. Among women, symptoms of cosmetic-induced subjective sensory skin discomfort occurred more commonly in the sensitive-skin cohort (53%) than in those who regarded themselves as being nonsensitive (17%) [2].

II. WHAT IS SUBJECTIVE/SENSORY IRRITATION?

Erythema and dryness are visible signs of irritation. These signs are typically associated with subjective responses such as tightness, stinging, burning, and a feeling of dryness. However, epidemiologic studies show that many consumers experience these sensations in the absence of visible cutaneous manifestations. At least 10% of women complain of stinging with certain facial products. These sensations often determine whether the consumer continues to use the product [1]. Thus, further work

0-8493-1443-7/05/$0.00+$1.50

is needed to develop a strategy to overcome this type of discomfort, which is found to be greater in the winter months and in persons with sensitive skin [3].

Those cutaneous responses that have no clinically evident, objective expressions have been termed "subjective/sensory irritation." In some of these situations, histologic damage may be substantial even if the skin is grossly normal [4]. Various methods have been devised to experimentally elicit these sensations.

Because of the lack of clinical signs, it has been difficult to document the sensory irritation syndrome. With the advent of new methods for sensory testing, more definite and objective information has been gathered to provide insights regarding the pathomechanisms of this phenomenon.

"Stinging" is elicited by applying hydrosoluble substances such as aqueous lactic acid on the nasolabial fold of subjects. The lactic acid sting test (LAST) has been developed as a model for nonvisible skin irritation. It provides dermatologists with the most widely accepted clinical method for confirming susceptibility to chemically induced neurosensory skin irritation. Typical LAST protocol involves induction of facial sweating followed by application of 5% lactic acid solution to one side of the face and water as a placebo to the other side. Intensity of skin sensations are graded and used to classify patients as stingers or nonstingers. "Stingers" are individuals who experience intense ascending stinging, peaking after 5 to 10 minutes [5]. Of the general population, 10% feel a stinging after application of lactic acid on sweating facial skin [6].

"Burning" is measured by determining the reaction time to a mixture of chloroform and methanol applied to the face [7]. Capsaicin and methyl salicylate are other chemicals used to quantitatively evaluate subjective responses.

"Itching" is experimentally induced by topical or intradermal injections of substances such as proteolytic enzymes, mast cell degranulators, and vasoactive agents. The intensity and duration of the itch is then evaluated. The goal of most studies is to determine whether there is indeed a subset of patients, the so-called "stingers," who are more susceptible to skin irritation.

III. PATHOMECHANISM

Although the pathomechanism remains unclear, several factors are believed to influence sensory irritation. The quality and concentration of the exposing agent and regional variations in reactivity [8], as well as microanatomical differences such as density of hair follicles and sweat glands, elaborate network of nerves [9], and skin sebum content [10] appear to be contributory. Physiological differences such as hyperreactive neural pathways, blood vessel hyperreactivity [9,11], and altered mechanisms for inflammatory mediator release [5] have all been implicated. Some sensory irritation is believed actually to be a form of subclinical contact urticaria [12].

IV. CORRELATION OF SELF-ASSESSED EFFECTS AND SUBCLINICAL CHANGE

Simion et al. [15] showed that self-assessed sensations, such as feelings of dryness, stinging, and burning, correlate with and predict visible effects like dryness and

TABLE 12.1
Studies Correlating Sensory Irritation and Skin Irritation

Authors	Assessment of Sensory Irritation	Parameters Measured	Results
Coverly 1998 [13]	86 volunteers were tested with 10% lactic acid in the nasolabial fold to assess their ability to sting: 44 were stingers, 42 were nonstingers	Capacity to develop nonimmunologic contact urticaria, evaluated by different urticants including methyl nicotinate, benzoic acid, cinnamic acid, cinnamaldehyde, and dimethylsulfoxide; skin irritation was also measured with dimethylsulfoxide	Sensory reactions and skin irritation response to dimethylsulfoxide were similar in stingers and nonstingers
Basketter and Griffiths 1993 [14]	Lactic acid 10% was used on nasolabial folds to detect stinging response: 25 of the subjects were stingers, 25 were nonstingers	Skin irritation: response to 0.3% sodium dodecyl sulfate patch test; visual assessment of erythema and dryness	Patch test responses using erythema and dryness scores were similar for stingers and nonstingers
Simion et al. 1995 [15]	30 women had been previously shown to be easily irritated by surfactants	Exaggerated arm-washing methods were used to induce skin irritation; subjective sensory evaluation; visual assessment for erythema and dryness	Significant correlations between all self-assessed effects and observed scored features were found

erythema. However, some studies suggest that there is no correlation between the susceptibility to stinging response and subsequent objective irritation (see Table 12.1). Coverly et al. [13] showed that sensory reactions and irritant responses were, in fact, similar for stingers and nonstingers. Basketter et al. [14] also showed that there was no difference in either the strength or the pattern of the irritant response between stingers and nonstingers as assessed by subjective erythema and dryness scores. Note that these three studies employed visual assessment of clinical irritation in evaluating responses.

Bioassays have used visible degrees of edema and erythema as indices of irritancy. Over the years, however, the development of physical techniques for measuring subtle degrees of noninflammatory skin damage has improved our understanding of barrier function in early preclinical disease. It is now generally accepted that alterations in skin impedance [16], the amount of transepidermal water loss (TEWL), skin thickness, dielectric characteristics [17], blood flow velocity [18], and skin pH [19] do occur in subclinical cutaneous irritation. It is desirable that we evaluate the correlation of self-assessed effects and subclinical change as measured by these methods. In other words, can we prove that stingers are a distinct subset from nonstingers in terms of cutaneous parameters measurable by objective methods?

V. SPECIAL ATTRIBUTES OF SKIN OF STINGERS

Various studies that have employed bioengineering methods have noted important differences in the skin of stingers compared with that of nonstingers. The skin of stingers may have higher baseline pH values and lower sebum content [10]. Stingers also appear to have a more reactive cutaneous vasculature. Studies demonstrating higher laser Doppler flowmetry values on areas of benzoic acid–induced nonimmunologic urticaria [9] and on skin areas applied with methyl nicotinate [18] support this claim. Trends toward higher chromometric a* values likewise provide evidence that indicates hypersensitive vascular responses in stingers [10]. The stratum corneum of stingers seems to be less hydrated, as evidenced by lower electrical capacitance values [10,20].

VI. BARRIER FUNCTION OF STINGERS

It is hypothesized that stingers have an inherent defect in barrier function that allows increased penetration of exogenous materials through the skin. Many of the differences in pH levels, vasodilatory responses, and capacitance observed in stingers may be explained by an overly permeable barrier. Because the relationship between TEWL and percutaneous penetration of chemicals has been established [21], it is worthwhile to examine the current evidence regarding TEWL in the skin of stingers.

VII. CORRELATION OF TEWL AND SENSORY IRRITATION

TEWL can be considered a determinant indicative of the functional state of the cutaneous barrier. One of the characteristics of healthy skin is that the relationship between TEWL and hydration remains directly proportional. Changes in TEWL for both physiologic and pathologic skin conditions have been extensively studied. It has been experimentally proven that TEWL is increased in diseases with dry scaly skin, such as atopic dermatitis, psoriasis, and irritant dermatitis.

TEWL often increases in association with erythematous irritation, which makes this method useful in studies of more acute irritant responses. However, TEWL is also increased in cumulative irritation [6], which can be demonstrated using bioengineering methods even before visible changes like erythema and dryness appear [22].

The results of several studies show a certain relationship between sensory irritation and the skin barrier, as manifested by changes in TEWL. Trends toward higher baseline TEWL values measured on the nasolabial fold [20] and cheek [10] were reported in stingers compared with nonstingers. Stingers also showed higher LAST

TABLE 12.2
Studies Correlating Sensory Irritation and Transepidermal Water Loss

Reference	Subjects	Methods	Results
Lammintausta et al. [6]	Eight stingers vs. 15 nonstingers	Transepidermal water loss measured after 24-hour occlusive-patch test with 1% sodium lauryl sulfate on the upper back	Trend toward a greater increase in transepidermal water loss in stingers vs. nonstingers; not statistically significant
Seidenari et al. [10]	26 subjects in the "sensitive-skin" group vs. 26 subjects in the "control" group	Lactic acid sting test scores; baseline transepidermal water loss values measured on cheek and arm	Higher lactic acid sting test scores and trend toward higher baseline transepidermal water loss values in the "sensitive-skin" group; not statistically significant
Wu et al. [20]	50 subjects	Transepidermal water loss was measured at the nasolabial fold before and 8 minutes after application of the 3% and 5% lactic acid test solutions	Increase in transepidermal water loss resulting from 3% and 5% lactic acid, $P < .05$; negative correlation between lactic acid sting test score and percentage increase in transepidermal water loss resulting from application of lactic acid, $P = .003$

scores and a tendency toward a more exaggerated increase in TEWL at sites of experimentally induced acute irritant dermatitis [6].

Although it is tempting to attribute sensory irritation to an inherent barrier defect, note that these data indicating that baseline TEWL of the skin of stingers is greater than that of nonstingers are not statistically significant [6,10,20]. There is also no published direct evidence indicating that the skin of stingers is, in fact, more permeable than that of nonstingers. Even to come to the conclusion that the ability to perceive skin stinging is indicative of an individual's susceptibility to irritation is difficult. Although some correlation has been made, other studies have proved otherwise [14].

Taken together, we believe that this common, albeit not typically serious, clinical entity needs more refined observation so that, by clarifying the mechanisms involved, solutions may be found.

REFERENCES

1. De groot, A., Nater, J., and Van der Hende, R., Adverse effects of cosmetics and toiletries: a retrospective study in the general population, *J. Cosmetic Sci.*, 9, 255, 1988.
2. Willis, C. et al., Sensitive skin: an epidemiological study, *Br. J. Dermatol.*, 145, 258, 2001.
3. Agner, T. and Serup, J., Sodium lauryl sulphate for irritant patch testing: a dose response study using bioengineering methods for determination of skin irritation, *J. Invest. Dermatol.*, 95, 543, 1990.
4. Kligman, A., The invisible dermatoses, *Arch. Dermatol.*, 127, 1375, 1991.
5. Frosch, P. and Kligman, A., A method for appraising the stinging capacity of topically applied substances, *J. Soc. Cosmetic Chem.*, 28, 197, 1977.
6. Lammintausta, K., Maibach, H., and Wilson, D., Susceptibility to cumulative and acute irritant dermatitis: an experimental approach in human volunteers, *Contact Dermatitis*, 19, 84, 1988.
7. Grove, C., Duncan, S., and Kligman, A., Effect of aging on the blistering of human skin with ammonium hydroxide, *Br. J. Dermatol.*, 107, 393, 1982.
8. Rougier, A. et al., Regional variation in percutaneous absorption in man: measurement by stripping method, *Arch. Derm. Res.*, 278, 465, 1986.
9. Lammintausta, K., Maibach, H., and Wilson, D., Mechanisms of subjective (sensory) irritation: propensity to non-immunologic contact urticaria and objective irritation in stingers, *Derm. Beruf. Umwelt*, 36, 45, 1988.
10. Seidenari, S., Francomano, M., and Mantovani, L., Baseline biophysical parameters in subjects with sensitive skin, *Contact Dermatitis*, 38, 311, 1998.
11. Berardesca, E. et al., *In vivo* transcutaneous penetration of nicotinates and sensitive skin, *Contact Dermatitis*, 25, 35, 1991.
12. Amin, S., Lahti, A., and Maibach, H., Contact uritcarua and the contact urticaria syndrome, in *Dermatotoxicology*, 5th ed., Marzulli, F. and Maibach, H., Eds., Taylor and Francis, Washington, D.C., 1995, pp. 485–503.
13. Coverly, J. et al., Susceptibility to skin stinging, non-immunologic contact urticaria and skin irritation: is there a relationship? *Contact Dermatitis*, 38, 90, 1998.
14. Basketter, D. and Griffiths, H., A study of the relationship between susceptibility to skin stinging and skin irritation, *Contact Dermatitis*, 29, 185, 1993.
15. Simion, F. et al., Self-perceived sensory responses to soap and synthetic detergent bars correlate with clinical signs of irritation, *J. Am. Acad. Dermatol.*, 32, 205, 1995.
16. Nicander, I. et al. Electrical impedance measured to five skin depths in mild irritant dermatitis induced by sodium-lauryl sulphate, *Br. J. Dermatol.*, 132, 718, 1995.
17. Maibach, H., Bronaugh, R., and Guy, R., Noninvasive techniques for determining skin function, in *Cutaneous Toxicity*, Drill, V.A. and Lazar, P., Eds., Raven Press, New York, 1984, pp. 63–97.
18. Guy, R., Tur, E., and Bjerke, S., Are there age and racial differences to methyl nicotinate-induced vasodilation in human skin? *J. Am. Acad. Dermatol.*, 12, 1001, 1985.
19. Grice, K. et al., Skin temperature and transepidermal water loss, *J. Invest. Dermatol.*, 57, 108, 1971.
20. Wu, Y. et al., Correlation between stinging, TEWL, and capacitance, *Skin Res. Technol.*, 9, 90, 2003.

21. Rougier, A., Lotte, C., and Maibach, H., *In vivo* relationship between percutaneous absorption and transepidermal water loss, in *Absorption of Dermatological Products*, 4th ed., Bronaugh, R. and Maibach, H., Eds., Marcel Dekker, New York, 2002, pp. 115–128.
22. Van der Valk, P., Nater, J., and Bleumik, E., Vulnerability of the skin to surfactants in different groups of eczema patients and controls as measured by vapor loss, *Clin. Exp. Derm.*, 10, 98, 1985.

13 Transepidermal Water Loss and Barrier Function of Aging Human Skin

Klaus-P. Wilhelm, Marianne Brandt, and Howard I. Maibach

CONTENTS

Abstract 143
I. Introduction 143
A. Stratum Corneum as a Physical Barrier 144
B. TEWL Measurements to Examine Skin Barrier Properties 144
II. Cutaneous Permeability Barrier in the Elderly 145
A. TEWL of Aging Skin 145
B. Percutaneous Penetration and Skin Aging 147
C. Proclivity to Skin Irritation of the Elderly 149
III. Discussion 151
A. Physicochemical Interpretation 152
IV. Conclusions 155
References 155

ABSTRACT

Today there is evidence that baseline transepidermal water loss (TEWL) is reduced in aged individuals compared with TEWL values from mid-adulthood. Several studies indicate that the permeability barrier *in vivo* is increased in aged subjects, not only for the egress of water but also for topically applied hydrophilic compounds. Regarding the irritation response of human skin, an age-dependence is documented in several publications. Possible explanations for these phenomena are discussed in this review. The exact mechanisms, however, are still to be described.

I. INTRODUCTION

It is generally accepted that preterm infants have a deficient cutaneous permeability barrier compared with that of term infants.[1–3] This has been demonstrated by

0-8493-1443-7/05/$0.00+$1.50

increased TEWL as well as by increased percutaneous penetration of topically applied compounds. Differences in TEWL, skin hydration, and skin pH between infants and adults have also been documented, showing higher values for the young that correspond to earlier findings.[4–6] During the first 2 to 4 weeks of life, the skin barrier gradually matures, and with increasing chronological age TEWL successively becomes comparable to adulthood levels. At the other end of the age spectrum (i.e., from adulthood through senescence), the age-dependence of the cutaneous permeability barrier function and, hence, the baseline TEWL, are still controversial.

A. Stratum Corneum as a Physical Barrier

It is established that the stratum corneum (SC), the outermost skin layer at the environment–individual interface, is the principle permeability barrier to TEWL and is a major barrier to percutaneous absorption of topically applied compounds.[7–9] SC is typically 6 to 20 μm thick, except for the palms and soles, where the thickness is approximately 400 to 600 μm.[10,11]

As a part of the epidermis, SC is constantly renewed from the granular layer, and the outermost corneocytes are gradually desquamated from the surface. The internal structure of SC is well organized and has often been schematically described by a "brick-and-mortar model."[9,12] Terminally differentiated, keratin-filled corneocytes of polyhedral shape, arranged in interdigitating vertical columns, represent the "bricks," and the intercellular lipoidal material in a multilamellar bilayer arrangement represents the "mortar." However, various other models of the SC structure are still discussed in addition to the described "brick-and-mortar model," including the "sandwich model"[13] and the "single gel phase model."[14] Lipid metabolism within the SC has been documented in all models, and TEWL seems to play a role in the regulation of lipid synthesis via regulation of HMG-CoA-reductase activity.[15,16]

Although there is circumstantial evidence that SC is not homogeneous throughout its thickness,[17,18] initial claims that the true barrier layer resides at the base of the SC[18–20] have been shown to be an inappropriate interpretation of experiments in which SC was removed, layer by layer, by adhesive tape stripping. More appropriate studies have demonstrated the contrary: The barrier properties are more evenly distributed across the entire thickness of the membrane[21,22] and are related to the phase behavior of the intercellular lipids (ceramides, cholesterol, and fatty acids).[23] The functional competence of each stratum corneum layer remains subjudicial.

B. TEWL Measurements to Examine Skin Barrier Properties

The permeability barrier function of human skin can be assessed by measuring the percutaneous absorption of xenobiotics or by evaluating the skin's proclivity to primary irritants. Alternatively, noninvasive TEWL measurements provide a generally accepted parameter for the skin permeability barrier function.[24–33] Different methodology has been used for assessing TEWL, as reviewed by Wilson and Maibach[34] and by Grice,[35] as well as in Chapter 22 of this book. Today, the majority of clinical studies are performed with the open, unventilated chamber technique and with instrumental calculation of the water evaporation gradient developing from the

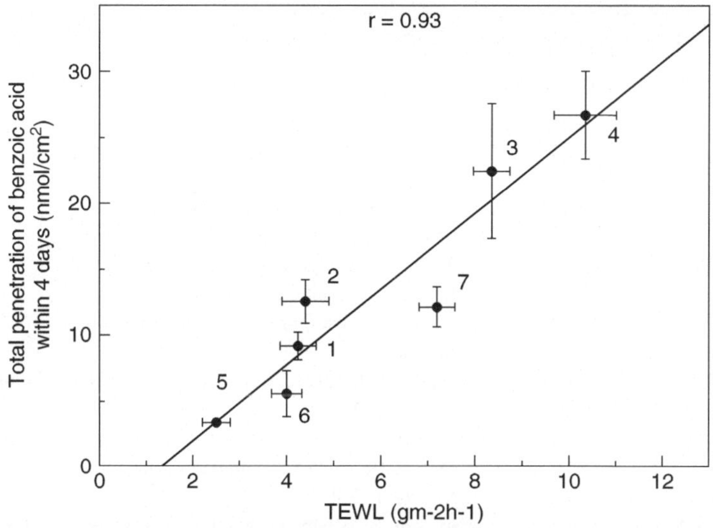

FIGURE 13.1 Correlation between percutaneous penetration and transepidermal water loss. *In vivo* relationship between transepidermal water loss and percutaneous penetration of benzoic acid according to anatomic site in man: 1, arm (upper outer); 2, abdomen; 3, postauricular; 4, forehead; 5, forearm (ventral elbow); 6, forearm (ventral mid); 7, forearm (ventral wrist). Mean values ± SD of n = 7 to 8 males (20 to 30 years of age). Drawn according to the data in Reference 38.

skin surface to atmospheric humidity.[36,37] The relationship between TEWL and percutaneous absorption has been clearly demonstrated in preterm infants[1] and in respect to anatomic variability (Figure 13.1).[31,38,39]

II. CUTANEOUS PERMEABILITY BARRIER IN THE ELDERLY

A. TEWL of Aging Skin

Although the fact that the cutaneous permeability barrier function of the newborn is incomplete is accepted without any doubt, no consensus has yet been established about the skin permeability barrier and baseline TEWL at the opposite end of the age spectrum; that is, from adulthood through senescence (Table 13.1). The majority of published studies,[39–47] however, note a significant decrease of TEWL with age, especially after the age of 60 to 70 years. A study performed by Leveque[40] on 145 healthy volunteers reported a significant decrease of TEWL on the forearm during the first 20 years of life and a second decrease after the age of 70 years as compared with adulthood levels (Figure 13.2). The decreased baseline TEWL in the elderly was confirmed by Wilhelm et al.[41] The authors demonstrated that TEWL was significantly lower in an aged group than in young individuals on nine out of 11 anatomic sites

TABLE 13.1
Age Dependence of TEWL

Reference	Group Size (n)	Result	Anatomic Site
Baker[42]	87	Lower water loss in individuals older than 70 years of age	—
Grice and Bettley[44]	39	TEWL tended to decrease with increasing age; however, correlation was not statistically significant (age range of the group was not indicated)	Abdomen
Jemec et al.[45]	21	Significant negative correlation between TEWL and age (22–71 years)	Onychial
Kligman[47]	21	Lower TEWL in 66- to 81-year-old individuals than in 19- to 26-year-old participants	Leg and forearm
Leveque et al.[40]	145	Decreased TEWL after 60 years of age	Forearm
Roskos et al.[48,49]	33	No correlation between baseline TEWL and age (range: 19–85 years); however, four out of six radiolabeled drugs showed signs of decreased percutaneous penetration in aged individuals	Forearm
Rougier et al.[38]	23	No difference in transepidermal water loss between age groups (20–30 vs. 65–80 years); the percutaneous penetration of benzoic acid, however, was significantly reduced in aged individuals	Upper arm
Tagami[46]	22	TEWL decreased in the aged group (19–22 vs. 61–85 years)	Pretibial
Tupker et al.[30]	43	TEWL showed no correlation with age (age range: 20–48 years)	Forearm
Wilhelm et al.[41]	29	Significant decrease with age on most anatomical regions (mean age 26.7 vs. 70.5 years)	11 different regions
Yosipovitch et al.[4]	64	TEWL lower in infants (2 days) on forehead, palms, and soles; higher on forearms	Seven different regions
Fluhr et al.[5]	88	Children (3.5 years) vs. adults (35 years); no significant difference in TEWL, but slightly higher values in children	—
Giusti et al.[6]	100	No significant difference in TEWL (babies vs. adult women)	Volar forearm, buttock

Note: The results of some recently published studies (without intention of completeness) on the effect of aging on *in vivo* transepidermal water loss (TEWL) measurements in humans.

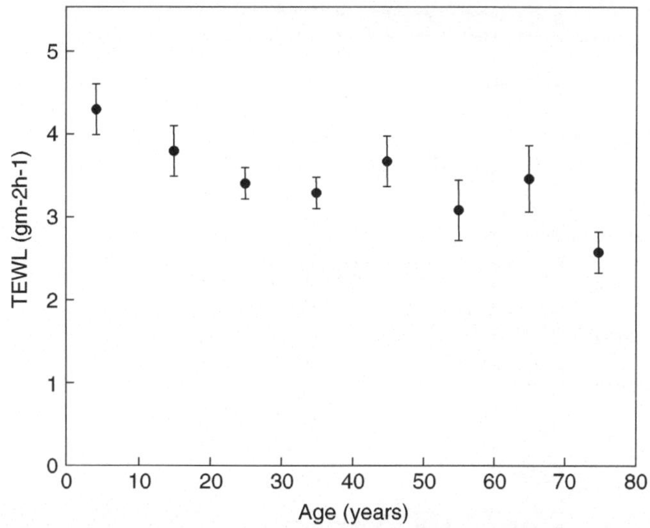

FIGURE 13.2 Increasing age and transepidermal water loss. Baseline transepidermal water loss on the volar forearm decreased during the first 20 years of life. A second decrease was noted after the age of 70 years. Mean values, n = 145. Modified from Reference 40.

(Figure 13.3). The changes were statistically and possibly biologically significant. However, in the same study, SC hydration as measured with a capacitance meter did not vary between the age groups (Figure 13.4).

There are also reports that failed to demonstrate any significant correlation between age and TEWL (Table 13.1).[5,6,30,38,44,48] The critical reader might wonder why results can be so conflicting, even though the method of measurement and group size were very similar. Although no information about the age of the study group was provided by Grice and Bettley[44] or Giusti et al.,[6] Tupker et al.[30] studied individuals aged from 20 to 48 years, Fluhr et al.[5] also included individuals of about 35 years of age. Both studies represented mid-adulthood levels. A correlation between age and TEWL could not be expected in their studies because TEWL apparently does not decrease significantly before the age of 60 to 70 years.[40] While studying individuals age 60 years and older, Roskos and Guy[48] and Rougier et al.[38] still failed to demonstrate significant age-related changes in TEWL, possibly because of high ambient temperature (23° ± 2°C) in the first study or high relative humidity (70%) in the second study. Unless TEWL is measured without contamination from eccrine sweat, false negative data may occur. Both groups, however, demonstrated an increase in the skin permeability barrier in aged individuals, using reduced percutaneous penetration of topically applied compounds (see below).[38,49]

B. Percutaneous Penetration and Skin Aging

A significant correlation between TEWL and percutaneous absorption of diverse drugs has been demonstrated by several studies.[31,38,39] It also appears that the

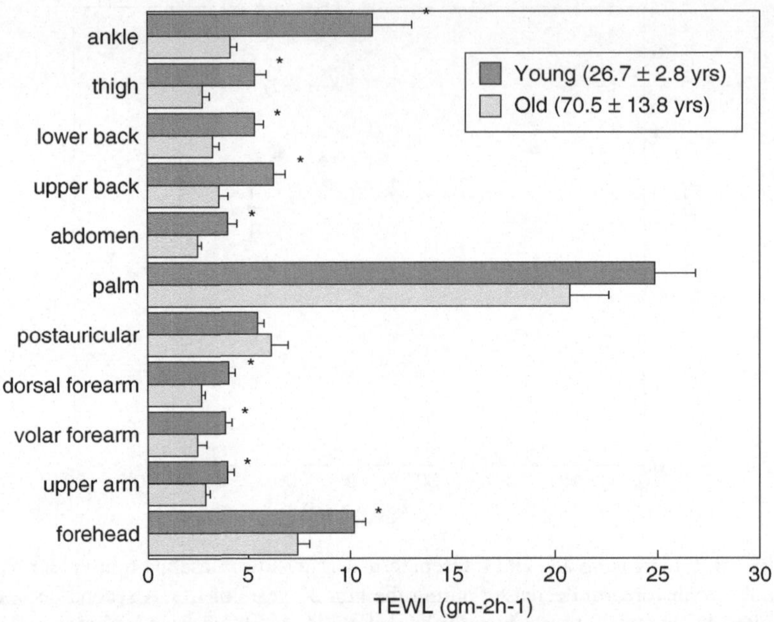

FIGURE 13.3 Anatomic variability of transepidermal water loss: influence of age. Age differences in transepidermal water loss rates at different anatomical locations. Mean values ± 1 SEM, n = 14 to 15. *Statistically significant difference between the two groups ($P \leq 0.05$). Modified from Reference 41.

decreased TEWL in aged individuals is revealing for a less permeable membrane to topically applied compounds.[38,39,49–51]

Rougier et al.[38] reported a significantly decreased percutaneous absorption of [^{14}C]-benzoic acid, a highly water soluble compound, in older subjects (aged 65 to 80 years). Roskos et al.[49] confirmed a significantly decreased penetration for four out of six radioisotope-labeled substances in older individuals (Figure 13.5). Only the percutaneous penetration of the two most lipophilic compounds considered was not significantly different in their experiments. Thus, like Rougier's benzoic acid data, hydrophilic compounds were less absorbed in the elderly. This was in agreement with earlier studies by Christophers and Kligman[50] and Tagami,[51] who concluded that the barrier function of human skin *in vivo* increases with increasing chronological age.

Another study, by Oriba et al.,[52] demonstrated lower percutaneous absorption of hydrocortisone in postmenopausal women than in premenopausal women. No difference was found in the absorption of testosterone. Other studies also observed changes in the skin barrier function and structure resulting from hormonal influences or occurring throughout the menstrual cycle of women.[43,53,54] However, whether the main cause for changes to the skin barrier is the aging procedure or a hormonal influence cannot be easily distinguished, because these effects are closely intertwined.

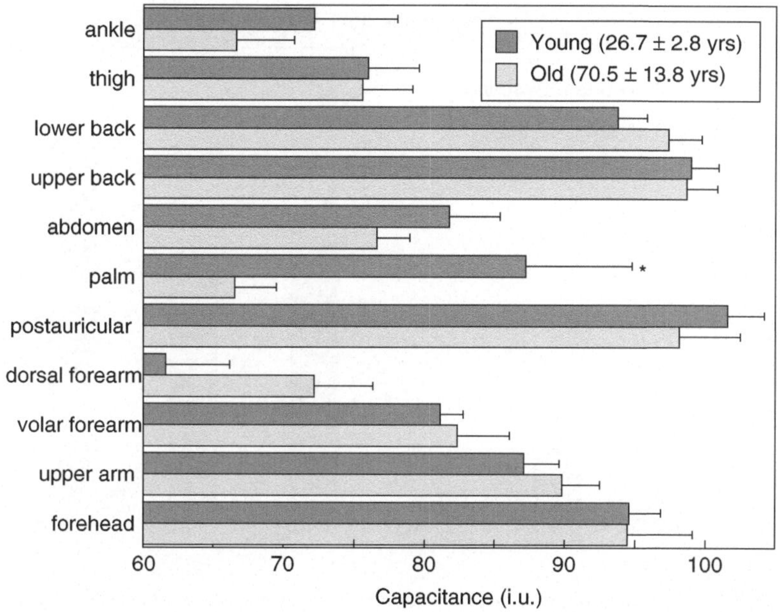

FIGURE 13.4 Anatomic variability of stratum corneum hydration: influence of age. Illustrated are age differences in stratum corneum hydration at different anatomical locations as evaluated by capacitance measurements. i.u., instrumental units. Mean values ± 1 SEM, n = 14 to 15. *Statistically significant difference between the age groups ($P \leq 0.05$). Modified from Reference 41.

C. Proclivity to Skin Irritation of the Elderly

Epidemiological data indicate a lower incidence of irritant contact dermatitis with increasing age.[55] Experimental studies also confirmed a decreased sensitivity to cutaneous irritants with increasing age.[43,56–61] Kwangsukstith et al.[62] proposed that increasing age resulted in a decreased ability to become sensitized to new allergens and a waning response in previously sensitized subjects.

Cua et al.[56] investigated the severity of the irritant response after application of sodium lauryl sulfate (SLS) on 11 anatomic sites in two groups. It was demonstrated that aged individuals had a significantly decreased irritant response on five out of 11 anatomic sites. The severity of SLS-induced skin irritation was quantified by visual scores and by TEWL measurements (Figure 13.6). Interestingly, aged individuals failed to demonstrate erythematous reactions at some anatomic sites completely (e.g., on the thigh and on the forearm), although, in contrast, TEWL measurements showed a significant barrier damage present in the aged group (Figure 13.6). Thus, there is disparity between visually documented inflammation and non–visually related function (i.e., TEWL).

Our own results, summarizing the reaction to repeated occlusive SLS patching of 195 volunteers (aged between 20 and 65 years), who participated in irritation studies according to a modified soap chamber test design from 1999 to 2002, describe

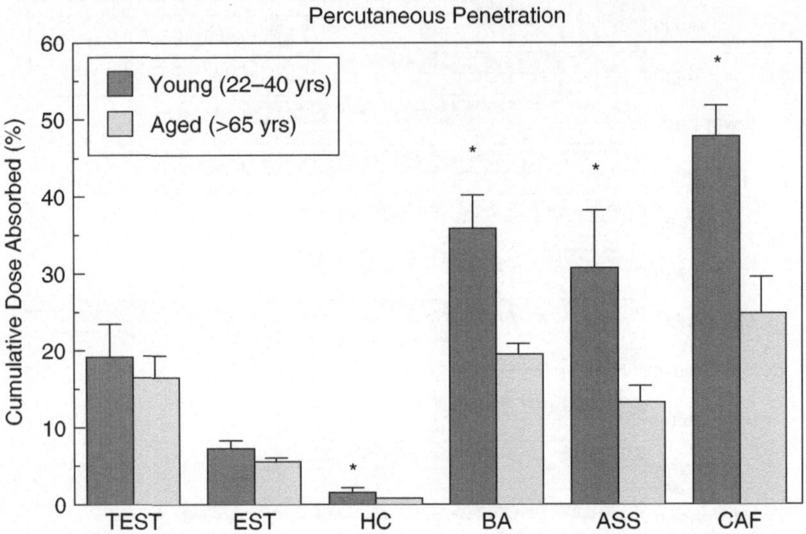

FIGURE 13.5 Age dependence of percutaneous absorption. Shown is the cumulative dose percutaneously absorbed within 7 days in percentage of the applied dose. The percutaneous absorption was decreased in the elderly group for hydrocortisone (HC), benzoic acid (BA), acetylsalicylic acid (ASS), and caffeine (CAF). Only the percutaneous absorption of the two most hydrophilic compounds testosterone (TEST) and estradiol (EST) was not significantly different between the age groups. All absorption data were corrected for incomplete renal elimination, using the appropriate population intravenous control. Mean values ± SEM. *Statistically significant difference between the age groups ($P \leq 0.05$). Drawn according to the data provided in Reference 49.

a significantly lower reaction (lower TEWL increase) of the older subjects to SLS as a standard positive irritant (Figure 13.7). These findings are in good correlation with the majority of publications confirming a decreased reactivity of older people to irritants.

Other studies[60,63] determined the recovery rate of human skin after irritating the skin by tape stripping or by applying irritants. They found that skin of the elderly (older than 70 years) shows a slower and more prolonged recovery. This result is confirmed by our own evaluations, represented in Figure 13.8. A summary of several studies investigating the barrier recovery after perturbing the skin with 0.25% SLS for 24 hours demonstrates higher TEWL values 7 days after skin perturbation for the older volunteers. This finding represents a slower barrier recovery for older skin. However, a significant difference was not documented. The reasons might be the small group size (n = 44) and missing representation of aged skin (from participants who were older than 65 years) in the panel, but a clear tendency of prolonged barrier recovery is given.

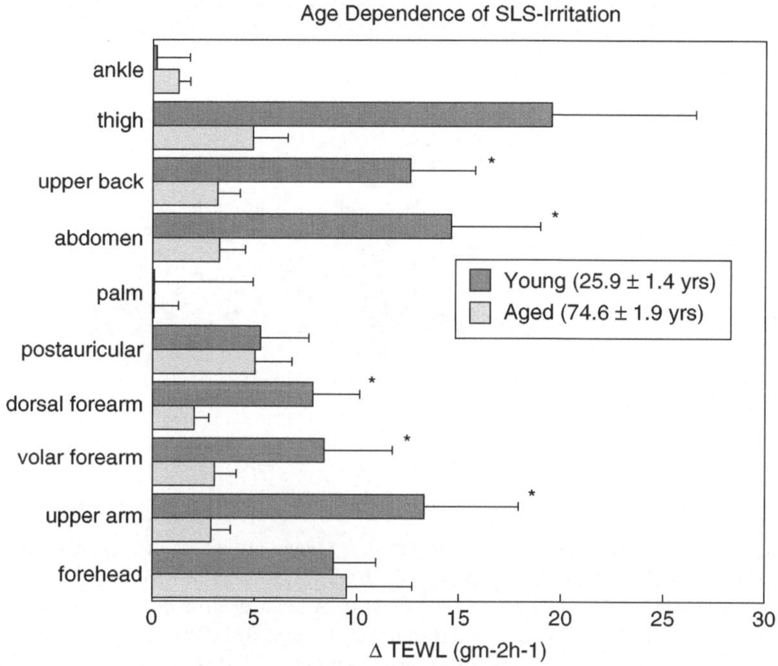

FIGURE 13.6 Age dependence of proclivity to sodium lauryl sulfate (SLS) irritation. Shown are increases in transepidermal water loss (TEWL; Δ TEWL = $TEWL_{SLS}$ – $TEWL_{control}$) after exposure to 0.25% SLS for 24 hours. On most anatomic sites, SLS induced lower increases in TEWL in elderly individuals. Mean ± SEM, n = 7 to 8. *Statistically significant difference between the age groups ($P \leq 0.05$). Drawn according to the data provided in Reference 56.

III. DISCUSSION

The clinical aspect of dry skin encountered by a large proportion of individuals aged 65 years and older has not been associated with a perturbed barrier to water permeability, as evaluated by TEWL measurements. In contrast, the majority of studies demonstrated lower TEWL measurements in individuals aged 60 to 70 years and older.[40,43,45–47]

Skin anatomy, physiology, and biochemistry change in many regards with increasing age.[64] It is not to be expected that the barrier function would be excluded from this modification (Table 13.2).

Reduced sweating associated with the aging process[65] probably does not contribute to the observed decrease in TEWL, because sweating does not significantly influence baseline TEWL at rest in a standard climatized environment.[66]

Decreased density and efficiency of the skin microvasculature[67] resulting in decreased skin temperature might explain reduced TEWL in elderly individuals, although Wilhelm et al.[41] corrected their TEWL measurements to a standard skin reference temperature of 30°C and still noted lower TEWL values in the elderly.

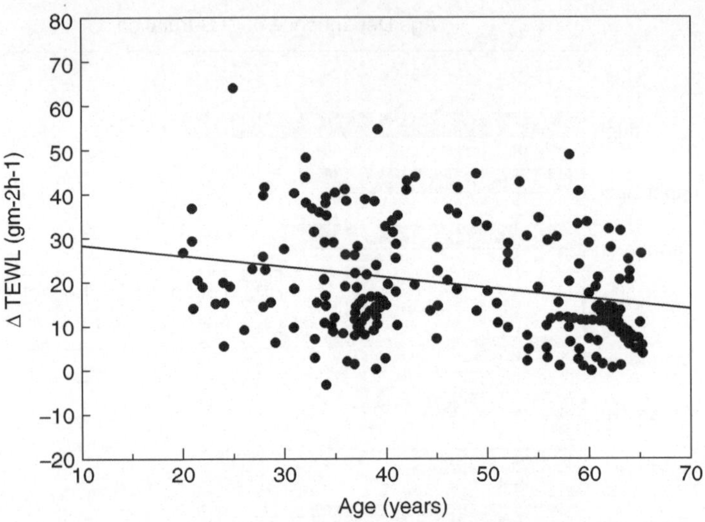

FIGURE 13.7 Age dependence of skin irritability. Shown are different transepidermal water loss (TEWL) values (Δ TEWL = $TEWL_{Day5} - TEWL_{Baseline}$) after repetitive exposure to 0.2% sodium lauryl sulfate (SLS) on 5 consecutive days. A clear decline of the fitting curve from the young to the elderly is demonstrated by a lower increase in TEWL. Single values, n = 195. Statistically significant reduced irritability of the aged ($P \leq 0.05$).

The thickness of the SC is not altered by age, but its renewal time is greatly prolonged: In young adults, SC transit time as estimated by the dansyl chloride staining method is about 20 days, whereas in older adults it is more than 30 days.[68] Whether the increased SC renewal time in aged skin is of relevance to its barrier properties is not yet known.

It has been shown that there is an inverse relationship between corneocyte size and TEWL.[69] In the case of skin senescence, there seems to be no clear correlation between these parameters;[54,70] however, they might be mutually dependent (see below). Another possible contribution to the decreased TEWL values in elderly individuals may be an altered composition of SC lipids.[71,72]

A. Physicochemical Interpretation

The simplest way of modeling both the process of percutaneous absorption and TEWL is to assume that Fick's first law of diffusion is applicable, although the SC is not an inert membrane. The form of the equation often quoted is

$$\frac{\mathbf{dQ}}{\mathbf{dt}} = \frac{\mathbf{DK_p\,c}}{\mathbf{h}}$$

where **dQ/dt** is the rate of skin penetration (TEWL), **D** is the effective diffusion coefficient of drug (water) in SC, **Kp** is the partition coefficient of drug (water)

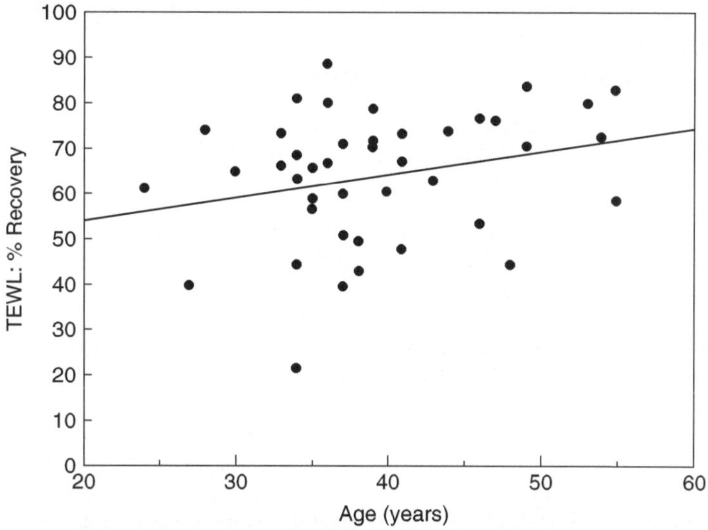

FIGURE 13.8 Age dependence of barrier recovery. Shown are the percentage transepidermal water loss (TEWL) values: % Recovery = $TEWL_{Recovery}/TEWL_{Maximum\ Irritation}$ [%], demonstrating the barrier recovery after exposure to 0.25% sodium lauryl sulfate (SLS) for 24 hours (where Recovery = TEWL values 7 days after taking off the SLS patch, and Maximum Irritation = TEWL values 1 day after taking off the SLS patch). Lower percentage recovery values represent less remaining skin barrier damage and therefore indicate a better barrier recovery. The values of barrier recovery of the older group were higher, indicating a slower recovery of older skin. Single values, n = 44. Statistically not significant.

between membrane and solution, **c** is the concentration gradient of drug (water), and **h** is the effective thickness of the skin barrier

For water diffusing through the SC in the opposite direction, d**Q**/dt is measurable at the skin surface as TEWL. Thus, all factors relevant to d**Q**/d**t** apply to TEWL. SC thickness is not altered by age[10,11] and may, therefore, not account for the increased *in vivo* barrier. One may also assume that **c** is constant, although epidermal atrophy and a reduced tissue hydration in the aged might decrease the concentration gradient for water.[67]

It has been demonstrated that the composition of SC lipids clearly changes with increasing age.[72] The relevance of this finding to barrier properties, however, has not yet been clearly elucidated. Roskos[73] confirmed a changed composition of SC lipids in aged individuals. In addition, she found an overall diminution of SC epidermal lipid content with increasing age in humans *in vivo* by means of attenuated total reflectance infrared spectroscopy.[73] A reduction of the intercellular lipids was also confirmed by Rogers et al.[71] A decreased epidermal lipid content in the elderly would reduce the partitioning (**Kp**) of water and of hydrophilic compounds in SC[74] and would, hence, explain the reduced TEWL and the decrease of the percutaneous absorption of preferentially hydrophilic compounds.

Because the corneocytes have been found to become larger with increasing age,[54,75] a reduction of the intercellular space follows.[76] The consequence is a smaller

TABLE 13.2
Summary of Age-Dependent Changes Relevant to the Permeability Barrier

Structure/Parameter	Change with Skin Aging	Influence on Permeability Barrier	References
SC thickness	Unchanged	—	10,11
SC intercellular lipids	Changed composition (decrease of sterol esters and triglycerides)	Possible influence on partioning and diffusivity	
	Decrease of quantity, decreased synthesis	Reduced absorption of hydrophilic substances, reduced water binding capacity	63,75
SC water content	Decrease	Decreased penetration of lipophilic compounds into SC	46,77
SC turnover/renewal	Prolonged	Unknown	68
SC/corneocytes	Increase of size	Reduction of intercellular space, smaller reservoir for penetrating substances; reduction of effective diffusion coefficient	75,76
Epidermis	Atrophy	Decreased water reservoir	67
Dermo-epidermal junction	Flattening	Reduced percutaneous absorption of substances	75

Note: SC, stratum corneum.

reservoir for penetrating substances. Water, as well as, for example, SLS, penetrates the intercellular space. Due to the reduction of intercellular space, the penetration decreases, which would result in a reduced TEWL and diminished irritability.[75] In the equation above, this would correspond to a reduction of the effective diffusion coefficient **D**.

In addition to SC lipids and structural changes, the water content of SC is an important variable influencing **Kp**. Significant differences in SC water content between young and old individuals were demonstrated by capacitance measurements,[4,6] although some other studies could not differentiate between young and old.[46,56] Using more sensitive attenuated total reflectance infrared spectroscopy instrumentation, it has been undoubtedly demonstrated that the SC of the elderly is drier than the young adult equivalent.[77] In addition, by use of a "water sorption–desorption" test, Tagami[46] confirmed the lower water binding capacity of old SC. A lower water content could be reasoned from a structural reduction of the intercellular space in combination with a change in the lipids. A reduced presence of water in the SC of old subjects would imply that the environment of aged skin is less attractive to hydrophilic molecules and to water, resulting in a decreased **Kp** and, hence, decreased d**Q**/d**t**.

All these findings give an overview of the changes in the skin barrier occurring in the aged human skin. The described reasons for these changes are most possibly closely correlated, but the mechanisms and processes are still undoubtedly not defined.

IV. CONCLUSIONS

The majority of recent studies investigating the influence of aging on the cutaneous permeability barrier by means of TEWL measurements, percutaneous penetration, and skin irritation models give clear evidence that the permeability barrier function changes with increasing chronological age, which is demonstrated by lower percutaneous penetration of hydrophilic substances. Reduced irritability and slower recovery of the skin barrier in the aged has been proved by various studies. Age-associated structural and compositional changes in the SC as the principle permeability barrier are most probably responsible for this effect. Further investigations are of interest, with improvements in the measuring techniques and clarification of the changes of biochemical processes in the aged skin.

REFERENCES

1. Rutter, N., The immature skin, *Br. Med. Bull.*, 44, 957, 1988.
2. Hammerlund, K., Sedin, G., and Stromberg, B., Transepidermal water loss in newborn infants, *Acta Paediatr. Scand.*, 72, 721, 1983.
3. Wilson, D.R. and Maibach, H.I., An *in vivo* comparison of skin barrier function, in *Neonatal Skin, Structure and Function*, Maibach, H.I. and Boisits, E.K., Eds., Marcel Dekker, New York, 1982, pp. 101–110.
4. Yosipovitch, G. et al., Skin barrier properties in different body areas in neonates, *Pediatrics*, 106, 105, 2000.
5. Fluhr, J.W., Pfisterer, S., and Gloor, M., Direct comparison of skin physiology in children and adults with bioengineering methods, *Pediatr. Dermatol.*, 17/6, 436, 2000.
6. Giusti, F. et al., Skin barrier, hydration, and pH of the skin of infants under 2 years of age, *Pediatr. Dermatol.*, 18/2, 93, 2001.
7. Blank, I.H. and Scheuplein, R.J., Transport into and within the skin, *Br. J. Dermatol.*, 81, 4, 1969.
8. Scheuplein, R.J. Percutaneous absorption after twenty-five years: or "old wine in new wineskins," *J. Invest. Dermatol.* 67, 31, 1976.
9. Elias, P.M., Epidermal lipids, membranes, and keratinization, *Int. J. Dermatol.* 20, 1, 1981.
10. Holbrook, K.A. and Odland, G.F., Regional differences in the thickness (cell layers) of the human stratum corneum: an ultrastructural analysis, *J. Invest. Dermatol.*, 62, 415, 1974.
11. Plewig, G. et al., Thickness of corneocytes, in *Stratum Corneum*, Marks, R. and Plewig, G., Eds., Springer, Berlin, 1983, pp. 171–174.
12. Michaels, A.S., Chandrasekaran, S.K., and Shaw, J.E., Drug permeation through human skin, theory and *in vitro* experimental measurement, *J. Am. Inst. Chem. Eng.*, 21, 985, 1975.
13. Bouwstra, J. et al., New aspects of the skin barrier organization, *Skin Pharmacol. Appl. Skin Physiol.*, 14, 52, 2001.

14. Norlen, L., Skin barrier structure and function: the single gel phase model, *J. Invest. Dermatol.*, 117/4, 830, 2001.
15. Grubauer, G., Elias, P.M., and Feingold, K.R., Transepidermal water loss: the signal for recovery of barrier structure and function, *J. Lipid Res.*, 30, 323, 1989.
16. Proksch, E., Elias, P.M., and Feingold, K.R., Regulation of 3-hydroxy-3-methylglutaryl-coenzyme A reductase activity in murine epidermis. Modulation of enzyme content and activation state by barrier requirements, *J. Clin. Invest.*, 85, 874, 1990.
17. Bommannan, D., Potts, R.O., and Guy, R.H., Examination of stratum corneum barrier function *in vivo* by infrared spectroscopy, *J. Invest. Dermatol.*, 95, 403, 1990.
18. Eriksson, G. and Lamke, L., Regeneration of human epidermal surface and water barrier function after stripping, *Acta Derm. Venereol. (Stockh.)*, 51, 169, 1971.
19. Bowers, P.A. and White, R.J., Isolation, barrier properties and lipid analysis of stratum compactum, a discrete region of the stratum corneum, *Br. J. Dermatol.*, 112, 1, 1985
20. Blank, I.H., Further observations on factors which influence the water content of the stratum corneum, *J. Invest. Dermatol.*, 21, 259, 1953.
21. Blank, I.H. and Gould, E., Location and reformation of the epithelial barrier to water vapor, *Arch. Dermatol.*, 78, 702, 1962.
22. Blank, I.H. and Gould, E., Study of mechanisms which impede the penetration of synthetic anionic surfactants into skin, *J. Invest. Dermatol.*, 37, 311, 1962.
23. Wertz, P.W., Lipids and barrier function of the skin, *Acta Derm. Venereol.*, 208(suppl), 7, 2000.
24. Thune, P., Evaluation of the hydration and the water-holding capacity in atopic skin and so-called dry skin, *Acta Derm. Venereol. (Stockh.)*, 144(suppl), 133, 1989.
25. Frost, P. et al., Ichthyosiform dermatoses. III. Studies of transepidermal water loss, *Arch. Dermatol.*, 98, 230, 1968.
26. Silverman, R.A., Lender, J., and Elmets, C.A., Effects of occlusive and semiocclusive dressings on the return of barrier function to transepidermal water loss in standardized human wounds, *J. Am. Acad. Dermatol.*, 20, 755, 1989.
27. Grice, K., Sattar, H., and Baker, H., The cutaneous barrier to salts and water in psoriasis and in normal skin, *Br. J. Dermatol.*, 88, 459, 1973.
28. Serup, J. and Staberg, B., Differentiation of allergic and irritant reactions by transepidermal water loss, *Contact Dermatitis*, 16, 129, 1987.
29. van der Valk, P.J.M., Nater, J.P., and Bleumink, E., Skin irritancy of surfactants as assessed by water vapor loss measurements, *J. Invest. Dermatol.*, 82, 291, 1984.
30. Tupker, R.A. et al., Baseline transepidermal water loss (TEWL) as a prediction of susceptibility to sodium lauryl sulphate, *Contact Dermatitis*, 20, 265, 1989.
31. Lotte, C. et al., *In vivo* relationship between transepidermal water loss and percutaneous penetration of some organic compounds in man: effect of anatomic site, *Arch. Dermatol. Res.*, 279, 351, 1987.
32. Onken, H.D. and Moyer, C.A., The water barrier in human epidermis, *Arch. Dermatol.*, 87, 90, 1963.
33. Murahata, R.I., Crowe, D.M., and Roheim, J.R., The use of transepidermal water loss to measure and predict the irritation response to surfactants, *Int. J. Cosmetic Sci.*, 8, 225, 1986.
34. Wilson, D.R. and Maibach, H.I., Transepidermal water loss: a review, in *Cutaneous Investigation in Health and Disease: Noninvasive Methods and Instrumentation*, Leveque, J.L., Ed., Marcel Dekker, New York, 1989, pp. 113–134.
35. Grice, K.A., Transepidermal water loss, in *The Physiology and Pathophysiology of the Skin*, Jarret, A., Ed., London, Academic Press, 1980, Vol. 6, pp. 2115–2127.

36. Nilsson, G.E., Measurement of water exchange through skin, *Med. Biol. Eng. Comput.*, 15, 209, 1977.
37. Pinnagoda, J. et al., Guidelines for transepidermal water loss (TEWL) measurement, *Contact Dermatitis*, 22, 164, 1990.
38. Rougier, A. et al., Relationship between skin permeability and corneocyte size according to anatomic site, age, and sex in man, *J. Soc. Cosmetic Chem.*, 39, 15, 1988.
39. Dupuis, D. et al., *In vivo* relationship between percutaneous absorption and transepidermal water loss according to anatomic site in man, *J. Soc. Cosmetic Chem.*, 37, 351, 1986.
40. Leveque, J.L., Measurement of transepidermal water loss, in *Cutaneous Investigation in Health and Disease: Noninvasive Methods and Instrumentation*, Leveque, J.L., Ed., Marcel Dekker, New York, 1989, pp. 135–152.
41. Wilhelm, K.P., Cua, A.B., and Maibach, H.I., The effect of skin aging on transepidermal water loss, stratum corneum hydration, skin surface pH, and casual sebum content, *Arch. Dermatol.*, 127, 1806, 1991.
42. Baker, H., Deperdition d'eau par voie trans-epidermique, *Ann. Dermatol. Syphiligr.*, 98, 289, 1971.
43. Elsner, P., Wilhelm, D., and Maibach, H.I., Sodium lauryl sulfate-induced irritant contact dermatitis in vulvar and forearm skin of premenopausal and postmenopausal woman, *J. Am. Acad. Dermatol.*, 23, 648, 1990.
44. Grice, K.A. and Bettley, F.R., Skin water loss and accidental hypothermia in psoriasis, ichthyosis and erythrodermia, *Br. Med. J.*, 4, 195, 1967.
45. Jemec, G., Agner, T., and Serup, J., Transonychial water loss. Relation to sex, age and nailplate thickness, *Br. J. Dermatol.*, 121, 443, 1989.
46. Tagami, H., Aging and the hydration state of the skin, in *Cutaneous Aging*, Kligman, A.M. and Takase, Y., Eds., University of Tokyo Press, Tokyo, 1988, pp. 99–109.
47. Kligman, A.M., Perspectives and problems in cutaneous gerontology, *J. Invest. Dermatol.*, 73, 39, 1979.
48. Roskos, K.V. and Guy, R.H., Assessment of skin barrier function using transepidermal water loss — effect of age, *Pharm. Res.*, 6, 949, 1989.
49. Roskos, K.V., Maibach, H.I., and Guy, R.H., The effect of aging on percutaneous absorption in man, *J. Pharmacokinet. Biopharm.*, 17, 617, 1989.
50. Christophers, E. and Kligman, A.M., Percutaneous absorption in aged skin, in *Advances in Biology of Skin. 6. Aging*, Montagna, W., Ed., Pergamon Press, Oxford, 1965, pp. 163–175.
51. Tagami, H., Functional characteristics of aged skin. Percutaneous absorption, *Acta Dermatol. (Kyoto)*, 66/67, 19, 1972.
52. Oriba, H.A., Bucks, D.A., and Maibach, H.I., Percutaneous absorption of hydrocortisone and testosterone on the vulva and forearm: effect of the menopause and site, *Br. J. Dermatol.*, 134/2, 229, 1996.
53. Agner, T., Damm, P., and Skouby, S.O., Menstrual cycle and skin reactivity, *J. Am. Acad. Dermatol.*, 24, 566, 1991.
54. Fluhr, J.W. et al., Differences in corneocyte surface area in pre- and postmenopausal women. Assessment with noninvasive videomicroscopic imaging of corneocytes method (VIC) under basal conditions, *Skin Pharmacol. Appl. Skin Physiol.*, 14, 10, 2001.
55. Malten, K.E. et al., Occupational dermatitis in five European dermatological departments, *Berufsdermatosen*, 19, 1, 1971.
56. Cua, A.B., Wilhelm, K.P., and Maibach, H.I., Cutaneous sodium lauryl sulfate irritation potential: age and regional variability, *Br. J. Dermatol.*, 123, 607, 1990.

57. Grove, G.L. et al., Use of nonintrusive tests to monitor age-associated changes in human skin, *J. Soc. Cosmetic Chem.*, 32, 15, 1981.
58. Soschin, D. and Kligman, A.M., Adverse subjective responses, in *Safety and Efficacy of Topical Drugs and Cosmetics*, Kligman, A.M., Ed., Grune and Stratton, New York, 1982, pp. 377–387.
59. Coenraads, P.J., Bleumink, E., and Nater, J.P., Susceptibility to primary irritants, *Contact Dermatitis*, 1, 377, 1975.
60. Schwindt, D.A. et al., Cumulative irritation in older and younger skin: a comparison, *Acta Derm. Venereol.*, 78/4, 279, 1998.
61. Robinson, M.K., Population differences in acute skin irritation responses. Race, sex, age, sensitive skin and repeat subject comparisons, *Contact Dermatitis*, 46/2, 86, 2002.
62. Kwangsukstith, C. and Maibach, H.I., Effect of age and sex on the induction and elicitation of allergic contact dermatitis, *Contact Dermatitis*, 33/5, 289, 1995.
63. Ghadially, R. et al., The aged epidermal permeability barrier. Structural, functional, and lipid biochemical abnormalities in humans and a senescent murine model, *J. Clin. Invest.*, 95/5, 2281, 1995.
64. Gilchrest, B.A., Skin aging and photoaging: an overview. *J. Am. Acad. Dermatol.*, 21, 610, 1989.
65. Sato, K. and Timm, D.E., Effect of aging on pharmacological sweating in man, in *Cutaneous Aging*, Kligman, A.M. and Takase, Y., Eds., University of Tokyo Press, Tokyo, 1988, pp. 111–126.
66. Pinnagoda, J. et al., Transepidermal water loss with and without sweat gland inactivation, *Contact Dermatitis*, 21, 16, 1989.
67. Miyake, I., Histological aging of facial skin, in, *Cutaneous Aging*, Kligman, A.M. and Takase, Y., Eds., University of Tokyo Press, Tokyo, 1988, pp. 571–588.
68. Grove, G.L. and Kligman, A.M., Age-associated changes in human epidermal cell renewal, *J. Gerontol.*, 38, 137, 1983.
69. Marks, R., Nicholls, S., and King, C.S., Studies on isolated corneocytes, *Int. J. Cosmetic Sci.*, 3, 251, 1981.
70. Leveque, J.L. et al., *In vivo* studies of the evolution of physical properties of the human skin with age, *Int. J. Dermatol.*, 23, 322, 1984.
71. Rogers, J. et al., Stratum corneum lipids: the effects of ageing and the seasons, *Arch. Dermatol. Res.*, 288, 765, 1996.
72. Saint Leger, D. et al., Age-associated changes in stratum corneum lipids and their relation to dryness, *Dermatologica*, 177, 159, 1988.
73. Roskos, K.V., The Effect of Skin Aging on the Percutaneous Penetration of Chemicals through Human Skin, Doctoral dissertation, University of California–San Francisco, 1989.
74. Surber, C. et al., Optimization of topical therapy: partitioning of drugs into stratum corneum, *Pharm. Res.*, 7, 1320, 1990
75. Schwindt, D.A., Nicht-invasive Untersuchungen zur Physiologie und Pathophysiologie der menschlichen Epidermis unter besonderer Berücksichtigung der Penetrationseigenschaften des Stratum Corneum. Dissertation, Medizinische Universität zu Lübeck, 1998.
76. Marks, R., Measurement of biological aging human epidermis, *Br. J. Dermatol.*, 104, 627, 1981
77. Potts, R.O. and Buras, E.M., *In vivo* changes in the dynamic viscosity of human stratum corneum as a function of age and ambient moisture, *J. Soc. Cosmetic Chem.* 36, 169, 1985.

14 Atopic Dermatitis and Other Skin Diseases

Manigé Fartasch

CONTENTS

Abstract 159
I. Introduction 160
II. TEWL in Atopic Dermatitis 160
A. The Eczematous Skin in Atopic Dermatitis 160
B. The Noneczematous Skin in Atopics 161
C. TEWL as Predictor for the Development of Irritant Contact Dermatitis in Atopics? 163
III. TEWL in Psoriasis and in Different Forms of Congenital Keratinization Disorders 164
A. Psoriasis 164
B. Congenital Keratinization Disorders 165
References 166

ABSTRACT

Lesional skin of inflammatory diseases (e.g., eczema, psoriasis) or congenital keratinization disorders with hyperkeratosis/parakeratosis show an elevated TEWL as a marker for their impaired barrier function. In nonlesional skin of atopics and in atopics with a history of eczema the mean baseline TEWL was found to be increased. This was not the case in psoriasis or patients with a history of contact dermatitis. Recent studies on various types of ichthyosis showed an increase of TEWL, which could be two- to fourfold higher than in controls. Even in clinically normal looking skin of patients with autosomal dominant ichthyosis the TEWL was still elevated, stressing the fact that in generalized skin diseases (e.g., atopic dermatitis) subclinical alterations of barrier capacity may exist in apparently normal skin. Another factor is that an enhancement of TEWL may be accompanied by an accelerated penetration of various substances. This might be of importance for the occupational exposure to various work-related chemical fluids and may be an explanation for the existing interindividual variation in systemic uptake.

0-8493-1443-7/05/$0.00+$1.50

I. INTRODUCTION

Changes in biophysical properties of the stratum corneum (SC), especially of the water-permeability barrier, appear in a variety of diseases of different etiologies. Defective barrier function with increased transepidermal water loss (TEWL) might be seen in inflammatory reactions of the epidermis, which influences epidermal proliferation and leads to incomplete cornification (like eczema or psoriasis) or in disease entities with the disturbance of terminal differentiations (keratinization disorders like different types of ichthyosis). In both cases the distribution, structure, and biochemistry of the intercellular lipid compartment of the SC, which are vital for the barrier function, are influenced leading to an impairment of the water diffusion barrier of the skin and probably facilitate penetration of various substances (Rougier et al., 1988; Kompaore et al., 1991; Oestermann et al., 1993; Lavrijsen et al., 1993).

Another point is that the clinically normal looking skin may be functionally abnormal in some cases, which can only be assessed by the measurements of TEWL. This is particularly the case in the noneczematous atopic skin. All of these issues will be addressed in the following chapter.

II. TEWL IN ATOPIC DERMATITIS

A. The Eczematous Skin in Atopic Dermatitis

The clinical and morphological changes in this inflammatory eczematous skin condition are severe. While clinically erythema, vesicles, papules, and scaling are seen, histologically eczematous disorders are characterized by epidermal spongiotic microvesicles (intercellular edema) with oozing and acanthosis, and the SC may be parakeratotic and contain aggregates of coagulated plasma, the substrate of crusts (Lever and Schaumburg-Lever, 1990). These inflammatory conditions show severe alterations of the SC.

Ultrastructural studies showed dilatation of the intercellular spaces of the SC, which depict an irregular distribution of lipid structures with disturbance of the normally lamellar-arranged epidermal lipid bilayers (Madison et al., 1987; Hou et al., 1991; Fartasch et al., 1993). These seemed to be intermingled with exudates (M. Fartasch, unpublished observations). Additionally, there is an increase of parakeratotic corneocytes. Alterations of the chemically bound lipid envelope of the corneocyte and its relation to the lamellar intercellular lipid bilayers (Schwartendr et al., 1987) in acute and chronic phases of the disease have not yet been studied. Due to all these changes it is therefore not astonishing that the TEWL values were reported to be increased in scaly eczematous lesions (Blichman and Serup, 1987). The chronic inflammation of atopic skin is associated with changes of atopic skin color, decrease of skin capacitance, and increase of TEWL and blood perfusion (van Neste et al., 1991). Increased TEWL has been recorded with instruments in diseased skin of atopic dermatitis (AD), i.e., on moderately to severely inflamed lesions (Rajka, 1997; Abe et al., 1978).

B. The Noneczematous Skin in Atopics

A common finding in patients with AD is the high incidence of dry skin with a finely scaling nonerythematous, noninflamed skin surface which feels rough to the touch, often with a perifollicular accentuation (Uehara and Harumitsu, 1984). Among the diagnostic criteria of AD defined by Hanifin and Rajka, xerosis or "dry skin" (DS) is a minor feature. In our prospective study comparing the frequency of major and minor criteria in normal population and atopic patients with flexural involvement, more than 91% compared to 26% showed DS (Diepgen and Fartasch, 1992). The term DS just refers to the superficial part of the SC that is harsh and looks like dry soil (Pierard, 1989). It should be noted that there are different types of DS and that many spongiotic and parakeratotic disorders are labeled dry skin, thus a single biological feature responsible for rough skin does not exist. Probably totally unrelated conditions induced by a number of biological pathways may show this feature.

There are conflicting data as to the cause of DS in noneczematous skin and the reason for the enhancement of TEWL. Some morphological studies have reported that atopic DS showed increased intercorneocytic cohesion (Finlay et al., 1980; Al-Jaberi and Marks, 1984; Uehara and Miyauchi, 1984; Watanabe et al., 1991), increased epidermal thickness with focal parakeratosis, and, in places, slight hypergranulosis or hypogranuloses, and it was suggested that the atopic DS reflects a subclinical eczema. Other studies could not confirm the previously proposed thesis that the persistent dry skin of atopics is the result of subclinical eczema. In some cases the epidermis of nonaffected atopic skin shows signs of suppressed synthesis of keratohyalin with a histological reduction of granular layer thickness. This feature when clinically accompanied by hyperlinear palms ("ichthyotic" palms) and keratosis pilaris was believed to be evidence for the coexistence of AD with autosomal dominant ichthyosis (ADI) and was suggested as the cause of DS in as many as 30 to 40% of atopics (Uehara and Hayashi, 1981). This would have been an explanation for the findings of why noneczematous dry skin and normal-looking skin show elevated TEWL. However, we have previously shown that only few atopics have concomitant ADI (4 to 6%) and that the dry condition in AD is structurally distinguishable from DS in dominant ichthyosis (Fartasch et al., 1989). Several lines of evidence indicate that the process of lipid translocation and transformation might be disturbed in DS (Schäfer and Kragballe, 1991; Melnik et al., 1990; Yamamoto et al., 1991; Imokawa et al., 1991). Biochemical, ultrastructural, and functional findings support the view that impaired biosynthesis of ceramides and acylceramides probably due to immunologically induced alterations of epidermal differentiation (especially increased epidermopoiesis) may be the cause of the atopic DS and the impaired barrier function. There are several studies that postulate abnormalities in the total amount of ceramides (Imokawa et al. 1991, Di Nardo et al. 1998, Bleck et al. 1999). In addition, some studies have found a different composition regarding the seven ceramides that partially form the intercellular lipids of the horny layer in noninvolved skin of atopics (Bleck et al. 1999). Others postulate a disturbance of lipid metabolism by dysfunction of enzymes (Murata et al. 1996, Ohnishi et al. 1999) or specific, newly found enzymes (Hara et al 2000).

TABLE 14.1
TEWL of Noneczematous Skin of Atopics

Reference	TEWL
Shahidulla et al. (1969)	↑
Rajka (1974)	↑
Werner and Lindberg (1985)	↑
van der Valk et al. (1985a)	↑
Thune (1989)	↑
Tupker et al. (1990)	↑
Imokawa et al. (1991)	↑
Fartasch et al. (1992)	↑
Loden et al. (1992)	↑

Ultrastructurally we did, however, find some evidence of disturbed maturation of the water-permeability barrier in atopic noneczematous DS, which was induced by a delayed and probably incomplete extrusion of lamellar bodies, resulting in a diminished delivery of their "probarrier" polar lipids in the SC intercellular spaces (Fartasch et al., 1992). The delayed lamellar body exocytosis may additionally impair the formation of the water permeability by disturbing other, presumably lamellar body-dependent processes, such as lipid transformation and extrusion of enzymes. While healthy skin has the ability to regulate disturbed barrier function by increased lipid synthesis (Proksch et al., 1990) because of the altered lipid-extruding mechanism, which may be at least partly responsible for the deficient barrier and the enhancement of TEWL and probably the resulting susceptibility of the atopic skin to irritants, the ability to regulate the barrier function is reduced. Another factor for the increased mean baseline TEWL is the observed significantly reduced mean corneocyte area in the uninvolved skin of patients with eczema (Al-Jaberi et al., 1984).

Many scaling diseases have been shown to be characterized by abnormally small corneocytes and an altered barrier. This variation in barrier function in the presence of small corneocytes may be related to the increased intercellular volume per unit volume of SC (Marks et al., 1981) or the different tortuosity of the intercellular pathway.

In numerous studies it has been reported that the DS in patients with AD shows an increase of the baseline TEWL (Table 14.1). However, it is difficult to accomplish a direct comparison since the results among different investigators vary either according to the techniques used and/or the different definition of the degree of skin involvement and the different anatomic sites examined as well as different opinions about the criteria for assessment of the clinical appearance of DS (Werner Linde, 1992; Thune, 1989). Additionally, the different ways to interpret "DS" have never been defined in a repeatable way (Pierard, 1989).

Dryness of the skin (e.g., rough skin) is not always correlated with altered barrier functions. Studies by Léveque et al. (1987) and Thune (1989) indicated that in those with ordinary winter xerosis and the xerotic skin of old patients (localization upper arm) the dryness was only poorly correlated with barrier function. Studies by Werner and Lindberg (1985) indicate that not only the clinically noneczematous DS but also the normal looking skin shows an elevated mean baseline TEWL. This has been confirmed by others examining the uninvolved skin of patients with AD on the forearm (Valk van, 1985; Tupker, 1990) and on the upper arm (Agner, 1991, 1992). Tupker et al. (1990) showed that subjects with a past history of healed AD had a higher mean baseline TEWL (forearms) compared to subjects with a past history of allergic contact dermatitis and the control group. On the other hand, basal TEWL measured on the upper arm in hand eczema patients with a childhood history of AD, but without atopic manifestations in adult life, was found to be normal. It was speculated that TEWL values may undergo changes related to the time and course of the disease (Agner, 1991).

This has also been confirmed by studies (Löffler and Effendy 1999, Matsumoto et al. 2000, Sakurai et al. 2002) showing that skin barrier function (baseline TEWL, water content) was not disturbed in patients with healed atopic dermatitis and that only atopics with clinical signs of eczema (active phase) showed a higher baseline TEWL.

C. TEWL as Predictor for the Development of Irritant Contact Dermatitis in Atopics?

Generally, there seems to be a relationship between a high baseline TEWL and skin susceptibility to detergents (Tupker, 1989; Tupker et al., 1990; Murahata et al., 1986; Pinnagoda et al., 1989; Fartasch, 1997b; Fartasch et al., 1998; Tupker et al., 1997). There is ample evidence that the unaffected atopic skin with higher baseline TEWL is more prone to develop irritant contact dermatitis than other skin (Bäurle et al., 1985; Lammintausta and Kalimo, 1998; Rystedt, 1985; Nilsson, 1985; Shmunes and Keil, 1984).

Whether or not chronic irritant contact dermatitis will develop in a particular individual may be explained by the disturbance of the balance between the "repair capacity" of the skin and the sum of the damaging factors. Individuals with AD not only have a lower threshold of irritation but also a slower healing. Dry skin in AD patients has been recognized as a risk factor for the development of hand eczema (Rystedt, 1985). Additionally, in the study of Tupker et al. (1990), dryness of the skin was a significant factor in the susceptibility to irritants. If tested in a quiet phase of the disease, patients with AD were found to react more severely to sodium lauryl sulfate (SLS) than healthy controls as assessed by measurements of TEWL (Valk van der et al., 1985). Agner (1992) stated that higher baseline TEWL values were found in AD patients, but the increase in TEWL after irritation was not significantly different from that in controls. In our time-course analysis, measurements of TEWL were performed before and after a single 0.5% SLS exposure in patients with atopic skin diathesis and controls (Fartasch et al., 1991). In controls the baseline TEWL was reached after 7 days; a normalization of TEWL was reached after 11 to 13 days in atopics. These findings stress the observation that the repair capacity of the atopic noneczematous skin seems to be impaired, which predisposes the development of irritant contact dermatitis.

III. TEWL IN PSORIASIS AND IN DIFFERENT FORMS OF CONGENITAL KERATINIZATION DISORDERS

A. PSORIASIS

The sharply demarcated brownish red papules and plaques, which are usually covered with layers of fine, silvery scales histologically show elongation of the epidermal rete ridges with thickening in their lower portion with elongation and edema of the papillae, and thinning of the suprapapillary portion of the epidermis with the occasional presence of very small spongiform pustules. The horny layer in some instances consists entirely of parakeratotic cells, and, since, in the epidermis a direct relationship exists between the absence of keratohyalin granules and the development of parakeratosis, there is concomitantly an absence of stratum granulosum (Lever and Schaumberg-Lever, 1985). Accumulation of pyknotic nuclei of neutrophils are located within parakeratotic areas of the horny layer (Munro's microabscess). Ultrastructural studies have shown that the epidermal barrier structures that are responsible for the water permeability of the SC show severe alterations (Menon and Elias, 1991). The lamellar body lipids persisted to higher layers within the SC interstices (Menon and Elias, 1991; Fartasch, 1997b). Although an abundance of lamellar material is present in the intercellular spaces, complete bilayers do not form; instead, foreshortened lamellae accumulate in lacunae at all levels of the SC interstices. TEWL measurements have shown that the psoriatic plaques with the pathological horny layer show consequent increased water passage (Marks et al., 1981; Frodin et al., 1988; Berardesca et al., 1990; Tagami and Yoshikuni, 1985; Serup and Blichmann, 1987; Rajka and Thune, 1976; Grice, 1980).

Evaporimetry has been used in monitoring the healing process of this disease, and the initially high TEWL value in a psoriatic plaque returns to normal at a rate paralleling the rate of healing (Frödin et al., 1988). In the studies of Tagami and Yoshikuni (1985) it has been stated that the amount of TEWL appeared to be proportional to clinical severity, i.e., thick, scaly lesions were accompanied by higher TEWL levels than were thin, scaly lesions. In contrast Serup and Blichmann (1987) did not find a direct correlation to the amount of scaling. Other authors stated that the increase of TEWL on the investigated lesional psoriatic skin can be used to predict a relapse occurring on the investigated lesional psoriatic area, since the TEWL change had preceded the clinical deterioration of psoriasis. Still, the method could not predict a relapse distant from the plaque investigated (Rajka and Thune, 1976).

While the nonlesional skin of AD shows an increased mean baseline TEWL suggesting that clinically normal skin may be functionally abnormal, it results in a defective barrier and a susceptibility to develop irritant dermatitis. Comparison of uninvolved normal-looking skin of psoriatic patients compared with normal subjects (Berardesca et al., 1990) showed that the TEWL in unaffected skin of psoriatic skin is variable since a hyperkeratotic psoriatic plaque is usually more resistant to irritation than normal skin (Rajka and Thune, 1976). During the therapy with dithranol the inflammation usually occurs in adjacent normal skin earlier than in lesional skin (Wilhelm and Maibach, 1990). On the other hand, psoriasis may be provoked in nonlesional skin by various irritant stimuli (Koebner's phenomenon) (Wilhelm and

Maibach, 1990). Despite normal baseline TEWL, former studies showed that the proliferative response of uninvolved psoriatic skin to fluid injection was increased and showed an abnormal proliferative behavior compared to controls (Grove, 1980). In psoriasis, TEWL measurement in clinical and pathological skin does not seem be a predictor that determines susceptibility to irritants. Reactivity of cellular targets underneath and components of the dermis may be important in this disease, which shows that differences in pathogenetic mechanism exist.

B. Congenital Keratinization Disorders

Despite the most hyperkeratotic SC in various keratinization disorders, an increased TEWL has been demonstrated in different forms of congenital ichthyosis. These studies were mostly been performed in the 1970s by using different techniques for the measurement of TEWL. Studies on four distinct ichthyosiform dermatoses (lamellar ichthyosis, ADI, X-linked ichthyosis, epidermolytic hyperkeratosis) with (Sage electric hygrometer) have been performed by Frost et al. (1968) who showed an increase of TEWL in all of these forms. The observed TEWL was highest in epidermolytic hyperkeratosis. Grice and Bettley (1967), who used the same technique showed that TEWL in ADI and the X-linked form of ichthyosis was found to be approximately double that of normal subjects. Using evaporimetry Kiistala et al. (1981) have shown that the basal TEWL in ichthyotic patients was almost threefold compared with that in controls, stressing that properties other than thickness of the SC (e.g., the lipid structure or the tortuosity of the intercellular pathway) determine its behavior.

Recent studies of Buyse et al. (1993) have shown that so-called "collodion babies" are at risk for developing a hypernatremic dehydration and hypothermia due to their severely impaired barrier function. The collodion baby showed a highly increased TEWL (on day four: 112 g/m^2X h vs. 18 g/m^2X h in controls). Hypernatremic dehydration is also a common finding in babies with Netherton's syndrome who display a generalized erythroderma covered by fine translucent scales. The course during infancy may be complicated by hypernatremic dehydration, hypothermia, failure to thrive, and enteropathy leading to death in infancy or childhood. Our TEWL measurements in an erythrodermic baby with Netherton's syndrome showed a fourfold enhancement in comparison to a control baby of the same ago. The TEWL measurements in these babies who seem to be at risk for developing dehydration are an important technique for a more accurate assessment of water loss (Fartasch et al., 1993, 1999).

In the elaborate study of Lavrijsen et al. (1993) on various keratinization disorders (ADI vulgaris, X-linked recessive ichthyosis, recessive congenital ichthyosis, dyskeratosis follicularis Darier, erythrokeratodermia variabilis), the measurement of water diffusion through the skin was performed by evaporimetry. This study additionally stressed the fact that these diseases show an accelerated penetration of hexyl nicotinate due to an impaired barrier of the skin. In Darier and erythokeratodermia variabilis only clearly affected skin showed increased TEWL, while in ADI, comparable to the noneczematous atopic skin, the TEWL values were significantly increased and penetration time accelerated on the volar site indicating an impaired barrier, despite the fact that some of the ADI patients showed normal-appearing skin on this location. These findings could be important for an occupational risk assessment and vocational guidance. An enhanced TEWL would simultaneously mean an easier and accelerated penetration of various chemical work-related substances. Consequently, direct skin contact with industrial chemicals can result in a varying extent of skin absorption and therefore in systemic uptake. Since ADI is a quite frequent disorder (1:250 to 320) (Wells and Kerr, 1965, 1966), this subgroup might be confronted with a higher percutaneous absorption, especially in occupations with chemical contact to large areas of the body.

REFERENCES

Abe, T., Okido, M., and Yamamoto, K., Studies on the skin surface barrier functions — skin surface lipids and transepidermal water loss in atopic skin during childhood, *J. Dermatol.,* 5, 223, 1978.

Agner, T., Susceptibility of atopic dermatitis patients to irritant dermatitis caused by sodium lauryl sulphate, *Acta Derm. Venereol. (Stockholm),* 71, 296, 1991.

Agner, T., Skin susceptibility in uninvolved skin of hand eczema patients and healthy controls, *Br. J. Dermatol.,* 125, 140, 1991.

Agner, T., Noninvasive measuring methods for the investigation of irritant patch test reactions: a study of patients with hand eczema, atopic dermatitis and controls, *Acta Derm. Venereol. (Stockholm),* 71 (Suppl.), 173, 1, 1992.

Al-Jaberi, H. and Marks, R., Studies of the clinically uninvolved skin in patients with dermatitis, *Br. J. Dermatol.,* 111, 437, 1984.

Bäurle, G., Hornstein, O. P., and Diepgen, T. L., Professionelle Handekzeme und Atopie. Eine klinische Prospektivstudie zur Frage des Zusammenhangs, *Dermatosen,* 33, 161, 1985.

Berardesca, E., Fideli, D., Bororni, G., Rabbiosi, G., and Maibach, H. I., *In vivo* hydration and water-retention capacity of stratum corneum in clinically uninvolved skin in atopic and psoriatic patients, *Acta Derm. Venereol. (Stockholm),* 70, 400, 1990.

Bleck, O. et al, Two ceramide subfractions detectable in Cer(AS) position by HPTLC in skin surface lipids of non-lesional skin of atopic eczema, *J. Invest. Dermatol.*, 113, 894, 1999.

Blichmann, C. and Serup, J., Hydration studies on scaly hand eczema. *Contact Dermatitis,* 16, 155, 1987.

Buyse, L., Graves, C., Marks, R., Wijeyesekera, K., Alfaham, M., and Finlay, A. Y., Collodion baby dehydration: the danger of high transepidermal water loss, *Br. J. Dermatol.,* 129, 86, 1993.

Di Nardo, A. et al., Ceramide and cholesterol composition of the skin of patients with atopic dermatitis, *Acta Derm. Venereol.*, 78, 27, 1998.

Diepgen, T. L. and Fartasch, M., Recent epidemiological and genetic studies in atopic dermatitis, *Acta Derm. Venereol. (Stockholm),* 176, 13, 1992.

Diepgen, T. L. and Fartasch, M., General aspects of risk factors in hand eczema, in *Hand Eczema,* Menné, T. and Maibach, H. I., Eds., CRC Press, Boca Raton, FL, 1994, chap. 16.

Fartasch, M., Ultrastructure of the epidermal barrier after irritation, *Microsc. Res. Tech.*, 37, 193–199, 1997a.

Fartasch, M., The epidermis barrier in disorders of the skin, *Microsc. Res. Tech.*, 38, 361–372, 1997b.

Fartasch, M., Bassukas, I. D., and Diepgen, T. L., Disturbed extruding mechanism of lamellar bodies in dry non-eczematous skin of atopics, *Br. J. Dermatol.,* 127, 221, 1992.

Fartasch, M., Bassukas, I. D., and Diepgen, T. L., Structural relationship between epidermal lipids, lamellar bodies and desmosomes in human epidermis: an ultrastructural study, *Br. J. Dermatol.,* 128, 1–9, 1993.

Fartasch, M., Diepgen, T. L., and Bassukas, I. D., Ultrastrukturelle Untersuchungen zur Funktion der Hornschichtbarriere an der nicht-ekzematösen Haut von Atopikern, *Allergologie,* 14, 315, 1991.

Fartasch, M., Diepgen, T. L., and Hornstein, O. P., Atopic dermatitis — ichthyosis vulgaris — hyperlinear palms — an ultrastructural study, *Dermatologica,* 178, 202, 1989.

Fartasch, M., Hüner, A., Tepe, A., Funke, U., and Diepgen, T. L., Hautphysiologische Untersuchungsmethoden in der Berufsdermatologie, *Allergologie,* 16, 25, 1993.

Fartasch, M., Schnetz, E., and Diepgen, T.L., Characterization of detergent-induced barrier alterations — effect of barrier cream on irritation, *J. Invest. Dermatol.*, Suppl. 3, S121–S127, 1998.

Fartasch, M., Williams, M. L., and Elias, P. M., Alterations in barrier formation and epidermal differentiation in Netherton's syndrome, Second Joint Meeting of the Society of Cutaneous Ultrastructural Research, NARA, Japan October 24–27, 1993. (Abstract)

Fartasch, M., Williams, M. L., and Elias, P. M., Altered lamellar body secretion and stratum corneum membrane structure in Netherton's syndrome: Differentiation from other infantile erythrodermas and pathogenic implications, *Arch. Dermatol.*, 135, 329–334, 1999.

Finlay, A. Y., Nicholls, S., King, C. S., and Marks, R., The dry non-eczematous skin associated with atopic eczema, *Br. J. Dermatol.,* 102, 249, 1980.

Frödin, T., Helander, P., Molin, L., and Skogh, M., Hydration of human stratum corneum studied *in vivo* by optothermal infrared spectrometry, electrical capacitance measurement, and evaporimetry, *Acta Derm. Venereol. (Stockholm),* 68, 461, 1988.

Frost, P., Weinstein, G. D., Bothwell, J. W., and Wildnauer, R., Ichthyosiform dermatosis. III. Studies on transepidermal water loss, *Arch. Dermatol.,* 98, 230, 1968.

Grice, K. A., Measurement of transepidermal water loss in pathological skin, in *The Physiology and Pathophysiology of the Skin,* Jarrett, A., Ed., Academic Press, New York, 1980, 247.

Grice, K. A. and Bettley, F. R., Skin water loss and accidental hypothermia in psoriasis, ichthyosis and erythrodermia, *Br. Med. J.,* 4, 195, 1967.

Grove, G. L., Proliferative response to fluid injection in psoriatic and normal human epidermis, *Br. J. Dermatol.,* 102, 281, 1980.

Hanifin, J. M. and Rajka, G., Diagnostic features of atopic dermatitis, *Acta Derm. Venereol. (Stockholm),* 92, 44, 1980.

Hara, J. et al., High-expression of sphingomyelin deacylase is an important determinant of ceramide deficiency leading to barrier disruption in atopic dermatitis, *J. Invest. Dermatol.* 115, 406, 2000.

Hou, S. Y. E., Mitra, A. K., White, S. H., Menon, G. K., Ghadially, R., and Elias, P. M., Membrane structures in normal and essential fatty acid-deficient stratum corneum: characterization by ruthenium tetroxide staining and X-ray diffraction, *J. Invest. Dermatol.,* 96, 215, 1991.

Imokawa, G. et al., Decreased level of ceramides in stratum corneum of atopic dermatitis: an etiologic factor in atopic dry skin? *J. Invest. Dermatol.*, 96, 523, 1991.

Imokawa, G., Abe, A., Jin, K., Higaki, Y., Kawashima, M., and Hidano, A., Decreased levels of ceramides in stratum corneum of atopic dermatitis: an etiologic factor in atopic dry skin? *J. Invest. Dermatol.*, 96, 523, 1991.

Kiistala, R., Lauharanta, J., and Kanvera, L., Transepidermal water loss and sweat gland response in lamellar ichthyosis before and during treatment with etretinate: report of three cases, *Acta Derm. Venereol. (Stockholm)*, 62, 268, 1981.

Kompaore, F., Dupont, C., and Marty, J. P., *In vivo* evaluation in man by two noninvasive methods of the stratum corneum barrier function after physical and chemical modifications, *Int. J. Cosmet. Sci.*, 13, 293, 1991.

Lammintausta, K. and Kalimo, K., Atopy and hand dermatitis in hospital wet work, *Contact Dermatitis*, 7, 301, 1981.

Lavrijsen, A. P. M., Oestmann, E., Hermans, J., Boddé, H. E., Vermeer, B. J., and Ponec, M., Barrier function parameters in various keratinization disorders: transepidermal water loss and vascular response to hexyl nicotinate, *Br. J. Dermatol.*, 129, 547, 1993.

Léveque, J. L., Grove, G., Rigal de, J. et al., Biophysical characterization of facial dry skin, *J. Soc. Cosmet. Chem.*, 82, 171, 1987.

Lever, W. F. and Schaumberg-Lever, G., Eds., *Histopathology of the Skin*, J.B. Lippincott Company, Philadelphia, 1990, 7.

Loffler, H. and Effendy, I., Skin susceptibility of atopic individuals, *Contact Dermatitis*, 40, 239, 1999.

Madison, K. C., Swartzendruber, D. C., Wertz, P. W., and Downing, D. T., Presence of intact intercellular lipid lamellae in the upper layers of the stratum corneum, *J. Invest. Dermatol.*, 88, 714, 1987.

Marks, R., Nicholls, S., and King, C. S., Studies on isolated corneocytes, *Int. J. Cosmet. Sci.*, 3, 251, 1981.

Marks, J., Rogers, S., Chadkirk, B., and Shuster, S., Clearance of chronic plaque psoriasis by anthralin-subjective and objective assessment and comparison with photochemotherapy, *Br. J. Dermatol.*, 105, 96, 1981.

Matsumoto, M. et al., Difference in ceramide composition between "dry" and "normal" skin in patients with atopic dermatitis, *Acta Derm. Venereol.*, 79, 246, 1999.

Melnik, B., Hollmann, J., Hofmann, U., Yuh, M.-S., and Plewig, G., Lipid composition of outer stratum corneum and nails in atopic and control subjects, *Arch. Dermatol. Res.*, 282, 549, 1990.

Menon, G. K. and Elias, P. M., Ultrastructural localization of calcium in psoriatic and normal human epidermis, *Arch. Dermatol.*, 127, 57, 1991.

Murahata, R. J., Crowe, D. M., and Roheim, J. R., The use of transepidermal water loss to measure and predict the irritation response to surfactants, *Int. J. Cosmet. Sci.*, 8, 225, 1986.

Murata, Y. et al., Abnormal expression of sphingomyelin acylase in atopic dermatitis: an etiologic factor for ceramide deficiency? *J. Invest. Dermatol.*, 106, 1242, 1996.

Nilsson, G. E., Mikaelsson, B., and Andersson, S., Atopy, occupation and domestic work as risk factors for hand eczema in hospital workers, *Contact Dermatitis*, 13, 216, 1985.

Oestmann, E., Lavrijsen, A. P. M., Hermans, J., and Ponec, M., Skin barrier function in healthy volunteers as assessed by transepidermal water loss and vascular response to hexyl nicotinate: intra- and interindividual variability, *Br. J. Dermatol.*, 128, 130, 1993.

Ohnishi, Y. et al., Ceramidase activity in bacterial skin flora as a possible cause of ceramide deficiency in atopic dermatitis, *Clin. Diagn. Lab. Immunol.*, 6, 101, 1999.

Piérard, G. E., What do you mean by dry skin?, *Dermatologica,* 179, 1, 1989.

Proksch, E., Elias, P. M., and Feingold, K. R., Regulation of 3-hydroxy-3-methyl-glutaryl-coenzyme a reductase activity in murine epidermis, *J. Clin. Invest.,* 85, 874, 1990.

Pinnagoda, J., Tupker, R. A., Coenraads, P. J., and Nater, J. P., Transepidermal water loss with and without sweat gland inactivation, *Contact Dermatitis,* 21, 16, 1989.

Rajka, G., *Atopic Dermatitis,* W.B. Saunders, London, 1975.

Rajka, G. and Thune, P., The relationship between the course of psoriasis and transepidermal water loss, photoelectric plethysmography and reflex photometry, *Br. J. Dermatol.,* 94, 253, 1976.

Rougier, A., Lotte, C., Corcuff, P., and Maibach, H. I., Relationship between skin permeability and corneocyte size according to anatomical site, age and sex in man, *J. Soc. Cosmet. Chem.,* 39, 15, 1988.

Rystedt, I., Hand eczema and long-term prognosis in atopic dermatitis (thesis), *Acta Derm. Venereol. (Stockholm),* 117, 1, 1985.

Rystedt, I., Factors influencing the occurrence of hand eczema in adults with a history of atopic dermatitis in childhood, *Contact Dermatitis,* 12, 185, 1985.

Sakurai, K. et al., Occurrence of patchy parakeratosis in normal-appearing skin in patients with active atopic dermatitis and in patients with healed atopic dermatitis: a cause of impaired barrier function of the atopic skin, *J. Dermatol. Sci.*, 30, 37, 2002.

Schäfer, L. and Kragballe, K., Abnormalities in epidermal lipid metabolism in patients with atopic dermatitis, *J. Invest. Dermatol.,* 96, 10, 1991.

Serup, J. and Blichmann, C., Epidermal hydration of psoriasis plaques and the relation to scaling, *Acta Derm. Venereol. (Stockholm),* 67, 357, 1987.

Shmunes, E. and Keil, J. E., The role of atopy in occupational dermatoses, *Contact Dermatitis,* 11, 174, 1984.

Swartzendruber, D. C., Wertz, P. W., Madison, K. C., and Downing, D. T., Evidence that the corneocyte has a chemically bound lipid envelope, *J. Invest. Dermatol.,* 88, 709, 1987.

Tagami, H. and Yoshikuni, K., Interrelationship between water-barrier and reservoir functions of pathologic stratum corneum, *Arch. Dermatol.,* 121, 642, 1985.

Thune, P., Evaluation of the hydration and the water-holding capacity in atopic skin and so-called dry skin, *Acta Derm. Venereol. (Stockholm),* 144 (Suppl.), 133, 1989.

Tupker, R. A., Coenraads, P. J., Pinnagoda, J., and Nater, J. P., Baseline transepidermal water loss (TEWL) as a prediction of susceptibility to sodium lauryl sulphate, *Contact Dermatitis,* 20, 265, 1989.

Tupker, R. A., Pinnagoda, J., Coenraads, P. J., and Nater, J. P., Susceptibility to irritants: role of barrier function, skin dryness and history of atopic dermatitis, *Br. J. Dermatol.,* 123, 199, 1990.

Tupker, R. A., Willis, C., Berrardesca, E., Lee, C. H., Fartasch, M., Agner, T., and Serup, J., Guidelines on sodium lauryl sulfate (SLS) exposure test, *Contact Dermatitis*, 37, 53–69, 1997.

Uehara, M. and Harumitsu, M., The morphologic characteristics of dry skin in atopic dermatitis, *Arch. Dermatol.,* 120, 1186, 1984.

Uehara, M. and Hayashi, S., Hyperlinear palms, *Arch. Dermatol.,* 117, 490, 1981.

Uehara, M. and Miyauchi, H., The morphologic characteristics of dry skin in atopic dermatitis, *Arch. Dermatol.,* 120, 1186, 1984.

van der Valk, P. G. M., Nater, J. P., and Bleumink, E., Vulnerability of the skin to surfactants in different groups of eczema patients and controls as measured by water vapour loss, *Clin. Exp. Dermatol.*, 10, 98, 1985.

Van Neste, D., Lyonnet, S., Brouwer de, B., and Thivolet, J., Noninvasive *in vivo* evaluation of biophysical parameters of atopic dermatitis, in *Immunological and Pharmacological Aspects of Atopic and Contact Eczema,* Czernielewski, J. M., ed., S. Karger, Basel, 1991, 128.

Watanabe, M., Tagami, T., Horii, I., Takahashi, M., and Kligman, A. M., Functional analyses of the superficial stratum corneum in atopic xerosis, *Arch. Dermatol.*, 127, 1689, 1991.

Wells, R. S. and Kerr, C. B., Genetic classification of ichthyosis, *Arch. Dermatol.*, 92, 1, 1965.

Wells, R. S. and Kerr, C. B., Clinical feature of autosomal dominant and sex-linked ichthyosis in an English population, *Br. Med. J.*, 1, 947, 1966.

Werner Linde, Y., Dry skin in atopic dermatitis, *Acta Derm. Venereol. (Stockholm),* Suppl. 177, 9, 1992.

Werner, Y. and Lindberg, M., Transepidermal water loss in dry and clinically normal skin in patients with atopic dermatitis, *Acta Derm. Venereol. (Stockholm),* 65, 102, 1985.

Wilhelm, K. P. and Maibach, H. I., Factors predisposing to cutaneous irritation, *Dermatol. Clin.,* 8, 17, 1990.

Yamamoto, A., Serizawa, M., Ito, M., and Sato, Y., Stratum corneum lipid abnormalities in atopic dermatitis, *Arch. Dermatol. Res.*, 283, 219, 1991.

15 Transepidermal Water Loss and Dry Skin

Marie Lodén

CONTENTS

I. Introduction 171
II. Characteristics of Dry Skin 172
  A. Clinical Symptoms 172
  B. Chemical Analysis 172
  C. Morphology and Function 173
III. Dry Skin and Defects in the Barrier Function 174
  A. Genetic Factors 174
  B. Environmental and Behavioral Factors 175
IV. Techniques for Barrier Function Assessment 176
V. Treatment Effects 176
VI. Conclusion 180
References 180

I. INTRODUCTION

The outer layer of the skin, the stratum corneum (SC), is produced by the basal layer in the epidermis. The SC consists of about 20 stacked layers of dry, flattened dead bodies of epidermal cells; the corneocytes. The protein-enriched corneocytes and the lipid-enriched intercellular domains make SC highly resistant to physical and chemical trauma.

The mechanical strength of the corneocytes derives from the tightly packed keratin bundles and the cross-linked proteins of the cornified envelopes. The corneocyte envelope constitutes a backbone for the intercellular barrier lipids, composed of cholesterol, free fatty acids and ceramides.[1] The content and organization of these lipids have broad implications for water retention, the permeability barrier function, and desquamation.

SC acts as a biosensor and regulates the metabolic response to a variety of stimuli.[2] Dryness is one reaction pattern to environmental stimuli and to internal abnormalities in the epidermis. Dryness is often related to impairment in the skin barrier function.[3–9] A defective barrier function promotes development of dermatitis by allowing entry of eczema-promoting substances and by allowing increased

0-8493-1443-7/05/$0.00+$1.50

transepidermal water loss (TEWL).[2] Increased TEWL can have pathological effects by overstimulating cytokines, which may result in cutaneous abnormalities.[2] Measurement of TEWL gives information about the function of SC as a diffusion barrier to water and may also serve as an indicator of the permeability of the skin to topically applied substances.[10,11]

II. CHARACTERISTICS OF DRY SKIN

A. Clinical Symptoms

There are several characteristics that give an impression of dry skin (Table 15.1). The degree of scaling, roughness, redness, and cracks (fissures) can be estimated by trained evaluators. One system includes measurement of the area involved and scoring of the intensity of the dryness signs on a five-point category scale.[12]

Sensory experiences in the skin can also be taken into account by the affected person. Sensory experiences are, for instance, that the skin feels tight, uncomfortable, painful, or itchy, and that it stings or tingles. The two systems may give different results, and more people will usually experience their skin as dry than are judged by the dermatologist to have dry skin.[13]

B. Chemical Analysis

The subjective appearance of dryness can be verified chemically. For example, in elderly patients with xerosis, the water content in SC is reduced,[14] and in winter in xerotic skin the water content is inversely related to clinical scores of dryness.[15] Furthermore, the dry-looking skin of patients with atopic dermatitis and psoriasis is less hydrated and less capable of binding water than normal skin.[6,14,16,17]

Not only water, but also the level of the osmotically active molecules, called natural moisturizing factor (NMF), may be decreased in dry skin. Urea, lactic acid, pyrrolidone carboxylic acid, and amino acids belong to the NMF. In ichthyosis vulgaris[14,18] and in psoriasis,[19] there is a virtual absence of NMF. In addition, the composition and organization of the intercellular lipids may be changed in dry skin.

TABLE 15.1
Dry Skin Can Display Several Characteristics

Analysis	Characteristics
Visual	Redness, a lackluster surface, dry white patches, flaky appearance, cracks, and even fissures
Tactile	Rough and uneven
Sensory	Feels dry, uncomfortable, painful, itchy, stings, and tingles
Chemical	Reduced water content, changed content of natural moisturizing factors and lipids
Morphological	Hyperkeratosis, decreased projected size of corneocytes area
Functional	Impaired barrier function to water and exogenous substances

The lipid membranes are organized in two crystalline lamellar phases with periodicities of approximately 6 and 13 nm.[20,21] In dry skin and in skin exposed to organic solvents, the composition and the structure of this bilayer are changed.[22–26]

Removal of the intercellular lipids will allow the NMF to easier leave the corneocytes and contribute to the appearance of dryness,[24,27,28] As early as in 1968, Middleton showed that powdering the SC destroyed the lipid membranes and made the skin more susceptible to drying-out effects.[28]

C. Morphology and Function

Dryness disorders, such as atopic eczema,[3,6,7] psoriasis,[29] and ichthyosis, are characterized by hyperkeratosis and scaliness. The elasticity of SC is often reduced, and the cohesive forces between the cells are increased.[7] These changes may explain why cracking and fissuring may be found in such conditions. Furthermore, the projected size of the corneocytes is smaller and the turnover time is shorter than in controls.[7,30]

The SC is the rate-limiting barrier to penetration of substances through skin. Although the thickness of SC is increased in hyperkeratotic skin, the rate of penetration is higher in dry atopic,[3,5–7] psoriatic,[8,29] and ichthyotic skin[25,31–33] than in normal SC. This increase in permeability may be the result of the smaller projected size of the corneocytes in hyperkeratotic skin than in normal skin. The major route of penetration of substances through the SC is considered to be the intercellular pathway. This highly convoluted and tortuous lipid pathway around the corneocytes will result in a distance for penetration longer than the actual thickness of the SC (Figure 15.1).[34,35] A decreased size of the projected area of the corneocytes reduces the tortuous permeability pathway and may be one of several reasons for the greater permeability in hyperproliferative diseases (Figure 15.1).[35,36] Another, probably more important, factor for restricting the penetration is the composition and organization of the lipids within the intercellular domains.[37]

A defect function of the intercellular lipid membranes may be compensated by an increased thickness of the SC. Hence, hyperkeratosis might indicate a failure of epidermis to produce a competent barrier with normal thickness. This may explain why TEWL is normal in certain dry skin conditions.[15] However, hyperkeratosis may also reflect an undesired inhibition of the desquamation process.

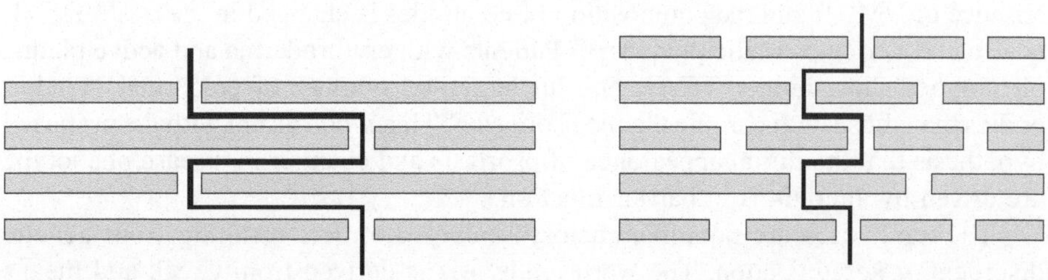

FIGURE 15.1 The tortuous pathway for penetration results in a longer penetration distance in normal skin (to the left) than in hyperkeratotic skin (right), with a smaller projected corneocyte size.

TABLE 15.2
Important Causes of Dry Skin

Factor	Characteristics
Genetic	Inherited disorders relating to the structure and function of the epidermis; for example, atopic dermatitis, ichthyosis
Environmental	Low humidity, low temperature
Behavioral	Exposure to solvents, cutting fluids, surfactants, acids, alkali, and so forth

III. DRY SKIN AND DEFECTS IN THE BARRIER FUNCTION

A. Genetic Factors

Dryness and impairment in the barrier function characterize several chronic disorders, such as atopic eczema and ichthyosis (Table 15.2). Furthermore, the repair process following acute deterioration of the barrier function can be delayed by normal aging and psychological stress in healthy individuals.[38,39]

Atopic eczema is one of the most common chronic inflammatory skin diseases, characterized by dry and itchy skin. Atopic eczema has a complex pathophysiology, including interactions between genetic predisposition and exogenous provocation factors. Its prevalence is increasing, and among school children in North Europe, values between 15 and 20% are reported.[40,41] The increasing prevalence during a rather short period of time suggests that there are factors in the environment that promote the rise, rather than genetic changes.[42] Higher TEWL is noted in dry skin in atopic patients,[5–7,16,43,44] whereas in completely healed atopic dermatitis, the barrier function is not disturbed.[45] Analyses of skin lipid composition show reduced amounts of ceramides along with a changed distribution of the different types.[26,46–48] The changed lipid composition may account for the aberrant lipid organization in atopic skin, with an increased frequency of hexagonal packaging.[49]

Psoriasis is another common inflammatory disorder with a prevalence of 2 to 3% in the general population. The disease is characterized by increased cell production with incomplete differentiation and parakeratotic plaques. There is a virtual absence of NMF,[19] and the composition of ceramides is changed in the SC.[50] TEWL is also elevated in psoriatic plaques.[8,29] Patients with erythroderma and active plaque phenotypes show higher TEWL and an increased number of epidermal lamellar bodies than those with chronic plaque psoriasis.[29] This is consistent with the proposed hypothesis that the initial appearance of psoriasis and changes in disease phenotype are driven by alterations in barrier function.[29]

The ichthyoses are noninflammatory scaling disorders resulting from genetic disorders of keratinization. The word ichthyosis is derived from Greek and means literally a fishy (ichthy-) condition (-osis). The disease vaguely resembles the scales or other surface characteristics of a fish. The abnormal formation and desquamation of the SC produce a persistently dry and scaly skin.

The most common form of ichthyosis is autosomal dominant ichthyosis vulgaris, which occurs at a frequency of about one in 300 people.[51] Ichthyosis vulgaris is a retention hyperkeratosis with normal rates of epidermal proliferation. The disease displays a relatively mild clinical spectrum and is usually not a major problem to the patient in a humid climate, whereas in a dry and cold climate, xerosis and hyperkeratosis may become severe. Scaling usually does not appear until late infancy or early childhood. In scaling, the stratum granulosum is thin or missing because of a defect in the processing of profilaggrin, which defect also is observed as tiny and crumbly keratohyalin granules.[51] This defect results in a deficiency of NMF.[18] TEWL is increased, and the barrier function to penetration of hexyl nicotinate is impaired.[32]

X-linked recessive ichthyosis occurs at a frequency of about one in 2500 people.[51] In this type of ichthyosis, the level of cholesterol sulfate is greatly increased because of a deficiency in steroid sulfatase, which hydrolyzes the ester.[52,53] The accumulation of cholesterol sulfate is supposed to alter the organization of the lamellar membranes, which will impair the barrier function.[54] In addition, cholesterol sulfate inhibits proteases, which delays dissolution of corneodesmosomes, thereby preventing desquamation.[55] Desquamation is linked to the conversion of cholesterol sulfate to free cholesterol.[56] Patients display both an abnormal barrier function under basal conditions and a delay in barrier recovery after acute perturbation, which correlate with minor abnormalities in membrane structure and extensive lamellar-phase separation.[32,54]

Lamellar ichthyosis (nonbullous ichthyosiform erythroderma) is rarer, with an incidence of around one in 100,000 people.[51] TEWL is increased,[25] which has been suggested to be the result of discontinuities in the extracellular membrane structure.[31] The hexagonal packing of the lipids was predominantly present, whereas the less permeable orthorhombic packing is observed only occasionally.[49] Crystalline cholesterol has been proposed to be present in the SC.[25]

B. Environmental and Behavioral Factors

Skin dryness and impaired barrier function can also be induced by factors in the environment, such as low temperature and low humidity (Table 15.2).[57,58] Already a short exposure to a low-humidity environment reduces the moisture content in human skin and induces changes in the skin surface pattern.[59] Expression of cytokines has been found in mice, indicating induction of inflammatory molecules.[58]

Exposure to solvents, such as water, may also cause barrier perturbations. Occlusion of the skin increases TEWL.[60] Water not only acts as a solvent but may also contain deleterious substances. For example, exposure to hard water, especially to calcium in domestic water, has been found to be associated with a higher prevalence of atopic eczema in primary-school children.[61] Moreover, irritant contact dermatitis is often the result of the continuous use of detergents by direct application to the skin surface. The hallmark of soap damage is "chapping," with a rough, dry, and cracked surface that is most evident on the hands, where the SC layer is thick.[62] The combination of surfactants and their inherent irritation potential contribute to differences in injuring effect among cleansers.[62,63] Other cosmetics, such as creams and mascara, can also contain substances that may cause dryness and increase TEWL.[64]

In addition, microorganisms present in dry skin in atopics may cause eczema by release of superantigenic exotoxins, which produce a massive release of cytokines.[65,66]

Acute and chronic perturbations of barrier function may lead to cutaneous inflammation and epidermal hyperplasia by increased production and secretion of cytokines.[2,67–69]

IV. TECHNIQUES FOR BARRIER FUNCTION ASSESSMENT

Noninvasive bioengineering techniques can be used to evaluate effects on skin barrier function.[70] Quantification of TEWL is a useful tool for monitoring the kinetics in the repair of a deteriorated barrier function. It has been suggested that the level of TEWL serves as an indicator of the permeability of the skin to topically applied substances.[10,11]

TEWL can be measured conveniently with an open-chamber evaporation gradient method.[71] Important factors to consider during TEWL measurements are room temperature and ambient humidity.[71,72] Sweating must be avoided, and a room temperature of 20° to 22°C is, therefore, recommended.[72] Furthermore, air convection in the room may disturb the readings, and some form of draught shield may be useful.[71,72] The rest time before TEWL readings also has to be considered. Moreover, volatile agents other than water might influence the readings if measurements are made immediately after application of moisturizers.[73] Application of aqueous formulations results in an immediate increase in the values, whereas application of water-free petrolatum will significantly reduce the values because of occlusion.[74]

However, TEWL may not necessarily reflect permeability to substances other than water. Therefore, evaluation of skin barrier function can also be made by the application of substances that cause biological responses.[75–78] Substances used to assess skin permeability are those inducing vasodilatation (e.g., nicotinates), irritation (surfactants such as sodium lauryl sulfate [SLS]), erosion (sodium hydroxide), whealing and flare (dimethyl sulfoxide), burning (chloroform:methanol), and stinging (lactic acid).[75,76]

V. TREATMENT EFFECTS

Symptoms of dryness may be diminished by treatment with emollients or moisturizing creams. Skin hydration will increase immediately after application of moisturizers to the skin.[74] Water will be absorbed into the skin, and lipid substances will occlude the skin and cause a build-up of water in the SC.[74]

Lipids are often used in products on the basis of their surface emolliency and occluding properties. Lipids may also exert deeper effects in the skin. For example, petrolatum has been shown to penetrate into the outer layer of delipidized SC and to reduce TEWL.[79] The addition of bath oils to water has also been shown to reduce TEWL in perturbed skin.[80] Topically applied lipids have been suggested to reduce

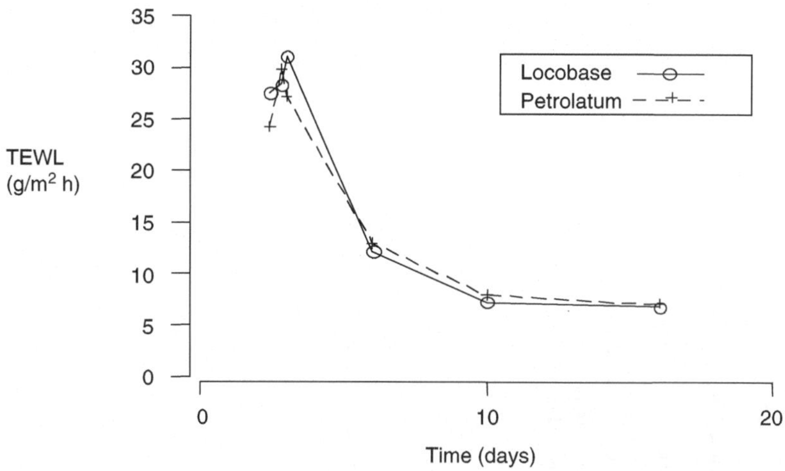

FIGURE 15.2 In surfactant-damaged skin, the rate of skin barrier recovery was almost the same during treatment with physiological lipids with ceramide 3 (Locobase® Repair) as with treatment with pure petrolatum. Modified from Reference 89.

the mitotic activity and to increase cell differentiation.[81] This application will result in a larger corneocyte area and in reduced skin permeability.[34,35]

In experimental models of dryness, moisturizers usually promote normalization of the skin.[82–84] The models used include barrier damage by successive tape strippings or by exposure to acetone or SLS. However, the treated skin abnormality and the composition of the cream may be crucial for the effect.[23,82,85–87] The ratios of physiological lipids have been suggested to be important.[87] One physiological lipid mixture was found to decrease TEWL and improve atopic dermatitis in an open study in children.[88] In addition, in a double-blind study, a commercially available physiological lipid mixture (containing ceramide 3) was found to promote barrier recovery in SLS-irritated and tape-stripped human skin, compared with the untreated control area.[84] However, the barrier recovery was not superior to its placebo (petrolatum) (Figure 15.2).[84,89]

Clinical studies in patients with xerosis also support the beneficial effect achieved by treatment with moisturizers (Table 15.3). For example, cleaners and kitchen workers show normalization of the skin texture during use of an emollient, whereas dryness increased during a nontreatment period.[90] However, the elevated TEWL was not normalized.[90]

Enhancement of the clinical appearance of dryness has often been found after inclusion of humectants in the moisturizer (Table 15.3). Only a few studies have monitored TEWL, but formulations with urea have shown barrier-improving effects.[77,91] One moisturizer with 5% urea reduced TEWL in atopic patients[92] and made skin less susceptible against irritation to SLS.[78] Another urea moisturizer was also superior to a glycerin moisturizer in lowering TEWL in recent double-blind study on atopic patients.[93] Furthermore, one urea-containing moisturizer reduced TEWL in ichthyosis,[94,95] whereas a lactic acid and propylene glycol–containing moisturizer actually increased TEWL in ichthyotic skin.[33] A moisturizer containing

TABLE 15.3
Effect of Moisturizer on Transepidermal Water Loss (TEWL) in Dry Skin Diseases

Active Substance	Control	Condition	Design/No. Patients	TEWL/Susceptibility	Effect on Dryness	Reference
12% ammonium lactate	No therapy	Dry skin including atopics	Open/24	No change in TEWL	Improved	96
15% glycolic acid	Water-in-oil emulsion	Dry xerotic legs	? bilateral/14	Increased TEWL, increased susceptibility in active, decreased in control	Both improved	76
Borage oil	Normal children	Infantile seborrheic dermatitis	Open/37	Decrease	Improved	101
3% and 10% urea	Untreated	Dry skin	Evaluator blind, bilateral/47	Decrease	Improved	91
10% urea cream	Essex base cream	Atopic skin	Open/40	Decrease	Improved, active better	102
Urea 4%	Glycerin 20%	Atopic skin	Parallel, double blind/197	Lower in urea group	Improved, active more	93
Urea 5%	Untreated	Atopic skin	Evaluator blind, bilateral/15	Improved barrier	Not evaluated	78
Ceramide-dominant barrier repair lipids	Untreated	Atopic children	Open/24	Decreased	Improved	88
Urea 10% + lactic acid 5% + betaine	Placebo	Ichthyosis	Bilateral/14	Lower in urea areas	Improved, active more	94
5% lactic acid + 20% propylene glycol in Locobase	5% urea in Locobase, 5% lactic acid + 20% propylene glycol in Essex cream, 20% propylene glycol in Locobase	Ichthyosis	Double blind, within patients/20	Increased	All improved but the active tended to be better	33

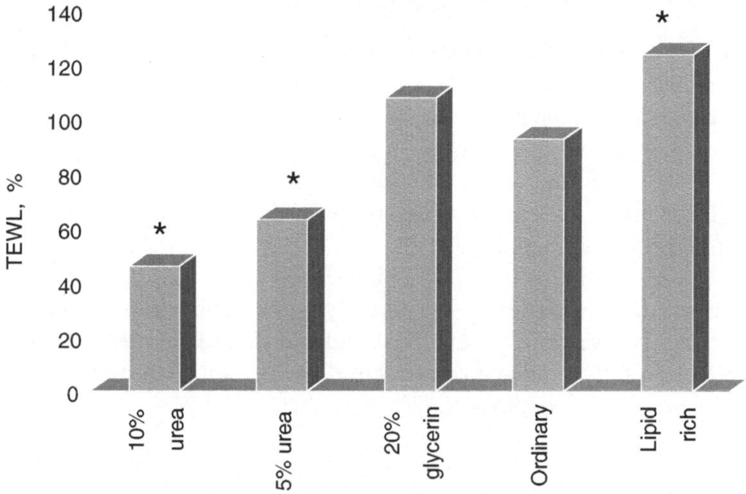

FIGURE 15.3 Skin susceptibility to irritation, measured as transepidermal water loss, after treatment of normal skin with different moisturizers. The value in the treated area is given in percent of the value in the control area. The graphs are compiled from data in the literature where similar experimental design has been used. Skin treated with 10%[77] and 5%[97] urea decreased skin susceptibility to sodium lauryl sulfate, whereas the moisturizer with 20% glycerin[103] and the ordinary[77] cream had no effect. The lipid-rich cream[98] increased the susceptibility. The asterisks show significant differences between treated and untreated area, $P < .05$.

ammonium lactate as humectant had no effect on TEWL in one study on dry skin in atopic and nonatopic individuals, although clinical appearance improved.[96] In addition, treatment of xerotic legs in the elderly with a lotion containing 15% glycolic acid increased both TEWL and the susceptibility to topically applied irritants.[76]

Treatment with moisturizers may also influence the barrier properties of normal skin. Repeated applications of urea-containing moisturizers (twice daily for 10 to 20 days) reduced TEWL and made skin less susceptible to SLS-induced irritation (Figure 15.3).[77,91,97] A lipid-rich cream without any humectant had no significant influence on TEWL, but increased skin susceptibility to SLS irritation compared with the susceptibility of untreated skin (Figure 15.3).[98] Increased skin reactivity was also found in a long-term study using benzyl nicotinate as a marker for permeability, where the time to maximum response was shorter for the cream-treated area compared with that of the untreated area.[99] In addition, the time to induce vasodilatation was shorter for the lipid-rich cream than for a moisturizer containing 5% urea.[99] Increased sensitivity to nickel was also found when nickel-sensitive participants treated their skin with moisturizers lacking humectant, compared with those who treated their skin with moisturizer containing humectant.[100] Areas treated with the glycerol-containing cream showed less reactivity to nickel than those treated with a cream without any humectant.[100]

VI. CONCLUSION

It has been hypothesized that permeability barrier abnormalities drive disease activity in inflammatory dermatoses, such as atopic dermatitis.[2] To relieve symptoms of dryness, emollients and moisturizers are frequently used. However, formulations may fail to strengthen the barrier function instead of lowering an increased TEWL.[23,33,76,90] Such formulations may sustain or even aggravate an existing barrier disease.[88] In addition, normal skin may react differently to environmental stimuli, depending on previous treatment.[77,98–100]

An increased understanding of skin biochemistry will facilitate the development of moisturizers that can address the underlying SC abnormality. The biological perspective can be even further explored and transferred into the development of moisturizers. Carefully conducted clinical trials will establish the usefulness of these moisturizers in specific conditions.

REFERENCES

1. Downing, D.T., Lipid and protein structures in the permeability barrier, in *Dry Skin and Moisturizers: Chemistry and Function*, Lodén, M. and Maibach, H.I., Eds., CRC Press, Boca Raton, Florida, 2000, pp. 59.
2. Elias, P.M., Wood, L.C., and Feingold, K.R., Epidermal pathogenesis of inflammatory dermatoses, *Am. J. Contact. Dermat.*, 10, 119, 1999.
3. Lodén, M. et al., Friction, capacitance and transepidermal water loss (TEWL) in dry atopic and normal skin, *Br. J. Dermatol.*, 126, 137, 1992.
4. Denda, M. et al., Stratum corneum sphingolipids and free amino acids in experimentally-induced scaly skin, *Arch. Dermatol. Res.* 284, 363, 1992.
5. Werner, Y. and Lindberg, M., Transepidermal water loss in dry and clinically normal skin in patients with atopic dermatitis, *Acta Derm. Venereol.*, 65, 102, 1985.
6. Thune, P., Evaluation of the hydration and the water-holding capacity in atopic skin and so-called dry skin, *Acta Derm. Venereol. (Stockh.)*, 144 (Suppl.), 133, 1989.
7. Finlay, A.Y. et al., The "dry" non-eczematous skin associated with atopic eczema, *Br. J. Dermatol.*, 103, 249, 1980.
8. Motta, S. et al., Abnormality of water barrier function in psoriasis, *Arch. Dermatol.*, 130, 452, 1994.
9. Takenouchi, M., Suzuki, H., and Tagami, H., Hydration characteristics of pathologic stratum corneum-evaluation of bound water, *J. Invest. Dermatol.*, 87, 574, 1986.
10. Aalto-Korte, K. and Turpeinen, M., Transepidermal water loss and absorption of hydrocortisone in widespread dermatitis, *Br. J. Dermatol.*, 128, 633, 1993.
11. Dupuis, D. et al., *In vivo* relationship between percutaneous absorption and transepidermal water loss according to anatomic site in man., *J. Soc. Cosmetic Chem.*, 37, 351, 1986.
12. Serup, J., EEMCO guidance for the clinical assessment of dry skin (xerosis) and ichthyosis: clinical scoring systems., *Skin Res. Technol.*, 1, 109, 1995.
13. Linde, Y. W., "Dry" skin in atopic dermatitis. A clinical study, *Acta Derm. Venereol. (Stockh.).*, 69, 311, 1989.
14. Horii, I. et al., Stratum corneum hydration and amino acid content in xerotic skin, *Br. J. Dermatol.*, 121, 587, 1989.

15. Leveque, J.L. et al., Biophysical characterization of dry facial skin, *J. Soc. Cosmetic Chem.*, 82, 171, 1987.
16. Berardesca, E. et al., *In vivo* hydration and water-retention capacity of stratum corneum in clinically uninvolved skin in atopic and psoriatic patients, *Acta Derm. Venereol.*, 70, 400, 1990.
17. Tagami, H. et al., Water sorption-desorption test of the skin *in vivo* for functional assessment of the stratum corneum, *J. Invest. Dermatol.*, 78, 425, 1982.
18. Sybert, V.P., Dale, B.A., and Holbrook, K.A., Ichthyosis vulgaris: identification of a defect in filaggrin synthesis correlated with an absence of keratohyaline granules, *J. Invest. Dermatol.*, 84, 191, 1985.
19. Marstein, S., Jellum, E., and Eldjarn, L., The concentration of pyroglutamic acid (2-pyrrolidone-5-carboxylic acid) in normal and psoriatic epidermis, determined on a microgram scale by gas chromatography, *Clin. Chim. Acta*, 43, 389, 1973.
20. White, S.H., Mirejovsky, D., and King, G.I., Structure of lamellar lipid domains and corneocyte envelopes of murine stratum corneum. An X-ray diffraction study, *Biochemistry*, 27, 3725, 1988.
21. Bouwstra, J.A., et al., The structure of human stratum corneum as determined by small angle X-ray scattering, *J. Invest. Dermatol.*, 96, 1006, 1991.
22. Imokawa, G. and Hattori, M., A possible function of structural lipids in the water-holding properties of the stratum corneum, *J. Invest. Dermatol.*, 84, 282, 1985.
23. Man, M.Q., Feingold, K.R., and Elias, P.M., Exogenous lipids influence permeability barrier recovery in acetone-treated murine skin, *Arch. Dermatol.*, 129, 728, 1993.
24. Imokawa, G., Kuno, H., and Kawai, M., Stratum corneum lipids serve as a bound-water modulator, *J. Invest. Dermatol.*, 96, 845, 1991.
25. Lavrijsen, A.P. et al., Reduced skin barrier function parallels abnormal stratum corneum lipid organization in patients with lamellar ichthyosis, *J. Invest. Dermatol.*, 105, 619, 1995.
26. Melnik, B. et al., Lipid composition of outer stratum corneum and nails in atopic and control subjects, *Arch. Dermatol. Res.*, 282, 549, 1990.
27. Laden, K., Natural moisturization factors in skin, *Am. Perfum. Cosmet.*, 82, 77, 1967.
28. Middleton, J.D., The mechanism of water binding in stratum corneum, *Br. J. Dermatol.*, 80, 437, 1968.
29. Ghadially, R., Reed, J.T., and Elias, P.M., Stratum corneum structure and function correlates with phenotype in psoriasis, *J. Invest. Dermatol.*, 107, 558, 1996.
30. Watanabe, M. et al., Functional analyses of the superficial stratum corneum in atopic dermtitis, *Arch. Dermatol.*, 127, 1689, 1991.
31. Elias, P.M. et al., Basis for the permeability barrier abnormality in lamellar ichthyosis, *Exp. Dermatol.*, 11, 248, 2002.
32. Lavrijsen, A.P. et al., Barrier function parameters in various keratinization disorders: transepidermal water loss and vascular response to hexyl nicotinate, *Br. J. Dermatol.*, 129, 547, 1993.
33. Gånemo, A., Virtanen, M., and Vahlquist, A., Improved topical treatment of lamellar ichthyosis: a double blind study of four different cream formulations, *Br. J. Dermatol.*, 141, 1027, 1999.
34. Potts, R.O. and Francoeur, M.L., The influence of stratum corneum morphology on water permeability, *J. Invest. Dermatol.*, 96, 495, 1991.
35. Rougier, A. et al., Relationship between skin permeability and corneocyte size according to anatomic site, age, and sex in man, *J. Soc. Cosmetic Chem.*, 39, 15, 1988.

36. Grove, G.L. and Kligman, A.M., Corneocytes size as an indirect measure of epidermal proliferative activity, in *Stratum Corneum*, Marks, R. and Plewig, G., Eds., Springer, New York, 1983, pp. 191.
37. Elias, P.M. and Menon, G.K., Structural and lipid biochemical correlates of the epidermal permeability barrier, *Adv. Lipid Res.*, 24, 1, 1991.
38. Ghadially, R. et al., The aged epidermal permeability barrier. Structural, functional, and lipid biochemical abnormalities in humans and a senescent murine model, *J. Clin. Invest.*, 95, 2281, 1995.
39. Garg, A. et al., Psychological stress perturbs epidermal permeability barrier homeostasis: implications for the pathogenesis of stress-associated skin disorders, *Arch. Dermatol.*, 137, 53, 2001.
40. Broberg, A. et al., Atopic dermatitis in 5-6-year-old Swedish children: cumulative incidence, point prevalence, and severity scoring, *Allergy*, 55, 1025, 2000.
41. Larsen, F.S., Diepgen, T., and Svensson, Å., The occurrence of atopic dermatitis in North Europe: an international questionnaire study, *J. Am. Acad. Dermatol.*, 34, 760, 1996.
42. Taieb, A., Hypothesis: from epidermal barrier dysfunction to atopic disorders, *Contact Dermatitis*, 41, 177, 1999.
43. Berardesca, E. and Maibach, H.I., Transepidermal water loss and skin surface hydration in the non invasive assessment of stratum corneum function, *Derm. Beruf. Umwelt*, 38, 50, 1990.
44. Loden, M. et al., Friction, capacitance and transepidermal water loss (TEWL) in dry atopic and normal skin, *Br. J. Dermatol.*, 126, 137, 1992.
45. Matsumoto, M., Sugiura, H., and Uehara, M., Skin barrier function in patients with completely healed atopic dermatitis, *J. Derm. Sci.*, 23, 178, 2000.
46. Imokawa, G. et al., Decreased level of ceramides in stratum corneum of atopic dermatitis: an etiologic factor in atopic dry skin? *J. Invest. Dermatol.*, 96, 523, 1991.
47. Matsumoto, M. et al., Difference in ceramide composition between "dry" and "normal" skin in patients with atopic dermatitis, *Acta Derm. Venereol.*, 79, 246, 1999.
48. Di Nardo, A., Ceramide and cholesterol composition of the skin of patients with atopic dermatitis, *Acta Derm. Venereol.*, 78, 27, 1998.
49. Pilgram, G.S. et al., Aberrant lipid organization in stratum corneum of patients with atopic dermatitis and lamellar ichthyosis, *J. Invest. Dermatol.*, 117, 710, 2001.
50. Motta, S. et al., Content of the different lipid classes in psoriatic scale, *Arch. Dermatol. Res.*, 287, 691, 1995.
51. Vahlquist, A., Ichthyosis — an inborn dryness of the skin, in *Dry Skin and Moisturizers: Chemistry and Function*, Lodén, M. and Maibach, H.I., Eds., CRC Press, Boca Raton, Florida, 2000, pp. 121.
52. Williams, M.L. et al., Ichthyosis induced by cholesterol-lowering drugs. Implications for epidermal cholesterol homeostasis, *Arch. Dermatol.*, 123, 1535, 1987.
53. Shapiro, L.J. et al., Enzymatic basis of typical x-linked ichthyosis, *Lancet*, 2, 756, 1978.
54. Zettersten, E. et al., Recessive x-linked ichthyosis: role of cholesterol-sulfate accumulation in the barrier abnormality, *J. Invest. Dermatol.*, 111, 784, 1998.
55. Sato, J. et al., Cholesterol sulfate inhibits proteases that are involved in desquamation of stratum corneum, *J. Invest. Dermatol.*, 111, 189, 1998.
56. Ranasinghe, A.W. et al., Lipid composition of cohesive and desquamated corneocytes from mouse ear skin, *J. Invest. Dermatol.* 86, 187, 1986.

57. Morris-Jones, R. et al., Dermatitis caused by physical irritants, *Br. J. Dermatol.* 147, 270, 2002.
58. Ashida, Y., Ogo, M., and Denda, M., Epidermal interleukin-1 alpha generation is amplified at low humidity: implications for the pathogenesis of inflammatory dermatoses, *Br. J. Dermatol.*, 144, 238, 2001.
59. Egawa, M. et al., Effect of exposure of human skin to a dry environment, *Skin Res. Technol.*, 8, 212, 2002.
60. Fluhr, J.W. et al., Effects of prolonged occlusion on stratum corneum barrier function and water holding capacity, *Skin Pharmacol. Appl. Skin Physiol.*, 12, 193, 1999.
61. McNally, N.J. et al., Atopic eczema and domestic water hardness, *Lancet*, 352, 527, 1998.
62. Frosch, P.J. and Kligman, A.M., The soap chamber test. A new method for assessing the irritancy of soaps, *J. Am. Acad. Dermatol.*, 1, 35, 1979.
63. Barany, E., Lindberg, M., and Lodén, M., Biophysical characterization of skin damage and recovery after exposure to different surfactants, *Contact Dermatitis*, 40, 98, 1999.
64. Lodén, M. and Wessman, C., Mascaras may cause contact dermatitis, *Int. J. Cosmetic Sci.*, 24, 1, 2002.
65. Cork, M.J., The role of *Staphylococcus aureus* in atopic eczema: treatment strategies, *J. Eur. Acad. Dermatol. Venereol.*, 7, S31, 1996.
66. Strange, P. et al., Staphylococal enterotoxin B applied on intact normal and intact atopic skin induces dermatitis, *Arch. Dermatol.*, 132, 27, 1996.
67. Wood, L.C. et al., Cutaneous barrier perturbation stimulates cytokine production in the epidermis of mice, *J. Clin. Invest.*, 90, 482, 1992.
68. Proksch, E. et al., Barrier function regulates epidermal DNA synthesis, *J. Clin. Invest.*, 87, 1668, 1991.
69. Denda, M. et al., The epidermal hyperplasia associated with repeated barrier disruption by acetone treatment or tape stripping cannot be attributed to increased water loss, *Arch. Dermatol. Res.*, 288, 230, 1996.
70. Serup, J. and Jemec, G.B.E., *Handbook of Non-Invasive Methods and the Skin*, CRC Press, Boca Raton, Florida, 1995.
71. Rogiers, V., EEMCO guidance for the assessment of transepidermal water loss in cosmetic science, *Skin Pharmacol. Appl. Skin. Physiol.*, 14, 117, 2001.
72. Pinnagoda, J. et al., Guidelines for transepidermal water loss (TEWL) measurement. A report from the Standardization Group of the European Society of Contact Dermatitis, *Contact Dermatitis*, 22, 164, 1990.
73. Morrison, B.M., ServoMed evaporimeter: precautions when evaluating the effect of skin care products on barrier function, *J. Soc. Cosmet. Chem.*, 43, 161, 1992.
74. Lodén, M., The increase in skin hydration after application of emollients with different amounts of lipids, *Acta Derm. Venereol.*, 72, 327, 1992.
75. Kolbe, L., Non-invasive methods for testing of the stratum corneum barrier function, in *Dry Skin and Moisturizers: Chemistry and Function*, Lodén, M. and Maibach, H.I., Eds., CRC Press, Boca Raton, Florida, 2000, pp. 393.
76. Kolbe, L., Kligman, A.M., and Stoudemayer, T., Objective bioengineering methods to assess the effects of moisturizers on xerotic leg of elderly people, *J. Derm. Treat.*, 11, 241, 2000.
77. Lodén, M., Urea-containing moisturizers influence barrier properties of normal skin, *Arch. Dermatol. Res.*, 288, 103, 1996.

78. Lodén, M., Andersson, A.-C., and Lindberg, M., Improvement in skin barrier function in patients with atopic dermatitis after treatment with a moisturizing cream (Canoderm®), *Br. J. Dermatol.*, 140, 264, 1999.
79. Ghadially, R., Halkier-Sorensen, L., and Elias, P.M., Effects of petrolatum on stratum corneum structure and function, *J. Am. Acad. Dermatol.*, 26, 387, 1992.
80. Hill, S. and Edwards, C., A comparison of the effects of bath additives on the barrier function of skin in normal volunteer subjects, *J. Dermatol. Treat.*, 13, 15, 2002.
81. Tree, S. and Marks, R., An explanation for the 'placebo' effect of bland ointment bases, *Br. J. Dermatol.*, 92, 195, 1975.
82. Held, E., Lund, H., and Agner, T., Effect of different moisturizers on SLS-irritated human skin, *Contact Dermatitis*, 44, 229, 2001.
83. Loden, M., Barrier recovery and influence of irritant stimuli in skin treated with a moisturizing cream, *Contact Dermatitis*, 36, 256, 1997.
84. Kucharekova, M. et al., Effect of a lipid-rich emollient containing ceramide 3 in experimentally induced skin barrier dysfunction, *Contact Dermatitis*, 46, 331, 2002.
85. Mao-Qiang, M. et al., Exogenous nonphysiologic vs physiologic lipids. Divergent mechanisms for correction of permeability barrier dysfunction, *Arch. Dermatol.*, 131, 809, 1995.
86. Zettersten, E.M. et al., Optimal ratios of topical stratum corneum lipids improve barrier recovery in chronologically aged skin, *J. Am. Acad. Dermatol.*, 37, 403, 1997.
87. Thornfeldt, C., Critical and optimal molar ratios of key lipids, in *Dry Skin and Moisturizers: Chemistry and Function*, Lodén, M. and Maibach, H.I., CRC Press, Boca Raton, Florida, 2000, pp. 337.
88. Chamlin, S.L. et al., Ceramide-dominant barrier repair lipids alleviate childhood atopic dermatitis: changes in barrier function provide a sensitive indicator of disease activity, *J. Am. Acad. Dermatol.*, 47, 198, 2002.
89. Lodén, M. and Barany, E., Skin-identical lipids versus petrolatum in the treatment of tape-stripped and detergent-perturbed human skin, *Acta Derm. Venereol.*, 80, 412, 2000.
90. Halkier-Sorensen, L. and Thestrup-Pedersen, K., The efficacy of a moisturizer (Locobase) among cleaners and kitchen assistants during everyday exposure to water and detergents, *Contact Dermatitis*, 29, 266, 1993.
91. Serup, J., A double-blind comparison of two creams containing urea as the active ingredient. Assessment of efficacy and side-effects by non-invasive techniques and a clinical scoring scheme, *Acta Derm. Venereol.*, 177, 34, 1992.
92. Andersson, A.-C., Lindberg, M., and Lodén, M., The effect of two urea-containing creams on dry, eczematous skin in atopic patients. I. Expert, patient and instrumental evaluation, *J. Dermatol. Treat.*, 10, 165, 1999.
93. Lodén, M. et al., Instrumental and dermatologist evaluation of the effect of glycerine and urea on dry skin in atopic dermatitis, *Skin Res. Technol.*, 7, 209, 2001.
94. Grice, K., Sattar, H., and Baker, H., Urea and retinoic acid in ichthyosis and their effect on transepidermal water loss and water holding capacity of stratum corneum, *Acta Derm. Venereol. (Stockh.)*, 54, 114, 1973.
95. Pope, F.M. et al., Out-patient treatment of ichthyosis: a double-blind trial of ointments, *Br. J. Dermatol.*, 86, 291, 1972.
96. Vilaplana, J. et al., Clinical and non-invasive evaluation of 12% ammonium lactate emulsion for the treatment of dry skin in atopic and non-atopic subjects, *Acta Derm. Venereol. (Stockh.)*, 72, 28, 1992.

97. Lodén, M., Barrier recovery and influence of irritant stimuli in skin treated with a moisturizing cream, *Contact Dermatitis*, 36, 256, 1997.
98. Held, E., Sveinsdottir, S., and Agner, T., Effect of long-term use of moisturizers on skin hydration, barrier function and susceptibility to irritants, *Acta Derm. Venereol. (Stockh.)*, 79, 49, 1999.
99. Duval, D. et al., Differences among moisturizers in affecting skin susceptibility to hexyl nicotinate, measured as time to increase skin blood flow, *Skin Res. Technol.*, 8, 1, 2002.
100. Hachem, J.P. et al., The effect of two moisturisers on skin barrier damage in allergic contact dermatitis, *Eur. J. Dermatol.*, 12, 136, 2002.
101. Tollesson, A. and Frithz, A., Transepidermal water loss and water content in the stratum corneum in infantile seborrhoeic dermatitis, *Acta Derm. Venereol.*, 73, 18, 1993.
102. Pigatto, P.D. et al., 10% urea cream (Laceran) for atopic dermatitis: a clinical and laboratory evaluation, *J. Dermatol. Treat.*, 7, 171, 1996.
103. Lodén, M. and Wessman, C., The influence of a cream containing 20% glycerin and its vehicle on skin barrier properties, *Int. J. Cosmetic Sci.*, 23, 115, 2001.

16 Bioengineering Correlates of the Sensitive Skin Syndrome: The Sensory Irritation Component

Robert Sinaiko and Howard I. Maibach

CONTENTS

I. Defining Sensitive Skin 187
II. Modeling the Sensory Irritation Component of the Sensitive Skin Syndrome: The Lactic Acid Sting Test (LAST) 188
III. Factors Determining Chemically Induced Stinging 189
IV. Stingers vs. Nonstingers: Differences Apparent from Bioengineering Measurements 190
A. Statistically Significant Correlations 190
B. Directional Trends with Less Clearly Established Statistical Significance 190
V. Discussion 191
References 195

I. DEFINING SENSITIVE SKIN

Approximately half of the general population, when asked, answer that they have "sensitive skin,"[1] but in the absence of rigorous definitional standards, that term is subject to different meanings and interpretations. In the most general sense, patients with sensitive skin are unusually susceptible to the induction of inflammatory or neurosensory symptoms by various exogenous triggers, including natural and synthetically derived chemical irritants, contact allergens, ingested foods, weather conditions, sun exposure, and incorrect skin care.

A proportion of patients who report that they have sensitive skin suffer from exogenously exacerbated inflammatory dermatitis. The development of inflammatory

0-8493-1443-7/05/$0.00+$1.50

signs may follow exacerbating exposures by minutes, as in the case of immunologic and nonimmunologic contact urticaria; after many hours, as in allergic contact dermatitis; or following longer periods of cumulative exposures or skin damage, and they may or may not be associated with unpleasant neurosensory symptoms such as itching, burning, or stinging.

Another group of patients, who make up the primary focus of this chapter, have a condition designated "sensory irritation." Best understood in terms of a neurosensory irritation model, this condition is closely related to the subjective (sensory) irritation variant of the cosmetic intolerance syndrome.[2] Beginning within several minutes after facial contact with a chemical trigger, usually a cosmetic or skin care product, these patients experience 5 to 10 minutes of intense facial discomfort, unaccompanied by any objective evidence of inflammation or other visible changes in the skin. Unable to ignore this crescendo of pain, they may frantically attempt to gain relief by washing the face. Having reached their peak, symptoms gradually fade, and they usually resolve completely within 30 minutes after the offending exposure.

II. MODELING THE SENSORY IRRITATION COMPONENT OF THE SENSITIVE SKIN SYNDROME: THE LACTIC ACID STING TEST (LAST)

A multicenter trial involving approximately 1000 subjects evaluated three chemical probes, comparing the correlations between their capacity to induce disagreeable facial symptoms and the self-reported skin sensitivity of each participant. Of these three chemical agents, 10% aqueous lactic acid, 10% balsam of Peru,* and 10:90 chloroform/methanol, the facial discomfort induced by lactic acid correlated most strongly with self-assessed sensitive skin.[3]

Although other chemical probes such as ammonium lactate are occasionally substituted, the LAST provides scientists with a widely accepted research method for confirming susceptibility to chemically induced neurosensory skin irritation as well as a useful tool for studying the pathophysiology of this aspect of the sensitive skin syndrome. In the hands of clinical investigators, the LAST has provided a useful experimental strategy for research seeking to elucidate the pathophysiology of the sensitive skin syndrome.

Details of the test methodology vary considerably. Sensitivity is increased and specificity is lost as the concentration of lactic acid, which ranges between reports from 3%[4] to 30%,[5] is raised.[6] Few publications provide information regarding the *p*H of the lactic acid solution used; *p*H is dependent on a number of factors, including the buffering capacity of the vehicle, and may exert important effects on test results. No published studies specifically address the possible advantages of including a diversified panel of chemical probes, though such a strategy might be expected to improve sensitivity.

* Because balsam of Peru contains 0.9% cinnamic aldehyde, a nonimmunologic contact urticant, burning, stinging, and itching triggered by this probe may represent urticaria of insufficient severity to produce a detectable wheal and flare response.

A typical published LAST protocol follows[7]:

> The subject is placed in a hot, humid environmental chamber until profuse sweating is achieved. Then a 5% solution of lactic acid is rubbed over the nasolabial folds and cheeks with a cotton-tipped applicator. The stinging sensation is scored on a scale at 10 seconds, 2.5 minutes, and 5 minutes.

Investigators are aware that, in addition to stinging, the LAST may induce other forms of discomfort, such as burning and intense itching,[8] and these sensations are often taken into account in evaluating and reporting the test results.

Many authors who have contributed to the sensitive skin literature have relied on inclusion criteria other than subjects' scores on the LAST. Some, for example, assign subjects to "sensitive" or "nonsensitive" groups by self-report (often using skin sensitivity questionnaires), or they equate skin sensitivity to an excessive propensity toward visible inflammatory reactions to chemical contactants.

To be sure, stingers generally do score higher on skin sensitivity questionnaires,[9] and because inflammatory mediators such as serotonin, histamine, and substance P can produce burning, itching, and stinging, it seems reasonable, although patients with sensitive skin syndrome show no visible inflammatory signs, to propose that neurosensory irritation may be an early step toward inflammation. However, many individuals who self-report sensitive skin are actually nonstingers,[3] and the reported observation that stingers do not appear to differ from nonstingers in their susceptibility to irritant-induced inflammation[10] indicates a bifurcated pathway, one of whose branches leads to visible inflammation, and the other toward neurosensory irritation.

III. FACTORS DETERMINING CHEMICALLY INDUCED STINGING

Characteristics of facial skin (and especially of the nasolabial fold) believed to give rise to the stinging response to chemicals such as lactic acid include

- Presumed thin stratum corneum (no measurements reported)
- An elaborate network of sensory nerves (no measurements reported)
- High density of skin appendages (hair follicles and sweat glands) (no measurements reported)
- High permeability[11] (no measurements reported)

Additional factors that may favor the likelihood of a positive LAST include

- Time of year (increased stinging is reported in winter months)[12]
- Physical trauma to stratum corneum, such as scratching or stripping with cellophane tape[5]
- Chemical delipidization of stratum corneum; for example, with acetone[5]
- Coexisting skin disease, such as rosacea[13]
- Coexistence of certain other pain syndromes, notably facial discomfort related to computer display use,[14] and possibly interstitial cystitis[15]

IV. STINGERS VS. NONSTINGERS: DIFFERENCES PRESUMED APPARENT FROM BIOENGINEERING MEASUREMENTS

A. Statistically Significant Correlations

- A 2003 report demonstrates that increases in nasolabial fold transepidermal water loss (TEWL) result from the application of 3% or 5% lactic acid to that site ($P = .003$). The same study demonstrates that this increase in TEWL is greater in subjects with lower LAST scores than in subjects with higher LAST scores (5% lactic acid).[4]
- Laser Doppler flowmetry reveals statistically significant exaggeration of local vasodilatory responses in stingers, compared with nonstingers, associated both with nonimmunologic contact urticaria induced on the upper back by 20 minutes of contact with benzoic acid 0.5% ($P < .05$) and with acute irritant dermatitis induced by 24-hour patch testing with 1% sodium lauryl sulfate ($P < .05$).[8]
- The rate at which skin surface *p*H returns toward normal after nasolabial fold acidification with lactic acid appears to be more rapid in stingers than in nonstingers ($P = .041$).[16]
- Chromometry reveals higher a* values (evidencing a stronger red component) in the skin color of subjects self-reporting sensitive skin who had higher LAST scores than in subjects self-reporting nonsensitive skin who had lower LAST scores ($P < .05$).[17]
- A 1998 publication demonstrated lower baseline electrical capacitance of the right cheeks of individuals self-reporting sensitive skin and with a propensity toward higher LAST scores compared with a self-reported nonsensitive group who had lower LAST scores ($P < .05$).[17] A 2003 report expands on that finding by similarly indicating that LAST score as determined by testing with 3% lactic acid solution is negatively correlated with baseline capacitance of the nasolabial skin ($P = .03$). This latter study also finds that 5% lactic acid application increases local capacitance to a lesser degree in subjects with higher LAST scores than in subjects with lower LAST scores ($P = .014$).[4]

B. Directional Trends with Less Clearly Established Statistical Significance

- A 2003 publication reports a statistically insignificant trend toward higher baseline TEWL measured at the nasolabial fold in subjects with higher LAST scores.[4] Earlier reports also indicated nonsignificant directional trends in a group self-reporting sensitive skin and with generally higher LAST scores (compared with a self-reported insensitive group with lower LAST scores) with the more sensitive subjects tending toward higher baseline TEWL values measured on the cheek[17] and toward more exaggerated increase in TEWL at sites of acute irritant dermatitis induced by a 24-hour occlusive patch test with 1% sodium lauryl sulfate on the upper back.[8]

- By Laser Doppler flowmetry, nonimmunologic contact urticaria induced by benzoic acid 1% and by sorbic acid 0.5% or 1% appears to produce a more pronounced local increase in skin blood flow velocity in a group of stingers than in nonstingers.[8] Laser Doppler flowmetry appears also to reveal that the topical application of methyl nicotinate induces a greater local vasodilatory effect in stingers than in nonstingers (the authors' assertion that this latter effect is "significant" appears to be supported by the graphical representation of their data, but no additional statistical analysis is provided).[18]
- A trend toward higher baseline skin surface *p*H was noted among a group of subjects with self-reported sensitive skin and with a propensity toward higher LAST scores, compared with a self-reported nonsensitive group with lower LAST scores.[17]
- Measured as transparency of an opaque band, skin sebum content may be lower in stingers than in nonstingers.[17]

V. DISCUSSION

The correlation between TEWL and the penetration of exogenous chemicals through the skin is well established,[11,19,20] so it is tempting to postulate that phenotypic stratum corneum hyperpermeability might be an important determinant of neurosensory irritation caused by chemicals applied to the skin. However, two important issues relative to this question remain to be resolved. First, data indicating that baseline TEWL of the skin of stingers is greater than that of nonstingers have thus far fallen short of statistical significance.[4,17] Second, there is no published direct evidence indicating that the skin of stingers is in fact more permeable than that of nonstingers.

Chemical provocation studies using laser Doppler flowmetry[8] and chromometry a* value[21] indicate that chemicals capable of dilating the skin's microvasculature on contact produce more pronounced vasodilation in stingers than in nonstingers. This exaggerated microvascular response to vasodilators or proinflammatory chemical contactants might reflect greater epidermal permeability to these agents, though it is also possible that the vasoregulatory mechanisms in the skin of stingers are hyperreactive to a given degree of chemical penetration.

The baseline skin surface pH of stingers may be slightly less acidic than that of nonstingers.[17] In addition, following acidification with lactic acid, the skin of the nasolabial fold appears to return more rapidly toward its normal pH in stingers than that in nonstingers.[16] If, indeed, the skin of stingers is constitutively more permeable than that of nonstingers, then both natural skin acidity and the duration of the hyperacidity induced by application of lactic acid might be reduced in stingers as a function of accelerated dilution and buffering, as more water and electrolytes migrate to the surface from the vascular bed of the skin, or as a function of the more rapid removal of the applied lactic acid from the surface as it is absorbed into the skin.[16]

TABLE 16.1
Summary of Published Studies Correlating *in vivo* Bioengineering Measurements with Susceptibility to Sensory Irritation as Determined by LAST

Reference	Study Design	Result	Statistical Significance of Differences between Stingers and Nonstingers
Lammintausta et al.[8]	Eight stingers were compared with 15 nonstingers in regard to the increase in laser Doppler flowmetry values associated with nonimmunologic contact urticaria induced by two different concentrations of sorbic acid and benzoic acid and in regard both to an increase in laser Doppler flowmetry and to an increase in TEWL after 24-hour occlusive patch test with 1% SLS	Laser Doppler flowmetry, greater increase in stingers than in nonstingers at sites of induction of nonimmunologic contact urticaria with sorbic acid and with benzoic acid and at site of the 24-hour SLS patch test; TEWL, greater increase in stingers than nonstingers with 24-hour SLS patch test.	$P < .05$ when contact urticaria is induced with benzoic acid 0.5%, but NS when contact urticaria is induced with sorbic acid 0.5% or 1%, or with benzoic acid 1%; $P < .05$ at site of 24-hour SLS patch test; difference in TEWL increase at SLS patch test site between stingers and nonstingers, NS
Issachar et al.[16]	Skin surface *p*H was measured at the nasolabial fold and volar forearm of 15 stingers and 15 nonstingers and was then serially determined at both areas after local application of 10% lactic acid solution	Serial skin *p*H measurement: increased rate of return of skin *p*H toward normal after acidification with lactic acid in stingers compared with nonstingers; no difference between stingers and nonstingers observed on volar forearm	$P = .041$ (nasolabial fold)
Issachar et al.[18]	A 0.5% solution of methyl nicotinate was applied to the forearms of 10 stingers and 10 nonstingers, and serial laser Doppler perfusion imaging measurements were conducted at 5- to 10-minute intervals for 1 hour	Laser Doppler flowmetry, for the first 35 minutes, topical application of methyl nicotinate induced a greater local vasodilatory effect in stingers than in nonstingers (maximal at 10 minutes)	The authors report that differences between stingers and nonstingers are "significant," an assertion that appears substantiated by the graphically represented data (error bars are nonoverlapping only at 10 minutes); no *P* value provided

Seidenari et al.[17]	Subjects were assigned to "sensitive skin" group (n = 26) and control group (n = 26) by questionnaire; LAST was administered to all subjects, but scores, though reported, had not been used as a criterion for group assignment and were not compared with results of bioengineering studies. Bioengineering measurements included: TEWL, cheek and forearm; stratum corneum hydration, cheek (as electrical capacitance); chromometry, cheek; sebum content, cheek (as transparency of an opaque band); skin surface *p*H, cheek	Compared with the control group, subjects reporting "sensitive skin" were found to have, in addition to higher LAST scores higher baseline TEWL values, lower baseline capacitance values, higher baseline a* values on chromometry (evidencing a stronger red component), lower baseline L* values on chromometry (indicating less overall reflective luminance); lower sebum content, and higher baseline skin surface *p*H	TEWL, NS; capacitance, $P < .05$; chromometric a* values, $P < .05$; chromometric L* values, NS; sebum content, NS; *p*H, NS
Wu et al.[4]	LAST tests were administered to 50 subjects; TEWL (by evaporimetry) and stratum corneum hydration (as skin capacitance) were measured at the nasolabial LAST sites before and again 8 minutes after application of the 3% (on one side) and 5% (on the other side) lactic acid test solutions	TEWL, increased both by 3% lactic acid and 5% lactic acid application—subjects with higher baseline TEWL tended to have higher LAST scores; LAST scores were negatively correlated with the percentage increase in nasolabial fold TEWL that followed application of the 5% lactic acid test solution; capacitance, increased by 3% (but not 5%) lactic acid; LAST scores using 3% lactic acid were negatively correlated with baseline capacitance; LAST scores were negatively correlated with percentage increase in capacitance that followed application of the 5% lactic acid test solution	Increase in TEWL caused by 3% and 5% lactic acid, $P < .05$; positive relationship between baseline TEWL and LAST score, NS; negative correlation between LAST score and percentage increase in TEWL resulting from application of lactic acid, $P = .003$; increase in capacitance resulting from 3% lactic acid, $P < .05$; negative correlation between LAST score with 3% lactic acid and baseline capacitance, $P = .03$; negative correlation between LAST scores and percentage increase in capacitance resulting from application of 5% lactic acid, $P = .014$

Note: LAST, lactic acid sting test; NS, not statistically significant; TEWL, transepidermal water loss; SLS, sodium lauryl sulfate.

Chromometric evidence that the skin of stingers is constitutively more vasodilated than that of nonstingers[17] suggests the intriguing possibility that stingers differ from nonstingers in their tissue concentrations of, or their vascular receptor sensitivity to, endogenous mediators that regulate the tone of microvascular smooth muscle. Greater permeability or sensitivity to more or less ubiquitous but inadvertent exogenous vasodilatory or proinflammatory environmental factors might also be hypothesized to explain the greater baseline vasodilation observed in stingers compared with nonstingers.

As evidenced by its comparatively reduced electrical capacitance,[4,17] the skin of stingers is less well hydrated than that of nonstingers. Returning again to the hypothesis that stingers have greater stratum corneum permeability than nonstingers, one might speculate that the skin of stingers becomes dehydrated as water migrates more rapidly out through the stratum corneum to the surface, where it is lost to evaporation.

Sebum content of the skin of stingers appears to be lower than that of nonstingers.[17] In addition to increasing the TEWL of skin, delipidization with acetone lowers its threshold for sensory perception of electrical current, an effect that is reversible by treatment with petrolatum.[5] Intercellular lipid content in the stratum corneum may be an important determinant of skin permeability, with highly lipid-laden skin manifesting the greatest resistance to transepidermal penetration of both hydrophilic and lipophilic contactants.[22] These findings indicate that the oft-postulated (but unproven) higher skin permeability of stingers, as well as their high neurosensory sensitivity, may both be related to deficient lipid content.

As more is learned of the correlations between the results of various bioengineering tests and the sensitive skin syndrome, bioengineering measurements will almost certainly be of help in deciphering the diverse physiologic factors involved in the genesis of this complex clinical entity (Table 16.2). Animal testing by bioengineering techniques may lead to safer product formulations, and the adoption of bioengineering tests into the clinical armamentarium might facilitate categorization of sensitive skin patients into diagnostic subsets distinguished by differing responses to a spectrum of therapeutic measures.

TABLE 16.2
Future Directions for Bioengineering Investigations

Parameter	Questions for Further Bioengineering Study
Features of nasolabial fold	How does the neurosensory irritability of the nasolabial fold compare with that of other anatomical areas? If the nasolabial area is more sensitive than other anatomical regions, what features might be identified to explain its greater susceptibility to sensory irritation?
Penetration of inducers	Will direct measurements of actual penetration of chemicals applied to the skin reveal that, because of such factors as phenotypic stratum corneum permeability, larger quantities of potentially triggering substances attain proximity to sensory nerve endings of stingers, compared with nonstingers?
Chemical properties of inducers	What specific characteristics of chemical substances determine their propensity to trigger neurosensory skin irritation? If more than one chemical property is found to correlate with triggering, what measurable features of the skin might predict individual susceptibilities to different classes of neurosensory irritants?
Seasonality/effects of weather conditions	What measurable features of the skin can be shown to change as a function of atmospheric conditions or season, and can a relationship between those features and neurosensory irritability be confirmed?
Age	In what way does the neurosensory irritability of the skin change as a function of age?
Gender	Is the neurosensory irritability of female facial skin greater than that of males, or is this apparent disproportion an artifact of sampling or reporting bias? If the gender difference is real, to what anatomic or physiologic characteristics of female facial skin might it be attributed?
Ethnicity	Are stingers disproportionately represented in certain ethnic groups? If so, can ethnically related features of skin anatomy or physiology be identified that might explain that disproportion?

REFERENCES

1. Willis, R.A. et al., Sensitive skin: an epidemiological study, *Br. J. Dermatol.*, 145, 258, 2001.
2. Maibach, H.I. and Engasser, P., Management of cosmetic intolerance syndrome, *Clin. Dermatol.*, 6, 102, 1988.
3. Bowman, J.P. et al., The use of chemical probes to assess the facial reactivity of women, comparing their self-perception of sensitive skin, *J. Cosmetic Sci.*, 51, 267, 2000.
4. Wu, Y. et al., Correlation between stinging, TEWL and capacitance, *Skin Res. Technol.*, 9, 90, 2003.

5. Kobayashi, H. et al., Measurement of electrical current perception threshold of sensory nerves for pruritus in atopic dermatitis patients and normal individuals with various degrees of mild damage to the stratum corneum, *Dermatology*, 206, 204, 2003.
6. Oh, B.M. et al., Comparison of results of several lactic acid sting tests on sensitive skin, *Kor. J. Dermatol.*, 41, 578, 2003.
7. Chew, A. and Maibach, H., Sensitive skin, in *Dry Skin and Moisturizers: Chemistry and Function*, Loden, M. and Maibach, H., Eds., CRC Press, Boca Raton, 2000, 435.
8. Lammintausta, K., Maibach, H.I., and Wilson, D., Mechanisms of subjective (sensory) irritation/propensity to non-immunologic contact urticaria and objective irritation in stingers, *Dermatozen*, 36, 45, 1988.
9. Muizzuddin, N., Marenus, K.D., and Maes, D.H., Factors defining sensitive skin and its treatment, *Am. J. Contact Dermat.*, 9, 170, 1998.
10. Basketter, D.A. and Griffiths, H.A., A study of the relationship between susceptibility to skin stinging and skin irritation, *Contact Dermatitis*, 29, 185, 1993.
11. Rougier, A., Lotte, C., and Maibach, H.I., *In vivo* percutaneous penetration of some organic compounds related to anatomic site in humans: predictive assessment by the stripping method, *J. Pharm. Sci.*, 76, 451, 1987.
12. Soschin, D. and Kligman, A.M., Adverse subjective responses, in *Safety and Efficacy of Topical Drugs and Cosmetics*, Kligman A.M. and Leyden J.J., Eds., Grune & Stratton, New York, 1982, 381.
13. Lonne-Rahm, S., Fischer, T., and Berg, M., Stinging and rosacea, *Acta Derm Venereol*, 79, 460, 1999.
14. Berg, M., Lonne-Rahm, S., and Fischer, T., Patients with visual display unit-related facial symptoms are stingers, *Acta Derm. Venereol. (Stockh.)*, 78, 44, 1998.
15. Madhu, A. et al., Interstitial cystitis: unexplained associations with other chronic disease and pain syndromes, *Urology*, 49(suppl. 5A), 52, 1997.
16. Issachar, N. et al., pH measurements during lactic acid stinging test in normal and sensitive skin, *Contact Dermatitis*, 36, 152, 1997.
17. Seidenari, S., Francomano, M., and Mantovani, L., Baseline biophysical parameters in subjects with sensitive skin, *Contact Dermatitis*, 38, 311, 1998.
18. Issachar, N. et al., Correlation between percutaneous penetration of methylnicotinate and sensitive skin, using laser Doppler imaging, *Contact Dermatitis*, 39, 182, 1998.
19. Dupuis, D. et al., *In vivo* relationship between percutaneous absorption and transepidermal water loss according to anatomic site in man, *J. Soc. Cosmetic Chem.*, 37, 351, 1986.
20. Rougier, A. et al., Regional variation in percutaneous absorption in man: measurement by the stripping method, *Arch. Dermatol. Res.*, 278, 465, 1986.
21. Berardesca, E. et al., *In vivo* transcutaneous penetration of nicotinates and sensitive skin, *Contact Dermatitis*, 25, 35, 1991.
22. Elias, P.M. et al., Percutaneous transport in relation to stratum corneum structure and lipid composition, *J. Invest. Derm.*, 76, 297, 1981.

17 Barrier Recovery

Silke M. Fuchs, Joachim W. Fluhr, and Howard I. Maibach

CONTENTS

Abstract 197
I. Introduction 197
II. Protective Mechanism of the Epidermis 198
III. Barrier Disruption and Repair 198
IV. Different Influences on Barrier Repair 199
A. Environmental Humidity 199
B. Electric Potential 201
C. Psychological and Physiological Aspects 202
D. Cosmetics 203
E. Racial Differences 205
V. Summary 205
References 205

ABSTRACT

The main demand on the human skin barrier is the prevention of the loss of water and, in addition, the prevention of the entry of water and matter from the environment into the body. Physical, chemical, and pathological factors can disturb barrier function. Disruption of skin barrier function is associated with skin diseases such as atopic dermatitis, psoriasis, and contact dermatitis. Barrier repair includes the immediate secretion of lamellar bodies and the formation of new lamellar bodies with further secretion and extracellular processing of secreted lamellar body–derived lipids. Several factors influence barrier repair; for example, environmental humidity, electronic potential, cosmetics, and more.

I. INTRODUCTION

An epidermal barrier is located in the stratum corneum. The stratum corneum is composed of two major components: protein-rich nonviable cells and intercellular lipid domains.[1] The physical barrier resides in long chain lipids, which are organized as bilamellar structures stacked on top of each other and filling the intercellular

0-8493-1443-7/05/$0.00+$1.50

spaces between the horny layer corneocytes, which give the barrier a brick (corneocytes)-and-mortar (tight lamellar structures) organization.[2]

To account for such barrier properties and for the hydrophilic and hydrophobic pathways through the skin barrier, an alternative model, namely, the domain mosaic model of the skin barrier,[3,4] has been proposed. This model envisages the barrier lipids as existing predominantly in crystalline — and therefore water-impermeable — domains surrounded by grain borders of lipids in a liquid crystalline state. This arrangement could provide an effective barrier that still allows a minute but controlled water loss through the liquid crystalline interdomains. This controlled water loss is sufficient to keep the stratum corneum keratin hydrated.[5] The liquid character of interdomain grain borders would allow passage of hydrophilic and hydrophobic molecules on downhill gradients. It is in these regions that permeation enhancers may induce structural transformations of the bilamellar structure that promotes transepidermal delivery of pharmacological substances.[6,7]

Physical, chemical, and pathological factors can disturb barrier function. Maintenance of the stratum corneum structural integrity is critical to barrier function. Increasing the hydration of stratum corneum can progressively reduce its barrier efficacy.[8–12] Disruption of skin barrier function is associated with skin diseases such as atopic dermatitis, psoriasis, and contact dermatitis.[13] Transepidermal water loss (TEWL) may act as an indicator of the recovery of barrier function and as a marker for the structural changes of the stratum corneum.[14,15]

II. PROTECTIVE MECHANISM OF THE EPIDERMIS

The outer layers of the epidermis mediate protective functions, including water repellence, integrity, cohesion, desquamation, antimicrobial defense, ultraviolet filtration, antioxidant defense, and resistance to xenobiotics, as as shown in Table 17.1.

III. BARRIER DISRUPTION AND REPAIR

Disruption of the barrier by organic solvents, detergents, or tape stripping elicits a homeostatic repair response in the epidermis, which rapidly results in restoration of barrier function[7,16] to its original level.[1] The degree of epidermal hyperplasia increases as the duration of barrier disruption is extended,[17] and decline of skin barrier function induces epidermal proliferation and an inflammatory response.[18] Epidermal injury results in a generalized increase in protein prenylation, which may play an important role in diverse cellular functions.[19]

Barrier repair includes the immediate secretion of lamellar bodies and the formation of new lamellar bodies with further secretion[20]; it also includes an increase in epidermal lipid synthesis,[21] extracellular remodeling of secreted lamellar body-derived lipids,[22] and enhanced cytokine production.[19] These enhancements will all lead ultimately to barrier restoration.

In addition, sphingolipids play an important role in the homeostasis and barrier function of human stratum corneum.[23–25] A disturbance of sphingolipid formation is presumed to be a crucial factor for the increased TEWL in common skin diseases like atopic eczema or psoriasis.[26] The low lipid content of the palm of the hand may

TABLE 17.1
Protective Functions of the Epidermis Including Water Repellence, Resistance to Xenobiotics, Antimicrobial Defense and More

Inflammatory Mediators	Antioxidants	Ultraviolet-Absorbing Molecules	Xenobiotic-Metabolizing Enzymes
Prostaglandins	Glutathione	Melanin	Glucuronidation
Eicosanoids	Oxidases	*Trans*-urocanic acid	Sulfation
Leukotrienes	Catalase	Vitamins C+D	Hydroxylation
Histamine	Cytochrome P450		
Cytokines	Vitamins C+E		
Antimicrobial systems	Natural moisturizing factors		
Lipids	Sebaceous glands		
Iron-binding proteins	Filaggrin breakdown products		
Complement	Free amino acids		
Peptides			
Acidic pH			
Free fatty acids			

explain its increased sensitivity to the development of dermatitis induced by soap and hot water, as it has an intrinsic barrier defect associated with reduced lipids.[27]

Acute disruption of the barrier results in an increase in epidermal DNA synthesis 16 to 20 hours after treatment,[28] and it also increases cytokine production in the epidermis, especially of interleukin IL1α expression.[29–31] Also involved in skin barrier repair function and the cutaneous condition of epidermal hyperplasia are the histamine H1 and H2 receptors. Ashida et al.[32] ascertained that histamine receptor antagonists may not only accelerate barrier repair but also inhibit scratching, which can break the skin barrier.

Even if the damage to the barrier is relatively small, when it is repeated[17] or occurs under low environmental humidity, the damage inflicted can induce obvious epidermal hyperplasia and inflammation (see below). Taljebini[33] demonstrated that barrier repair responses are similar regardless of the etiology or extent of barrier disruption. Skin occlusion obstructs the normal ventilation of the skin surface and increases stratum corneum hydration, hence compromising skin barrier function.[14]

IV. DIFFERENT INFLUENCES ON BARRIER REPAIR

A. Environmental Humidity

Skin conditions change at different environmental humidity levels (see Table 17.2): The severity of skin disorders has been reported to increase in winter[34,35] as a result of the decreased humidity.[36–38] There is no remarkable difference in the skin barrier recovery of aged skin in comparison with young skin in a dry environment, however.[39]

TABLE 17.2
Main References Concerning the Influence of Environmental Humidity on Skin Barrier Repair

First Author	Title	Result	Comment	Ref.
Ashida	Epidermal interleukin 1α-generation is amplified at low humidity.	Interleukin 1α in the epidermis was higher in animals kept in a low-humidity environment than that of animals in a high-humidity environment	Interleukin 1α is an important factor in mediating the relationship between environmental humidity and epidermal pathology	40
Choi	The skin barrier state of aged hairless mice in a dry environment	The number of stratum corneum layers increased in the dry environment. The recovery rate of the aged skin barrier was neither accelerated nor delayed in the dry environment	No remarkable difference was found in the skin barrier recovery of aged hairless mice in a dry environment	39
Denda	Exposure to dry environment enhances epidermal permeability barrier function	Prolonged exposure of normal murine skin to low humidity for 1 to 2 weeks increased stratum corneum weight with a reduction of transepidermal water loss	Environmental conditions influence epidermal structure and function and suggest inducing or exacerbating cutaneous disorders	43
Denda	Low humidity stimulates epidermal DNA synthesis and amplifies the hyperproliferative response to barrier disruption	This provides evidence that changes in environmental humidity contribute to the seasonal exacerbations of cutaneous disorders	Exacerbation of many dermatoses with a defective barrier by low humidity environment may result from this cutaneous response to decreased humidity	41
Egawa	Effect of exposure of human skin to a dry environment	Short exposure of skin to low-humidity induced changes in the moisture contents in the stratum corneum	A dry environment might make fine wrinkles related to lack of water in stratum corneum	45
Sato	Dry condition affects desquamation of stratum corneum in vivo	Dry environment perturbs desmosome degradation in intact stratum corneum by decreasing water content	This study suggests that a dry environment has a rather rapid effect on normal skin in vivo and induces scaliness by impairing desquamation in stratum corneum	42

Denda et al.[41] discussed the effect of environmental humidity on barrier recovery. Low environmental humidity stimulates DNA synthesis and amplifies the hyperproliferate response to barrier disruption, leading to more proinflammatory cytokines being present in the skin.[40,41] Dry environment perturbs desmosome degradation in intact stratum corneum by decreasing the water content.[42] In contrast, longer exposure to low humidity enhanced cutaneous barrier function. Denda et al.[43] demonstrated that the prolonged exposure of normal murine skin to low humidity for 1 to 2 weeks increased stratum corneum weight and thickness, with a commensurate reduction in basal TEWL.[43] They also noted that environmental humidity influences the sensitivity of skin to the topical application of detergents.[44] Egawa et al.[45] mentioned that having a dry environment in daily life would cause fine wrinkles related to lack of water in the stratum corneum. A sudden change from a very humid to very dry environment results in an abnormality in permeability barrier function: Formerly "wet" keratinocytes upregulate neither epidermal lamellar body secretion nor DNA synthesis sufficiently, allowing a barrier defect to develop until a new layer of keratinocytes reaches the stratum granulosum/stratum corneum interface.[46] There was a significant increase in TEWL noted on the cheek in winter, as opposed to in summer, suggesting that a cold dry environment stimulates the skin and provokes subclinical inflammation.[43] Halkier-Sorensen et al.[47] observed that cold temperatures inhibit barrier repair. Changes in the environmental humidity contribute to the seasonal exacerbations of cutaneous disorders (characterized by a defective barrier), such as atopic dermatitis and psoriasis.[41] Alterations in our living and working conditions (e.g., air-conditioning) might alter skin function, thus exposing our body to a harmful environment.[48]

B. Electric Potential

Healthy skin surface has a negative potential against the inner layers of skin (Table 17.3). After barrier disruption, the potential decreases immediately.[49,50] Baker et al.[51] indicated in 1982 that the negative potential experienced was induced by epidermal living layers. Ion gradients of calcium and magnesium in the epidermis play a crucial role in skin barrier homeostasis[52,53]: They appear to be synergistic in inhibiting barrier recovery,[52] and both were localized in the stratum corneum. Within 30 minutes after barrier disruption, the gradation of calcium and magnesium disappeared.[53,54] Menon et al.[55] demonstrated that low calcium concentration accelerates lamellar body secretion, and they suggested that an increase of Ca^{2+} in the stratum corneum perturbed lamellar body secretion and delayed skin barrier repair.

Topical application of magnesium and calcium on mice skin accelerated barrier recovery.[56] Denda and Kumazawa[57] suggested that the external electric potential affects the ionic gradients in the epidermis and also influences skin barrier homeostasis; they observed that the external negative electric potential induced the secretion of epidermal lamellar bodies at the stratum corneum/stratum granulosum interface.[57]

In addition, Said et al.[58] found that epidermal calcium release is a useful indicator of skin barrier function, although it appears mainly to illustrate damage to epidermal cells and functions.[58] Recent studies suggest that various types of skin diseases, characterized by barrier abnormality (e.g., atopic dermatitis),[59] might be associated

TABLE 17.3
Main References Concerning the Influence of Electric Potential on Skin Barrier Repair

First Author	Title	Result	Comment	Ref.
Denda	Some magnesium salts and a mixture of magnesium and calcium salts accelerate skin barrier recovery	Aqueous 10 mM magnesium chloride and calcium chloride hastened barrier recovery more effectively than a solution of 10 mM magnesium chloride	This indicates that the effects of these metal ions are different depending on application method	56
Denda	P2X purinergic receptor antagonist accelerates skin barrier repair and prevents epidermal hyperplasia induced by skin barrier disruption	Barrier homeostasis can be regulated by cation flux through a P2X3-like ATP receptor	This indicates a new way to regulate calcium dynamics in the epidermis and an effective therapeutic strategy to improve epidermal ion dynamics and skin condition	60
Lo	Transepidermal potassium ion, chloride ion, and water flux across delipidized and cellophane tape-stripped skin	Cellophane stripping of normal SC resulted in an increase in outward transepidermal potassium and chloride ion flux	The potassium electrode would be useful for investigating and assessing the epidermal electrolyte barrier	50
Said	Epidermal calcium release (ECR) *in vivo* sampled with a simple washout chamber technique	Epidermal calcium release is showed to be a barrier marker of high reproducibility	Calcium release can have major clinical relevance and may become an important barrier marker	58

with abnormal ion dynamics.[60] For example, the well-known high mineral content of tap water could interfere with barrier recovery in patients with a damaged skin barrier.[61]

C. Psychological and Physiological Aspects

Animal and human studies have shown that psychological stress is another important aspect relating to the epidermal barrier, as it affects the cutaneous immune system (Table 17.4).[62,63] Denda et al.[64,65] demonstrated that either crowding or immobilization of mice delays permeability barrier recovery after acute barrier insults. They observed the same effect of immobilization stress on barrier homeostasis in male and female rats, indicating that sex hormones may not be crucial mediators of the stress on the alteration in barrier homeostasis. Moreover, the delay in repair induced by immobilization was further linked to psychological stress, as treatment with

TABLE 17.4
Main References Concerning the Influence of Cosmetics on Skin Barrier Repair

First Author	Title	Result	Comment	Ref.
Chamlin	Ceramide-dominant barrier repair lipids alleviate childhood atopic dermatitis: changes in barrier function provide a sensitive indicator of disease activity	Stratum corneum integrity and hydration improved significantly during therapy; stratum corneum revealed extracellular lamellar membranes (no vehicle control)	Ceramide-dominant emollient represents an adjunct to possible childhood atopic dermatitis treatment	72
Denda	γ-aminobutyric acid (A) receptor agonists accelerate cutaneous barrier recovery and prevent epidermal hyperplasia induced by barrier disruption	Positive effect of topical applied γ-aminobutyric acid and γ-amino-butyric acid receptor–specific agonists on barrier recovery after disruption	These results suggested that the γ-aminobutyric acid-like receptor is associated with skin barrier homeostasis	71
Fluhr	Glycerol accelerates recovery of barrier function in vivo	Glycerol creates a stimulus for barrier repair and improves the stratum corneum hydration even 7 days after treatment	Glycerol can be regarded as a barrier, stabilizing and moisturizing compound	70

sedative drugs before immobilization restored the normal kinetics of barrier recovery.[64,65] Furthermore, Denda et al.[16] showed that an odorant that has a sedative effect could prevent the deterioration of homeostasis that is induced by psychological stress in peripheral tissue such as the skin.[16]

Fluhr et al.[66] documented that skin at different body sites shows distinguished patterns in barrier repair that are likely to be related to structural and physiological differences. Forehead and back (lipid-rich skin areas) were the most vulnerable to barrier disruption with tape stripping and with acetone extraction of stratum corneum lipids.[66] There is a circadian rhythm in the stratum corneum barrier homeostasis;[67] with ageing, the stratum corneum barrier becomes fragile, and recovery is delayed.[62] The subjects as a group demonstrated a decline in permeability barrier recovery kinetics after barrier disruption by cellophane tape stripping, in parallel with an increase in perceived psychological stress during the higher- vs. the lower-stress occasions.[68]

D. Cosmetics

Barrier problems can occur with cosmetic use (Table 17.5). Compounds in cosmetics that should enhance the barrier function may also present risks.[69] However, barrier regeneration can be enhanced through the use of topical products (e.g., effective

TABLE 17.5
Main References Concerning the Influence of Psychological and Physiological Aspects on Skin Barrier Repair

First Author	Title	Result	Comment	Ref.
Denda	Stress alters cutaneous permeability barrier homeostasis	Transfer to a new environment affects skin permeability barrier homeostasis; an increase in endogenous glucocorticoid production is essential for the barrier abnormalities that occur in response to psychological stress.	It may be possible to improve certain cutaneous disorders by specifically blocking the pathways by which stress produces defects in permeability barrier homeostasis	64
Denda	Immobilization-induced and crowded environment-induced stress delay barrier recovery in murine skin	Barrier recovery was delayed in rats exposed to stress in comparison with untreated controls	Objective quantification of the influence of stress on the skin would allow further progress; this stress model might be useful in studying stress and cutaneous functions	65
Fluhr	Impact of anatomical location on barrier recovery, surface pH and stratum corneum hydration after acute barrier disruption	Lipid-rich skin areas were the most vulnerable to barrier disruption	Skin at different body sites shows distinct patterns of barrier recovery that are likely to be related to structural and physiological differences	66
Garg	Psychological stress perturbs epidermal permeability barrier homeostasis	After barrier disruption there was a decline in permeability barrier recovery kinetics with an increase in perceived psychological stress	This provides a link between psychological status and cutaneous function in humans	68

moisturizers of appropriate composition). Glycerol creates a stimulus for barrier repair and improves the stratum corneum hydration.[70] Topical application of γ-aminobutyric acid accelerated skin barrier recovery of hairless mice and improved epidermal hyperplasia under low environmental humidity.[71] Chamlin et al.[72] found out that a ceramide-dominant barrier repair emollient represents a useful adjunct to the treatment of childhood atopic dermatitis. The cholesterol-dominant mixture of physiologic lipids, including ceramides and free fatty acids, also can be effective in restoring normal function of aged epidermis.[73]

E. Racial Differences

Stratum corneum of lumbar skin requires more cell layers and more tape strips for removal and barrier disruption in Blacks and Asians than in Whites.[74] In addition, in Asians and Blacks, the TEWL was higher than in Whites.[75] Finally, Blacks and Asians more than Whites have an increased TEWL after irritation with sodium lauryl sulfate.[76]

V. SUMMARY

This review has given an overview regarding the mechanisms involved in barrier homeostasis and barrier repair. Factors influencing barrier repair, such as environmental humidity, electrical potential, and psychological, physiological, cosmetic, and racial factors, were also discussed. Other factors may also be involved. We are aware of the discrepancy of some of the data, which may be, at least partially, due to the chosen experimental paradigm (e.g., rodent vs. human skin). Analogies are discovered frequently, but an extrapolation from rodent to human data has to based on experimental data. This area of research is still experiencing a rapid increase of knowledge.

REFERENCES

1. Feingold, K.R. and Elias, P.M., The environmental interface: regulation of permeability barrier homeostasis, in *Dry Skin and Moisturizers. Chemistry and Function*, Loden, M. and Maibach, H.I., Eds., CRC Press, Boca Raton, Florida, 2000, pp. 45–58.
2. Williams, M.L. and Elias, P.M., From basket weave to barrier. Unifying concepts for the pathogenesis of the disorders of cornification, *Arch. Dermatol.*, 129, 626, 1993.
3. Forslind, B., A domain mosaic model of the skin barrier, *Acta Derm. Venereol.*, 74, 1, 1994.
4. Forslind, B., The skin: upholder of physiological homeostasis. A physiological and (bio) physical study program, *Thromb. Res.*, 80, 1, 1995.
5. Norlen, L., Emilson, A., and Forslind, B., Stratum corneum swelling. Biophysical and computer assisted quantitative assessments, *Arch. Dermatol. Res.*, 289, 506, 1997.
6. Forslind, B., Engstrom, S., Engblom, J., et al., A novel approach to the understanding of human skin barrier function, *J. Dermatol. Sci.*, 14, 115, 1997.
7. Elias, P.M. and Feingold, K.R., Lipids and the epidermal water barrier: metabolism, regulation, and pathophysiology, *Semin. Dermatol.*, 11, 176, 1992.
8. Berardesca, E. and Maibach, H.I., The plastic occlusion stress test (POST) as a model to investigate skin barrier function, in *Dermatologic Research Techniques*, Maibach, H.I., Ed., CRC Press, Boca Raton, Florida, 1996, pp. 179–186.
9. Bucks, D., Guy, R., and Maibach, H.I., Effects of occlusion, in *In Vitro Percutaneous Absorption: Principles, Fundamentals and Applications*, Bronaugh, R.L. and Maibach, H.I., Eds., CRC Press, Boca Raton, Florida, 1991, pp. 85–114.

10. Bucks, D. and Maibach, H.I., Occlusion does not uniformly enhance penetration in vivo, in *Percutaneous Absorption: Drug-Cosmetics-Mechanisms-Methodology*, Bronaugh, R.L. and Maibach, H.I., Eds., Marcel Dekker, New York, 1999, pp. 81–105.
11. Kligman, A.M., Hydration injury to human skin, in *The Irritant Contact Dermatitis Syndrome*, van der Valk, P.G.M. and Maibach, H.I., Eds., CRC Press, Boca Raton, Florida, 1996, pp. 187–194.
12. Fluhr, J.W., Lazzerini, S., Distante, F., et al., Effects of prolonged occlusion on stratum corneum barrier function and water holding capacity, *Skin Pharmacol. Appl. Skin Physiol.*, 12, 193, 1999.
13. Grice, K.A., Transepidermal water loss in pathologic skin, in *The Physiology and Pathophysiology of the Skin*, Jarrett, A., Ed., Academic Press, London, 1980, pp. 2147–2155.
14. Zhai, H. and Maibach, H.I., Occlusion vs. skin barrier function, *Skin Res. Technol.*, 8, 1, 2002.
15. Grubauer, G., Elias, P.M., and Feingold, K.R., Transepidermal water loss: the signal for recovery of barrier structure and function, *J. Lipid Res.*, 30, 323, 1989.
16. Denda, M., Tsuchiya, T., Shoji, K., et al., Odorant inhalation affects skin barrier homeostasis in mice and humans, *Br. J. Dermatol.*, 142, 1007, 2000.
17. Denda, M., Wood, L.C., Emami, S., et al., The epidermal hyperplasia associated with repeated barrier disruption by acetone treatment or tape stripping cannot be attributed to increased water loss, *Arch. Dermatol. Res.*, 288, 230, 1996.
18. Denda, M., Kitamura, K., Elias, P.M., et al., trans-4-(Aminomethyl)cyclohexane carboxylic acid (T-AMCHA), an anti-fibrinolytic agent, accelerates barrier recovery and prevents the epidermal hyperplasia induced by epidermal injury in hairless mice and humans, *J. Invest. Dermatol.*, 109, 84, 1997.
19. Denda, M., Brown, B.E., Elias, P.M., et al., Epidermal injury stimulates prenylation in the epidermis of hairless mice, *Arch. Dermatol. Res.*, 289, 104, 1997.
20. Menon, G.K., Feingold, K.R., and Elias, P.M., Lamellar body secretory response to barrier disruption, *J. Invest. Dermatol.*, 98, 279, 1992.
21. Feingold, K.R., The regulation and role of epidermal lipid synthesis, *Adv. Lipid Res.*, 24, 57, 1991.
22. Holleran, W.M., Takagi, Y., Menon, G.K., et al., Processing of epidermal glucosylceramides is required for optimal mammalian cutaneous permeability barrier function, *J. Clin. Invest.*, 91, 1656, 1993.
23. Holleran, W.M., Gao, W.N., Feingold, K.R., et al., Localization of epidermal sphingolipid synthesis and serine palmityl transferase activity: alterations imposed by permeability barrier requirements, *Arch. Dermatol. Res.*, 287, 254, 1995.
24. Holleran, W.M., Man, M.Q., Gao, W.N., et al., Sphingolipids are required for mammalian epidermal barrier function. Inhibition of sphingolipid synthesis delays barrier recovery after acute perturbation, *J. Clin. Invest.*, 88, 1338, 1991.
25. Holleran, W.M., Feingold, K.R., Man, M.Q., et al., Regulation of epidermal sphingolipid synthesis by permeability barrier function, *J. Lipid. Res.*, 32, 1151, 1991.
26. Stachowitz, S., Alessandrini, F., Abeck, D., et al., Permeability barrier disruption increases the level of serine palmitoyltransferase in human epidermis, *J. Invest. Dermatol.*, 119, 1048, 2002.
27. Berardesca, E., Vignoli, G.P., Distante, F., et al., Effects of water temperature on surfactant-induced skin irritation, *Contact Dermatitis*, 32, 83, 1995.

28. Proksch, E., Holleran, W.M., Menon, G.K., et al., Barrier function regulates epidermal lipid and DNA synthesis, *Br. J. Dermatol.*, 128, 473, 1993.
29. Wood, L.C., Jackson, S.M., Elias, P.M., et al., Cutaneous barrier perturbation stimulates cytokine production in the epidermis of mice, *J. Clin. Invest.*, 90, 482, 1992.
30. Nickoloff, B.J. and Naidu, Y. Perturbation of epidermal barrier function correlates with initiation of cytokine cascade in human skin, *J. Am. Acad. Dermatol.*, 30, 535, 1994.
31. Wood, L.C., Elias, P.M., Calhoun, C., et al., Barrier disruption stimulates interleukin-1 alpha expression and release from a pre-formed pool in murine epidermis, *J. Invest. Dermatol.*, 106, 397, 1996.
32. Ashida, Y., Denda, M., and Hirao, T., Histamine H1 and H2 receptor antagonists accelerate skin barrier repair and prevent epidermal hyperplasia induced by barrier disruption in a dry environment, *J. Invest. Dermatol.*, 116, 261, 2001.
33. Taljebini, M., Warren, R., Mao-Qiang, M., et al., Cutaneous permeability barrier repair following various types of insults: kinetics and effects of occlusion, *Skin Pharmacol.*, 9, 111, 1996.
34. Sauer, G.C. and Hall, J.C., Seasonal skin diseases, in *Manual of Skin Diseases*, Sauer, G.C. and Hall, J.C., Eds., Lippincott-Raven, Philadelphia, 1996, pp. 23–28.
35. Wilkinson, J.D. and Rycroft, R.J., Contact dermatitis, in *Textbook of Dermatology*, Vol. 5, Champion, R.H., Burton, J.L., and Ebeling, F.J.G., Eds., Blackwell Scientific, Oxford, 1992, pp. 614–615.
36. Rycroft, R.J. and Smith, W.D., Low humidity occupational dermatoses, *Contact Dermatitis*, 6, 488, 1980.
37. Hosoi, J., Hariya, T., Denda, M., et al., Regulation of the cutaneous allergic reaction by humidity, *Contact Dermatitis*, 42, 81, 2000.
38. Denda, M., Influence of dry environment on epidermal function, *J. Dermatol. Sci.*, 24(Suppl 1), S22, 2000.
39. Choi, E.H., Kim, M.J., Ahn, S.K., et al., The skin barrier state of aged hairless mice in a dry environment, *Br. J. Dermatol.*, 147, 244, 2002.
40. Ashida, Y., Ogo, M., and Denda, M., Epidermal interleukin-1 alpha generation is amplified at low humidity: implications for the pathogenesis of inflammatory dermatoses, *Br. J. Dermatol.*, 144, 238, 2001.
41. Denda, M., Sato, J., Tsuchiya, T., et al., Low humidity stimulates epidermal DNA synthesis and amplifies the hyperproliferative response to barrier disruption: implication for seasonal exacerbations of inflammatory dermatoses, *J. Invest. Dermatol.*, 111, 873, 1998.
42. Sato, J., Denda, M., Nakanishi, J., et al., Dry condition affects desquamation of stratum corneum in vivo, *J. Dermatol. Sci.*, 18, 163, 1998.
43. Denda, M., Sato, J., Masuda, Y., et al., Exposure to a dry environment enhances epidermal permeability barrier function, *J. Invest. Dermatol.*, 111, 858, 1998.
44. Denda, M., Epidermal proliferative response induced by sodium dodecyl sulphate varies with environmental humidity, *Br. J. Dermatol.*, 145, 252, 2001.
45. Egawa, M., Oguri, M., Kuwahara, T., et al., Effect of exposure of human skin to a dry environment, *Skin Res. Technol.*, 8, 212, 2002.
46. Sato, J., Denda, M., Chang, S., et al., Abrupt decreases in environmental humidity induce abnormalities in permeability barrier homeostasis, *J. Invest. Dermatol.*, 119, 900, 2002.

47. Halkier-Sorensen, L., Menon, G.K., Elias, P.M., et al., Cutaneous barrier function after cold exposure in hairless mice: a model to demonstrate how cold interferes with barrier homeostasis among workers in the fish-processing industry, *Br. J. Dermatol.*, 132, 391, 1995.
48. Sato, J., Katagiri, C., Nomura, J., et al., Drastic decrease in environmental humidity decreases water-holding capacity and free amino acid content of the stratum corneum, *Arch. Dermatol. Res.*, 293, 477, 2001.
49. Edelberg, R., Relation of electrical properties of skin to structure and physiologic state, *J. Invest. Dermatol.*, 69, 324, 1977.
50. Lo, J.S., Oriba, H.A., Maibach, H.I., et al., Transepidermal potassium ion, chloride ion, and water flux across delipidized and cellophane tape-stripped skin, *Dermatologica*, 180, 66, 1990.
51. Barker, A.T., Jaffe, L.F., and Vanable, J.W., Jr., The glabrous epidermis of cavies contains a powerful battery, *Am. J. Physiol.*, 242, R358, 1982.
52. Lee, S.H., Elias, P.M., Proksch, E., et al., Calcium and potassium are important regulators of barrier homeostasis in murine epidermis, *J. Clin. Invest.*, 89, 530, 1992.
53. Mauro, T., Bench, G., Sidderas-Haddad, E., et al., Acute barrier perturbation abolishes the Ca2+ and K+ gradients in murine epidermis: quantitative measurement using PIXE, *J. Invest. Dermatol.*, 111, 1198, 1998.
54. Denda, M., Hosoi, J., and Asida, Y., Visual imaging of ion distribution in human epidermis, *Biochem. Biophys. Res. Commun.*, 272, 134, 2000.
55. Menon, G.K., Price L.F., Bommannan, B., et al., Selective obliteration of the epidermal calcium gradient leads to enhanced lamellar body secretion, *J. Invest. Dermatol.*, 102, 789, 1994.
56. Denda, M., Katagiri, C., Hirao, T., et al., Some magnesium salts and a mixture of magnesium and calcium salts accelerate skin barrier recovery, *Arch. Dermatol. Res.*, 291, 560, 1999.
57. Denda, M. and Kumazawa, N., Negative electric potential induces alteration of ion gradient and lamellar body secretion in the epidermis, and accelerates skin barrier recovery after barrier disruption, *J. Invest. Dermatol.*, 118, 65, 2002.
58. Said, L., Serup, J., and Rebel, C., Epidermal calcium release (ECR) in vivo sampled with a simple washout chamber technique, *Skin Res. Technol.*, 8, 219, 2002.
59. Forslind, B., Werner-Linde, Y., Lindberg, M., et al., Elemental analysis mirrors epidermal differentiation, *Acta Derm. Venereol.*, 79, 12, 1999.
60. Denda, M., Inoue, K., Fuziwara, S., et al., P2X purinergic receptor antagonist accelerates skin barrier repair and prevents epidermal hyperplasia induced by skin barrier disruption, *J. Invest. Dermatol.*, 119, 1034, 2002.
61. Warren, R., Ertel, K.D., Bartolo, R.G., et al., The influence of hard water (calcium) and surfactants on irritant contact dermatitis, *Contact Dermatitis*, 35, 337, 1996.
62. Ghadially, R., Brown, B.E., Sequeira-Martin, S.M., et al., The aged epidermal permeability barrier. Structural, functional, and lipid biochemical abnormalities in humans and a senescent murine model, *J. Clin. Invest.*, 95, 2281, 1995.
63. Hosoi, J., Tsuchiya, T., Denda, M., et al., Modification of LC phenotype and suppression of contact hypersensitivity response by stress, *J. Cutan. Med. Surg.*, 3, 79, 1998.
64. Denda, M., Tsuchiya, T., Elias, P.M., et al., Stress alters cutaneous permeability barrier homeostasis, *Am. J. Physiol. Reg. Integr. Comp. Physiol.*, 278, R367, 2000.
65. Denda, M, Tsuchiya, T., Hosoi, J., et al., Immobilization-induced and crowded environment-induced stress delay barrier recovery in murine skin, *Br. J. Dermatol.*, 138, 780, 1998.

66. Fluhr, J.W., Dickel, H., Kuss, O., et al., Impact of anatomical location on barrier recovery, surface pH and stratum corneum hydration after acute barrier disruption, *Br. J. Dermatol.*, 146, 770, 2002.
67. Denda, M. and Tsuchiya, T., Barrier recovery rate varies time-dependently in human skin, *Br. J. Dermatol.*, 142, 881, 2000.
68. Garg, A., Chren, M.M., Sands, L.P., et al., Psychological stress perturbs epidermal permeability barrier homeostasis: implications for the pathogenesis of stress-associated skin disorders, *Arch. Dermatol.*, 137, 53, 2001.
69. Elsner, P., Trends in stresses on barrier integrity, *Cutis*, 70, 4, 2002.
70. Fluhr, J.W., Gloor, M., Lehmann, L., et al., Glycerol accelerates recovery of barrier function in vivo, *Acta Derm. Venereol.*, 79, 418, 1999.
71. Denda, M., Inoue, K., Inomata, S., et al., gamma-Aminobutyric acid (A) receptor agonists accelerate cutaneous barrier recovery and prevent epidermal hyperplasia induced by barrier disruption, *J. Invest. Dermatol.*, 119, 1041, 2002.
72. Chamlin, S.L., Kao, J., Frieden, I.J., et al., Ceramide-dominant barrier repair lipids alleviate childhood atopic dermatitis: changes in barrier function provide a sensitive indicator of disease activity, *J. Am. Acad. Dermatol.*, 47, 198, 2002.
73. Elias, P.M. and Ghadially, R., The aged epidermal permeability barrier: basis for functional abnormalities, *Clin. Geriatr. Med.*, 18, 103, 2002.
74. Weigand, D.A. and Gaylor, J.R., Irritant reaction in Negro and Caucasian skin, *South Med. J.*, 67, 548, 1974.
75. Kompaore, F., Marty, J.P., and Dupont, C., In vivo evaluation of the stratum corneum barrier function in blacks, Caucasians and Asians with two noninvasive methods, *Skin Pharmacol.*, 6, 200, 1993.
76. Berardesca, E. and Maibach, H.I., Racial differences in sodium lauryl sulphate induced cutaneous irritation: black and white, *Contact Dermatitis*, 18, 65, 1988.

18 Testing the Efficacy of Moisturizers

Bernard Gabard

CONTENTS

I. Introduction 212
II. Definitions 212
III. TEWL, Hydration of the SC, and Mode of Action of Moisturizers 213
  A. Water in the SC 213
  B. TEWL 214
  C. Effect of Hydration, Relationship with TEWL 214
    1. Effects on Healthy Skin with Intact Barrier Function 214
    2. Effects on Damaged/Diseased Skin or Skin with Modified Barrier Function 214
  D. Mode of Action of Moisturizers 215
IV. Experimental Measurement of TEWL and Hydration 216
  A. Measurement of TEWL 216
  B. Measurement of Hydration 216
V. Experiments on Healthy Skin 216
  A. Effects of a Single Application 216
    1. Testing Moisturizers 216
    2. Testing Humectants 219
  B. Effects of Repeated Applications 220
VI. Experiments on Diseased Skin 221
  A. Experimental Studies 221
    1. Single Irritation 221
    2. Cumulative Irritation 223
    3. Discussion and Conclusion 228
  B. Clinical Studies 231
VII. Conclusion 231
References 232

0-8493-1443-7/05/$0.00+$1.50

I. INTRODUCTION

Among the beneficial properties claimed for dermatological and cosmetic products, "moisturizing" is possibly the most widely used. This term has been coined after the now classic observations of Blank,[1,2] who discovered the plasticizing effect of water in the stratum corneum (SC). A common skin surface disturbance, xerosis, or so-called dryness of the skin, is experienced by most persons at some time, by a few persons all the time, and by all individuals increasingly as they grow older.[3] Thus, as consumers advance in age, concern about dry skin increases. In addition, moist, clean, soft, and wrinkle-free skin is perceived as youthful, and for this reason, moisturizers are widely used, and skin care with these products is regarded as a dominant growth area in cosmetics and toiletries.

II. DEFINITIONS

The term "emollient" is another ill-defined term that is often used as a synonym to moisturizer in connection with dermatological or cosmetic products.[3,4] Emollient is derived from the Latin adjective for soft ("molle," i.e., soft or smooth), and in fact, keeping the skin soft, "molle," smooth, and supple is a main claim of moisturizers. Idson[5,6] defines moisturizers as actively increasing the water content of the skin and emollients as lubricating or occluding the skin with water-insoluble materials. Thus, moisturizer means any water-miscible material externally applied that hydrates the skin, and emollient means a material that imparts a smooth, soft feeling to the skin but increases the water content in the SC passively by occluding the surface. "Occlusive moisturizers" combine the softening effects of emollients with the water-retaining action of moisturizers and the coating activity of lubricants.[6] Lodén[4] defines emollients in a more general fashion, as having a mode of action similar to the occlusive moisturizers. Moisturizing also may be considered in a more general way, including occlusion, humectancy, and restoration of possibly deficient materials[7] ("humectants" being defined as compounds that will attract moisture to the skin).[6,7] Emolliency is considered to be associated only with imparting smoothness and a general sense of well-being to the skin, as determined by touch.[7]

There is no general agreement about these definitions. Moisturizers are considered further as compounds imparting water-binding properties to a cream,[8] whereas Ummenhofer defines compounds with this property as humectants.[9] Moreover, this author considers moisturizers to be materials impairing the evaporation of water from the cream after application to the skin, and emollients as simply creams containing moisturizers. Wehr and Krochmal[10] define emollients as systems that smooth the roughened surface of the SC but do not occlude the skin and that usually do not show any effect on the transepidermal water loss (TEWL) unless applied in great quantities. Furthermore, he considers humectants to be compounds that will attract water when applied to the skin, but the source of this water is transepidermal, not coming from the atmosphere.[10] Idson[6] expands this concept and makes the source of the attracted water dependent on the relative ambient humidity.

There is now a general trend toward using the name "moisturizer" to define a product that will be applied to the skin (not to describe the compound or compounds contained in that product); mostly an emulsion — containing or not an active substance — such as a lotion, a cream, an ointment, or a bath oil with the purpose of hydrating the SC and of regenerating a dry, rough, or scaly surface resulting from xerosis, irritation, or any other cause.[4,10–16] In an introduction to a recent book, Kligman[17] vividly pointed out how difficult it is to precisely define what a moisturizer is, partly because of the absence of an accepted definition of what constitutes dry skin.

Given the contradictions and possible confusions outlined above, the term "moisturizer" will be used throughout the following pages in reference to the product applied to the skin — emolliency will not be considered a separate entity — and humectants will be defined as any compound or material that is intended to bind or attract water in or to the SC, and that is incorporated in the moisturizer.

The action of moisturizers is considered only on the SC, even if for convenience or stylistic clearness the terms "skin" or "skin surface" are used. The primary barrier against the physical and biological environment is the SC, the thin coherent membrane of keratinized epithelial cells making up the dead surface layer of the epidermis.[2,18] The SC ensures most of the mechanical strength of the epidermis and remains soft and flexible under various conditions, allowing free body movement without fissuring, provided a constant flux of water from the deeper layers, and providing an intact barrier function. Hydration, defined as the content of water in the SC, and the barrier function (determined by, e.g., measurement of the TEWL) are very important factors influencing the biophysical properties of the SC. Maintaining an appropriate water content or barrier function is a significant clinical and cosmetic concern.[19]

III. TEWL, HYDRATION OF THE SC, AND MODE OF ACTION OF MOISTURIZERS

It is important, for proper interpretation of the TEWL changes resulting from the use of a moisturizer, to recall some basic knowledge on water in the SC and the relationship between hydration state and TEWL.

A. Water in the SC

Normal, healthy SC contains about 10 to 20% water.[1,2,20,21] The water content depends on

1. The quantity and flux of water passing from the inferior layers to the SC — there is always a water supply from the underlying living tissue *in vivo*, and the water concentration in the lowest SC layer is about 49 M, which is equal to that of the epidermis[22]
2. The flux of water that is eliminated by evaporation (TEWL)
3. The ability of the SC to retain the water; the previous elements determine the value of the TEWL, which is about 2 to 5 g/m^2 per hour on the volar forearm of healthy persons,[20] which corresponds to the passive diffusion of water vapor through epidermis and SC and depends on the gradient of

water concentration between both sides of the SC: As the water concentration in the upper layer of the SC is about 12 M (at 40% relative humidity and 31°C temperature), the concentration gradient amounts to 37 M[22]

B. TEWL

The mathematical principle governing the diffusion of water through SC is Fick's law.[18] As the SC is not an inert membrane and shows a certain affinity for water, Fick's law needs to be modified:[7,22,23]

$$J = K_m \times D(C_s/e),$$

where J is the flow per unit area (mol/cm^2 per hour); K_m is the partition coefficient [(concentration in membrane)/(concentration in solution)]; D is the diffusion coefficient of water (cm^2/h), which itself depends on the water content; C_s is the concentration gradient of water across the SC (mol/cm^3), and e is the SC thickness (cm).

C. Effect of Hydration, Relationship with TEWL

1. Effects on Healthy Skin with Intact Barrier Function

Hydration has two main effects: it increases SC thickness e and the diffusion constant D,[22–24] and K_m may increase or decrease depending on the concentration in either compartment. For example, occlusion may increase the water content of the SC up to 50%, whereby 37% more surface area will be found.[21] Thus, the net result of a change in SC hydration on TEWL is not always predictable. Given no change in the partition coefficient, an increase of the diffusion constant will lead to an increase of TEWL that will be antagonized by the increase in thickness. However, in healthy skin (with a normal barrier function not influenced by application of the moisturizer), D predominates and TEWL increases.[1,25]

The recent availability of modern devices to measure the water content of the SC with some degree of reliability allows us to draw some general rules governing the relationship between hydration and barrier function of the SC in health and disease[19,20,26,27] and to confirm the direct relationship between TEWL and hydration in healthy skin. TEWL is low in newborns of 37 weeks' gestation or more who possess an efficient SC.[19] This relationship also holds on different anatomical locations in infants and adults.[19,20,28–30] There is at least circumstantial evidence that skin aging as a physiological event is characterized by a decrease of both TEWL and SC hydration, whereby both maintain their directly proportional relationship.[19,31] This shows that the notion of "dry skin" does not apply for the same phenomenon in aging or pathological conditions, as in the elderly the barrier function is intact.

2. Effects on Damaged/Diseased Skin or Skin with Modified Barrier Function

Because of the stringent dependence of TEWL on the barrier function of the SC, the relations defined above do not hold for when the skin is diseased or when the

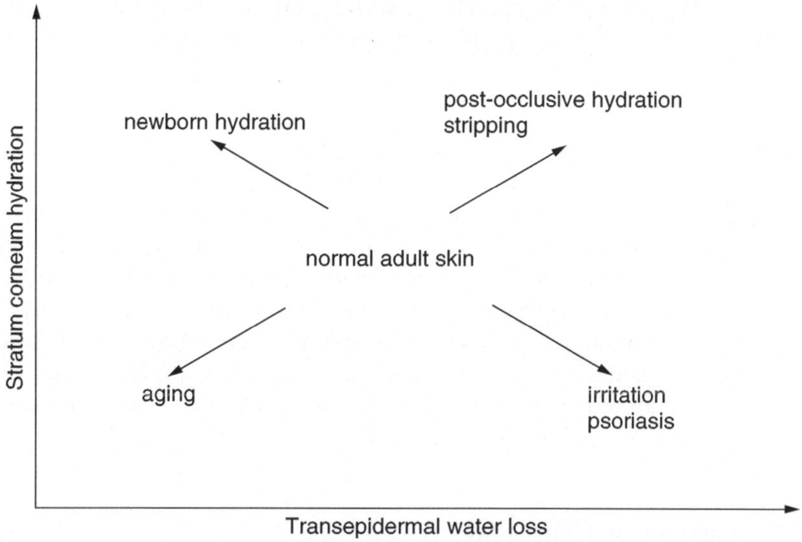

FIGURE 18.1 Relationship between SC hydration and TEWL in health and disease. (From Reference 19; with permission.)

applied product influences the barrier function in either way (e.g., decrease through irritation or increase through occlusion).

Indeed, in pathological SC, there may be an inverse relationship between TEWL and hydration: the lower the skin hydration, the higher the TEWL. This may be found in newborns of 32 weeks' gestation or less in whom a high TEWL is accompanied by a decreased water-holding capacity of an incomplete SC.[19] The same paradoxical results are consistently found in various scaly dermatoses.[20] Thus, the pathological skin becomes dry and hard because the SC lacks the capacity or presents a reduced ability to bind water. In this case, TEWL cannot be regarded as an exact reflection of the water content in the SC (Figure 18.1).[19,20]

D. Mode of Action of Moisturizers

Given these general considerations, moisturizers may basically act on the water concentration gradient in the SC in two ways: by restoring, retaining, or increasing moisture in the SC allowing the addition of water from the dermis and stratum Malpighi or from the applied product, and by occluding the skin surface, impeding TEWL, and causing a buildup of moisture in the SC.

These modes of action are often combined in the same product. There are, however, exceptions such as petrolatum or floating (spreading) bath oils that are considered to work only by forming an occlusive layer on the SC.[10,25,32,33] One may again quote Albert Kligman[17] asking himself, "How do moisturizers work?" and answering, "we really do not know and much of what we think we know is wrong."

IV. EXPERIMENTAL MEASUREMENT OF TEWL AND HYDRATION

A. Measurement of TEWL

An important consequence of the mathematical formula describing TEWL is the need to perform measurements under carefully controlled conditions. It is easy to understand that if, for example, the relative humidity is not constant, the water content of the upper SC layers will change correspondingly, as will the concentration gradient, which almost reaches zero as the relative humidity approaches 100%.[22] In addition, the diffusion constant is highly dependent on temperature.[22,23] Sweating because of heat or emotional or sympathetic activation should be carefully avoided. Applied products themselves or their ingredients may greatly influence the outcome of the TEWL measurements.[34] Unfortunately, the conditions of measurements are not always indicated in published work. Published guidelines should be imperatively followed.[35,36]

B. Measurement of Hydration

Several techniques may be considered to evaluate the water content of the SC *in vivo*. Among the most popular are those based on electrical methods, as convenient-to-use devices are now available (see Part III of this book).[20] A recent published compilation shows the diversity of available techniques for the measurement of SC hydration.[37] In any case, measurements under controlled temperature and relative humidity conditions are strongly recommended, particularly during experiments that extend over several hours or days.

V. EXPERIMENTS ON HEALTHY SKIN

A. Effects of a Single Application

1. Testing Moisturizers

Alter a single application of a moisturizer on healthy skin, two distinct phases can be defined on the basis of measured TEWL changes.[11]

a. *Evaporation Phase*

The evaporation phase is of short duration, usually 10 to 15 minutes, during which TEWL initially shows an abrupt increase followed by a gradual decrease to baseline values.[4,11,38,39] This phenomenon is caused by the evaporation of the water contained in the applied product on contact with the skin surface and by temperature adjustment.

If pure water is applied on the skin, evaporation as measured by TEWL changes is complete within 2 minutes, showing the rapidity of this process.[25,40] The fact that on application of a moisturizer, evaporation of water will be measured for a longer period of time is a consequence of the nature of the applied product and of the interaction of the moisturizer water with the SC.

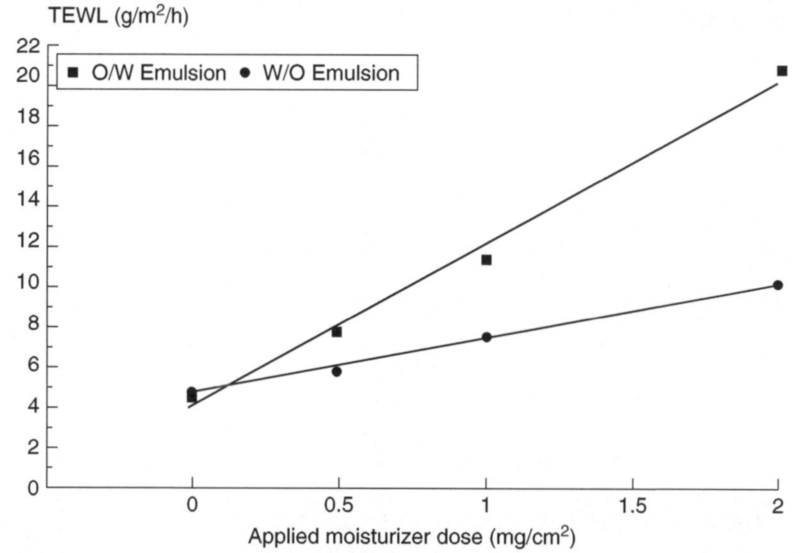

FIGURE 18.2 Relationship between TEWL 2 min after application and the applied amount of the moisturizing emulsion. (Redrawn from References 38 and 39; with permission.)

Different authors[4,38–40] demonstrated that the maximal TEWL value attained immediately after application of a moisturizer depends on the quantity of water the product contains. This is illustrated in Figure 18.2, which shows that the application of an oil-in-water (O/W) emulsion leads to higher TEWL values than the application of a water-in-oil (W/O) emulsion. Similar results were reported by Lodén[4] using an ordinary cream (71% water) and a lipid-rich moisturizer (30% water). In addition, Lodén confirmed the results of Rietschel[25] by showing that application of water-free petrolatum did not lead to an increase of TEWL. In fact, petrolatum decreased TEWL because of its occlusive properties.[4,25] Peak TEWL depends on the quantity of product applied to the skin (Figure 18.2), confirming that the measured value represents the evaporation of water contained in the product.

In addition, if the applied products are wiped off from the surface of the skin after only 5 minutes to eliminate their possible influence on the barrier function of the SC, a significant increase in TEWL could be measured (Figure 18.3).[4] This increase is the result of the release of excess water already absorbed in the SC. Here, too, the magnitude of the measured evaporation changes depend on the water content of the moisturizers: the more water the products contain, the more water is absorbed into the skin. Thus, an increase in TEWL was not observed after removal of petrolatum applied to the skin for 5 minutes.[4] It is known that the upper SC takes up water from the environment avidly and almost immediately.[1,41] Thus, it is not surprising that even a short application of a moisturizer leads to an increase of water in the SC that is measurable by TEWL under carefully controlled conditions after wiping off the product.

b. *Lipidization Phase*

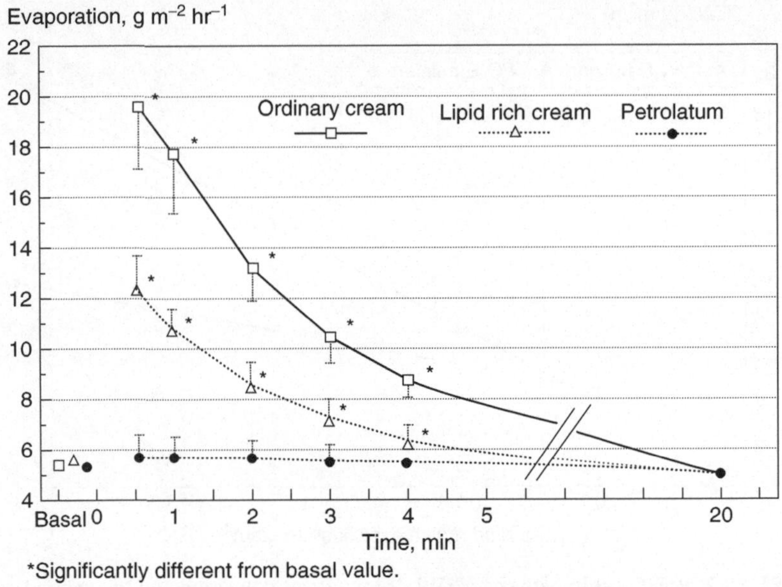

FIGURE 18.3 TEWL after removal of moisturizer residue (mean ± SE). The products were wiped off the skin after 5 min. (From Reference 4; with permission.)

Following the evaporation phase, a much longer (order of magnitude up to several hours) lipidization phase is found, characterized by the slow clearing of the emulsion lipids respective of the nonvolatile components from the surface of the skin.[11,33] An increase in the water content of the SC may be seen, but only slight or no TEWL changes will be detected in comparison to a control area, depending on the nature of the applied moisturizer.

Lipids on the skin surface usually occlude the SC and impair water evaporation, depending on the amount applied and the capacity of these lipids to build an isotropic oily phase after evaporation of the water from the emulsion.[42] During the evaporation phase, a "new" emulsion is formed on the skin, containing a small quantity of nonevaporated water and other nonvolatile components such as surfactants. Indeed, no change[11] or slight (16%)[4] or important (45%)[4,25] decrease of TEWL may be detected during the lipidization phase, depending on the formulation applied on the skin: a lotion containing 80% water in the first case, a cream with 34% or 71% water in the second, and petrolatum in the third.

Again, wiping off the applied products during the lipidization phase allows the excess water absorbed in the SC to be measured.[4,25] An increase in TEWL occurs that, contrary to the results obtained during the evaporation phase, is inversely related to the water content of the moisturizers applied: The more lipids in the formulation, the higher the TEWL increase after removal of the remaining components of the products, with petrolatum leading to the greatest changes (Figure 18.4). This, of course, emphasizes the role of the occlusive properties of the moisturizer.

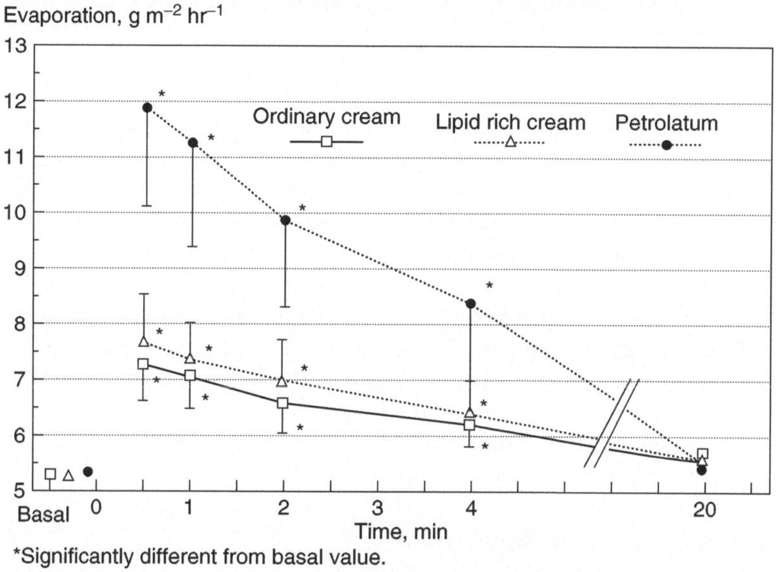

FIGURE 18.4 TEWL after removal of moisturizer residue (mean ± SE). The products were wiped off the skin after 40 min. (From Reference 4; with permission.)

Thereafter, TEWL returns to the basal values within 10 to 20 minutes.[4] An analysis of the water concentration profile in the SC during the lipidization phase shows that the water is not uniformly distributed across the tissue and accumulates for the greatest part in the upper tenth of the SC.[24] It is, therefore, not surprising that the evaporation of this excess water takes only a short time.

2. Testing Humectants

Humectants, as defined earlier, encompass a wide range of products such as glycerol, urea, lipids, α-hydroxyacids, and so forth.[43]

Similar to the TEWL changes described earlier, two phases were observed after a single application of an aqueous humectant solution (containing 5 or 15% glycerol) to the skin: an initial increase resulting from the evaporation of water from the solution, with the TEWL returning to basal values within minutes, followed by a marked decrease in TEWL for hours.[44] As an aqueous solution was used, no lipids were applied to the skin that could decrease TEWL by occlusion. Thus, the diminution measured during the second (lipidization) phase was the result of an increased retention of water in the SC. A parallel augmentation of the SC hydration was indeed detected. Identical results were obtained if the glycerol was applied in emulsion form (at 10% concentration).[44]

These results should not be generalized, as other humectants may lead to different results. For example, Rietschel showed an increase of TEWL during the lipidization phase after applying urea (10% in a lotion) to the skin.[25,45] Different humectants may create different water retention and diffusion conditions in the SC that will ultimately lead to different TEWL changes. The following points should be borne in mind:

1. The skin naturally contains different hydrosoluble substances in a balanced mixture controlling the hydration state of the SC (natural moisturizing factors; NMFs).[46] Urea is one of these substances. Externally applied humectants may interact with or modify the concentration of one or more of these NMF components, leading to modification of the water-retention capacity of the final mixture.
2. Humectants play an important role in the formation of an isotropic oily phase from the applied emulsion on the skin, as outlined previously.[42,47] Their effects are variable. Thus, finally, the occlusive properties of an emulsion during the lipidization phase will also depend on the interaction of the humectant or humectants contained in it with other nonvolatile components of the formulation, in addition to the quantity of lipids contained within and the amount applied.

B. Effects of Repeated Applications

Frödin and colleagues[48] investigated the effect of an O/W cream on the hydration of normal skin (test area was the volar forearm) in a group of seven women. The cream was applied four times during 2 days, and hydration of the SC, as well as TEWL, was measured before and 2 hours after the application. A statistically significant increase of hydration was measured, but changes in TEWL were absent.

Serup, Winther, and Blichmann[12] treated the volar forearm of 16 healthy women twice daily with a lotion containing 80.4% water. Measurements of SC hydration, TEWL, skin surface lipids, and cutaneous blood flow were performed at regular intervals starting 12 hours after the application for 14 days. A significant increase in SC hydration was found, but there were no changes in other parameters.

Serup[49] compared two moisturizing lotions containing humectants such as urea (3%, respectively 10%), lactic acid (1.5%, respectively 5%), and betaine (1.5%, respectively 5%). The water content was 86% for the first lotion, and 94% for the second. Twenty-three volunteers tested the first ("light") moisturizer, and 24 the other. They were volunteers with healthy skin but were selected out of 72 participants because of the low hydration values of their SC. Test sites were delimited on the volar side of one forearm, with the other arm serving as control. The moisturizers were applied to the test areas twice daily for 3 weeks, and measurements were made 12 h after the last application. Clinical scoring, skin color, hydration of the SC, TEWL, skin surface lipids, scaling, and epidermal desquamation were assessed in a blind fashion for the investigator. In addition, a questionnaire was completed by the participants. Although the two lotions appeared equally effective, the "light" one was found cosmetically more acceptable by the users. There was a definite increase in SC hydration, a reduced scaling, and overall "clinical" improvement. TEWL decreased in the group using the 10% urea lotion compared with control values and with the values in the 3% urea group.

These studies call for a short comment: They were designed to test or demonstrate the reliability of SC hydration measurements. They were not designed to record the TEWL changes that depend on the action of the applied moisturizer. In fact, as described earlier, TEWL measurements performed several hours after the application (i.e., during the late lipidization phase) may show no change or only a slight decrease that is difficult to detect if the measurements are performed at sporadic intervals.[12,48] Here, TEWL was considered merely as an indicator of the barrier function that may have been modified by the application of the product and not as an indicator of the moisturizer effect on the hydration state of the SC. The results obtained with the 10% urea lotion are different from those reported by Rietschel[25,45] and mentioned earlier. As a decrease of TEWL was measured, it was concluded that urea improved the barrier function, but it could have been an increase in the water-retaining capacity of the SC as well. This emphasizes once more that there are no general rules concerning the effects of humectants on TEWL and the hydration state of the SC.

VI. EXPERIMENTS ON DISEASED SKIN

A. Experimental Studies

Two kinds of experimental studies can be distinguished. In the first one, the studies are designed to test the efficacy of a treatment with moisturizers on recovery after the irritative challenge, and the skin is irritated once.[50] This implicates a relatively strong irritative stimulus, as a reaction should be clearly seen after a short time. In the second kind of study, a mild, standardized irritation is repeated at regular intervals. A reaction of the skin may appear only after several consecutive stimuli ("cumulative" irritation).[51] This is thought to better mimic an everyday situation. In this latter case, treatment with moisturizers may be conducted together with or after the irritation to obtain information on the prevention of the irritation or on the recovery from the irritated state.

A further distinction may be made depending on the nature of the irritative stimulus: chemical, such as organic solvents or sodium lauryl sulfate (SLS), or physical, such as stripping the SC with Sellotape. In the majority of cases in which a chemical irritant is used, an aqueous SLS solution under occlusion, usually in a large Finn chamber, is the "standard" irritative challenge used in single as well as cumulative irritation studies.[52] Removal of the SC with tape stripping is another irritative stimulus often used, but exact standardization of this procedure is needed.[53]

1. Single Irritation

a. Chemical Irritation

Blanken and colleagues[54] irritated the skin of the forearms with SLS (2.5%) under occlusion in a group of 30 healthy volunteers for 12, 24, or 48 hours. After this period, four different moisturizers ("after-work emollient creams") were applied to the irritated sites twice daily for 5 days. Recovery rate from the irritation was quantified with visual scoring of erythema and scaling according to Frosch and Kligman,[55] with TEWL, and with cutaneous blood flow measurements. There were no differences in recovery between any of the moisturizer-treated sites and the control sites.

Van Neste and Antoine[56] applied SLS 1, 5, and 10% under occlusion for 48 hours on the skin of the lower back of 10 healthy female volunteers who were otherwise complaining from "dry" skin. Skin reactions were followed for 10 days by clinical scoring of roughness and erythema and measurements of hydration of the SC, TEWL, and skin blood flow (laser Doppler). Irritated sites were left untreated (control) or treated twice daily with a moisturizer containing urea and lactic acid (10 and 5%, respectively) or with the same vehicle without the humectants. On the control sites, the measured effects corresponded with the concentration of SLS (0 to 10%), but no differences could be observed between the two highest concentrations (5 and 10%). Increased TEWL values and decreased hydration values were recorded, with the decrease in hydration being delayed in comparison with the course of TEWL changes. Treated sites showed statistically significant higher hydration values, but there were no differences between the humectants containing moisturizer and their vehicle. TEWL was increased by treatment compared with the TEWL of irritated control sites, but again, no statistical significance could be found. The rate of recovery was identical on all sites.

A similar experiment was conducted by Berardesca et al.[57] in 10 healthy volunteers. SLS (1%) was applied under occlusion on the volar forearm for 24 hours. A moisturizing emulsion containing 4.5% NMF or a lotion containing 0.01% ceramide 1 among other agents as applied daily for 3 days. The rate of recovery was assessed by measurements of TEWL and hydration of the SC. Peak TEWL was lower and recovery faster (TEWL not significantly different from baseline) after treatment with the ceramide moisturizer compared with the NMF moisturizer, but only the NMF-containing emulsion increased the water content of the SC.

More recent publications emphasize the value of the TEWL measurements in the single SLS-application model for claim substantiation of moisturizer efficacy.[58,59]

The skin may also be irritated by organic solvents such as acetone, toluene, or a combination of different solvents.[33,60] SLS and solvents may be combined to create an irritation that is more representative of daily conditions.[61]

b. Physical Irritation

Tape-stripped skin is most sensitive to occlusion and, as such, is most suitable to test the occlusive properties of moisturizers. As already described, these occlusive properties are, among others, dependent on the lipids contained in the emulsion.

The SC of the forearm of six healthy volunteers was tape stripped until TEWL attained values between 40 and 50 g/m^2 per hour. The stripped sites (1.5 × 2 cm) were located in the middle of larger areas (2.5 cm), to which 2.5 mg/cm^2 of the test products were carefully applied after a rest period of 1 hour. TEWL was measured before and at different times after application (Figures 18.5, 18.6). Percentage changes related to time were calculated as

$$\% \text{ change} = 100 \times [(\text{corr. TEWL}_{\text{test, start}})\ (\text{TEWL}_{\text{test, }t})]/(\text{corr. TEWL}_{\text{test, start}}),$$

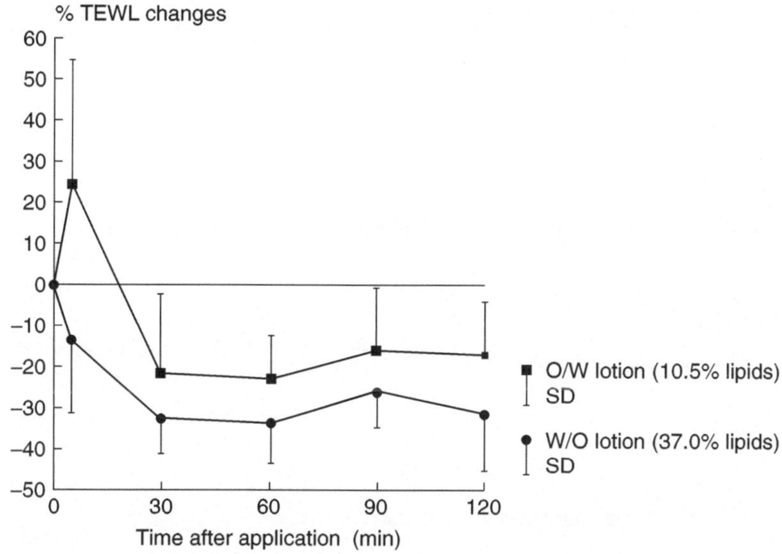

FIGURE 18.5 Percent changes of TEWL of stripped skin after application of two different moisturizers (lotions; mean ± SD). Baseline values were 53 ± 13 (O/W), respectively, 55 ± 11 (W/O) $g/m^2/h$. First measurement at t = 5 min.

where (corr. $TEWL_{test, start}$) = ($TEWL_{test, start}$) × [($TEWL_{contr., t}$)/($TEWL_{contr., start}$)], corr. stands for corrected; test is the site treated with a moisturizer; contr. stands for control, untreated site; start is the values measured before application; and *t* is the values measured at time *t*.

These calculations are needed because the small but steady decrease of TEWL of stripped skin during the experiment must be taken into account. Figures 18.5 and 18.6 show the results with moisturizers containing different amounts of lipids. It can be seen that for the O/W lotion, an evaporation phase may be shown, and that otherwise this model is extremely sensitive to the lipid content of the applied moisturizers.

This model was also used to test the effect of lipids deposited on the skin surface by bath oil after a bath under standardized conditions.[32] More lipids could be recovered from the skin surface after a bath with bath oils of the floating type than with dispersible bath oils. TEWL was efficiently reduced after the bath with the duration, not the extent, of this reduction being related to the amount of lipids deposited on the skin. Tape stripping may also be used to study the effects of moisturizers or humectants on barrier function recovery measured with TEWL.[62]

2. Cumulative Irritation

In the experiment of Hannuksela and Kinnunen,[15] frequent skin washing with a liquid dishwashing detergent was used as the irritative stimulus. Twelve healthy female volunteers washed their upper arms with the detergent for 1 minute twice a day for 1 week. Eight moisturizers (three O/W creams containing 65 to 70% water;

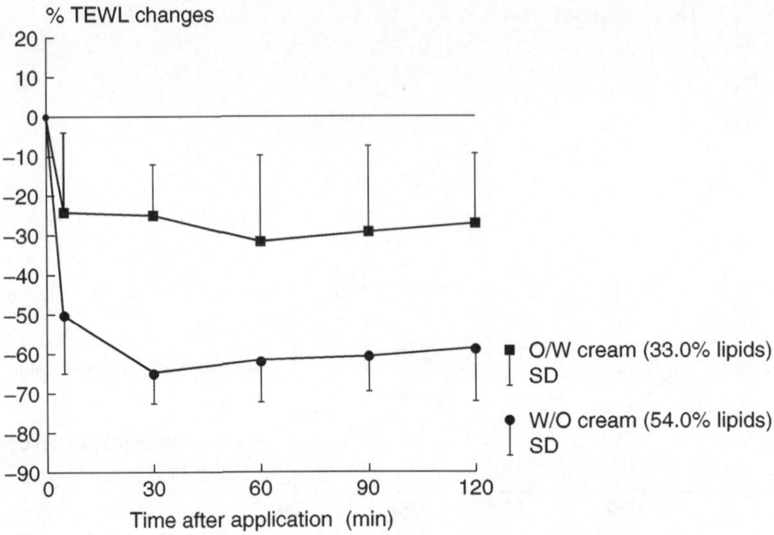

FIGURE 18.6 Percent changes of TEWL of stripped skin after application of two different moisturizers (creams; mean +/– SD). Baseline values were 53+/–3 (O/W) respectively, 60+/–7 (W/O) g/m2/h. First measurement at t = 5 min.

one skin oil; and four emulsions containing 48 to 67% water) were applied to the areas of the left arm after each washing. The right arm was washed in a similar fashion but not treated with the moisturizers. On all test sites, TEWL and cutaneous blood flow (laser Doppler) were measured 4 hours after application of the moisturizers both before the start of the experiment and on days 3 and 7. During the second week, the irritative stimulus was stopped but treatment with the moisturizers and regular measurements were continued until day 15. Another experiment was conducted 6 months later with the same subjects, who washed their upper arms in the same manner for 1 week. Treatment with one of the moisturizers was then continued for another week on one site, and the rate of recovery from the irritation was compared between both sites. The same parameters as in the first part of the study were measured. Regular washing with the dishwashing detergent induced a local irritant dermatitis characterized by a progressive increase of TEWL and cutaneous blood flow. The authors were able to show that treatment with the moisturizers prevented the development of the irritation regardless of the moisturizer used. The rate of recovery from the irritation, measured in the second part of the experiment, was only slightly faster on the treated site.

Frequent skin washing has now been standardized as a model to investigate the efficacy of moisturizers.[63] In this model, toxic influences of cleansing agents are maximized. Repetitive washings may be performed over 1 day or on several consecutive days. Many moisturizers containing, for example, glycerol, urea, or dexpanthenol as humectants have been already tested in this model using TEWL measurements as the efficacy parameter. The results have been recently summarized.[63]

TABLE 18.1
Composition of the Tested Formulations (short-term SLS irritation model)

Products	Type	Lipid Content (%)	Other
Lotio alba (NRF)	Hydrogel	0	ZnO 20; talc 20; glycerol 85% 30; water 30
Pasta zinci (DAB 6)	Fatty paste	100	ZnO 25; talc 25; petrolatum 50
Unguentum molle (DAB 6)	Onguent	98	Lanolin 50; petrolatum 50
Physiologic saline		0	Control

The effects of three mineral oil– and linoleic acid–containing W/O emulsions were assessed by Blanken and coworkers[64] on the course of an irritation induced on the forearm skin of nine healthy volunteers by a twice-daily application of occlusive SLS patches (0.5%) for 45 minutes. The moisturizers were rubbed into the irritated skin 10 minutes after removal of the SLS patches three times daily. Visual scoring and TEWL measurements of the test areas were made at regular intervals for 15 days. The mineral oil–containing ointment was shown to counteract to some extent the SLS-induced TEWL increase. Between the other two emulsions, one reduced the TEWL changes in the first week only, and the other showed no influence at all.

An interesting model for testing moisturizers in cumulative irritation was developed by Elsner[14] (the "short-term model"). A subclinical irritant dermatitis was induced on the forearms of 15 healthy volunteers by occlusive application of SLS (0.5%) for 24 hours. Daily postirritative treatment with different test products (Table 18.1) was begun after removal of the occlusion and continued for 2 days. On day 4, a new SLS challenge was performed exactly as on day 1. Visual scoring[55] and measurement of TEWL were performed immediately after the first and last irritative challenge. The results (Figure 18.7) showed that among the formulations tested, two (Lotio alba and Pasta zinci) were capable of keeping the TEWL on the level measured on day 1.

This model was further developed and extended over a 3-week period in collaboration with our laboratories ("chronic irritation model"). Twenty healthy volunteers aged 30 ± 8 years (mean ± SD; range 21 to 47; 6 female and 14 male) participated in the study after informed consent. Approval was given by the ethical committee of the University Hospital Zurich. For the induction of the cumulative irritation, eight test areas were delimited on the volar surface of each forearm (including one occluded and one nonoccluded control area). The contact dermatitis was induced by the occlusive application of SLS (1%; 30 minutes) on one arm and of toluene (100%; 10 minutes) on the other for a total of 9 days over a period of 2 weeks, with a break on the weekend (Figure 18.8). The irritants were applied on the skin as filter paper disks soaked with the solutions and placed in large Finn chambers (ø11 mm). After removal of the Finn chambers, the skin was gently blotted dry with a paper towel, and treatment was performed with moisturizers of different types, as listed in Table 18.2. Then 3 mg/cm^2 of product were carefully spread with

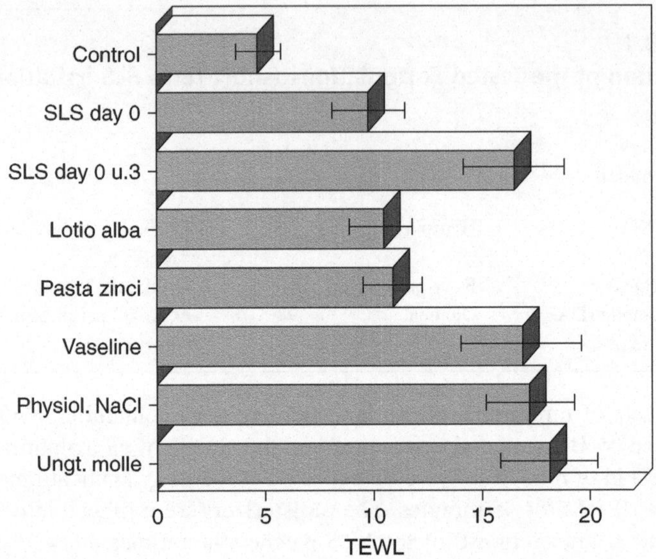

FIGURE 18.7 TEWL (g/m2/h; mean +/– SD) after postirritative treatment with different products (short-term model).

TABLE 18.2
Composition of the Tested Formulations (chronic SLS/toluene irritation-model)

Products	Type	Lipid Content (%)	Other
Lotion A	O/W	10.5	Urea 2%
Lotion B	O/W	10.5	
Lotion C	W/O	37.0	Urea 4%
Lotion D	W/O	37.0	
Petrolatum	Onguent	100.0	
LAA (FH 1991)	Hydrogel	0	ZnO 15; talc 15; PEG 15; bentonite 5; water 50

a metal spatula over the test area. The order of application was randomized, but the same products were applied to the corresponding areas on either forearm. Measurements were performed before the start (baseline) and again 24 hours later before the next irritant and subsequent test product application (Figure 18.8). They included visual scoring according to Frosch and Kligman,[55] TEWL, skin color,[65] and SC hydration (NOVA DPM 9003, as described in Chapter 15). The data were analyzed by analysis of variance and Duncan multiple test procedure with SPSS-PC+ 4.0.

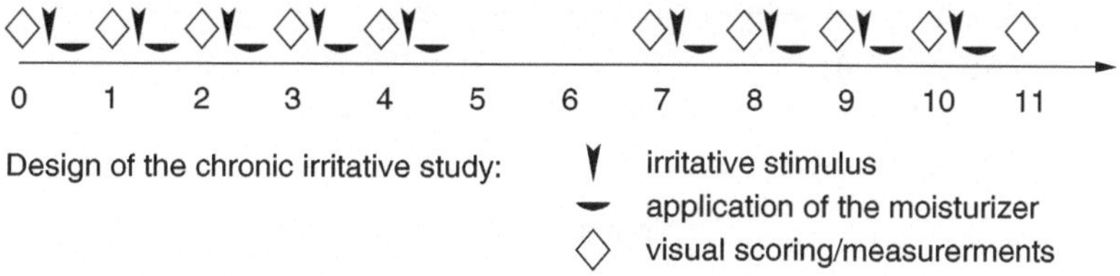

FIGURE 18.8 Design of the chronic irritation study (numbers = days).

TABLE 18.3
Chronic Irritation Model: Visual Score for Erythema, Number of Volunteers with Positive Reactions (score range)

	Day 4		Day 11	
Test Product	**SLS**	**Toluene**	**SLS**	**Toluene**
No (control)	4 (1–2)	2 (1)	11 (1–2)	2 (1)
Lotion A	5 (1)	0	7 (1)	0
Lotion B	4 (1)	0	4 (1)	0
Lotion C	4 (1)	0	5 (1–3)	0
Lotion D	2 (1)	0	8 (1)	0
Petrolatum	6 (1)	0	9 (1–3)	0
LAA	11 (1–3)	1 (1)	16 (1–3)	10 (1–2)

Note: SLS — sodium lauryl sulfate (1%, 30 min daily) and toluene — 100%, 10 min daily. For test products see Table 18.2.

The results are shown in Tables 18.3 to 18.5. There were no differences between occluded and nonoccluded control areas. Toluene was definitely less irritating than SLS in our experiments, although all volunteers experienced a strong erythema and some stinging during occlusion and immediately after removal of the patch. This reaction faded within half an hour. By visual scoring assessment, toluene provoked a minimal change of the skin surface in only two volunteers. At the end of the test period, 10 volunteers showed a distinct irritation after SLS application (SLS control; Table 18.3). Treatment with the moisturizers did not change the visual scores significantly, with the exception of Lotio alba aquosa (LAA), which significantly enhanced the irritation caused by both substances. Skin color measurements confirmed these results (data not shown). TEWL was significantly enhanced compared with the occluded control area after SLS (Table 18.4). After toluene, only a weak, statistically insignificant enhancement was noticed. None of the test products influenced these TEWL changes after both irritants, with the exception of LAA, which further significantly enhanced TEWL. This is illustrated in Figure 18.9. Both irritants significantly reduced the hydration of the SC compared with the control areas

TABLE 18.4
Chronic Irritation Model: Algebric Sums of TEWL Changes, Days 1 and 11

	Algebraic Sums of TEWL Changes ($g/m^2/h$)	
Test Product	**SLS**	**Toluene**
No (control)	+26.8 ± 18.6[a]	+10.2 ± 11.4
Lotion A	+24.7 ± 14.0	+3.0 ± 14.7
Lotion B	+22.1 ± 15.5	+9.8 ± 9.2
Lotion C	+26.0 ± 14.6	+6.2 ± 11.7
Lotion D	+24.3 ± 15.9	+8.7 ± 7.1
Petrolatum	+27.3 ± 18.6	+5.4 ± 10.8
LAA	+37.3 ± 18.4[b]	+14.9 ± 11.9[c]

Note: Mean ± SD ($g/m^2/h$). SLS — sodium lauryl sulfate (1%, 30 min daily) and toluene — 100%. 10 min daily. For test products see Table 18.2.

[a] $p < 0.05$ with control.
[b] Significantly different from all other groups.
[c] Significantly different from lotion A, lotion C, and petrolatum.

(Table 18.5). This decrease was partly counteracted by the moisturizer treatments in the case of SLS irritation only, but these data were not statistically significant. Moisturizers containing urea showed a slightly better effect, but again, this effect did not reach statistical significance. LAA showed a strong drying effect. In the case of toluene irritation, only lotion A significantly inhibited the hydration decrease of the SC, which was greatly enhanced by LAA.

The chronic irritation model has been expanded by using further irritants such as sodium hydroxide, lactic acid, and solvents such as toluene.[66] This model was refined and is now used in many laboratories to investigate the efficacy of moisturizers for protecting the skin against irritative stimuli, as well as for their capacity to support or enhance regeneration of the skin after and between irritations.[67,68] It has now been shown that a 1-week period is sufficient for evaluating the efficacy of moisturizers. A multicenter study has been designed with the purpose of standardizing the 5-day test procedure,[69] and a further refinement has now been achieved by showing that two different irritant stimuli applied concomitantly may cause a stronger irritation than if the same irritants are applied twice daily on their own.[61] The so-called "Tandem Repeated Irritation Test" may now be considered to investigate protective as well as regenerative properties of moisturizers.[70]

3. Discussion and Conclusion

The single and repeated irritation experiments show that moisturizers may lead to different results depending on the test model. However, one general conclusion may be drawn: Moisturizers do not seem to influence the development or the rate of

TABLE 18.5
Chronic Irritation Model: Algebric Sums of SC Hydration-Changes between Days 1 and 11 (arbitrary DPM-units)

Test Product	SC-Hydration: Algebric Sums of Changes from Baseline (DPM-Units)	
	SLS	Toluene
No (control)	−10.5 ± 21.3[a]	−18.5 ± 15.7[a]
Lotion A	+3.5 ± 18.5	−4.9 ± 15.1[c]
Lotion B	+0.2 ± 21.0	−18.6 ± 19.8
Lotion C	+3.6 ± 21.4	−11.5 ± 14.3
Lotion D	+0.1 ± 16.7	−9.5 ± 22.9
Petrolatum	−4.3 ± 21.8	−17.1 ± 22.4
LAA	−37.6 ± 23.8[b]	−26.2 ± 16.4[d]

Note: Means ± SD, SLS — sodium lauryl sulfate (1%, 30 min daily) and toluene — 100%, 10 min daily. For test products, see Table 18.2.

[a] $p < 0.05$ with control.
[b] Significantly different from all other groups.
[c] Significantly different from control.
[d] Significantly different from lotion A, lotion C, and lotion D.

recovery from an irritative stimulus, as measured with TEWL in the test models, thus far. The only study showing a clear-cut positive effect[15] was conducted with a commercially available dishwashing product that may differ in its irritative properties from the standard stimuli used in the other experiments. Often, commercially available detergents are mixtures of different compounds and are optimized regarding skin tolerance. The last described study using SLS and toluene as irritants showed that different irritation mechanisms are involved: SLS-treated sites showed increased TEWL and slightly decreased hydration, whereas toluene-treated sites revealed slightly increased TEWL and decreased hydration. It is known that different irritants such as SLS, acetone, petroleum ether, and different solvent mixtures all influence the barrier function of the SC by different mechanisms, as characterized by analysis of SC lipids, for example.[43,71–73] Furthermore, recovery of the SC barrier function may greatly depend on the occlusive properties of the applied products.[74] As a consequence, moisturizers with different modes of action or with action mechanisms better adapted to the nature of the irritative stimuli are needed. Use of the wrong product, such as LAA in SLS or toluene irritation, produced deleterious effects and aggravated the dermatitis. This is also illustrated by the effect of petrolatum, as reported by Ghadially et al.,[33] after irritation of the skin with acetone, a result that contradicts the effects of the same compound in other models such as SLS or toluene irritation. Further, the results of Berardesca et al.[57] or of Imokawa et al.[75] illustrate the need for a specific moisturizer adapted to the nature of the irritative stimuli.

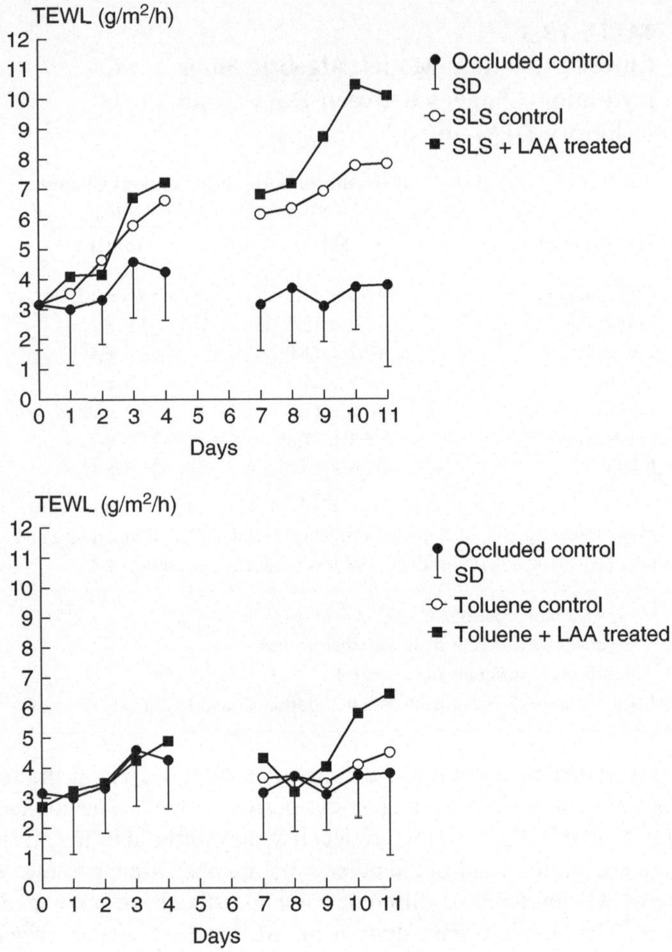

FIGURE 18.9 TEWL of occluded control sites, irritated control sites, and irritated sites treated by LAA (mean ± SD). For the sake of clarity, only 1/2 SD of control mean values are shown.

However, moisturizers alone may change the properties of the SC. Two recent publications[76,77] pointed out that the enhancement of irritation described with LAA above may not be fortuitous: Long-term[76] as well as short-term[77] use of moisturizers may lead to an increased response of the skin to a standard irritative stimulus. These studies also confirm that even in experimentally diseased (healthy) skin, the relationship between hydration of the SC and TEWL is not constant. For this reason, experiments designed to test a possible effect of moisturizers on diseased skin should always include both TEWL and hydration measurements; otherwise, the TEWL changes that are a result of changes in the barrier function of the SC cannot be related to real value barrier function changes.

B. Clinical Studies

Moisturizers with or without humectants are often used for the treatment of dry, sometimes pruritic skin. However, there are only a few reports of trials including TEWL and SC hydration measurements to control the effects of the treatment. Nevertheless, two examples will be given below.

Vilaplana and coworkers[16] tested the efficacy of a moisturizing emulsion containing ammonium lactate (12%). Twenty-four female patients entered the study, nine of whom were suffering from atopic dermatitis, and 15 being nonatopic. All patients showed dry, rough, scaling but noninflamed skin on the legs. The moisturizer was applied on the dry skin twice daily for a month as the only treatment allowed. Clinical evaluation and TEWL, hydration of the SC, surface lipids, and extensibility and firmness of the skin were measured on the first day (control value), after 14 days, and at the end of the test period. Replicas of the areas were evaluated by scanning electron microscopy and image analysis. Treatment already significantly reduced the clinical scores for dryness, desquamation, and pruritus after 15 days. The hydration of the SC showed a significant increase, but the TEWL did not change and stayed in the normal range, as before treatment. The other parameters showed changes consistent with the clinical improvement; in particular, a significant increase in skin surface lipids was measured 12 hours after application, after completion of the lipidization phase.

A study of similar length was conducted by Tollesson and Frithz[78] in children suffering from seborrheic dermatitis. Thirty-seven children 1 to 7 months of age were included, and 22 could be evaluated at the end of the trial. The children exhibited a normal hydration state of the SC, but an elevated TEWL. They were treated topically with pure borage oil twice daily. At the end of the treatment, the children had improved clinically and were free from skin symptoms, and the elevated TEWLs returned to normal values.

These studies illustrate two opposite clinical situations. In the first situation, TEWL was normal but SC hydration was low, and in the second study, TEWL was elevated in the presence of a normal hydration state of the SC. In both cases, carefully selected moisturizers could significantly improve the clinical situation. This confirms the conclusion drawn from the experimental studies: Always keep in mind the nature of the moisturizer to be used. In the first case, a normal TEWL indicated an intact barrier function, but the reduced hydration was the result of a low water-retaining capacity of the SC. In the second case, the water-binding capacity was not altered, but the barrier function was impaired. In addition to the restoration of the barrier properties of the SC, the fact that borage oil contains γ-linolenic acid might have played an important role.

VII. CONCLUSION

This short review of the recent literature and of the reports of several of the author's experiments demonstrates that TEWL measurements should be carefully planned when moisturizers are to be tested; otherwise, results of limited value will be obtained.

The determination of TEWL alone is indicated when the occlusive properties of a product are to be measured. Stripped skin is the model of choice. Wiping off the products during the lipidization phase, as suggested by Rietschel[24] and Lodén,[4] is another possibility, but the measured variations are not as dependent on the lipid content of the formulations as they are on stripped skin. In contrast, the results are more representative of the behavior of moisturizers in an *in vivo* healthy skin situation in which the direct relationship between TEWL and SC hydration is constant, unless humectants are included in the formulation.

In other investigational or clinical situations dealing with irritated skin, TEWL measurements are of little value if they are not accompanied by measurements of the SC hydration. This is particularly true if humectants are included in the formulations. It has been shown that experimental irritation, although leading to "dry" skin, modifies the relationship of TEWL hydration in a way that depends on the irritative stimulus used. Therefore, moisturizers may show no effect at all, a deleterious effect, or an effect on the SC hydration but not on the TEWL or reciprocally. The relationships between water-binding capacity, concentration gradient of water, and barrier function of the SC are complex, and moisturizers adapted to the different modifications of these relationships are needed. This ambitious goal will only be attained by integrating and combining results from different parameters that allow different properties of the moisturizing formulations to be optimized.

REFERENCES

1. Blank, I.H., Factors which influence the water content of the stratum corneum, *J. Invest. Dermatol.*, 18, 433, 1952.
2. Blank, I.H., Further observations on factors which influence the water content of the stratum corneum, *J. Invest. Dermatol.*, 21, 259, 1953.
3. Marks, R., Methods to evaluate effects of skin surface texture modifiers, in *Principles of Cosmetics for the Dermatologist*, Frost, P. and Horwitz, S.N., Eds., Mosby, St. Louis, 1982, chap. 8.
4. Lodén, M., The increase in skin hydration after application of emollients with different amounts of lipids, *Acta Derm. Venereol. (Stockh.)*, 72, 327, 1992.
5. Idson, B., Moisturizers, emollients, and bath oils, in *Principles of Cosmetics for the Dermatologist*, Frost, P. and Horwitz, S.N., Eds., Mosby, St. Louis, 1982, chap. 5.
6. Idson, B., Dry skin: moisturizing and emolliency, *Cosmet. Toiletr.*, 107, 69, 1992.
7. Wilkinson, J.B. and Moore, R.J., *Harry's Cosmetology*, 7th ed., Chemical Publishing, New York, 1982, 62.
8. Charlet, E., *Kosmetik für Apotheker*, Wissentschaftliche Verlagsgesellschaft, Stuttgart, 1989, 77.
9. Ummenhofer, B., Hautschutz durch Dermatika, in *Externe Therapie von Hautkrankheiten: Pharmazeutische und Medizinische Praxis*, Hornstein, O.P. and Nürnberg, E., Eds., Georg Thieme, Stuttgart, 1985, chap. 15.
10. Wehr, R.F. and Krochmal, L., Considerations in selecting a moisturizer, *Cutis*, 39, 512, 1987.

11. Blichmann, C.W., Serup, J., and Winther, A., Effects of single application of a moisturizer: evaporation of emulsion water, skin surface temperature, electrical conductance, electrical capacitance, and skin surface (emulsion) lipids, *Acta Derm. Venereol. (Stock.)*, 69, 327, 1989.
12. Serup, J., Winther, A., and Blichmann, C.W., Effects of repeated application of a moisturizer, *Acta Derm. Venereol. (Stockh.)*, 69, 457, 1989.
13. Lodén, M. and Lindberg, M., The influence of a single application of different moisturizers on the skin capacitance, *Acta Derm. Venereol. (Stockh.)*, 71, 79, 1991.
14. Elsner, P., Objektivierung und Quantifizierung von Externa-Wirkungen mittels nichtinvasiver biophysikalischer Messverfahren, *Zentralbl. Haut*, 161, 493, 1992.
15. Hannuksela, A. and Kinnunen, T., Moisturizers prevent irritant dermatitis, *Acta Derm. Venereol. (Stockh.)*, 72, 42, 1992.
16. Vilaplana, J., Coll, J., Trulläs, C., Azön, A., and Pelejero, C., Clinical and noninvasive evaluation of 12% ammonium lactate emulsion for the treatment of dry skin in atopic and non-atopic subjects, *Acta Derm. Venereol. (Stockh.)*, 72, 28, 1992.
17. Kligman, A., Introduction, in *Dry Skin and Moisturizers, Chemistry and Function*, Lodén, M. and Maibach, H.I., Eds., CRC Press, Boca Raton, Florida, 2000, chap. 1.
18. Marks, R., Lévêque, J.-L., and Voegeli R., *The Essential Stratum Corneum*, Martin Dunitz, London, 2002.
19. Berardesca, E. and Maibach, H.I., Transepidermal water loss and skin surface hydration in the non-invasive assessment of stratum corneum function, *Dermatosen*, 38(2), 50, 1990.
20. Tagami, H., Impedance measurement for evaluation of the hydration state of the skin surface, in *Cutaneous Investigation in Health and Disease: Noninvasive Methods and Instrumentation*, Lévêque, J.-L., Ed., Marcel Dekker, New York, 1989, chap. 5.
21. Schaefer, H., Zesch, A., and Stüttgen, G., *Skin Permeability*, Springer, Berlin, 1982, 733.
22. Wilson, D.R. and Maibach, H.I., Transepidermal water loss: a review, in *Cutaneous Investigation in Health and Disease: Noninvasive Methods and Instrumentation*, Lévêque, J.-L., Ed., Marcel Dekker, New York, 1989, chap. 6.
23. Lévêque, J.-L., Measurement of transepidermal water loss, in *Cutaneous Investigation in Health and Disease: Noninvasive Methods and Instrumentation*, Lévêque, J.-L., Ed., Marcel Dekker, New York, 1989, chap. 7.
24. Wu, M.S., Yee, D.J., and Sullivan M.E., Effect of a skin moisturizer on the water distribution in human stratum corneum, *J. Invest. Dermatol.*, 81, 446, 1983.
25. Rietschel, R.L., A method to evaluate skin moisturizers *in vivo*, *J. Invest. Dermatol.*, 70(3), 152, 1978.
26. Chilcott, R.P., Patel, A., Ashley, Z., Hughes, J.N., and Parkes, J.A., The effects of chemical damage on TEWL, in Marks R., Lévêque, J.-L., and Voegeli R., Eds., *The Essential Stratum Corneum*, Martin Dunitz, London, 2002, chap. 45.
27. Chilcott, R.P., Dalton C.H., Emanuel, A.J., Allen, C.E., and Bradley, S.T., The effects of physical damage on TEWL, in Marks R., Lévêque, J.-L., and Voegeli R., Eds., *The Essential Stratum Corneum*, Martin Dunitz, London, 2002, chap. 46.
28. Elsner, P. and Maibach, H.I., The effect of prolonged drying on transepidermal water loss, capacitance and pH of human vulvar and forearm skin, *Acta Derm. Venereol., (Stockh.)*, 70, 105, 1990
29. Clarys, P., Manou, I., and Barel, A., Relationship between anatomical site and response to halcinonide and methylnicotinate studied by bioengineering techniques, *Skin Res. Technol.*, 3, 161, 1997

30. Schnetz, E., Kuss, O., Schmitt, J., Diepgen, T.L., Kuhn, M., and Fartasch, M., Intra- and inter-individual variations in transepidermal water loss on the face: facial locations for bioengineering studies, *Contact Dermatitis*, 40, 243, 1999.
31. Wilhelm, K.P. and Maibach, H.I., Influence of aging on the barrier function of human skin evaluated by *in vivo* transepidermal water loss measurements, in *Aging Skin: Properties and Functional Changes*, Lévêque, J.-L. and Agache, P.G., Eds., Marcel Dekker, New York, 1993, chap. 17.
32. Gabard, B. and Bieli, E., Measurement of lipid deposition on the skin of the forearm: comparison of different bath oils, *J. Soc. Cosmet. Chem.*, 42, 299, 1991.
33. Ghadially, R., Halkier-Sorensen, L., and Elias, P.M., Effects of petrolatum on stratum corneum structure and function, *J. Am. Acad. Dermatol.*, 26, 387, 1992.
34. Morrison, B.M., Jr., ServoMed evaporimeter: precautions when evaluating the effect of skin care products on barrier function, *J. Soc. Cosmet. Chem.*, 43, 161, 1992.
35. Pinnagoda, J., Tupker, R.A., Agner, T., and Serup J., Guidelines for transepidermal water loss measurement: a report from the standardization group of the European Society of Contact Dermatitis, *Contact Dermatitis*, 22, 164, 1990.
36. Rogiers, V. and the EEMCO Group, EEMCO Guidance for the assessment of transepidermal water loss in cosmetic sciences, *Skin Pharmacol. Appl. Skin Physiol.*, 14, 117, 2001.
37. Bernengo J.-C. and De Rigal, J., Techniques physiques de mesure de l'hydratation du stratum corneum *in vivo*, in Agache, P., Ed., *Physiologie de la peau et explorations fonctionnelles cutanées*, EM Inter, Cachan, 2000, chap.13.
38. Marty, J.P., Vincent, C.M., and Fiquet, E., Etude des proprietés hydratantes de la crème hydratante visage Neutrogena, *Réalités Ther. Derm. Vénéréol.*, 15, 1, 1992.
39. Vincent, C.M., Fiquet, E., Cohen-Letessier, A., and Marty, J.R., Evaluation des proprietés hydratantes de la crème hydrophile lipophile (Effadiane®), *Nouv. Dermatol.*, 11, 419, 1992.
40. Blichmann, C.W. and Serup, J., Assessment of skin moisture: measurement of electrical conductance, capacitance and transepidermal water loss, *Acta Derm. Venereol. (Stockh.)*, 68, 284, 1988.
41. Tagami, H., Ohi, M., Iwatsuki, K., and Yamada, M., Electrical measurement of the hydration state of the skin surface *in vivo*, in *Stratum Corneum*, Marks, R. and Plewig, G., Eds., Springer, Berlin, 1983, 252.
42. Wepierre, J., Adrangui, M., and Marty, J.P., Factors in the occlusivity of aqueous emulsions, *J. Soc. Cosmet. Chem.*, 33, 157, 1982.
43. Lodén, M. and Maibach, H.I., *Dry Skin and Moisturizers, Chemistry and Function*, CRC Press, Boca Raton, Florida, 2000.
44. Batt, M.D., Fairhurst, D.E., Gerrard, W.A., and Ridge, B.D., Changes in the physical properties of the stratum corneum following treatment with glycerol, *J. Soc. Cosmet. Chem.*, 39, 367, 1988.
45. Rietschel, R.L., A skin moisturization assay, *J. Soc. Cosmet. Chem.*, 30, 369, 1979.
46. Jung, L. and Pelzer, I., Effects of moisturizing agents: *in vivo* measurements, in *Principles of Cosmetics for the Dermatologist*, Frost, P. and Horwitz, S.N., Eds., Mosby, St. Louis, 1982, chap. 6.
47. Choudhury, T.H., Marty, J.P., Oreechioni, A.M., Seiller, M., and Wepierre, J., Factors in the occlusivity of aqueous emulsions. Influence of humectants, *J. Soc. Cosmet. Chem.*, 36, 255, 1985.
48. Frödin, T., Helander, P., Molin, L., and Skogh, M., Hydration of human stratum corneum studied *in vivo* by optothermal infrared spectrometry, electrical capacitance measurement, and evaporimetry, *Acta Derm. Venereol. (Stockh.)*, 68, 461, 1988.

49. Serup, J., A double-blind comparison of two creams containing urea as the active ingredient: assessment of efficacy and side-effects by non-invasive techniques and a clinical scoring scheme, *Acta Derm. Venereol. (Stockh.)*, 177, 34, 1992.
50. Van Neste, D., Masmoudi, M., Leroy, B., Mahmoud, G., and Lachapelle, J.M., Regression patterns of transepidermal water loss and of cutaneous blood flow values in sodium lauryl sulfate induced irritation: a human model of rough dermatitic skin, *Bioeng. Skin*, 2, 103, 1986.
51. Malten, K.E. and den Arend, J.A.C.J., Irritant contact dermatitis: traumiterative and cumulative impairment by cosmetics, climate, and other daily loads, *Dermatosen*, 33(4), 125, 1985.
52. Tupker, R.A., Willis, C., Berardesca, E., Lee, C.H., Fartasch, M., Agner, T., and Serup, J., Guidelines on sodium lauryl sulfate (SLS) exposure tests, *Contact Dermatitis*, 37, 53, 1997.
53. Surber, C., Schwarb, F.P., and Smith, E.W., Tape-stripping technique, in *Percutaneous Absorption: Drugs, Cosmetics, Mechanisms, Methodology*, 3rd ed., Bronaugh, R.L. and Maibach, H.I., Eds., Marcel Dekker, New York, 1999, chap. 23.
54. Blanken, R., van der Valk, P.G.M., Nater, J.P., and Dijkstra, H., After-work emollient creams: effects on irritant skin reactions, *Dermatosen*, 35(3), 95, 1987.
55. Frosch, P.J. and Kligman, A.M., The soap chamber test, *J. Am. Acad. Dermatol.*, 1, 35, 1979.
56. Van Neste, D. and Antoine, J.L., A vehicle controlled study of the effects of hydrating agents in a human model of rough dermatitic skin, *Bioeng. Skin*, 4, 243, 1988.
57. Berardesca, E., Vignoli, G.P., Borroni, G., Oresajo, C., and Rabbiosi, G., Surfactant damaged skin: which treatment? in *The Environmental Threat to the Skin*, Marks, R. and Plewig, G., Eds., Martin Dunitz, London, 1992, chap. 49.
58. De Paepe, K., Derde, M.-P., Roseeuw, D., and Rogiers, V., Claim substantiation and efficiency of hydrating body lotions and protective creams, *Contact Dermatitis*, 42, 227, 2000.
59. De Paepe, K., Hachem, J.-P., Vanpee, E., Goossens, A., Germaux, M.-A., Lachapelle, J.-M., Lambert, J., Matthieu, L., Roseeuw, D., Suys, E., Van Hecke, E. and Rogiers, V., Beneficial effects of a skin tolerance-tested moisturizing cream on the barrier function in experimentally-elicited irritant and allergic contact dermatitis, *Contact Dermatitis*, 44, 337, 2001.
60. Gloor, M., Gabard, B., Fluhr, J.W., and Lehmacher W., Action of an aluminium chlorohydrate and glycerol containing skin protection cream in experimental skin irritation produced by sodium lauryl sulphate and solvents, *Occup. Environ. Dermatol.*, 49, 67, 2001
61. Wigger-Alberti, W., Krebs, A., and Elsner, P., Experimental irritant contact dermatitis due to cumulative epicutaneous exposure to sodium lauryl sulfate and toluene: single and concurrent application, *Br. J. Dermatol.*, 143, 551, 2000.
62. Fluhr, J.W., Gloor, M., Lehmann, L., Lazzerini, S., Distante, F., and Berardesca, E., Glycerol accelerates recovery of barrier function *in vivo*, *Acta Derm. Venereol. (Stockh.)*, 79, 418, 1999.
63. Gehring, W., The repetitive washing test as a model for the evaluation of barrier creams, in *Trends in Dermatopharmacy*, Wohlrab, J., Neubert, R., and Marsch, W., Eds., *Trends Clin. Exp. Dermatol.* Schaker, Aachen, Vol. 1, 47, 2003
64. Blanken, R., van Vilsteren, M.J.T., Tupker, R.A., and Coenraads, P.J., Effect of mineral oil and linoleic acid-containing emulsions on the skin vapour loss of sodium lauryl sulphate-induced irritant skin reactions, *Contact Dermatitis*, 20, 93, 1989.

65. Wilhelm, K.P. and Maibach, H.I., Skin color reflectance measurements for objective quantification of erythema in human beings, *J. Am. Acad. Dermatol.*, 21, 1306, 1989.
66. Frosch, P.J. and Kurte, A., Efficacy of skin barrier creams (IV). The repetitive irritation test (RIT) with a set of 4 standard irritants, *Contact Dermatitis*, 34, 278, 1994.
67. Schlüter-Wigger, W. and Elsner, P., Efficacy of 4 commercially available protective creams in the repetitive irritation test (RIT), *Contact Dermatitis*, 34, 278, 1996.
68. Wigger-Alberti, W., Caduff, L., Burg, G., and Elsner, P., Experimentally induced chronic irritant dermatitis to evaluate the efficacy of protective creams *in vivo*, *J. Am. Acad. Dermatol.*, 40, 590, 1999.
69. Schnetz, E., Diepgen T. L., Elsner, P., Frosch, P.J., Klotz, A.J., and Kresken, J. Multicentre study for the development of an *in vivo* model to evaluate the influence of topical formulations on irritation, *Contact Dermatitis*, 42, 336, 2000.
70. Spoo, J., Wigger-Alberti, W., Schliemann-Willers, S., Klotz, A.J., and Elsner, P., The tandem repeated irritation test (TRIT), in *Trends in Dermatopharmacy*, Wohlrab, J., Neubert, R., and Marsch, W., Eds., *Trends Clin. Exp. Dermatol.* Schaker, Aachen, Vol. 1, 58, 2003.
71. Deffond, D., Saint-Leger, D., Lévêque, J.L., and Agache, P., *In vivo* measurement of epidermal lipids in man, *Bioeng. Skin*, 2, 71, 1986.
72. Fulmer, A.W. and Kramer, G.J., Stratum corneum lipid abnormalities in surfactant induced dry scaly skin, *J. Invest. Dermatol.*, 86, 598, 1986.
73. Grubauer, G., Feingold, K.R., Harris, R.M., and Elias, P.M., Lipid content and lipid type as determinant of the epidermal permeability barrier, *J. Lipid Res.*, 30, 89, 1989.
74. Proksch, E., Holleran, W.M., Menon, G.K., Elias, P.M., and Feingold, K.R., Barrier function regulates epidermal lipid and DNA synthesis, *Br. J. Dermatol.*, 128, 473, 1993.
75. Imokawa, G., Akasaki, S., Minematsu, Y., and Kawai, M., Importance of intercellular lipids in water-retention properties of the stratum corneum: induction and recovery study of surfactant dry skin, *Arch. Dermatol. Res.*, 281, 45, 1989.
76. Held, E., Sveinsdóttir, S., and Agner, T., Effect of long-term use of moisturizer on skin hydration, barrier function and susceptibility to irritants, *Acta Derm. Venereol. (Stockh.)*, 79, 48, 1999.
77. Held, E. and Agner, T., Effect of moisturizers on skin susceptibility to irritants, *Acta Derm. Venereol. (Stockh.)*, 81, 104, 2001.
78. Tollesson, A. and Frithz, A., Transepidermal water loss and water content in the stratum corneum in infantile seborrhoeic dermatitis, *Acta Derm. Venereol. (Stockh.)*, 73, 18, 1993.

19 Dynamic Measurements: The Plastic Occlusion Stress Test, Moisture Accumulation Test, and Sorption–Desorption Test

Grazia Primavera and Enzo Berardesca

CONTENTS

I. Introduction 237
II. MAT 238
  A. MAT and Skin Irritation 239
  B. MAT and Psoriatic Skin 240
III. POST 240
  A. POST in Dry and Irritated Skin 241
  B. POST in Psoriatic Skin 242
IV. SDT 242
  A. SDT in Psoriatic Skin 242
  B. SDT in Infant Skin 243
  C. SDT in Delipidized Skin 243
V. Conclusions 244
References 244

I. INTRODUCTION

The stratum corneum (SC) is physiologically poor in water and is one of the tissues in the human body with the lowest water content. In its natural state, the SC contains about 30% water, nonuniformly distributed. A decreasing water gradient exists from the deeper SC layers (40%) toward the outer SC (15%) [1]. The SC of the human

0-8493-1443-7/05/$0.00+$1.50

skin enables the integrity of the organism and is the major constituent of barrier function.

The exceptional poverty of water in the SC (hence its hygroscopicity) has encouraged the development of three tests used to take measurements relating to the degree of water retention in the SC: the Sorption–Desorption Test (SDT), the Moisture Accumulation Test (MAT), and the Plastic Occlusion Stress Test (POST). These are dynamic "additive" hydration tests that assess the behavior of SC when in contact with water.

There are many factors affecting the water-holding capacity (WHC) of the SC and, hence, the desorption values of the SC. They may be classified as intrinsic to the SC [2], extrinsic [3], and environmental. The information that has been collected in the SC is obtained *in vivo* and *in vitro,* using noninvasive methods of an electrical type that are usually designed to assess the passive electric characteristics of SC [4,5]. The factors that influence the WHC of the SC include the depth of the survey level carried out on the SC, the thickness of the SC, the survey area, the lipid component, and the presence of pathological conditions. Several epidermal components modulate WHC. In the corneocyte, filaggrin rapidly disintegrates, releasing water and low-molecular-weight substances that contribute to the so-called natural moisturizing factors (NMFs). These substances are formed during epidermal differentiation and may represent up to 10% of the corneocyte mass. They are principally amino acids, carboxylic pyrrolidone acid, lactic acid, urea, glucose, and mineral ions. Keratinization plays an important part in the formation of NMFs that exhibit strong osmotic potential, attracting the water molecules. The binding of water to NMFs is the static aspect of cutaneous hydration. The second, dynamic, aspect is related to the selective permeability of the stratum corneum and to its lipid barrier properties, the permeability of which depends on the integrity and nature of the intercorneocyte lipids and on the lamellar organization between their cells [6–8].

In fact, once in the intercellular spaces, the water is free or very weakly bound to polar groups such as ceramides or other amphiphilic lipids [9]. This water is attracted to the surface by the usually lower relative humidity of the atmosphere. The transepidermal water loss (TEWL) constitutes water diffusing through the SC from the viable epidermis and water from the physiological dehydration of corneocytes. This natural SC hydration should not be confused with the added endogenous hydration arising from skin surface occlusion, a passive phenomenon that is achieved in the MAT and POST. In that case, one may assume that the blocked TEWL leads to a filling and dilatation of intercellular spaces throughout the SC thickness, and then to a slowing down of the water loss from deepest corneocytes and to passive hydration with swelling of the stratum compactum and stratum disjunctum corneocytes.

II. MAT

The MAT measures the increase in SC water content following occlusion using an impedance meter. It concerns the rate of progression of this hydration during an artificial occlusion. When the TEWL is blocked at the surface by an impermeable

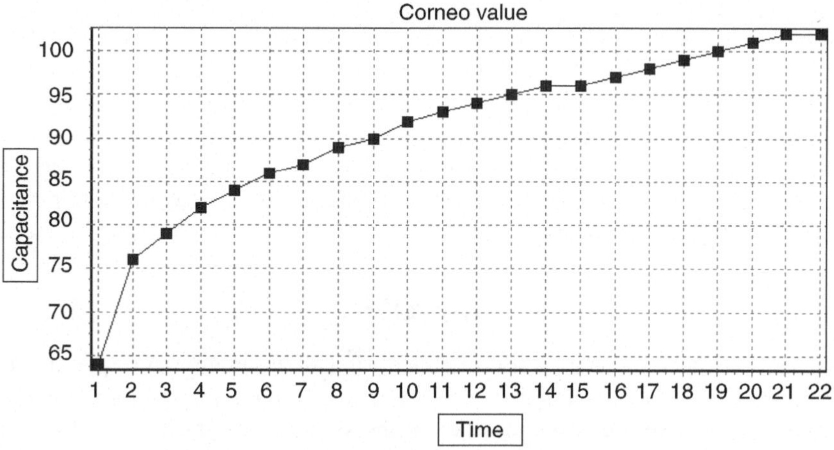

FIGURE 19.1 Typical Moisture Accumulation Test curve: There is a steep increase of capacitance within the first 20 seconds that levels off afterward, indicating an exponential function.

layer, SC hydration rapidly increases. The water accumulation probably depends on the osmotic potential of the SC. This dynamic test was described by Van Neste in 1990 [10], suggesting use of an early model of the capacitance-based Corneometer (Courage & Khazaka, Cologne, Germany). The probe head is covered with a water-impermeable foil, and the probe is placed on the skin, causing an occlusion. Measurements are taken in regular intervals over a total of 45 seconds. Van Neste proposed 3-second intervals, whereas we measured intervals of 1.5 seconds. Nowadays, it is possible to record the capacitance up to 20 times per second using the Corneometer CM 825 (Courage + Khazaka), together with a computer.

A typical MAT curve shows a steep increase of capacitance within the first 20 seconds that levels off afterward, indicating an exponential function (Figure 19.1). In particular, during the first second or two, capacitance increases rapidly and then more slowly. This increase appears to be bilinear or biexponential, and it probably indicates rapid filling of the intercellular spaces by an interrupted TEWL (initial phase), followed by progressive cell hydration of the SC. The latter may arise from osmotic hydration, simple diffusion, or both. Van Nest proposed calculating the area under the curve (AUC) of the MAT registration curve as an indicator of water influx into the SC.

A. MAT and Skin Irritation

Van Nest showed the usefulness of the MAT as a dynamic test compared with a static baseline value in the assessment of irritant contact dermatitis [10]. Although baseline capacitance was not significantly different in skin sites of healthy volunteers treated with H_2O (control sites) and sodium lauryl sulfate (SLS) in increasing concentrations for 48 hours, AUC rose in correlation with SLS concentrations. Furthermore, a close correlation between AUC and log TEWL was observed.

B. MAT and Psoriatic Skin

With psoriasis, we have found an increase of the corneal accumulation of unbound water, as expressed by MAT [11]. The overall increase was three times greater than that in "normal" skin, which may imply SC dehydration. Both the MAT and the POST may be useful dynamic parameters to assess therapeutic effects in psoriasis. Because no preocclusion is needed for the MAT, and as the measuring period is substantially lower for the MAT than for the POST, the MAT may be more practical for use in larger populations.

III. POST

The POST is a method that uses skin occlusion and allows noninvasive quantification of some functional parameters of the SC. The test consists of positioning a plastic chamber (or whatever occlusive device) on the skin surface and sealing it firmly to induce an occlusion to increase the water content underneath the dressing. Hydration is proportional to the occlusion time, and in our experience we have seen that 24 hours is a very practical period of time for our work giving good results in terms of water accumulation and of the discriminating power of the decay curves that were obtained. When the occlusion is removed, the excess water on the skin starts to evaporate and can be recorded using an evaporimeter. We prefer to call this evaporation skin surface water loss (SSWL) rather than transepidermal water loss (TEWL), representing the evaporation of water trapped within and over SC. Immediately after removal of occlusion the SSWL is monitored every minute (or every 5 minutes) for 25 to 30 minutes [12].

Figure 19.2 shows a typical dehydration curve obtained with the POST technique. The amount of water evaporating during the first few minutes is directly proportional to skin hydration. The higher the SC water content, the higher the degree of evaporation in the air. At the end of the dehydration time, SSWL is greatly reduced and the evaporation detected by the evaporimeter probe is mainly as a result of TEWL. Therefore, changes induced in this part of the curve are caused by damage to the water barrier and reflect the condition of the barrier function. The shape and the slope of the curve describe the total amount of water evaporated and, therefore, describe the WHC of the SC. The evaluation of the slope is rather difficult, because the curve shows a biexponential pattern of decay and can be fitted using biexponential regression analysis. However, an easier approach consists of transforming the data logarithmically. The second slope, which appears after 10 minutes, is subtracted from the overall log curve, allowing the initial decay curve to be obtained (0 to 9 min). The parameters obtained are the slope and the y-intercept of the SSWL/time log regression lines. The slopes (SSWL decrease rate constant) are expressed as the percentage of SSWL decrease per unit time. The y-intercept of the initial slope is the log of the initial SSWL (not to be mistaken for the total desorbed water). This parameter is less valuable than the SSWL rate. An important piece of information that can be obtained by analyzing the POST curves is the estimate of the free- and bound-water compartments of the SC (Figure 19.1). In the first part of decay, the curve is steeper, although the evaporation of the free water is recorded during the

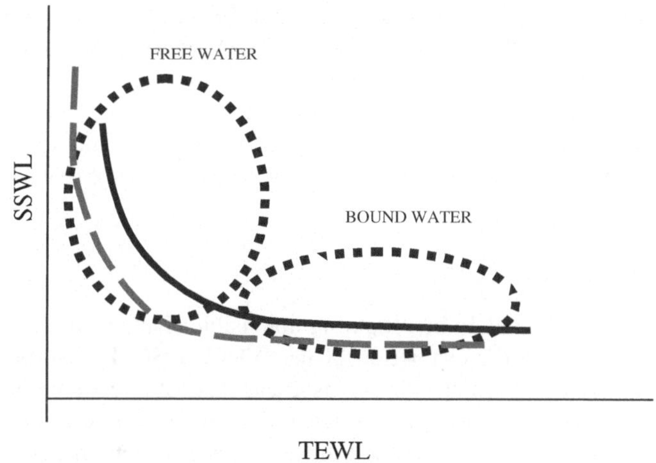

FIGURE 19.2 Typical dehydration curve obtained with the POST technique in normal skin (—) and irritated skin (——). The amount of water evaporating during the first few minutes is directly proportional to skin hydration. The shape of the curve changes during the course of pathological conditions of the skin.

first 10 minutes. Later, after 10 minutes, the bound-water compartment is evaluated [13,14]. The initial rapid desorption is caused by the free water located in the intercellular spaces, and the late phase is caused by the water most probably located within the corneocytes. The area under the SSWL curve (AUC), which indicates the amount of desorbed water over the measurement period, can also be calculated. The POST gives information on SC hydration, integrity of the barrier function, and SC WHC and is useful in assessing SC changes in various dermatological conditions.

A. POST in Dry and Irritated Skin

The shape of the curve changes during the course of pathological conditions of the skin (Figure 19.2). A flat curve with high SSWL levels, during the deconvolution time with high TEWL levels at the end, is detectable during acute irritant dermatitis and is the result of the dramatic increase of TEWL secondary to barrier function impairment [12]. In contrast, dry skin shows a curve, with lower SSWL levels existing throughout the deconvolution time.

In the uninvolved skin of atopic dermatitis, we have found that the initial SSWL and the AUC were markedly decreased [13], whereas the rate constant remained normal, demonstrating a reduced amount of water stored during occlusion.

The POST has also been used to investigate and quantify visible and nonvisible skin irritation induced by surfactants, to measure functional properties of the SC in dry pathological skin, and to assess the influence of some classes of lipids on the WHC of the SC. We have found that SLS decreases [12,15] the initial SSWL and also steadily decreases the SSWL decrease rate constant. This means that the accumulated water was retained longer, somewhat surprisingly for a compound well known to alter the SC intercellular spaces and increase the TEWL. Chloroform–methanol 2:1 decreased the initial SSWL and increased the SSWL decreasing rate constant [16]. This solvent removes lipids and free amino acids and may modify both the intercellular spaces and NMFs. Ether–acetone 1:1 did not change the initial SSWL and slightly increased the SSWL decreasing rate constant [16]. This less potent lipid solvent seems to have a similar action to chloroform–methanol, but the effect was less pronounced.

B. POST in Psoriatic Skin

The usefulness of the POST in investigating nonvisible changes in skin function has been further assessed in the evaluation of the WHC of SC, hydration, and barrier function in apparently normal skin in psoriasis and atopic dermatitis [13]. No significant differences were detected between uninvolved psoriatic skin and normal controls. However, although evaporation of free water is exactly the same in the two groups, the bound-water compartment shows reduced hydration in psoriatic skin because of defective keratinization and SC formation. This reduced hydration was also present in apparently normal skin of psoriatic subjects. Therefore, the decay constants show a higher rate of decay in psoriasis and, thus, a decreased WHC. The atopic curve shows a shape similar to control skin, but with lower SSWL values, corresponding to a drier skin condition. Indeed, basal TEWL indicates that the atopic group differs from controls and psoriatics being characterized by high TEWL and low water-content values. The slope is not different from control; thus, we can assume that WHC in these two groups is normal. However, the WHC expressed by POST was increased in lesional psoriatic skin [11].

IV. SDT

The SDT measures the hydration of the SC following external water contact (hygroscopicity) and the spontaneous loss of this hydration when this water is removed. This test consists of the application of water to the SC *in vivo*, then monitoring of the decrease in SC water content. An area of the skin is marked out, and the stratum corneum hydration is measured. Distilled water (20 μL) is left on the skin for 10 seconds before being dried off with soft tissue paper. SC hydration is measured immediately after drying and then every 30 seconds for 2 minutes. A sorption peak, indicating maximal hydration, is observed immediately after applying the water (sorption phase) — within 10 seconds. This is followed, at the end of 30 seconds, by a rapid fall in hydration values (desorption) [5]. Return to baseline values may take over 2 minutes. The increase in hydration resulting from the artificial application of water is the "hygroscopicity," and its rate of decrease represents the WHC.

A. SDT in Psoriatic Skin

The contact time with water, the quantity of water, and the instrument used are crucial when defining the hygroscopicity of a pathological and hyperkeratotic SC. Indeed, if relatively high quantities of water (1 mL) are used on a surface of 50 cm^2

and water is allowed to be adsorbed on the keratinic substrate for a relatively long period (1 minute) before measuring the capacitance values, the hygroscopicity is notable even on psoriatic SC, in particular after 1 minute. The WHC of the SC of the psoriatic lesion is not significantly different from the skin of the controls in the first 10 minutes, although it differs significantly after 20 minutes [17]. Rather than concluding that the pathological horny layer is incapable of binding water, we speculate that the water-binding function is deficient over a fairly long period. Thus, using different methods for a period of time proportional to the quantity of water used and to the thickness of the SC. The SC of the psoriatic lesion is highly hygroscopic and "greedy" for water but is defective regarding the water-binding capacity. Hence, psoriasis presents desorption times that are shorter than those of the normal SC. If the quantity of water used in the experiment is minimal (20 μL a droplet), and the period in question is relatively short (10 seconds), the hygroscopity of the psoriatic SC is much lower than that of perilesional skin [18]. This is probably the result of a disproportion between the water quantity and the SC volume that is not water tied to polar groups of the keratinic substrate but, rather, water secondarily bound to the weaker hydrogen link.

B. SDT in Infant Skin

The water sorption–desorption tests were also performed by Visscher et al. as an assessment of surface hydrophobicity in infant skin [19]. The desorption curves obtained over the first 28 days of life, for both the diapered and nondiapered sites, exhibited a significant increase over the first few weeks, followed by a plateau period. A more positive desorption slope is, by hypothesis, indicative of a slower rate of water desorption. Measurement of the corresponding maternal forearm over the first 2 weeks showed no change. The data of this study are consistent with the immediate newborn period, exhibiting specialized properties with respect to water handling and barrier properties. In particular, the more positive slope for the water desorption curves indicates a diminished ability to shed water quickly.

C. SDT in Delipidized Skin

Skin lipids play a critical role in maintaining the efficacy of the barrier function and the water-holding properties of the SC. In fact, when the hygroscopicity and the SDT are assessed after delipidization of the SC, using a surfactant and different associations of organic solvents, hydration and desorption modalities vary according to the lipid-depleting substances applied to the skin, and thus to the nature of the extracted lipids. The acetone/ether solution (1:1) causes a reduction in WHC after 3 minutes from contact with water [20]. Imokawa et al. [21] have demonstrated that this impairment can be repaired by topical application of isolated intercellular lipids, among which sphingolipids play the most significant role. A 5% SLS solution, after 5 minutes of contact, does not seem to play a crucial role in affecting WHC, indicating a capacity to elute the hydrating hydrosoluble substances and, probably, the surface lipids of sebaceous origin. A 5-minute contact with chloroform/ethanol (2:1) causes a rapid and significant impairment in WHC. Because chloroform/methanol can separate ceramides from other SC lipids [21], it may be inferred that chloroform/ethanol is also able to extract substances that are crucial in the water-binding function. Moreover, most volunteers showed severe inflammatory changes in the contact sites for chloroform/ethanol that lasted several weeks. This indicates that this solution, under the test conditions used, caused damage to viable cells of the malpighian layer, leading to clinically evident inflammation. A further argument in favor of the fundamental role of epidermal lipids in the SC hydration is the effect of delipidization of structural lipids from the SC obtained with acetone/ether. This extraction might be counterbalanced by the application of ceramides, glycolipids, and fractions of cholesterol, but not by application of free fatty acids and cholesterol ester fractions [21]. Moreover, sphingolipids have been demonstrated to play an important role in restoring the WHC in the SC after 5% sodium dodecyl sulfate treatment of SC *in vivo* [22]. Regarding the relationship between hygroscopicity and SDT, the middle layers of the SC have been demonstrated *in vivo* to have a higher WHC than the outermost layers [5].

V. CONCLUSIONS

The dynamic hydration tests are useful in investigative dermatology for studying skin physiology because they increase the dynamic of skin behavior and skin responses. To study the SC modifications induced by skin irritation or hydration, the SDT and the MAT appear to be the best choices. In fact, these two techniques are both quick (3 minutes for the MAT and 2.5 minutes for the SDT) and easy to use. The MAT is also the least prone to technical variability, which in fact only concerns the device used and its reliability. The POST is complementary to the MAT. A comparison is necessary to better understand the relationship among the water's accumulation velocity, the total amount of desorbed water, and the WHC. However, this test allows us to differentiate smaller degrees of SC changes subsequent to modifications of the intercellular spaces or cell membranes. The main inconvenience of the POST is the time required to carry out the test: two measurement sessions separated by 24 hours, with the final session lasting 30 minutes.

REFERENCES

1. Warner, R.R., Stone, K.J., and Boissy, Y.L., Hydration disrupts human stratum corneum ultrastructure, *J. Invest. Dermatol.*, 120(2), 275, 2003.
2. Clar, E., Her, E.P., and Sturelle, C.G., Skin impedance and moisturization, *J. Soc. Cosm. Chem.*, 26, 337, 1975.
3. Borroni, G., Bernardi, L., Berardesca, E., Tassi, G., Bellosta, M., and Rabbiosi G., Impedenza della cute senile e modificazioni indotte dal trattamento idratante topico, *Boll. Soc. Med. Chir. Pavia*, 95, 629, 1979.
4. Tagami, H., Ohi, M., Iwatsuki, K., Kanamaru, Y., Yamada, M., and Ichijo B., Evaluation of the skin surface hydration *in vivo* by electrical measurements, *J. Invest. Dermatol.*, 75, 500, 1980.

5. Tagami, H., Kanamaru, Y., Inoue, K., Suehisa, S., Inoue, F., Iwatsuki, K., Yoshikuni, K., and Yamada, M., Water sorption-desorption test of the skin *in vivo* for functional assessment of the stratum corneum, *J. Invest. Dermatol.*, 78, 425, 1982.
6. Elias, P.M., Lipids and epidermal permeability barrier, *Arch. Dermatol. Res.*, 270, 95, 1981.
7. Yardley, H.J. and Summerly, R., Lipid composition and metabolism in normal and diseased epidermis, *Pharmacol. Ther.*, 13, 357, 1981.
8. Marty, J.P., NMF and cosmetology of cutaneous hydration, *Ann. Dermatol. Venereol.*, 129, 131, 2002.
9. Yamamura, T. and Tezuka, T., The water holding capacity of the stratum corneum measured by H-NMR, *J. Invest. Dermatol.*, 93, 160, 1989.
10. Van Neste, D., *In vivo* evaluation of unbound water accumulation in stratum corneum: the influence of acute skin irritation induced by sodium laurylsulfate, *Dermatologica*, 181, 197, 1990.
11. Elsner, P. and Burg, G., *Dynamic functional properties of psoriatic skin assessed by non-invasive bioengineering methods in vivo*, Fifth International Psoriasis Symposium, San Francisco, July 1991.
12. Berardesca, E. and Maibach, H.I., Monitoring the water-holding capacity in visually non-irritated skin by plastic occlusion stress test (POST), *Clin. Exp. Dermatol.*, 15, 107, 1990.
13. Berardesca, E., Fideli, D., Borroni, G., Rabbiosi, G., and Maibach, H., *In vivo* hydration and water-retention capacity of stratum corneum in clinically uninvolved skin in atopic and psoriatic patients, *Acta Derm. Venereol. (Stockh.)*, 70, 400, 1990.
14. Werner, Y., Lindberg, M., and Forslind, B., The water-binding capacity of stratum corneum in dry noneczematous skin of atopic eczema, *Acta Derm. Venereol. (Stockh.)*, 62, 334, 1982.
15. Berardesca, E., Fideli, D., Gabba, P., Cespa, M., Rabbiosi, G., and Maibach, H.I., Ranking of surfactant skin irritancy *in vivo* in man using the plastic occlusion stress test (POST), *Contact Dermatitis*, 23, 1, 1990.
16. Berardesca, E., Herbst, R., and Maibach, H.I., Plastic Occlusion Stress Test as a model to investigate the effects of skin delipidization on the stratum corneum water holding capacity *in vivo*, *Dermatology*, 187, 91, 1993.
17. Borroni, G., et al., Changes in the water holding capacity of psoriatic stratum corneum *in vivo*, *Acta Derm. Venereol. (Stockh.)*, 146, 192, 1989.
18. Tagami, H. and Yosikuni, K., Interrelationship between water-barrier and reservoir functions of pathologic stratum corneum, *Arch. Dermatol.*, 121, 642, 1985.
19. Visscher, M.O., et al., Changes in diapered and nondiapered infant skin over the first month of life, *Pediatric Dermatol.*, 17, 45, 2000.
20. Borroni, G., et al., The sorption-desorption test in the evaluation of skin hygroscopicity in physiological and pathological conditions, *Skin Pharmacol.*, 5, 237, 1992.
21. Imokawa, G., Akasaki, S., Hattori, M., and Yoshizuka, N., Selective recovery of deranged water-holding properties by stratum corneum lipids, *J. Invest. Dermatol.*, 87, 758, 1986.
22. Imokawa, G., Akasaki, S., Minematsu, Y., and Kawai, M., Importance of intercellular lipids in water retention properties of the stratum corneum: induction and recovery study of surfactant dry skin, *Arch. Dermatol. Res.*, 281, 45, 1989.

Part III

Water Content

20 Assessment of Stratum Corneum Hydration: Corneometer CM 825

Gabriel Khazaka

CONTENTS

I. Introduction 249
II. History of the Development of the Corneometer 250
III. Measuring Principle 250
IV. Structure of the Sensor 253
V. Calibration of the Corneometer Probe 255
A. Production 255
B. Check Calibration by the User 255
C. Important Notes for the User 255
VI. Technical Data 256
VII. Advantages of the Development of the CM 825 256
VIII. Corneometer in Comparison with Other Bioengineering Instruments 256
IX. Linearity of the Probe CM 825 257
A. Amount of Water 257
B. Measurement of Substances with Different Dielectricity *In Vitro* 257
X. Penetration Depth of the Measurement with the Corneometer CM 825 257
XI. Classification of the Skin Type with the Corneometer CM 825 259
XII. Software 259
References 260

I. INTRODUCTION

The assessment of skin moisture is one of the first and most important measurements for testing the efficacy of cosmetic products on the skin surface. The quantity of literature worldwide dealing with this topic indicates the significance of this measurement. Numerous studies about the advantages, disadvantages, and comparisons among the different commercially available devices have been published. The main topics of the scientific discussion are the measuring principle, measuring frequency,

0-8493-1443-7/05/$0.00+$1.50

FIGURE 20.1 The first Corneometer probe: laboratory prototype of the Corneometer, 1980.

penetration depth of the measurement into the skin, influence of different substances on the measurement, and handling of the probe. The Corneometer is one of the most used instruments for stratum corneum hydration measurement. Because of its unique measuring principle, accuracy, and international reputation, this method of measurement is referred to as the Corneometer method in the literature.

II. HISTORY OF THE DEVELOPMENT OF THE CORNEOMETER

In 1980, the first study with a prototype of the Corneometer was published by Tronnier [1], and the idea for manufacturing and selling this device came from Karl Heinz Schrader [2]. The stages of development are shown in Figures 20.1 to 20.6.

III. MEASURING PRINCIPLE

Two metal plates electrically isolated by a so-called dielectric (e.g., vacuum, air, glass, plastic) are defined as a capacitor. When connecting a supply point (e.g., a battery) to the capacitor, electrons flow from one metal plate via the supply point into the other metal plate. As a consequence, one plate has an excess of electrons (negative charge) and the other has a lack of electrons (positive charge). This condition remains even when the supply point is removed. The ability of the capacitor to store a certain quantity of electric charge is called "capacity." An electric field forms between the plates, visualized by lines of electric flux, as the opposite charges of the plates attract each other. These forces of attraction also influence the dielectric reaction of the medium measured. The electron sheaths are attracted by the plate, with the positive charge and the atomic cores attracted by the plate with the negative charge. The electrically neutral molecules get both a positive and a negative pole.

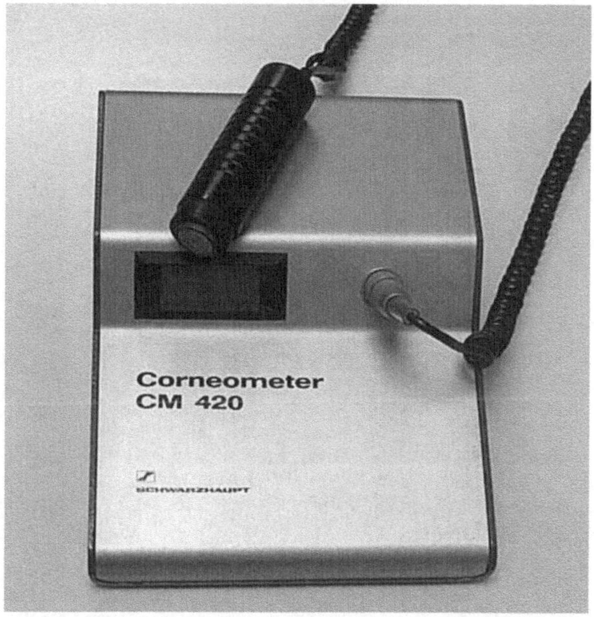

FIGURE 20.2 The first commercial Corneometer CM 420 probe with plastic foil, 1982.

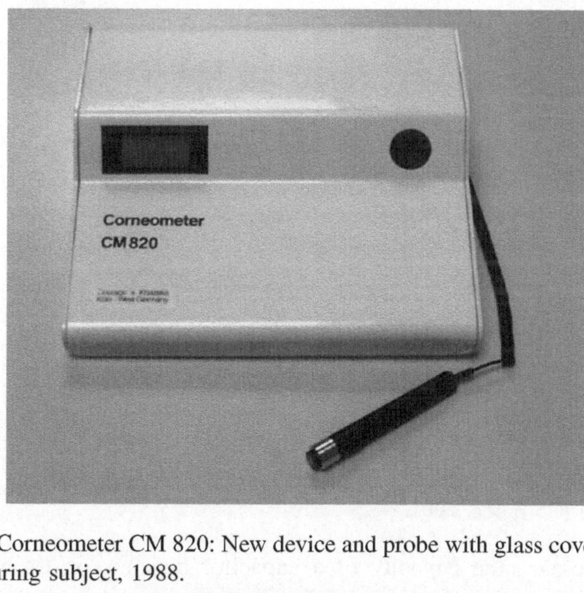

FIGURE 20.3 Corneometer CM 820: New device and probe with glass cover between probe head and measuring subject, 1988.

The molecules are attracted by the opposite charges of the plates and orientate themselves accordingly (polarization) (Figure 20.7).

The charge of the plates is partially compensated by this. Therefore, the capacitor can store a higher charge, which means an increase in its capacity. The ability of a

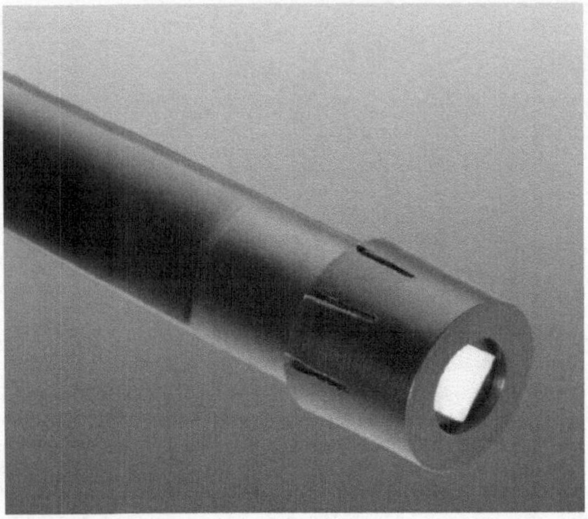

FIGURE 20.4 Corneometer CM 825 probe with new electronic deisgn and special captor continuous measuring mode, 1998.

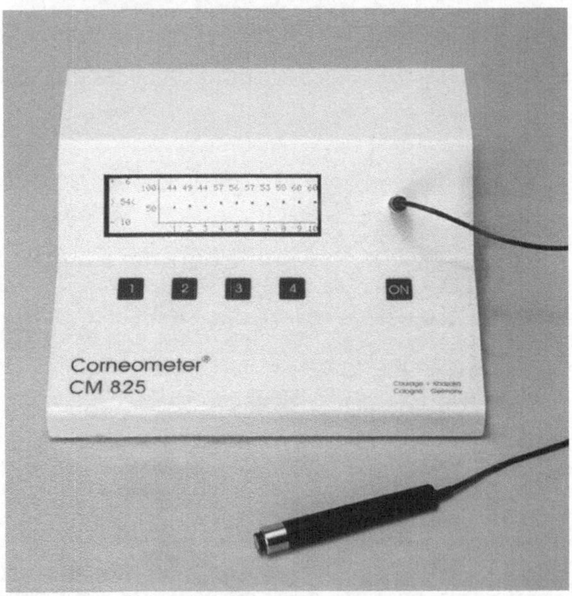

FIGURE 20.5 Corneometer CM 825 single device with new graphic display and new digital Corneometer CM 825 probe, 2001.

dielectric to increase the capacity of a capacitor depends on the material. Most materials increase the capacity compared with that of a vacuum by a factor of more than seven (dielectricity constant); water, however, increases the capacity by a factor of 81. This means that for a mixture of different substances, such as found in the skin, a change in water content leads to large changes in the capacity of the measuring capacitor.

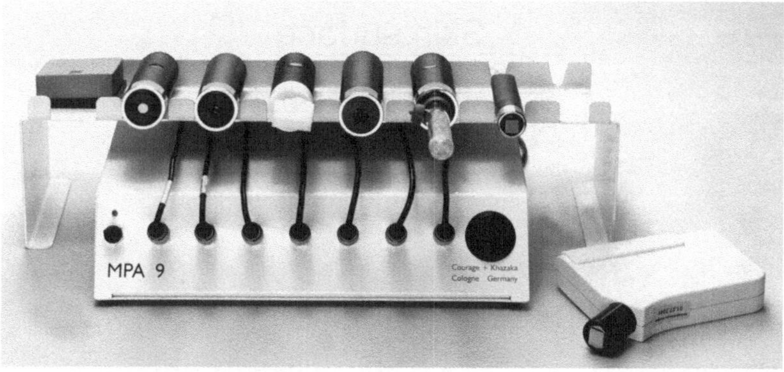

FIGURE 20.6 New Multi Probe Adapter MPA 9-system: Basic station for free combination of different Courage + Khazaka probes, 2002.

FIGURE 20.7 Capacitor scheme.

IV. STRUCTURE OF THE SENSOR

The Corneometer probe was developed according to a measuring principle described previously [3]. The probe consists of a 7 × 7-mm ceramic tile with many gold lines that are close to each other, functioning like capacitor plates. The parallel gold lines have such a small distance between them that they are only visible under a magnifying glass. The even lines, as well as the odd lines, are linked and, thus, form a coupling in parallel with the individual capacitors (Figures 20.8 and 20.9).

The ideal capacitor forms a homogeneous electrical field between the plates. In the case of a real capacitor, there is an additional scatterfield at the edges of the plates decreasing approximately reciprocally, proportional with the distance. Normally this is an undesired effect, but the Corneometer method makes use of this effect. The lines are very thin compared with their breadth, and therefore, this

FIGURE 20.8 Conductor track of the Corneometer, front view of the electrode.

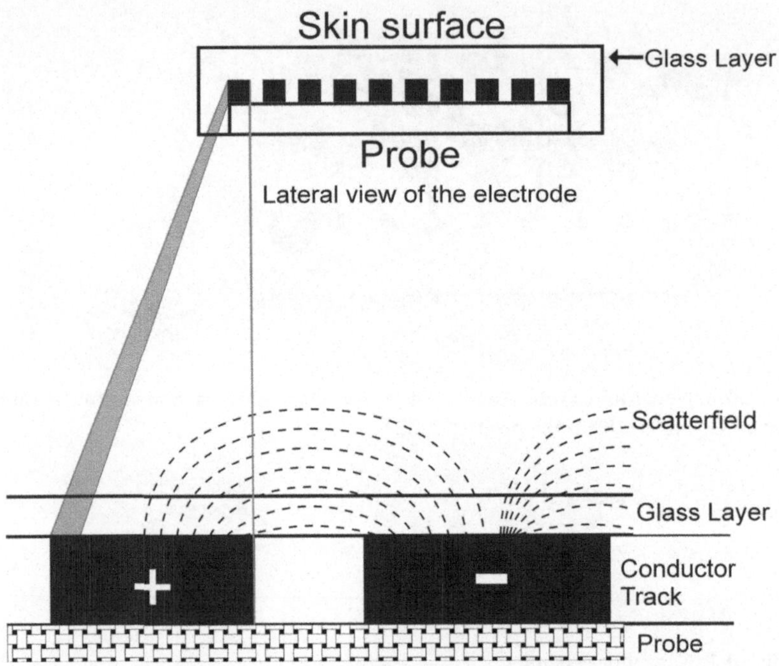

FIGURE 20.9 Conductor track of the Corneometer, lateral view of the electrode.

scatterfield is spread broadly. If skin containing water is brought into the scatterfield of the sensor, the capacity of the system is increased.

For protection of the fragile gold lines, the probe head is sealed with a very thin glass layer. This isolating layer also prevents a conductive connection to the skin, so that no charge or current can flow away through the skin while measuring on the skin surface.

By choosing the appropriate conductor spacing and thickness of the glass layer, the penetration depth of the electric field into the skin, and therefore the measuring depth, can be controlled. However, it is difficult to determine the exact measuring depth, as the electric field and the sensitivity decrease reciprocally, proportional with the distance.

If a capacitor is charged with a constant current I until a voltage U, the time t necessary is directly proportional to the capacity C of the capacitor:

$$t = U/I \times C.$$

The electronics in the probe head are responsible for alternating the charging of the capacitor. This means that the current flows into the capacitor first through one and then through the other connection of the capacitor. For dry skin or air, 10 microseconds are needed for a change; for moist skin or water, 20 microseconds are needed. This process prevents polarization effects both on the surface of the sensor and in the skin.

The sensor is flexible within the probe. During measurement, it is pressed onto the skin by a spring. Measurement is triggered at a pressure of 0.7 to 0.8 N, and the total pressure is approximately 1.3 N (new CM 825). The processor calculates the results from the information of the measuring values and from the calibration data.

V. CALIBRATION OF THE CORNEOMETER PROBE

A. Production

The Corneometer probe CM 825 contains a memory for storing the calibration values. With this technology, the Corneometer probe is independent from the device itself. The probe can be used with different devices from Courage + Khazaka electronic GrahH (Cologne, Germany).

For calibration, the following steps are necessary:

- Testing the imperviousness of the sensors with respect to water
- Checking the electric capacity of the sensors
- Checking the measurement pressure of the probe

The calibration itself consists of the following steps:

- Zero calibration with air (dry material)
- Determination of the maximum value with saltwater
- Determination of the penetration depth with a foil of 10 μm in thickness
- Comparison of measurement on the skin surface using a Corneometer reference device

B. Check Calibration by the User

A check calibration of the Corneometer CM 825 probe can be easily done at any time. A kit containing a special salt solution, filters, and a 15-μm foil is supplied with each probe. As long as the measured values correspond with the values of the manufacturer, the probe can be used without problem.

C. Important Notes for the User

- Always put the probe vertically onto the skin surface.
- Do not measure on hairy skin areas.
- Start measurement with constant pressure of the probe onto the skin.
- Always clean the sensor.
- Do the check calibration two to three times a year.
- Send the probe to the manufacturer for cleaning and calibration at least every second year.

VI. TECHNICAL DATA

The measuring principle is one of capacitance, with nonelectric contact with the skin. The surface of the sensor is 7 × 7 mm in size. Data processing is done digitally, inside the probe. Check calibration can be done by the user with a special check calibration kit delivered with each probe. The frequency of the is 1.2 to 0.9 MHz. The measuring time is 1 second. Finally, the current consumption is 6 mA.

VII. ADVANTAGES OF THE DEVELOPMENT OF THE CM 825

According to experiences with the Corneometer during the last twenty years, the technology has steadily increased. The advantages of the new Corneometer CM 825 are as follows:

- High linearity corresponding to the water content
- Very low penetration depth
- Influences of ground capacity are minimized
- Influences of salts and other substances on the measurement are less important
- All calibration values and data processing are stored in the probe itself
- Check-calibration function
- Continuous measurement with or without direct skin contact
- Measuring time of 1 second
- Probe can be used with various devices from Courage + Khazaka electronic GmbH

VIII. CORNEOMETER IN COMPARISON WITH OTHER BIOENGINEERING INSTRUMENTS

In the last 20 years, the Corneometer was used not only for efficacy testing but also for comparisons with instruments measuring other parameters in the field of biophysical assessment of the skin, as well as with instruments using other moisture-measuring techniques. Examples of this bioassessment are skin roughness [4], transepidermal water loss [5], desquamation [6], laser Doppler [7], skin thickness measurement, and nuclear magnetic resonance spectroscopy [8]. The user gets an overview about the advantages and disadvantages of the different systems when looking at the comparisons between the Corneometer CM 820 and the CM 825 [9], and between those devices using the conductance method [10–14]. In this context, the European Group for Efficacy Measurements on Cosmetics and Other Topical Products (EEMCO) group has to be mentioned, having issued an important guideline for doing skin moisture measurements [15].

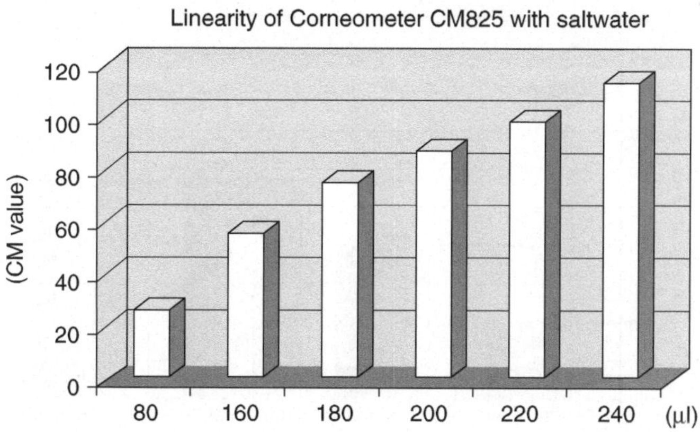

FIGURE 20.10 Linearity of Corneometer CM 825 measuring values with different amounts of saltwater: 80 µl of saltwater was added on a filter. After the initial measurement, the amount was doubled and then increased by 20 µl in every step.

IX. LINEARITY OF THE PROBE CM 825

A. Amount of Water

The following experiment shows the linearity of the Corneometer probe with respect to the measured amount of water. Eighty microliters of saltwater was applied on a filter. After the first measurement, the amount of saltwater was doubled and then increased by 20 µL in every step (Figure 20.10).

B. Measurement of Substances with Different Dielectricity *in Vitro*

Bielfeldt et al. [4] show impressively how the Corneometer probe CM 825 reacts to the different dielectricities of some cosmetic raw materials. The following substances, characterizing the different skin types, were used (Figure 20.11):

- 2-Ethyl-1-Hexanol (very dry skin)
- Isopropanol 99% (dry)
- Ethanol 90% (normal to moisturized skin)
- Ethanol/water 1:1 (well-moisturized skin)

X. PENETRATION DEPTH OF THE MEASUREMENT WITH THE CORNEOMETER CM 825

One of the most important demands concerning skin moisture measurement is a low penetration depth of the measurement into the skin. The most important skin layer

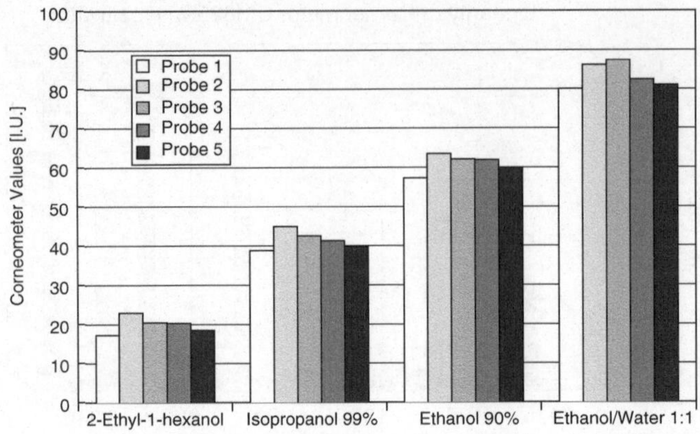

FIGURE 20.11 Comparison of Corneometer CM 825 measuring values on same cosmetic raw materials with different dielectricities.

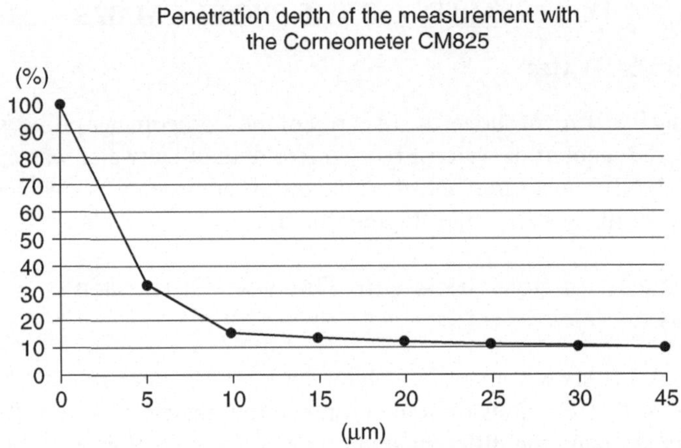

FIGURE 20.12 Penetration depth of the Corneometer CM 825 measurement.

here is the stratum corneum. It consists of corneocytes without nuclei and forms the barrier and the storage system (i.e., for water) of the skin. The thickness is approximately 100 μm at the palms of the hands and soles of the feet, and at other skin sites it is approximately 10 μm. The geometry of the Corneometer sensor allows a very low penetration depth of the electric measuring field into the skin surface. For the determination of the penetration depth, measurement was carried out on a filter soaked with saltwater and covered with a 5-μm polyurethane foil. Table 20.1 shows that 80% of the measuring value corresponds with a thickness of the foil of 10 μm. This is approximately the normal thickness of the stratum corneum (Figure 20.12).

TABLE 20.1
Definition of Different Skin Types with the Corneometer CM 825

Very dry skin	<30 CM 825 arbitrary units
Dry skin	30–40 CM 825 arbitrary units
Normal skin	>40 CM 825 arbitrary units

XI. CLASSIFICATION OF THE SKIN TYPE WITH THE CORNEOMETER CM 825

A multicenter study on 349 persons in six locations in Germany was recently published [5]. The values listed in Table 20.1, measured with the Corneometer CM 825, were taken to define the different skin types.

XII. SOFTWARE

The latest development of the Corneometer probe allows its operation with various devices from Courage + Khazaka electronic. They are equipped with a USB interface. In addition, a sensor for room temperature and relative humidity can be used at the same time; the room condition data are stored together with the measuring values. The sophisticated software allows single measurement (Figure 20.13) and continuous measurement (occlusion) with or without contact to the skin (Figure 20.14). More than one Corneometer probe can be operated with one device, and all data can be transferred to Excel together with ambient conditions (room temperature and relative humidity), measured with a sensor supplied with the device (Figure 20.15).

For more scientific literature, please visit http://www.courage-khazaka.de and download a comprehensive list of studies.

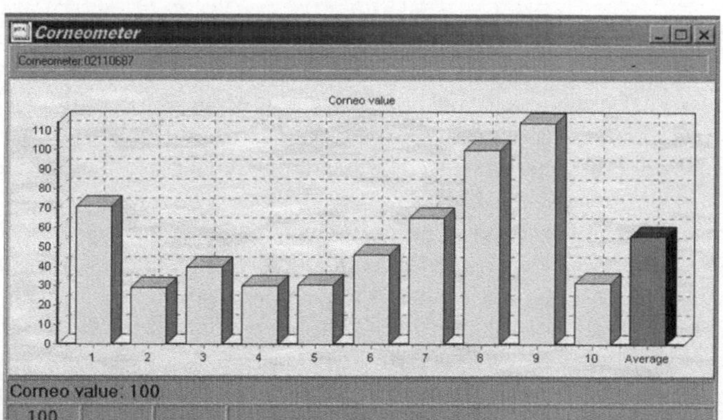

FIGURE 20.13 Single measurements carried out with the Corneometer CM 825 probe and calculation of the average value.

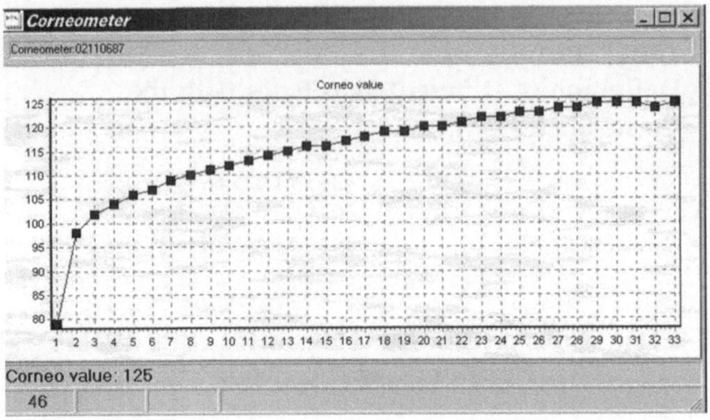

FIGURE 20.14 Measuring curve of continuous measurement with the Corneometer CM 825 probe with direct contact to the skin (occlusion). As a result of the occlusion, the measuring values increase with every measurement.

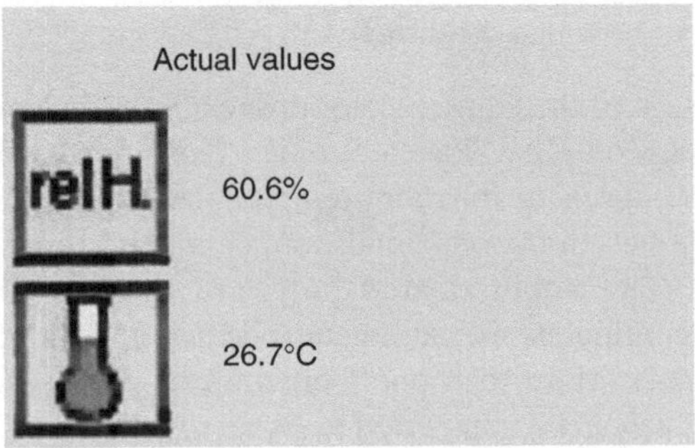

FIGURE 20.15 The measurement of ambient conditions (room temperature and relative humidity) with the Corneometer CM 825 measurement using a digital sensor provided with the device. Ambient conditions are stored with the measuring values.

REFERENCES

1. Tronnier, H., Dermatologisch-pharmakologische Methoden zur Prüfung kosmetischer Präparate und Grundstoffe, *Ärztliche Kosmetol.,* 10, 361, 1980.
2. Schrader, K.H., Untersuchungen wasserretinierender Kosmetika auf der Haut, *Parfümerie Kosmetik*, 62, 1981.
3. Courage, W., Hardware and measuring principle: Corneometer, in *Bioengineering of the Skin: Water and the Stratum Corneum*, Elsner, P., Berardesca, E., and Maibach, H.I., Eds., CRC Press, Boca Raton, Florida, 1994, pp. 171–175.
4. Rohr, M. and Schrader, A., *FOITS-Corneometry Influenced by Peripheral Experimental Conditions, the Essential Stratum Corneum*, Martin Dunitz, 2002.
5. Matta, A., Muret, P., and Humbert, P., In vivo *Calibration of the Corneometer CM 820 and the Evaporimeter EP1 Servomed for the Assessment of the Water Content of the Upper Part of the Epidermis Using a Water Sorption Desorption Test*, 13th ISBS Congress, Jerusalem, March 2000.

6. De Paepe, K., Janssens, K., Hachem, J.P., Roseeuw, D., and Rogiers, V., Squamometry as a screening method for the evaluation of hydrating products, *Skin Res. Technol.*, 7, 2001.
7. Black, D., Del Pozo, A., Diridollou, S., Lagarde, J.M., and Gall, Y., *Stratum Corneum Barrier Function Assessment Using the Sorption-Desorption Test and Laser Doppler Flowmetry*, 13th ISBS Congress, Jerusalem, March 2000.
8. Wiechers, J.W., et al., *Factors Influencing Skin Moisturisation Signal Using Near Infrared Spectroscopy*, Proceedings of the 22nd IFSCC Congress, Edinburgh, September 2002.
9. Heinrich, U., et al., Multicentre comparison of skin hydration in terms of physical-, physiological- and product-dependent parameters by the capacitive method (Corneometer CM 825), *Int. J. Cosmetic Sci.*, 25, 45, 2003.
10. Fluhr, J.W., Gloor, M., Lazzerini, S., Kleesz, P., Grieshaber, R., and Beradesca, E., Comparative study of five instruments measuring stratum corneum hydration (Corneometer CM 820 and CM 825, Skicon 200, Nova DPM 9003, DermaLab) Part II: *In vivo*, *Skin Res. Technol.*, 5, 1999.
11. Rodrigues, L., Salgueiros, Y.R., Galego, N., Pinto, P., Silva, N., and Ferro, I.Z., Study on the *in vivo* performance of two capacitance systems: assessment of the experimental reproducibility and sensitivity, *Skin Res. Technol.*, 5, 1999.
12. Barel, A.O. and Clarys, P., *In vitro* calibration of the capacitance method (Corneometer CM 825) and conductance method (Skicon 200) for the evaluation of the hydration state of the skin, *Skin Res. Technol.*, 3, 1997.
13. Barel, A., Evaluation of stratum corneum hydration: comparison between electrical capacitance (Corneometer CM 825) and conductance (Skicon 200) measurements, *Skin Res. Technol.*, 2, 1996.
14. Distante, F. and Berardesca, E., *Hydration, Bioengineering of the Skin: Methods and Instrumentation*, CRC Press, Boca Raton, Florida, 1995.
15. Berardesca, E., EEMCO guidance for the assessment of stratum corneum hydration: electrical methods, *Skin Res. Technol.*, 3, 1997.
16. Bielfeldt, S., et al., Capacitance *Measurement of Skin Moisture: Sophisticated Calibration of Instruments*, poster presented at the 14th ISBS Congress, Hamburg, Germany, May 2003.

21 Hardware and Measuring Principles: The NOVA Dermal Phase Meter

Randy Wickett

CONTENTS

I. Introduction ... 263
II. Hardware ... 264
A. Description of the DPM 9003 ... 264
1. Measurement Modes ... 264
2. Probes for the DPM 9003 ... 264
3. Mode of Operation for the DPM 9003 ... 265
B. The NOVA Petite ... 265
C. The Edwina Software ... 265
III. Experimental ... 266
A. Static Measurements ... 266
B. The Water Sorption/Desorption Test ... 267
C. The Moisture Accumulation Test ... 268
D. Correlations with Other Methods ... 270
E. Correlations with Other Electrical Instruments ... 271
F. Correlations between the NOVA DPM 9003 and the NOVA Petite ... 271
G. Precautions When Using the Instrument ... 271
IV. General Conclusions ... 272
References ... 272

I. INTRODUCTION

Measurements of the impedance or capacitive reactance of the stratum corneum have been used to asses the hydration status of the skin for more than 20 years.[1,2] Over this time, three instruments have been the dominant tools for this purpose: the Skicon, the Corneometer, and the NOVA Dermal Phase Meter (DPM). The NOVA DPM 9003 will be the primary focus of this chapter, and the new NOVA Petite Instrument will also be discussed.

0-8493-1443-7/05/$0.00+$1.50

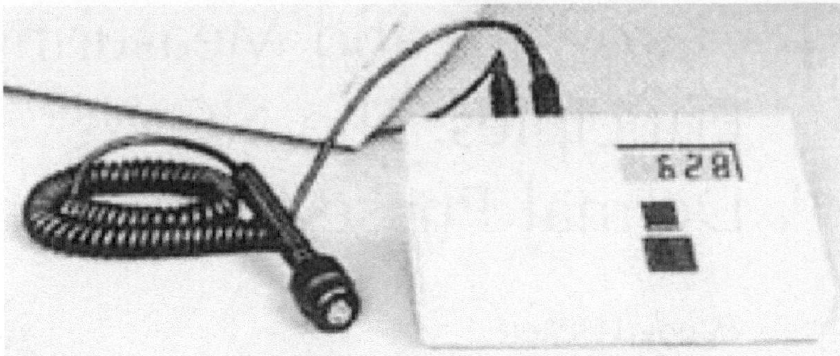

FIGURE 21.1 The NOVA Dermal Phase Meter 9003 with standard probe.

II. HARDWARE

A. Description of the DPM 9003

The NOVA DPM 9003 is a lightweight, portable, battery-operated device that can either be read from the liquid crystal display or connected to a laptop computer (Figure 21.1). The device is automatically calibrated when turned on, using an internal calibration. The measurements are activated by an automatic sensor and displayed on the liquid crystal display in whole numbers. The minimum reading is 90. Values lower than 90 result in zero readings.

The NOVA DPM 9003 runs on four 1.5-volt (AA) batteries. Its dimensions are 155 by 92 mm, and it is 33 mm tall. It weighs 227 g without batteries.

1. Measurement Modes

There are three modes of measurement. They can be selected by holding the Mode button down and waiting for the desired mode to appear on the display.

dl 0: In this mode, the reading is instantaneous when the probe touches the skin. The immediate state of the skin is measured with little or no contribution from water build-up resulting from probe occlusion.

dl 5: In this mode, the reading is "latched" after a 5-second reading interval. The reading will be slightly higher than the instantaneous reading because 5 seconds of probe occlusion occurs.

Con: This setting provides continuous readings. In this mode, the measurement continually increases when the probe is held against the skin surface as water builds up beneath the probe head because of occlusion.

2. Probes for the DPM 9003

A variety of interchangeable probes is available for this instrument. The standard probe that comes with the instrument is the DPM 9103. It has a flat circular surface with two concentric brass ring electrodes separated by a nonconducting polymer. The ring diameters are 8.76 and 4.34 mm, and the probe is 113.41 mm long. The DPM 9105 6-mm sensor has ring diameters of 5.08 and 2.54 mm. We have found the 9105 probe to be very useful for studies with neonatal rats[3,4] and athymic mice[5] because of its smaller size. Readings from 9105 probes will be lower than those from the 9103 because of the different geometry, but in our experience, the relative changes are consistent between the two probes. Other probes, including custom and disposable probes, are available. Specifications of all the probes can be found on NOVA Technology Corporation Web site (http://www.novatechcorp.com/dpm9003.html).

3. Mode of Operation for the DPM 9003

The device emits a 1-MHz span of simultaneously produced frequencies, producing a differential current source with a controlled rise time. Impedance is evaluated at several frequencies using a proprietary chip in the instrument. Impedance is a complex quantity with both imaginary and real parts. The electrical impedance of skin seems to be mostly capacitive.[6] In the past, we have referred to the DPM measurement as capacitive reactance, which is indeed the main component, but the overall impedance is what is actually measured. The reading is related to the inverse of the overall impedance and increases as the skin goes from very dry to very wet. Values range from 90 to 999. More information is available on the NOVA Technology Corporation Web site (http://www.novatechcorp.com/impedance.html).

B. The NOVA Petite

The NOVA Petite is a small, handheld probe that is interfaced directly to a personal digital assistant. The probe diameter is 6 mm. The device measures both skin surface temperature and the contact force of the probe against the skin. Data collected on the personal digital assistant can be exported directly to a spreadsheet. The principle of operation is the same as the DPM 9003, but the readings start at 0 rather than 90. Our preliminary results with the NOVA Petite indicate that it correlates well with the DPM at high readings but may be more sensitive to differences in the very low end of the scale (Figure 21.2).

C. The Edwina Software

The Edwina software package is available for both PC and Mac computer systems. The program can be set up to automatically record data when the probe contacts the skin and can be configured to record data continuously at predetermined intervals. This feature is particularly useful when performing dynamic measurements such as the Moisture Accumulation Test (see below). Data can be easily exported to commercial spreadsheets. The next generation of the software is scheduled for release in the summer of 2003 and is intended to provide additional rate-of-change analysis and capabilities.

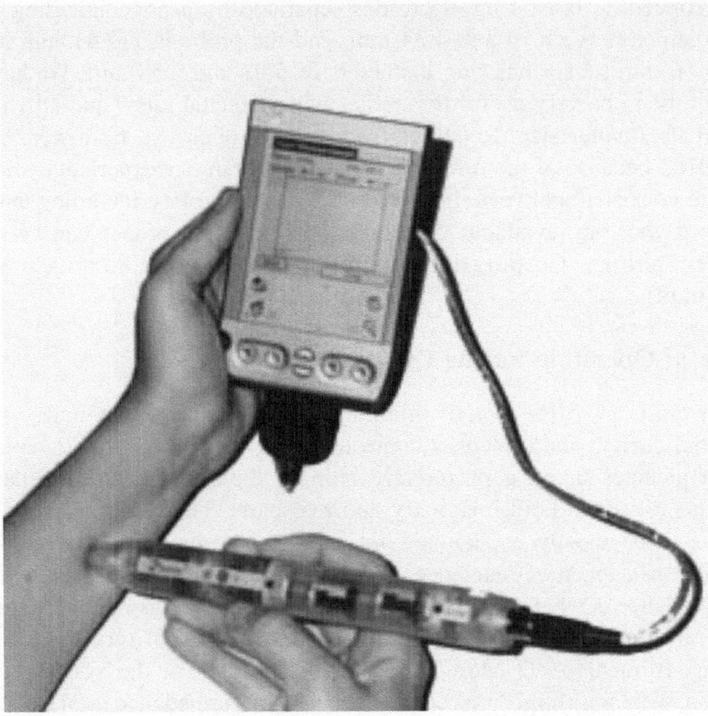

FIGURE 21.2 The NOVA Petite. Photo courtesy of NOVA Technology Corporation.

III. EXPERIMENTAL

A. Static Measurements

DPM measurements collected using the dl0 or dl5 modes can be used to estimate the relative water content of skin[7–9]; these methods have long been used to substantiate the efficacy of moisturizers.[10] In our laboratory, we have investigated the effects of simple formulations containing glycerin and sodium chloride on measurements taken 1 and 2 hours after application with the NOVA DPM 9003, the Skicon, and the Corneometer CM 825.[11,12]

The data in Figure 21.3 were taken 1 hour after treatment with otherwise identical formulations that contained glycerin concentrations ranging from 0 to 15% and with an untreated control site.

Li at al.[12] also investigated, in a 2-week study, the ability of these formulations to improve skin grades, and the correlation between instrumental readings at 1 hour and skin grade reductions at 1 and 2 weeks. Results are illustrated in Figure 21.4.

Similar correlations were found with the Skicon and Corneometer. The correlation between grade reductions and the 1-hour measurement with the DPM indicates that electrical measures are predictive of long-term moisturizing ability, at least with glycerin-based moisturizers.

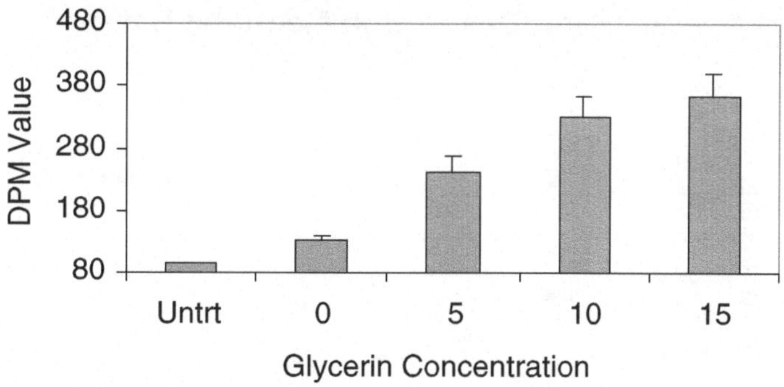

FIGURE 21.3 Dermal Phase Meter values 1 hour after treatment with formulations containing glycerin levels from 0 to 15%; $N = 9$, error bars are standard errors of the mean.

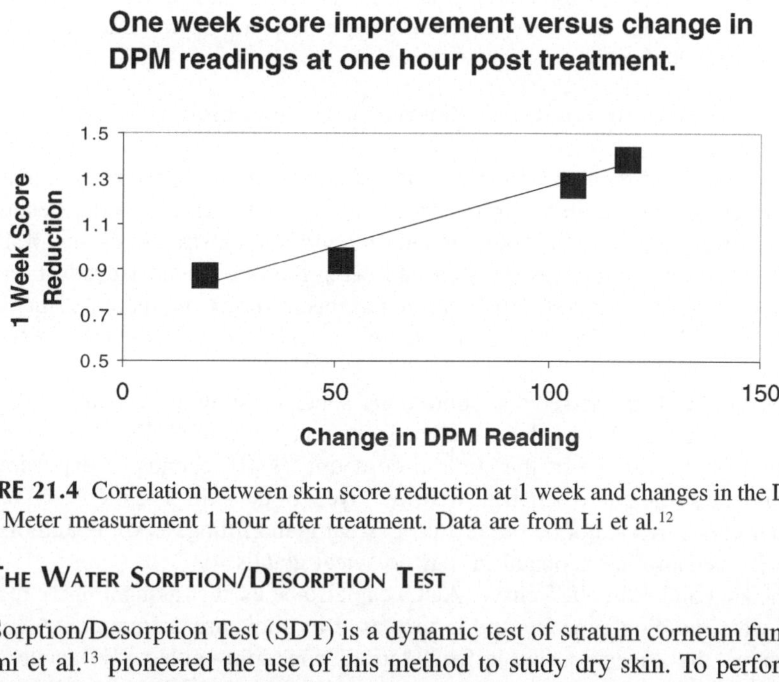

FIGURE 21.4 Correlation between skin score reduction at 1 week and changes in the Dermal Phase Meter measurement 1 hour after treatment. Data are from Li et al.[12]

B. The Water Sorption/Desorption Test

The Sorption/Desorption Test (SDT) is a dynamic test of stratum corneum function. Tagami et al.[13] pioneered the use of this method to study dry skin. To perform the SDT, water is placed on the skin for a defined time (usually 10 to 20 seconds) and then quickly blotted off. Measurements are then made with time until as the reading returns to baseline. This can be considered a measure of the water-holding capacity of the skin.

The neonatal rat has a thick stratum corneum with no sweat glands or hair follicles. At birth, the animal has a hydrophobic coating called the periderm. Removal of the periderm as the animal grows or by artificial means such as tape stripping leads to a dramatic increase in water sorption as judged by SDT curves.[3] Figure 21.5

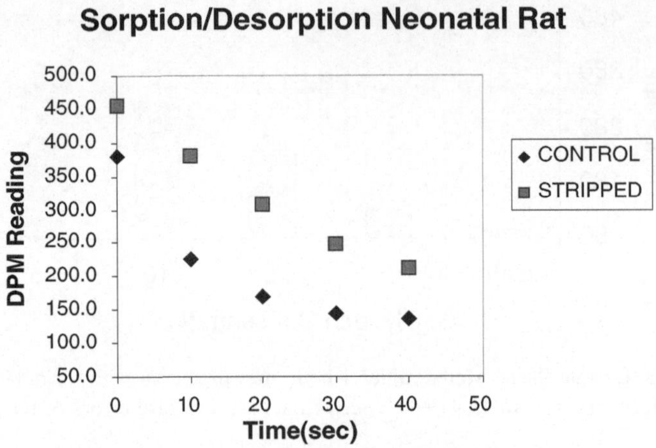

FIGURE 21.5 Sorption/Desorption Test curves after 10 seconds of water exposure for a neonatal rat pup 1 day after birth; effect of a single tape strip shown. Data were obtained with the model 9015 probe.

shows SDT curves from a neonatal rat pup 1 day after birth both before and after a single tape strip.

The removal of the hydrophobic periderm allows the stratum corneum to absorb more water, leading to higher initial values and a slower rate of water desorption. Water desorption curves for both neonatal rat and human stratum corneum fit well to a single exponential decay constant, allowing the rate of water desorption to be calculated.[3,13–15] Moisturizer application has been shown to increase the water-holding capacity of the stratum corneum, as calculated using this method of measurement.[16,17]

The neonatal rat presents a continuous sheet of stratum corneum. The SDT curves from neonatal rats show conclusively that the NOVA DPM is measuring hydration of the surface of the stratum corneum, as 10 seconds of exposure time would not be sufficient to penetrate to lower layers of the skin. The increased reading on water exposure cannot be the result of sweat gland filling, as some authors have claimed,[18] because the neonatal rat has no sweat glands.

The area under the SDT curve (AUC) can also be used as a quantitative measure of the water-holding capacity and is increased by moisturizer treatment.[8,16] Figure 21.6 shows the effects of simple formulations containing either 8% lactate or 8% glycerin on AUC 1 hour after treatment measured with the DPM 9003. Both active ingredients increased AUC compared with vehicle or untreated skin.[19]

C. The Moisture Accumulation Test

When the probe of an impedance or capacitance device is held against the skin surface, occlusion causes the measurement to change with time, as water that would have passed through the stratum corneum builds up under the probe. This is the Moisture Accumulation Test (MAT).[4,8,14,16,20]

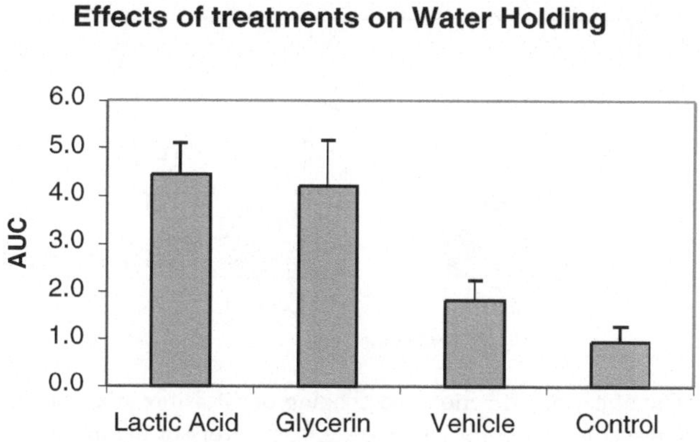

FIGURE 21.6 Area under the curve values for skin treated with moisturizers containing either 8% lactic acid or 8% glycerin and vehicle; 18 subjects' error bars are standard errors of the mean.

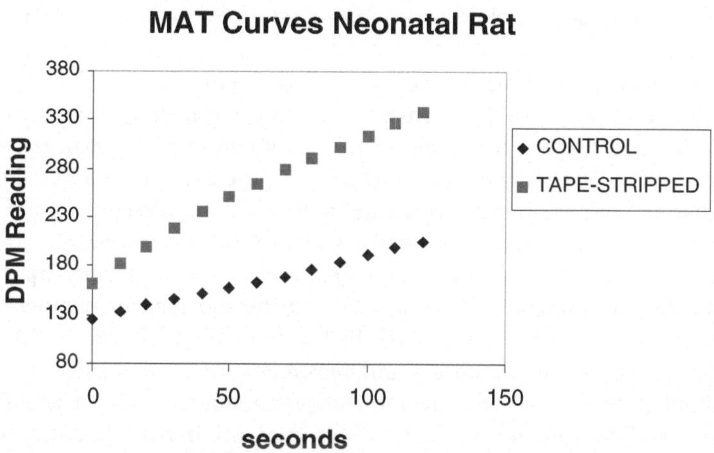

FIGURE 21.7 Moisture Accumulation Test curves from a neonatal rat, day 1.

When the stratum corneum barrier is disrupted, MAT increases. Figure 21.7 shows DPM vs. time curves for neonatal rats on day 1 of life, both before and after tape stripping. MAT data are usually presented in one of two ways. Either the actual curve of DPM reading vs. time is presented, as in Figure 21.7, or the rate of change is presented. Fluhr et al.[8] refer to the rate of change as the moisture accumulation velocity. Some authors have also referred to the measurements as transepidermal water movement.[21]

The MAT has proven to be useful tool to evaluate barrier function in situations with a disturbed barrier in which transepidermal water loss is difficult or inconvenient to measure, such as with cell culture systems[5,22] and with premature infants. The rate of moisture accumulation is very high in severely premature infants and

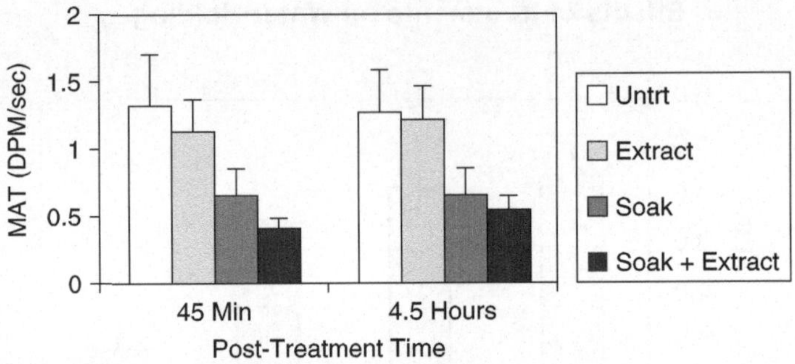

FIGURE 21.8 The effect of extraction and soaking on Moisture Accumulation Test slopes; 11 subjects were used, and the error bars are standard errors of the mean.

correlates to gestational age.[14] Moisture accumulation decreases dramatically over the first few days of life.

MAT measurements have also been used successfully to evaluate abnormal barrier development in transgenic mice that are defective in Kruppel-like growth factor 4[23] and, in several studies, to evaluate healing of human skin transplanted to athymic mice.[5,22,24,25]

In situations with a normal barrier but with dry skin, the rate of change in DPM values is low and the MAT is a sensitive parameter reflecting skin dryness.[26] We have hypothesized that the lower moisture accumulation in dry skin reflects lower levels of natural moisturizing factor (NMF).[26–28] Figure 21.8 shows MAT slope values for skin that had been untreated, extracted with 50/50 acetone/ether for 5 minutes, soaked in water for 10 minutes, or soaked in water following acetone/ether extraction.

Treatment of skin by acetone/ether, which removes some surface lipids, did not reduce moisture accumulation significantly, but the subsequent treatment by a 10-minute water soak to remove the NMF had a dramatic effect on the MAT. A 10-minute water soak alone caused a significant reduction in moisture accumulation, even without pretreatment with acetone/ether. Treatment with an artificial NMF mixture restored moisture accumulation.[27] Recent work in our laboratory has shown that the reduction in moisture accumulation on soaking and extraction are correlated with reductions in measured values of the amino acid components of NMF.[29]

D. Correlations with Other Methods

MAT slopes have been shown to correlate to transepidermal water loss in cases in which the skin barrier is disrupted.[4] In situations in which the skin is dry, the MAT and transepidermal water loss measurements do not correlate.

There is a correlation between increases in DPM readings and increased elasticity of the skin under some conditions[30]; however, DPM measurements are much more sensitive to both immediate hydration effects and long-term moisturization effects than are measurements with either the Cutometer[30–32] or the Dermal Torque Meter.[32]

E. Correlations with Other Electrical Instruments

There have been studies comparing the values obtained with various instruments.[7–9,11] We have found that NOVA DPM readings correlate well to Skicon readings over the entire range of measured values but only correlate well with Corneometer CM 825 readings when the CM 825 readings are about 40 or higher.[11] Our work is in agreement with that of Fluhr et al.,[8] finding a better correlation between the NOVA and the Skicon than that between either instrument and the Corneometer. This might be expected because both NOVA and the Skicon use higher frequencies than the Corneometer.

F. Correlations between the NOVA DPM 9003 and the NOVA Petite

We are not aware of any published studies with the NOVA Petite. M. Visscher and coworkers (unpublished observations) compared MAT values from the NOVA DPM 9003 on normal skin and on skin that had been tape stripped to increase moisture accumulation. Their results are shown in Figure 21.9: There was an excellent correlation between the MAT values from the two instruments.

G. Precautions when Using the Instrument

The NOVA DPM 9003 is very easy to use but, naturally, certain precautions must be taken. As with other bioengineering measurements it is important to control the environmental conditions. Humidity should be controlled in the range from 40% to

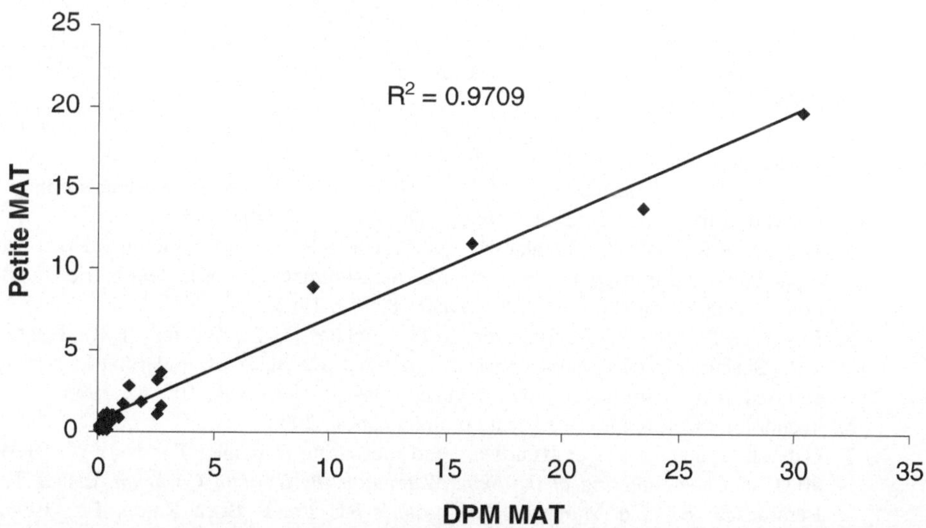

FIGURE 21.9 Comparison between MAT values measured using the NOVA Dermal Phase Meter 9003 and those measured using the NOVA Petite. Data courtesy of Marty Visscher, Skin Sciences Institute, Children's Hospital, Cincinnati, OH.

60%, and temperature should fall in the range from 20° to 22°C, in accordance with the EEMCO guidelines.[33] Subjects should generally be equilibrated for 30 minutes before measurements, with the site to be measured exposed to the air. Controlling contact pressure is very important to obtain consistent values. With the DPM 9003, we generally use the weight of the probe to provide a uniform contact pressure. Another approach is to use small "lead donuts" on the probe to get a precise probe weight. The NOVA Petite has a force transducer in the probe to allow control of contact pressure, and a temperature sensor to precisely measure skin surface temperature.

IV. GENERAL CONCLUSIONS

The NOVA DPM 9003 is a versatile instrument for the evaluation of skin hydration. It is very sensitive to moisturizer treatment and has been shown to be capable of predicting the efficacy of glycerin-based moisturizers by measuring within 1 hour of treatment.[12] The availability of different probes increases the flexibility of the instrument in nonstandard situations, such as measurements on small animals. Dynamic results using either sorption–desorption (SDT) or moisture accumulation (MAT) protocols can increase the sensitivity to such factors as barrier damage and the level of natural moisturizer in the stratum corneum. SDT and MAT results with the neonatal rat clearly demonstrate that the measurements are primarily sensitive to the hydration level of the superficial stratum corneum.

REFERENCES

1. Tagami, H., Ohi, M., Iwatsuki, K., Kanamaru, Y., Yamada, M., and Ichijo, B., Evaluation of the skin surface hydration *in vivo* by electrical measurement, *J. Invest. Dermatol.*, 75, 500, 1980.
2. Leveque, J.L. and de Rigal, J., Impedance methods for studying skin moisturization, *J. Soc. Cosmetic Chem.*, 29, 419, 1983.
3. Wickett, R.R., Mutschelknaus, J.L., and Hoath, S.B., Ontogeny of water sorption-desorption in the perinatal rat, *J. Invest. Dermatol.* 100, 407, 1993.
4. Wickett, R.R., Nath, V., Tanaka, R., and Hoath, S.B., Use of continuous electrical capacitance and transepidermal water loss measurements for assessing barrier function in neonatal rat skin, *Skin Pharmacol.*, 8, 179, 1995.
5. Boyce, S.T., Supp, A.P., Harriger, M.D., Pickens, W.L., Wickett, R.R., Hoath, S.B., Surface electrical capacitance as a noninvasive index of epidermal barrier in cultured skin substitutes in athymic mice, *J. Invest. Dermatol.*, 107, 82, 1996.
6. Technology Corporation, personal communication, 2003.
7. Gabard, B. and Treffel, P., Hardware and measuring principle: The NOVA™ DPM 9003, in *Bioengineering of the Skin: Water and the Stratum Corneum*, Elsner, P., Berardesca, E., and Maibach, H.I., Eds., CRC Press, Boca Raton, FL, 1994, pp. 177–196.
8. Fluhr, J.W., Gloor, M., Lazzerini, S., Kleesz, P., Grieshaber, R., and Berardesca, E., Comparative study of five instruments measuring stratum corneum hydration (Corneometer CM 820 and CM 825, Skicon 200, Nova DPM 9003, DermLab). Part I. *In vitro*, *Skin Res. Technol.*, 5, 161, 1999.

9. Clarys, P., Barel, A.O., and Gabard, B., Non-invasive electrical measurements for the evaluation of the hydration state of the skin: comparison between three conventional instruments—The Corneometer, Skicon and NOVA DPM, *Skin Res. Technol.*, 5, 14, 1999.
10. Loden, M., Biophysical methods of providing objective documentation of the effects of moisturizing creams, *Skin Res. Technol.*, 1, 101, 1995.
11. Li, F., Visscher, M., Conroy, E., and Wickett, R.R., The ability of electrical measurements to predict skin moisturization. I. Effects of salt and glycerin on short-term measurements, *J. Cosmetic Sci.*, 52, 13, 2001.
12. Li, F., Visscher, M., Conroy, E., and Wickett, R.R., The ability of electrical measurements to predict skin moisturization. II. Correlations between one hour measurements and long term results, *J. Cosmetic Sci.*, 52, 23, 2001.
13. Tagami, H., Kanamaru, Y., Inoue, K., Suehisa, S., Inoue, F., Iwatsuki, K., Yoshikuni, K., and Yamada, M., Water sorption-desorption test of the skin *in vivo* for functional assessment of the stratum corneum, *J. Invest. Dermatol.*, 78, 425, 1982.
14. Okah, F.A., Wickett, R.R., Pickens, W.L., and Hoath, S.B., Surface electrical capacitance as a noninvasive bedside measure of epidermal barrier maturation in the newborn infant, *Pediatrics*, 96, 688, 1995.
15. Okah, F.A., Wickett, R.R., Pompa, K., and Hoath, S.B., Human newborn skin: the effect of isopropanol on skin surface hydrophobicity, *Pediatr. Res.*, 35, 443, 1994.
16. Treffel, P. and Gabard, B., Stratum corneum dynamic function measurements after moisturizer or irritant application, *Arch. Dermatol. Res.*, 287, 474, 1995.
17. Sindhvananda, J., Gritiyarangsan, P., Rungrairatanaroij, P., and Kullavanijaya, P., Hygroscopicity and water-holding capacity of moisturizing agents: a single application *in vivo* study, *J. Soc. Cosmetic Chem.*, 44, 279, 1993.
18. Martinsen, O.G. and Grimnes, S., Facts and myths about electrical measurement of stratum corneum hydration state, *Dermatology*, 202, 87, 2001.
19. Wickett, R.R., Bidaye, A., and Visscher, M., Comparing the effects of lactic acid and glycerin on skin using simple formulations, in preparation.
20. Van Neste, D., *In vivo* evaluation of unbound water accumulation in stratum corneum, *Dermatologica*, 181, 197, 1990.
21. Visscher, M., Maganti, S., Munson, K.A., Bare, D.E., and Hoath, S.B., Early adaption of human skin following birth: a biophysical assessment, *Skin Res. Technol.*, 5, 213, 1999.
22. Supp, A.P., Wickett, R.R., Swope, V.B., Harriger, M.D., Hoath, S.B., and Boyce, S.T., Incubation of cultured skin substitutes in reduced humidity promotes cornification *in vitro* and stable engraftment in athymic mice, *Wound Repair Regen.*, 7, 226, 1999.
23. Segre, J.A., Bauer, C., and Fuch, E., Klf4 is a transcription factor required for establishing the barrier function of skin, *Nat. Genet.*, 22, 356, 1999.
24. Boyce, S.T., Skin substitutes from cultured cells and collagen-GAG polymers, *Med. Biol. Eng. Comput.*, 36, 791, 1998.
25. Boyce, S.T., Supp, A.P., Swope, V.B., and Warden, G.D., Vitamin C regulates keratinocyte viability, epidermal barrier, and basement membrane *in vitro*, and reduces wound contraction after grafting of cultured skin substitutes, *J. Invest. Dermatol.*, 118, 565, 2002.

26. Wickett, R.R., Tolia, G., Visscher, M., and Hoath, S.B., Bioengineering evaluation of the water handling capabilities of stratum corneum *in vivo*. Creation and Hope, Proceedings of the 2001 IFSCC International Conference, Taipei Taiwan, International Federation of Societies of Cosmetic Chemists, 2001.
27. Visscher, M., Tolia, G., Wickett, R.R., and Hoath, S.B., Effect of soaking and natural moisturizing factor on stratum corneum water handling properties, *J. Cosmetic Sci.*, 2003, in press.
28. Wickett, R.R., Tolia, G., and Visscher, M., The effect of extraction and NMF treatment on the water handling capability of stratum corneum as measured by TEWL and MAT, in *The Essential Stratum Corneum*, Marks, R., Leveque, J.L., and Voegeli, R., Eds., Martin Dunitz, Andover, 2002, pp. 243–250.
29. Robison, M. and Wickett, R.R., in preparation, 2003.
30. Murray, B.C. and Wickett, R.R., Sensitivity of Cutometer data to stratum corneum hydration, *Skin Res. Technol.*, 2, 167, 1996.
31. Wickett, R.R., Stretching the skin surface: skin elasticity, *Cosmetics Toiletries*, 116, 47, 2003.
32. Murray, B.C. and Wickett, R.R., Correlations between Dermal Torque Meter, Cutometer and Dermal Phase Meter measurements of moisturizer effects on human skin, *Skin Res. Technol.*, 3, 101, 1997.
33. Berardesca, E., EEMCO guidance for the assessment of stratum corneum hydration: electrical methods, *Skin Res. Technol.*, 3, 126, 1997.

22 Hardware and Measuring Principles: The Computerized DermaLab® Transepidermal Water Loss Probe

Gary L. Grove and Charles Zerweck

CONTENTS

I. Introduction 275
II. The Vapor Pressure Gradient Estimation Method of Nilsson 276
III. The Computerized Dermalab TEWL Probe 278
IV. Diagnostic Programs 279
V. Calibration Procedures 280
VI. Measurement of TEWL under Steady-State Conditions 281
VII. Measurement of TEWL under Dynamic Conditions 283
VIII. Measurement of TEWL Using Dual-Probe Instruments 283
IX. Retrospective Analysis of the Data 284
X. Concluding Remarks 284
References 285

I. INTRODUCTION

It is generally appreciated that *in vivo* measurements of transepidermal water loss (TEWL) rates through human skin can be used to noninvasively monitor changes in stratum corneum barrier function. In normal, healthy skin, the barrier is quite effective, and water loss rates are typically very low. If the barrier is not fully functional because of pathological processes, or if it is damaged by physical or chemical agents, then there will be a corresponding increase in water loss rates that directly relates to the degree of impairment. Conversely, there will be a corresponding

0-8493-1443-7/05/$0.00+$1.50

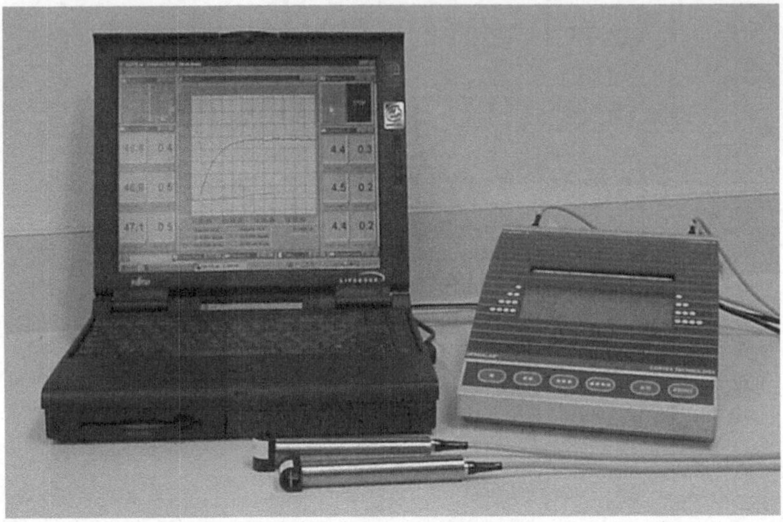

FIGURE 22.1 The computerized DermaLab Transepidermal Water Loss Probe showing a dual-probe configuration interfaced into a laptop computer.

decrease in TEWL as the barrier is restored. This means that monitoring changes in TEWL over time not only allows one to evaluate therapeutic response to different treatments but can also help determine the effectiveness of various prophylactic strategies that could be used to prevent or lessen the injury in the first place. Thus, it is not surprising that there is considerable literature dealing with TEWL measurements. Indeed, these methods were the very first to be reviewed by the Standardization Group of the European Society of Contact Dermatitis [1]. Guidelines for TEWL measurements have also been established by EEMCO, as well [2].

Over the years, a variety of instrumental methods have been devised to measure TEWL (for general reviews, see [3–5]). A convenient way to classify these instruments is based on whether the configuration of the probe is closed, ventilated, or an open-chamber design. By far the most popular instruments are open-chamber designs, which are based on the Vapor Pressure Gradient Estimation Method of Nilsson and that were first commercialized as the Servo Med Evaporimeter in 1979 [6]. More recently, Cortex Technology (Hadsund, Denmark) has introduced a new and improved version of an open-chamber TEWL probe as part of their DermaLab System [7,8]. This chapter will primarily deal with the computerized DermaLab TEWL Probe shown in Figure 22.1, but most of the comments will be equally applicable to other open chamber systems.

II. THE VAPOR PRESSURE GRADIENT ESTIMATION METHOD OF NILSSON

Open-chamber devices for measuring TEWL are based on the Vapor Pressure Gradient Estimation Method, which was first proposed by Nilsson [6] as part of his

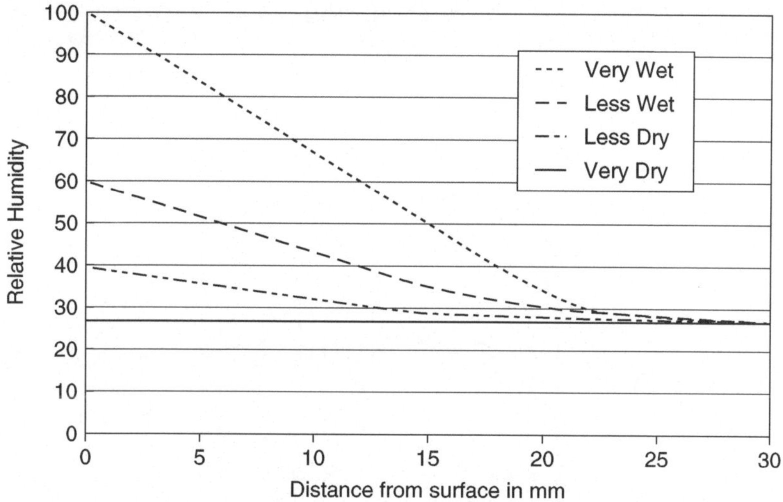

FIGURE 22.2 Hypothetical plot of the relative humidity as a function of distance from the surface in four different situations: bone dry, slightly damp, slightly wet, and soaking wet. The ambient room condition in this case was 27% relative humidity.

doctoral thesis at Linkoping University, Sweden. To understand the underlying principle on which these measurements are based, let us consider the following hypothetical situation, depicted in Figure 22.2. If we plot the water content of the air above a completely dry surface as a function of the distance from that surface, we would obtain a flat line at whatever level the ambient conditions might be. If, however, we plot the situation that exists above a completely wet surface, we now find a very steep profile that begins at 100% and then falls progressively until reaching an asymptote that corresponds to the ambient conditions. Plots from situations between these extremes would show intermediate plots that directly correspond to the degree of surface wetness and to how much water must be evaporating from it. It is easy to envision that by placing sensors at fixed points from the surface, one could measure the initial slope and, thus, determine the rate of evaporation from that surface.

A similar situation exists above the skin surface, and the resulting vapor pressure gradient is directly dependent on the rate of water exchange through the skin. Although this gradient is disturbed when the probe is first placed on the skin surface, with an open-chamber design, the gradient will be quickly reestablished. As shown in Figure 22.3, within this cylinder, which is placed perpendicular to the skin surface, there are sensors that are set at two fixed distances above the skin surface. At each point, the local relative humidity (RH) and temperature are measured by paired sensors, and the vapor pressure at each point is computed. The difference between the vapor pressure at these two points along the gradient is directly related to the rate of evaporative water loss from that skin site, as described above.

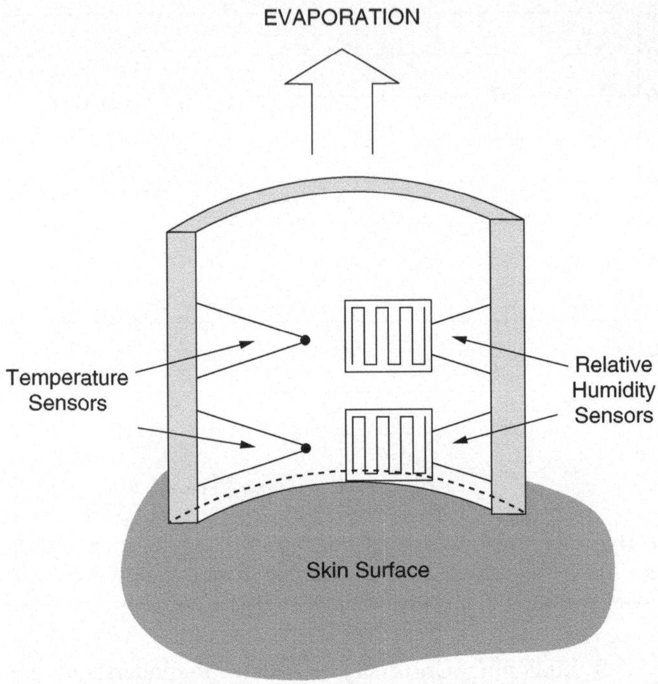

FIGURE 22.3 Schematic drawing of the DermaLab TEWL probe. The open chamber consists of a cylinder that is placed perpendicular to the skin surface being measured. Within this chamber, there are two pairs of sensors set at fixed distances above the skin surface. From the relative humidity and temperature readings obtained from these two levels, the water vapor pressure gradient can be determined and the amount of water evaporating from 1 square meter of the skin surface per hour calculated in grams. Based on Nilsson's Vapor Pressure Gradient Method [6].

III. THE COMPUTERIZED DERMALAB TEWL PROBE

The DermaLab is a modular system that can be configured in a variety of ways that include not only TEWL measurements but other noninvasive assessments of skin function such as skin surface hydration levels, sebaceous gland activity, and elasticity, depending on the module chosen. For TEWL measurements, the DermaLab can be configured as either a single-probe or a dual-probe instrument by attaching one TEWL probe to just channel A or two probes to both channels A and B, respectively.

Although the DermaLab System can be used as a stand-alone instrument, as will be shown below, optimal performance is achieved when it is interfaced with a PC. This is easily done using a built-in RS232c link to the serial port of the host PC. When placed in continuous mode, the DermaLab will transmit data in the form of ASCII text at a rate of four lines per second. Each line provides not only the computed water loss rate but also the values for the RH and temperature of the upper and lower sensors on which it is based. The fastest filter should be employed so that the least-damped signal is available for computer analysis.

The application software used with the DermaLab system is provided by cyberDERM, Inc. (Media, PA), and is based on DASYLab (DasyTec, Amherst, NH). The incoming line of ASCII text is parsed into the appropriate segments so that each of the values can subsequently be processed separately. Data processing includes generating a time-based list of the raw data inputs, computing means, and standard deviations for certain blocks of data, as well as reporting selected parameters in real-time visual displays using strip chart recorder graphs or digital meters. In addition to displaying data in real time, the graphic user interface also provides interactive switches that allow the operator to choose certain process options such as duration of measurement or type of display.

Retrospective analysis can be performed by exporting the raw data list into spreadsheets created with Microsoft Excel. Multitasking operations based on dynamic data exchange (DDE) allow the extracted values to be electronically transferred to appropriately linked spreadsheets in near real time.

IV. DIAGNOSTIC PROGRAMS

It should be appreciated that the real-time outputs of the DermaLab include not only the computed water loss rate but also the values for the RH and temperature of the upper and lower sensors on which it is based. This means that the condition of the probe can be easily checked using computerized evaporimetry and should be the keystone of your quality assurance program. If we allow the probe to sample just the environment, the upper and lower sensors are surrounded by the same air and, thus, the RH and temperature values obtained from them should be equal, as shown in Figure 22.4.

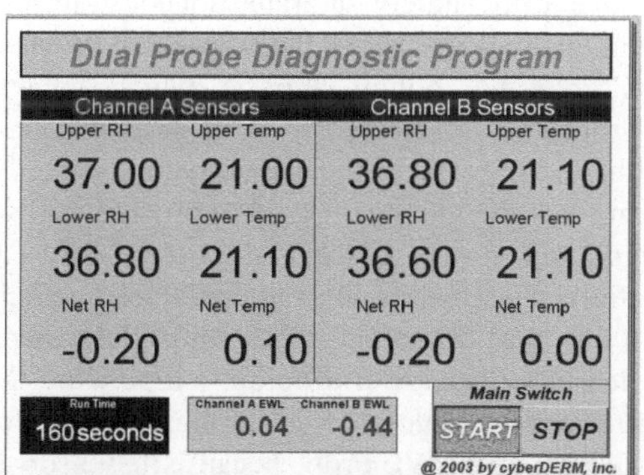

FIGURE 22.4 Graphic user interface showing typical results from a performance test of a computerized DermaLab. Note that there is good agreement in the upper and lower sensors for temperature and relative humidity both within and between the two probes.

Such diagnostic programs can be further refined by color coding the RH and temperature meters according to internationally recognized standards for the ideal environmental conditions for obtaining reliable evaporative water loss measurements from human volunteers [1]. Thus, if the ambient RH detected by either sensor falls outside the range of from 30 to 50%, or if the ambient temperature is not between 20° and 22°C, the respective value will be displayed in red or yellow instead of green. They are meant to be visual warnings that less than ideal conditions exist, but they will not prevent the measurements from being taken.

Keeping track of the values obtained under ambient conditions in a routine fashion makes it easy to distinguish any changes in instrument performance that would indicate that recalibration is needed. This is especially true if the studies are being done under carefully controlled conditions in an environmental chamber. Any deviations from the expected values are a clear-cut indication that the probe is no longer in calibration or that the environmental chamber conditions have been altered.

V. CALIBRATION PROCEDURES

It is recommended that the DermaLab TEWL probe not be calibrated in the field but, rather, only by the factory, which will issue a certificate that the probe conforms to very strict tolerances that are based on National Institute of Standards and Technology traceable standards. During calibration, which involves two saturated salt solutions (lithium chloride and sodium chloride), trim potentiometers are used to set the appropriate RH and temperature values for each standard. Before leaving the factory, each of the main units is adjusted to have a common reference standard as the default setting so that the probes can be freely exchanged.

Those familiar with the older Servo Med Evaporimeter will recall that they used a three-point calibration procedure. At first glance, it would seem that a three-point calibration would be better than a two-point calibration, especially as the reference standards are in the same form, namely, saturated aqueous salt solutions that generate a constant humidity in the air space above the solution. As reviewed by Brownawell [9], only lithium chloride and sodium chloride, which are used to calibrate the DermaLab TEWL Probe, are considered to be a good choice. Potassium sulfate, which is used to calibrate the Servo Med Evaporimeter, is greatly affected by traces of other salts requiring that the glassware used to prepare said solutions be scrupulously cleaned. Magnesium nitrate, which is the other salt used by Servo Med, is especially problematic because of its high temperature coefficient. Compounding this problem is the fact that because air has much less thermal mass than the water/salt solution, it can be used at a different temperature than the liquid. This difference will cause a considerable error in the observed RH reading. This is not a problem with the DermaLab TEWL Probe because its sensors are directly reading the temperature of the vapor phase and can be trimmed accordingly.

It is important to appreciate that even when the sensors are properly calibrated for both RH and temperature, there are other factors that can affect probe performance that need to be considered to guarantee that the probe measures TEWL reliably. Chief among these factors is the position of the sensors because they must be accurately positioned above the skin surface to sample the vapor pressure gradient correctly. The sensors must also be rapid and equally responsive to changes in either RH or temperature, especially when dynamic assessments are being made. If only steady-state assessments are being made, then slower-responding sensors can be tolerated with the understanding that it will take longer for the values to plateau. These additional concerns are best addressed by using a reference box that is covered with polymeric films that differ in water vapor transmission rates according to ASTM-96-95 [10]. This is similar to the constant water evaporation device that was constructed by Pinnagoda et al. [1] using a polyetherurethane Op-site membrane stretched over a Petri dish. Wheldon and Monteith [11] had earlier used the water vapor transmission rate (WVTR) analyzer of Celgard (Charlotte, NC) to evaluate the performance of a skin surface evaporimeter.

VI. MEASUREMENT OF TEWL UNDER STEADY-STATE CONDITIONS

In most cases, it is highly desirable that TEWL rates be measured under steady-state conditions. No matter how careful you are, the mere placement of the probe on the skin surface will disturb the water vapor gradient at that site. It will also take some time for the sensors to warm up and respond to the changes in RH that take place as the gradient is reestablished. As a consequence, after the probe is placed on the skin, the water loss rate will progressively increase over time and then plateau when equilibrium is restored. It is only under such steady-state conditions that meaningful measurements of stratum corneum barrier function can be made.

Blichmann and Serup [12], in their studies of the Servo Med EP1 Evaporimeter, found that TEWL values reached a linear state approximately 20 seconds after application of the probe to the flexor side of the forearm of healthy volunteers. Pinnagoda et al. [13] also found that stabilization of the TEWL value is usually reached 30 to 45 seconds after probe placement. Our experiences with the DermaLab TEWL probe are quite similar, and for routine baseline measurements we employ an application program in which the duration of the measurement run is fixed at 1 minute. Although the DermaLab is continually sending data to the PC, it is a simple matter to ignore the inputs during the first 40 seconds and use only those over the last 20 seconds to compute the mean and standard deviation of the water loss rates. Because the sampling rate is four inputs per second, the computed values are based on 80 individual data points. However, if one uses the noncomputerized DermaLab as a stand-alone device, the mean TEWL is based on only the last eight values, taken at a sampling rate of one per second.

With the computerized DermaLab, such processing occurs automatically and requires no operator action other than initially clicking a start button with the mouse. Nevertheless, the operator needs to be fully aware of any fluctuations or deviations that occur in the water loss values while the measurements are being taken. That is why the real-time display provided by the strip chart recorder is so important. Problems caused by air currents, probe movements, and so forth will be readily apparent, and their effect on the measurement will be immediately determined.

It should be emphasized that the recommended stabilization period of 30 to 45 seconds [1] is for baseline TEWL measurements. There are many situations such as

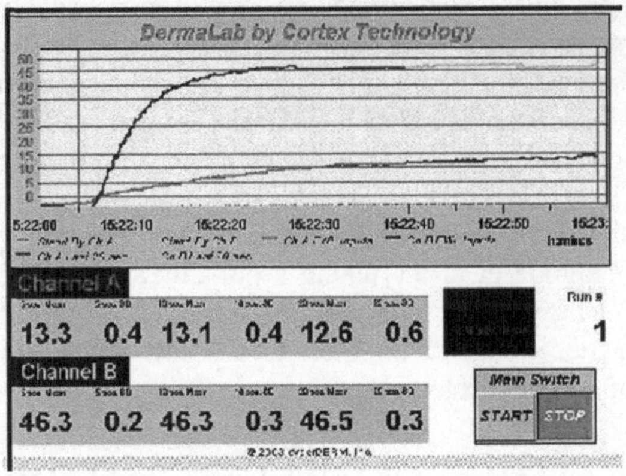

FIGURE 22.5 Graphic user interface showing typical results from a steady-state assessment using a computerized DermaLab with dual probes. Note that when steady-state conditions are achieved, the plot has plateaued; the mean values for the 5-, 10-, and 20-second time blocks are in good agreement; and the standard deviation is quite low.

diseased or wounded skin sites in which a longer time will be required. Indeed, Van Sam et al. [14] have found that as long as 15 minutes may be required to obtain stabilized TEWL in some of their human volunteers. This means that there can be no hard-and-fast rule as to how long the measurement will take. Thus, the question becomes how to best determine that true steady-state conditions have been reestablished. With computerized evaporimetry, this can easily be done by allowing the operator to decide, using the real-time display provided by the strip chart recorder. To facilitate the decision process, the operator can compare the mean values computed over 10-, 15-, and 20-second intervals, as reported on digital meters. At first, water loss rates will be increasing progressively and there will be considerable disagreement among the metered values. However, as steady-state conditions are approached, the strip chart plot will begin to plateau, and all meters will come into good general agreement at that level. By comparing these values, it becomes quite easy to determine when to "freeze" the measurements and record the appropriate values. Figure 22.5 provides representative results of steady-state determinations from the two sites on the volar forearm, one of which was tape stripped, and the other left intact.

Through computerized evaporimetry, it is possible to write programs that will automatically stop when certain conditions other than elapsed time have been met. At first glance, it would seem that termination based on achieving a certain low value for standard deviation would be quite an attractive way of letting the computer determine when steady-state conditions have been met. Unfortunately, in actual practice this is not so easy to do, because the cut-off criteria differ depending on the range of values likely to be encountered. To date, the most practical way to obtain reliable measurements is to allow the data collection to "run on" until the operator makes the decision that steady-state conditions exist and manually stops the program.

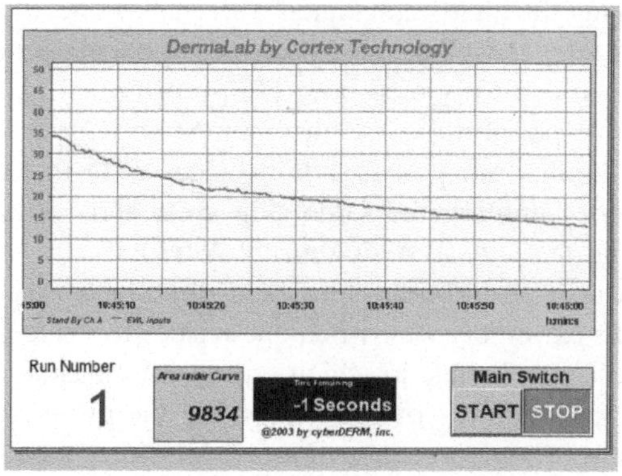

FIGURE 22.6 Graphic user interface showing typical results from a dynamic assessment using a computerized DermaLab with a single probe. In this situation the area under the curve is often used to evaluate these types of differences.

VII. MEASUREMENT OF TEWL UNDER DYNAMIC CONDITIONS

Although most measurements of TEWL are under steady-state conditions, there are certain situations in which the investigator needs to monitor rapid changes in evaporative water loss kinetics over a short period of time. A very good example is the use of computerized evaporimetry to measure skin surface wetness under occlusive covers such as diapers. It was first shown by Zimmerer et al. [15] that the elevated water loss rates observed after removing prewetted patches of diaper materials from the skin was inversely related to their absorbency. More recently, this adult forearm model has been refined so that the integrity of the diaper design is preserved [16]. The basic strategy entails placing an intact diaper around the forearm, loading it with saline to simulate urination, and then removing the test article 1 hour later. By immediately measuring water loss rates over the next 2 minutes, the relative degree of skin wetness can be determined by the area under the curve as shown in Figure 22.6.

With computerized evaporimetry, the programs can be written in such a way that the operator is provided with a delay from the time the PC is activated and the actual dynamic data collection begins. This feature allows for the operator's hands to be free for the removal of the test article and the subsequent placement of the TEWL probe on the uncovered site. This is especially useful in studies in which the diapers are being worn by babies [17].

VIII. MEASUREMENT OF TEWL USING DUAL-PROBE INSTRUMENTS

Because both channels of the DermaLab can be fitted with TEWL probes, it becomes a simple matter with computerized evaporimetry to simultaneously display and process the EWL values for each probe while the measurements are being taken.

To facilitate identifying the inputs associated with each probe, the respective tracings on the strip chart recorder as well as the associated digital meters can be color coded.

An instrument that can truly read both TEWL probes simultaneously opens up a number of attractive possibilities. At the Skin Study Center, we have devised a special fixture that holds both probes side by side. With this close spacing, both probes are essentially sampling the same skin area, and we can obtain duplicate measurements of that test site in a single run. Alternatively, you can separate the probes, place them on two different sites, and directly compare them during the same run. This works extremely well when one wishes to make a paired comparison between treated and control sites, in which sequential measurements may be problematic. Regardless of whether you are computing the difference or averaging the two values, the dual-probe approach will work for either steady-state or dynamic conditions. Indeed, both Figures 22.4 and 22.5 were done this way.

IX. RETROSPECTIVE ANALYSIS OF THE DATA

Although the primary advantage of computerized evaporimetry is that it allows the operator to conveniently monitor evaporative water loss rates in real time, one should not overlook that retrospective analysis is sometimes highly desirable. Perhaps after doing a series of measurements, you feel that a mean based on the water loss rates over the last 30 seconds instead of the last 20 seconds might be a more meaningful measure. By logging the data inputs during the measurement run, these values become available for export into Excel, where they can be replotted or analyzed using standard spreadsheet commands and formulas.

X. CONCLUDING REMARKS

Although Cortex Technology's DermaLab with TEWL can be used as a stand-alone instrument, there are a number of advantages to using computerized evaporimetry as described in this chapter. First and foremost is that it allows the operator to conveniently monitor evaporative water loss rates in real time so that any undesirable influences caused by air currents, probe movements, and so forth will be readily apparent and their effect on the measurement immediately determined. Although an X–Y recorder could be used for the same purpose, computerized evaporimetry also allows one to analyze the values in real time. This opens up many possibilities ranging from calculating mean and standard deviations for steady-state determinations to complex curve characterizations in dynamic situations with either single- or dual-probe instruments. Because the raw values are electronically logged during the measurement run, it is also possible to retrospectively analyze these data using other approaches, with the simplest approach being the spreadsheet operations available through Microsoft Excel.

REFERENCES

1. Pinnagoda, J., Tupker, R.A., Agner, T., and Serup, J., Guidelines for transepidermal water loss (TEWL) measurement. A report from the Standardization Group of the European Society of Contact Dermatitis, *Contact Dermatitis*, 22, 164, 1990.
2. Rogiers, V. and the EEMCO group, EEMCO guidance for the assessment of transepidermal water loss in cosmetic sciences, *Skin Pharmacol. Appl. Skin Physiol.*, 14, 117, 2001.
3. Wilson, D.R. and Maibach, H., Transepidermal water loss: a review, in *Cutaneous Investigation in Health and Disease: Noninvasive Methods and Instrumentation*, Leveque, J.L., Ed., Marcel Dekker, New York, 1989, pp. 113–133.
4. Leveque, J.L., Measurement of transepidermal water loss, in *Cutaneous Investigation in Health and Disease: Noninvasive Methods and Instrumentation*, Leveque, J.L., Ed., Marcel Dekker, New York, 1989, pp. 135–152.
5. Elsner, P., Berardesca, E., and Maibach, H.I., *Bioengineering of the Skin: Water and the Stratum Corneum*, CRC Press, Boca Raton, Florida, 1994.
6. Nilsson, G.E., On the measurement of evaporative water loss. Methods and clinical applications. Doctoral thesis, Linkoping Medical University, Linkoping, Sweden, 1977.
7. Grove, G.L., Grove, M.J., Zerweck, C., and Pierce, E., Comparative metrology of the evaporimeter and the DermaLab TEWL probe, *Skin Res. Technol.*, 5, 1, 1999.
8. Grove, G.L., Grove, M.J., Zerweck, C., and Pierce, E., Computerized evaporimetry using the DermaLab TEWL Probe, *Skin Res. Technol.*, 5, 9, 1999.
9. Brownawell, M., Interchangeability: key to choosing RH sensors, *Sensors*, April 1995.
10. American Society for Testing Materials, *ASTM E96-95, Standard Test Method for Water Vapor Transmission of Materials*, American Society for Testing Materials, West Conshohoken, Pennsylvania, 1997.
11. Wheldon, A.E. and Monteith, J.L., Performance of a skin evaporimeter, *Med. Biol. Eng. Comput.*, 18, 201, 1980.
12. Blichmann, C.W. and Serup, J., Reproducibility and variability of transepidermal water loss measurement. *Acta Derm. Venereol.*, 67(3), 206, 1987.
13. Pinnagoda, J., Tupker, R.A., Coenraads, P.J., and Nater, J.P., Comparability and reproducibility of the results of water loss measurements: a study of 4 evaporimeters, *Contact Dermatitis*, 20(4), 241, 1989.
14. Van Sam, V., Passet, J., Maillols, H., Guillot, B., and Guilhou, J.J., TEWL measurement standardization: kinetic and topographic aspects, *Acta Derm. Venereol.*, 74(3), 168, 1994.
15. Zimmerer, R.E., Lawson, K.D., and Calvert, C.J., The effects of wearing diapers on the skin, *Pediatr. Dermatol.*, 3, 95, 1986.
16. Akin, F.J., Lemmen, J.T., Bozarth, D.L., Garofalo, M.J., and Grove, G.L., A refined method to evaluate diapers for effectiveness in reducing skin hydration using the adult forearm, *Skin Res. Technol.*, 3, 173, 1997.
17. Davis, J.A., Leyden, J.J., Grove, G.L., and Raynor, W.J., Comparison of disposable diapers with fluff absorbent and fluff plus absorbent polymers: effects on skin hydration, skin pH, and diaper dermatitis, *Pediatr. Dermatol.*, 6(2), 102, 1989.

23 Electrical Assessment of Skin Hydration: Standardization of Measurements and Guidelines

Grazia Primavera, Joachim W. Fluhr, and Enzo Berardesca

CONTENTS

I. Introduction ... 287
II. Sources of Error and Associated Variables ... 288
  A. Instrument-Related Variables ... 288
    1. Start-Up and Use ... 288
    2. Use of the Probe Protection Covers ... 289
    3. Instrumental Variability ... 289
  B. Environment-Related Variables ... 289
    1. Ambient Air Temperature and Humidity ... 289
  C. Individual-Related Variables ... 290
    1. Anatomical Sites ... 290
    2. Age, Sex ... 290
    3. Skin Surface Temperature ... 291
    4. Sweating ... 291
III. Conclusions ... 291
References ... 291

I. INTRODUCTION

The importance of water to the proper functioning of the stratum corneum (SC) is well recognized. The reliable quantification of water in the corneum and its interaction with topically applied products is, in fact, essential for understanding skin physiology and developing efficient skin care formulation.

0-8493-1443-7/05/$0.00+$1.50

Thus, it is not surprising that a wide variety of techniques have been developed for measuring SC water content, including electrical measurements (Nova Dermal Phase Meter [DPM] 9003 [1,2], Skicon 200 [2–5], Corneometer [2,3]), mechanical measurements (Dia-Stron DTM Torque Meter [6]), spectroscopic measurements (Fourier-transform infrared spectroscopy [FTIR] [7], nuclear magnetic resonance spectroscopy [NMR] [8]), Transient Thermal Transfer [TTT] [8]), and direct imaging measurements (magnetic resonance imaging [MRI] [9–11]). Between these different methods, the electrical assessments have become very popular because of their relatively low cost and ease of use.

The use of these techniques has been documented in both normal and diseased skin [12–18]. However, it is clear that many factors may influence the measurements and can alter the final interpretation of results [19–21]. Furthermore, reproducibility and variability between different instruments and methods can make it difficult to adequately compare data obtained in different laboratories and experiments. To obtain reliable values, it is necessary to have a standardization of procedures (EEMCO [21]). In this chapter, therefore, attention is focused on the standardization of measurements by determining the variables associated with the method of measuring skin hydration, assessing the extent to which they influence the measured value, and elaborating the techniques by which they may be resolved.

II. SOURCES OF ERROR AND ASSOCIATED VARIABLES

The relationship between electrical conductance and water content is not linear, but depends on the binding state of water molecules to the keratin chains, which state is itself described by the water sorption isotherm [22]. It has been customary in the skin care literature to define three types of water according to their strength of binding to the keratin: "tightly-bound" water for water contents from 0 to around 7%, "bound water" between about 7 and 35%, and "free water" beyond. This division can be considered simplistic on the basis of more detailed theory [23] but is generally helpful. The consequence of variation in water-binding strength is that there is no direct proportionality between water content and electrical conductance. Three categories of factors and sources of variation exist in measurements, including instrumental, environmental, and individual-related variables. A detailed account of these influencing variables is given in this section.

A. Instrument-Related Variables

1. Start-Up and Use

It is very important, independent of the instrument used, that the measurement surface of the probe is absolutely free of dirt particles. Therefore, be sure to clean the measuring head before starting the measurement. In case of cleaning with alcohol, it is necessary to dry the measuring surface with a soft tissue, as alcohol and water residues can lead to errors in measurement. The probe should be placed perpendicularly on on a hair-free skin surface with slight pressure just sufficient to start the measurement process. Either an inclining position of the probe into the skin or an overabundance of hairs causes the measuring surface to have an incorrect pressure into the skin [24]. Each of the devices described above has a spring in the probe to ensure a reproducible pressure. When necessary, shaving or clipping the hairs is recommended 3 days before starting the study.

The instrument should be turned on at least 20 minutes before measurements are performed. Repeated measurements on the same site can cause occlusion, which results in an increase in the values displayed. Therefore, it is recommended to wait for at least 5 seconds before repeating a measurement on the same site. It is advisable to measure at least three times, once at each of three different but nearby sites, and to calculate the median to have more reproducible data.

2. Use of the Probe Protection Covers

The condition of the glass cover of the probe surface has an influence on the measurement results. If this cover has been damaged, it can allow a part of the current flow to reach the skin. This is not dangerous, as the current is only minor (>10 μÅ), but it will alter the recharge time of the capacitor, and therefore the measured value will increase.

3. Instrumental Variability

DPM and Skicon readings are usually higher than those of the Corneometer [2,3,13]. Furthermore, it should be remembered that, in general, different examples of the same instrument type might give variable readings [5]. Also, it appears that the DPM and Skicon are more useful in assessing relatively high water content levels and desorption kinetics, whereas the Corneometer is more sensitive when assessing relatively dry skin [4]. Furthermore, the Skicon probably measures the SC hydration of more superficial parts of the skin than does the Corneometer. The user should be aware that the electric-based instruments measure the integratum of a field covering a three-dimensional space within the upper parts of the skin. The DPM readings seem to be less variable than the Corneometer readings [25]. As the Corneometer readings have already been shown to be less variable than the Skicon readings [26,13], the DPM thus appears to be the least variable of the three instruments [2]. Other advantages of the DPM are that it is small, light, and portable and that it is not necessary to put it into an electric outlet, thus making it very easy to use.

B. Environment-Related Variables

1. Ambient Air Temperature and Humidity

Temperature and relative humidity are known to influence SC water content [27,28]. Under some conditions, a linear relationship has been found between "capacitance" and relative humidity [29]. In fact, high environmental humidities easily influence the hydration state of the skin surface because the SC is highly hygroscopic. Furthermore, chemical changes in environmental humidities have an influence on barrier related factors (e.g., water-holding capacity) and, thus, on SC hydration. Therefore, room humidity should be kept constant. In cases of variation of relative humidity, this variation should be included in the statistical analysis of variance and, if justified, some suitable normalization procedure such as proposed by Barel and Clarys [29] should be performed. Below 60% relative humidity, the influence of the environment is much smaller; however, it is preferable to keep the relative humidity low. Room temperature should be kept constant and at a level that will minimize sweat gland activity; hence, the range from 18° to 20°C is recommended. A steady increase in "capacitance" as room temperature rises above 22°C has been reported [19], which may arise from increased perspiration. Seasonal variations resulting from alterations in temperature and relative humidity occur [28,30–33], and particularly in summer, the room should be air conditioned. Day-to-day and even diurnal variations are also reported [34,35].

C. Individual-Related Variables

1. Anatomical Sites

Regional differences in SC water content have been reported [13,15,18,25–27,36,37], but when interpreting such results, it must be remembered that all these electrical devices give only relative, not absolute, indications of water content expressed in arbitrary units (except by Skicon: µm). The actual relationship between their outputs and true water content is highly complex and also variable. Site-to-site differences in the same anatomical region (e.g., the forearm) are reported by some [19] but denied by others [25]. High hydration values appear to exist on the forehead and palms; lower values are observed at the abdomen, thigh, and lower leg [29]. The hydration of contralateral pairs of sites is generally indistinguishable [19] except on the dorsum of the hands. Hence, comparative left–right studies of the same selected skin sites are recommended to avoid skin site differences. The influence of body hair should be noted [24], especially if comparison is made between sites that are not contralateral.

2. Age, Sex

No significant differences have been reported between males and females when matched by age and tested on the same skin sites [19,37,38]. However, during certain periods of life, significant differences may occur. For example, during the first days of postnatal life SC hydration is higher than in adults, indicating that the SC barrier is still in the process of adapting to extrauterine life [39,40]. On the basis of the almost identical values for the parameters of TEWL, SC hydration, and pH value, the skin physiology of the child differs very little in SC hydration and barrier function from that of adults [41]. However, skin aging appears to induce a slow decrease in hydration from around the age of 25 years onward [19,37,38]. In fact, with increasing age, significantly decreased levels of all major barrier lipids and changes in cutaneous proteoglycans' size have been described [42,43]. The observation of these dramatic age-related differences indicates that these changes may be involved in the age-related changes in the physical properties of skin.

3. Skin Surface Temperature

It is possible that skin that is moist and cool gives exactly the same electrical response to measurements made at a single frequency as skin that is dry and warm. To separate and specify potentially confounding influences such as water content, temperature change, and sweat gland activity, it is necessary to use some form of electrical spectroscopic technique; that is, stimulation at three or more different frequencies, or a time-domain approach followed by Fourier transformation [44–46].

4. Sweating

Thermal sweat gland activity is unlikely if the ambient air temperature is below 20°C, humidity is between 30 and 60%, and skin temperature is below 30°C, provided that the skin is not exposed to forced convection and no excessive body heat is produced (as a result of physical exercise) [47–49]. Subjects, in fact, should be relaxed and adapted to the measuring environment for at least 20 minutes before testing takes place. Therefore, skin should be exposed to the ambient air of the test environment for at least 10 minutes before taking measurements [21].

III. CONCLUSIONS

Water has an important effect on the physical properties of the SC, and reliable quantification of water in the corneum and its interaction with topically applied products is essential for understanding skin physiology and developing efficient skin care formulations. Electrical conductance offers a sensitive approach to assessing SC water, even though the exact relationship between instrument readings and actual SC water content is complex and variable, preventing accurate calibration and opening the possibility of misinterpretation.

New methods, more experimental and expensive, are emerging that could improve on the performance of electrical approaches, but for the foreseeable future, the electrical techniques used will remain the more basic and widespread methodologies and will remain useful provided their limitations are clearly understood and the experimental conditions are designed to overcome the associated difficulties as much as possible.

It should be remembered that no single type of instrumental test of skin is sufficient: A battery of methods (visual, tactile, instrumental) should always be used to guard against errors of interpretation.

REFERENCES

1. Li, F., Conroy, E., Visscher, M., and Wickett, R.R., The ability of electrical measurements to predict skin moisturization. I. Effects of NaCl and glycerin on short-term measurements, *J. Cosmetic Sci.*, 52(1), 13, 2001.
2. Fluhr, J., Gloor, M., Lazzerini, S., et al., Comparative study of five instruments measuring stratum corneum hydration (Corneometer CM 820 and CM 825, Skicon 200, Nova DPM 9003, DermaLab). Part I. In vitro, *Skin Res. Technol.*, 5, 161, 1999.

3. Barel, A.O. and Clarys, P., *In vitro* calibration of the capacitance method (Corneometer CM 825) and conductance method (Skicon-200) for the evaluation of the hydration state of the skin, *Skin Res. Technol.*, 3, 107, 1997.
4. Van Neste, D., Comparative study of normal and rough human skin hydration *in vivo*: evaluation with four different instruments, *J. Dermatol. Sci.*, 2, 119, 1991.
5. Hashimoto-Kumasaka, K., Takahashi, K., and Tagami, H., Electrical measurements of the water content of the stratum corneum *in vivo* and *in vitro* under various conditions: comparison between the Skin Surface Hygrometer and Corneometer in evaluation of the skin surface hydration state, *Acta Derm. Venereol. (Stockh.)*, 73, 335, 1993.
6. Boyce, S.T., Supp, A.P., Wickett, R.R., Hoath, S.B., and Warden, G.D., Assessment with the dermal torque meter of skin pliability after treatment of burns with cultured skin substitutes, *J. Burn Care Rehabil.*, 21(1 Pt 1), 55, 2000.
7. Brancaleon, L., Bamberg, M.P., Sakamaki, T., and Kollias, N., Attenuated total reflection — Fourier transform infrared spectroscopy as a possible method to investigate biophysical parameters of stratum corneum *in vivo*, *J. Invest. Dermatol.*, 116(3), 380, 2001.
8. Girard, P., Beraud, A., and Sirvent, A., Study of three complementary techniques for measuring cutaneous hydration *in vivo* in human subjects: NMR spectroscopy, transient thermal transfer and corneometry — application to xerotic skin and cosmetics, *Skin Res. Technol.*, 6(4), 205, 2000.
9. Szayna, M. and Kuhn, W., *In vivo* and *in vitro* investigations of hydration effects of beauty care products by high-field MRI and NMR microscopy, *J. Eur. Acad. Dermatol. Venereol.*, 11(2), 122, 1998.
10. Ozawa, T. and Takahashi, M., Skin hydration: recent advances, *Acta Derm. Venereol. Suppl. (Stockh.)*, 185, 26, 1994.
11. Querleux, B., Richard, S., Bittoun, J., et al., *In vivo* hydration profile in skin layers by high-resolution magnetic resonance imaging, *Skin Pharmacol.*, 7(4), 210, 1994.
12. Leveque, J.L. and de Rigal, J., Impedance methods for studying skin hydration, *J. Soc. Cosm. Chem.*, 34, 419, 1983.
13. Blichmann, C. and Serup, J., Assessment of skin moisture, *Acta Derm. Venereol. (Stockh.)*, 68, 284, 1988.
14. Serup, J. and Blichmann, C., Epidermal hydration of psoriasis plaques and the relation to scaling, *Acta Derm. Venereol. (Stockh.)*, 67, 357, 1987.
15. Salter, D.C., Instrumental methods for assessing skin moisturization, *Cosmet. Toilet.* 102, 103, 1987.
16. Leonardi, G.R., Gaspar, L.R., and Maia Campos, P.M., Application of a non-invasive method to study the moisturizing effect of formulations containing vitamins A or E or ceramide on human skin, *J. Cosmet. Sci.*, 53(5), 263, 2002.
17. Martinsen, O.G. and Grimnes, S., Facts and myths about electrical measurement of stratum corneum hydration state, *Dermatology*, 202(2), 87, 2001.
18. Black, D., Del Pozo, A., Lagarde, J.M., and Gall, Y., Seasonal variability in the biophysical properties of stratum corneum from different anatomical sites, *Skin Res. Technol.*, 6(2), 70, 2000.
19. Rogiers, V., Derde, M.P., Verleye, G., and Roseeuw, D., Standardized conditions needed for skin surface hydration measurements, *Cosmet. Toilet.*, 105, 73, 1990.
20. Hashimoto-Umasaka, K., Takahashi, K., and Tagami, H., Electrical measurement of the water content of the stratum corneum *in vivo* and *in vitro* under various conditions: comparison between skin surface hygrometer and Corneometer in evaluation of the skin surface hydration state, *Acta Derm. Venereol.*, 73(5), 335, 1993.

21. Berardesca, E., EEMCO guidance for the assessment of stratum corneum hydration: electrical methods, *Skin Res. Technol.*, 3, 126, 1997.
22. Liron, Z., Clewell, H.J., and McDougal, J.N., Kinetics of water vapor sorption in porcine stratum corneum, *J. Pharm. Sci.*, 83(5), 692, 1994.
23. Salter, D.C., Studies in the measurement, form and interpretation of some electrical properties of normal and pathological human skin *in vivo*, Doctoral thesis, University of Oxford.
24. Loden, M., Hagforsen, E., and Lindberg, M., The presence of body hair influences the measurement of skin hydration with the Corneometer, *Acta Derm. Venereol.*, 75(6), 449, 1995.
25. Yamamoto, T. and Yamamoto, Y., Analysis for the change of skin impedance, *Med. Biol. Eng. Comput.*, 15, 219, 1977.
26. Barel, A.O., Clarys, P., Wessels, B., and de Romsee, A., Non-invasive electrical measurements for evaluating the water content of the horny layer: comparison between capacitance and conductance measurements, in *Prediction of Percutaneous Penetration: Methods, Measurements, Modelling*, Vol. 2, Scott, R.C., Guy, R.H., Hadgraft, J., and Boode, H.E., Eds., IBC Technical Services, London, 1991, p. 46.
27. Loden, M., Olsson, H., Axell, T., and Linde, Y.W., Friction, capacitance and transepidermal water loss (TEWL) in dry atopic and normal skin, *Br. J. Dermatol.*, 126, 137, 1992.
28. Tagami, H., Impedance measurements for evaluation of the hydration state of the skin surface, in *Cutaneous Investigation in Health and Disease*, Leveque, J.L., Ed., Marcel Dekker, New York, 1989, chap. 5.
29. Barel, A.O. and Clarys, P., Measurements of epidermal capacitance, in *Handbook of Non-Invasive Methods and the Skin*, Serup, J. and Jemec, G.B.E., Eds., CRC Press, Boca Raton, Florida, 1995, pp. 165–170.
30. Prall, J.K., Theiler, R.F., Bowser, P.A., and Walsh, M., The effectiveness of cosmetic products in alleviating a range of skin dryness conditions as determined by clinical and instrumental techniques, *Int. J. Cosmetic Sci.*, 8, 159, 1986.
31. Leveque, J.L., Grove, G., de Rigal, J., Corcuff, P., Kligman, A.M., and Saint-Leger, D., Biophysical characterisation of dry facial skin, *J. Soc. Cosmetic Chem.*, 82, 171, 1987.
32. Wilhelm, K.P., Possible pitfalls in hydration measurements, *Curr. Probl. Dermatol.*, 26, 223, 1998.
33. Tupker, R.A., Coenraads, P.J., Fidler, V., et al., Irritant susceptibility and wheal and flare reactions to bioactive agents in atopic dermatitis. II. Influence of season, *Br. J. Dermatol.*, 133(3), 365, 1995.
34. Yosipovitch, G., Xiong, G.L., Haus, E., et al., Time-dependent variations of the skin barrier function in humans: transepidermal water loss, stratum corneum hydration, skin surface pH, and skin temperature, *J. Invest. Dermatol.*, 110(1), 20, 1998.
35. Le Fur, I., Reinberg, A., Lopez, S., et al., Analysis of circadian and ultradian rhythms of skin surface properties of face and forearm of healthy women, *J. Invest. Dermatol.*, 117(3), 718, 2001.
36. Fluhr, J.W., Dickel, H., Kuss, O., et al., Impact of anatomical location on barrier recovery, surface pH and stratum corneum hydration after acute barrier disruption, *Br. J. Dermatol.*, 146(5), 770, 2002.
37. Ya-Xian, Z., Suetake, T., and Tagami, H., Number of cell layers of the stratum corneum in normal skin — relationship to the anatomical location on the body, age, sex and physical parameters, *Arch. Dermatol. Res.*, 291(10), 555, 1999.

38. Leveque, J.L., Methodes experimentales d´etude du vieillisement cutane chez l´homme *in vivo, Acta Derm. Venereol. (Stockh.)*, 144, 1297, 1987.
39. Yosipovitch, G., Maayan-Metzger, A., Merlob, P., and Sirota, L., Skin barrier properties in different body areas in neonates, *Pediatrics*, 106(1 Pt 1), 105, 2000.
40. Hoeger, P.H. and Enzmann, C.C., Skin physiology of the neonate and young infant: a prospective study of functional skin parameters during early infancy, *Pediatr. Dermatol.*, 19(3), 256, 2002.
41. Fluhr, J.W., Pfisterer, S., and Gloor, M., Direct comparison of skin physiology in children and adults with bioengineering methods, *Pediatr. Dermatol.*, 17(6), 436, 2000.
42. De Paepe, K., Vandamme, P., Roseeuw, D., and Rogiers, V., Ceramides/cholesterol/FFA-containing cosmetics: the effect on barrier function, *SOFW J.*, 122, 199, 1996.
43. Carrino, D.A., Sorrell, J.M., and Caplan, A.I., Age-related changes in the proteoglycans of human skin, *Arch. Biochem. Biophys.*, 1, 373(1), 91, 2000.
44. Salter, D.C., *Monitoring Human Skin Hydration in vivo Using Electrical Impedance—A Model of Skin as a Solid Ionic Conductor.* Proceedings IX International Conference on Electrical Bio-Impedance and European Community Concerted Action on Impedance Tomography, Heidelberg 26–30 September 1995, pp. 17–20.
45. Salter, D.C., Further hardware and measurement approaches for studying water in the stratum corneum, in *Bioengineering of the Skin: Water and the Stratum Corneum*, Elsner, P., Berardesca, E., and Maibach, H., Eds., CRC Press, Boca Raton, Florida, 1994, pp. 205–215.
46. Egawa, M., Oguri, M., Kuwahara, T., and Takahashi, M., Effect of exposure of human skin to a dry environment, *Skin Res. Technol.*, 8(4), 212, 2002.
47. Shahidullah, M., Raffle, E.J., Rimmer, A.R., and Frain-Bell, W., Transepidermal water loss in patients with dermatitis, *Br. J. Dermatol.*, 81, 722, 1969.
48. Baker, H. and Kligman, A.M., Measurement of transepidermal water loss by electrical hygrometry. Instrumentation and responses of physical and chemical insults, *Arch. Dermatol.*, 96, 441, 1967.
49. Pinnagoda, J., Tupker, R.A., Coenraads, P.J., and Nater, J.P., Transepidermal water loss: with and without sweat gland inactivation, *Contact Dermatitis*, 21, 16, 1989.

24 Hydration and Moisturizers

Marie Lodén

CONTENTS

I. Introduction 295
II. Chemistry and Function of Dry Skin 296
III. Chemistry and Function of Moisturizers 297
IV. Treatment with Moisturizers 298
V. Evaluation Techniques 300
VI. Conclusion 301
References 302

I. INTRODUCTION

The stratum corneum (SC) covers the whole body surface and is able to stay soft and flexible in the usual ambient conditions. The SC is composed of protein-enriched corneocytes and lipid-enriched intercellular domains. During the life cycle of the keratinocytes, they differentiate — they gradually flatten out, extrude lipids by exocytosis, and eventually completely cornify to become corneocytes filled with keratin and amorphous matrix. The corneocytes also contain a water–soluble fraction called a natural moisturizing factor (NMF), which enhances their water-holding capacity.[1,2] In the intercellular region, lipids are arranged in a lamellar phase consisting of alternative layers of water and lipid bilayers.

Failure of the SC to retain water induces dryness of the skin. Scaling and decreased mechanical flexibility of the SC are associated with the impression of dryness. The condition is not fatal, but it is often linked with an impaired permeability barrier function.[3–5] Increased permeability facilitates development of contact eczema by allowing deleterious substances to enter into the skin. Furthermore, transepidermal water loss is increased, which also may result in cutaneous abnormalities by overstimulation of cytokines.[6]

A wide range of hydrating and emollient products is used for treatment or prevention of dry skin. Application of oily materials to the skin is almost intuitive and may be as old as mankind itself. The term "emollient" (from the Latin derivation) implies a material designed to soften the skin (i.e., a material that "smooths" the surface to the touch and makes it look smoother to the eye). The term "moisturizer"

0-8493-1443-7/05/$0.00+$1.50

is often used synonymously with emollient, but moisturizers usually contain humectants, aimed at further increasing the hydration of the SC.

Creams, ointments, lotions, and bath oils belong to the treatment arsenal, but gels and soap substitutes also may be used for dry skin. Treatment with emollients and moisturizers is able to break the dry skin cycle and maintain the smoothness of the skin. The products are also useful treatment adjuncts for the inflammatory dermatoses as steroid-sparing agents.[7] However, moisturizers and emollients vary considerably in their chemical composition, and few of them address the underlying SC abnormalities.

II. CHEMISTRY AND FUNCTION OF DRY SKIN

The impression of dry skin is a common disorder, but the lack of water may not explain all types of problems covered by the term "dry skin." The subjective self-assessment and the clinical assessment of the presence of dry skin show discrepancies in the frequency of dryness.[8,9] However, in studies of winter xerotic skin, the water content of the SC correlates inversely with clinical scores of dryness,[10,11] and in elderly patients with xerosis the SC contains less water than in normals.[12,13] Furthermore, the dry-looking skin of patients with atopic dermatitis and psoriasis is less hydrated and less capable of binding water than normal skin.[3,14–17]

Water is the plasticizer of keratin, allowing the SC layer to bend and stretch, avoiding cracking and fissuring. Furthermore, water increases the activity of enzymes involved in the desquamation process.[18] Increased corneocyte cohesion is found in the dry skin in atopic patients and in patients with senile dry pruritus.[12,19] This increased cohesion interrupts the normal shedding of SC cells. Failure of the corneocytes to separate from one another induces scaling and hyperkeratosis (i.e., shedding of clumps of SC cells and increased thickness of the SC, respectively). However, hyperkeratosis may also be the result of an increased production of keratinocytes. Hyperhydration of the skin by a long bath causes the SC surface to be easily rubbed off. This is because of progressive degradation of corneodesmosomes and disruption of the intercellular bilayer lipids.[20]

Three types of water with different molecular mobility can be found in SC. At a water content below 10% the primary water is tightly bound, presumably to the polar sites of the proteins or to the intercellular regions.[21] When the degree of hydration exceeds 10%, the secondary water is hydrogen-bonded around the protein-bound water, and above 40 to 50%, the water resembles the bulk liquid.[16] The amount of tightly bound water, which does not seem to have any plasticizing effect, is almost the same in different types of pathological skin, whereas the amount of secondary water, accounting for the elasticity, is much smaller in SC from psoriatic patients and from elderly persons with xerosis than in normal SC.[16] In normal SC from glabrous skin, the content of water is 38.2 mg/100 mg dry tissue, as compared with 31.7 mg in senile xerosis and 27.2 mg in psoriatic scales per 100 mg dry tissue.[16]

NMFs contribute to the content of water in the SC. The fraction of NMFs can make up about 15 to 20% of the total weight of the corneum.[22] Substances belonging to NMFs are amino acids, pyrrolidone carboxylic acid (PCA), lactates, and urea.[22] Lactic acid and urea show highest concentration in the uppermost layer of the SC, with a steep decline with depth.[23] PCA and the other components of NMFs show a more flattened profile in the uppermost layers and declining levels at the lowermost part of SC.[23]

In normal and affected SC of atopic patients, the amount of urea is substantially reduced.[24] In addition, other hygroscopic substances belonging to the NMFs is reduced in the SC of atopic skin as well as in other dry skin conditions.[13,25–27] There is a virtual absence of NMFs in ichthyosis vulgaris[13,28] and psoriasis.[29] Decreased profilaggrin production has been suggested to be responsible for the reduced level of PCA and water-soluble amino acids.[13]

The corneocytes in SC are embedded in lipid bilayers consisting primarily of three classes of lipids: free sterols, free fatty acids, and ceramides. An abnormal lipid composition has been observed in surfactant-irritated skin,[30] in experimentally induced scaly skin,[25] in normal winter dry skin,[31] in dry atopic skin,[32–34] in psoriatic plaques,[35] and in hereditary ichthyosis.[36] In particular, the content and distribution of the ceramides are changed.[25,30,32,33,35–37] In addition to the changes in lipid composition, the normal bilayer structure is perturbed in skin dryness[38] and in skin exposed to organic solvents.[39]

III. CHEMISTRY AND FUNCTION OF MOISTURIZERS

Moisturizers contain a great variety of ingredients and can execute multiple functions. The interplay between the ingredients is essential to get a stable and cosmetically attractive product with the desired effect on the skin.

Creams are the most common types of delivery system used for hydrating and emollient products. They enable a wide variety of ingredients to be quickly and conveniently delivered to the skin. The two phases are usually oil and water, producing either an oil-in-water (O/W) or a water-in-oil (W/O) emulsion. The droplet size is often between 1 and 100 μm. Emulsifiers provide stability by turning their nonpolar hydrocarbon end into the oil phase and their polar end into the water phase, thereby embedding the droplets. Emulsifiers may interact with the skin and change transepidermal water loss.[40]

Humectants widely used in moisturizers are glycerin, propylene glycol, PCA, lactic acid, and urea. Lactic acid belongs to the group of α-hydroxy acids (AHAs; Table 24.1). The water-binding capacity of the sodium salts of lactic acid and PCA appears to be higher than that of glycerin and sorbitol.[41,42] Urea also has strong osmotic activity.[43,44] However, which of these substances most efficiently increases the skin hydration is not known. In addition to differences in water-binding capacity, their absorption into the skin is important for the resultant hydration. The amount of urea[45,46] and glycerin[47] absorbed into normal SC can be determined using a tape-stripping technique.

After treatment with urea and glycerin, the water-holding capacity of normal SC and of scales from psoriatic and ichthyotic patients is substantially increased.[48–51] Likewise, PCA attracts water and increases the degree of hydration of solvent-damaged guinea pig footpad corneum.[52]

Although water is known to play an important role in maintaining skin suppleness and plasticity, other components may also affect its physical properties. For

TABLE 24.1
Alpha-Hydroxy Acids Used in Moisturizers

Substance	CAS-No	Mw	Other name	Natural source
Glycolic acid	79-14-1	76.05	2-hydroxyethanoic acid, hydroxyacetic acid	Sugar cane juice
Lactic acid	50-21-5	90.1	2-hydroxypropanoic acid	Sour milk and tomato juice
Malic acid	6915-15-7	134.1	Hydroxybutanedioic acid, N-hydroxysuccinic acid	Apples, pears
Tartaric acid	87-69-4	150.1	(+)-L-2,3-dihydroxybutanedioic acid, dihydroxysuccinic acid.	Grapes and wine

Note: CAS = Chemical Abstract Service.

example, topically applied lipids influence the visible and tactile properties of the skin.[53,54] Moisturizers may also influence the crystalline arrangement of the bilayer lipids.[55] In dry skin, the proportion of lipids in the solid state may be increased, and it has been suggested that certain substances, such as glycerin, maintain the lipids in a liquid crystalline state at low relative humidity.[55,56] It has also been proposed that glycerin may aid the digestion of the superficial desmosomes in subjects with dry skin and thereby ameliorate dry flaky skin.[18] Furthermore, AHAs, such as lactic acid and glycolic acid, might be useful in moisturizers because of their influence on the flexibility of SC and removal of superficial scales (Tables 24.1, 24.2).[57–60]

Dry skin disorders and experimentally induced dryness are improved by treatment with moisturizers, and clinical studies indicate enhanced effect from those containing humectants[50,52,61–63] (Table 24.2). Lipids in the moisturizers may increase skin hydration by several mechanisms. The most conventional one is occlusion, which implies a simple reduction of the loss of water from the outside of the skin. A hydrophobic material such as petrolatum can also decrease transepidermal water loss,[57,64] possibly by absorption into the skin.[64] Thus, lipids in moisturizers may interact with the intercellular lipids in SC and assist in retaining the moisture content in the corneocytes.[1,64,65]

IV. TREATMENT WITH MOISTURIZERS

Topical treatment is fundamental for patients with skin diseases, but many health care professionals and patients overlook the importance of moisturizers and consider them as not being "active" treatments. This may also be true if the moisturizers have an unsuitable composition or are used in too small quantities.

Adverse skin reactions from topical preparations may occur, but compared to traditional drugs used by dermatologists moisturizers can be considered as safe. Virtually any topical substance can disturb the skin, but some individuals are more susceptible than others. For example, those with dry skin with an impaired barrier function are more prone to adverse reactions than those with normal skin. Facial

TABLE 24.2
Moisturizers Improve Dry Skin Conditions

Active Substance	Control	Condition	Design/Number of Patients	Effect on Dryness	Ref.
12% Ammonium lactate	Petrolatum based cream	Xerosis	Double blind, bilateral/73	Active more effective	66
12% Ammonium lactate	5% lactic acid+2.5% PCA	Xerosis on legs	Double blind/41	Active more effective	67
12% Ammonium lactate	No therapy	Dry skin including atopics	Open/24	Active improved	68
12% Ammonium lactate	No therapy	Dry heels	Evaluator blind, bilateral/60	Active improved	69
12% Lactate	5% lactic acid/emollient lotion	Xerosis on legs	Double blind, bilateral/60	All improved, equally effective, but 12% lasted longer	63
7.5% Lactic acid	12% lactic acid	Xerosis on psoriatic patients	Double blind, bilateral/11	Improved, but no difference	70
5% Lactic acid	Eucerin lotion	Xerosis	Double blind, bilateral/56	Active, more effective	71
15% Glycolic acid	Eucerin cream	Xerosis on legs	?/bilateral/14	Active superior according to objective evaluation	57
5% Pyrrolidone carboxylic acid	Placebo and 10% urea	Xerosis	Cross-over/150	Active better than placebo, and equal as urea	52
3% and 10% urea	Untreated	Dry skin	Evaluator blind, bilateral/47	Improved	72
10% Urea cream	Essex base cream	Atopic skin	Open/40	Improved, active better	73
10% Urea	Placebo	Ichthyosis	Double blind, bilateral/60	Improved	74
10% Urea	Placebo	Senescent dryness on forearm	Double blind, bilateral/60	Improved	75
4% Urea + 4% sodium chloride	Placebo	Asteatosis, senescent dryness on leg	Double blind/26	Improved, active better	61
Urea 4%	Glycerin 20%	Atopic skin	Parallel, double blind/197	Improved, active more	76

skin is also more sensitive than other body regions,[77] possibly because of a less efficient barrier with smaller number of SC cell layers and because of the presence of large follicular pores.[78]

The most common adverse reactions are sensory or subjective sensations (no signs of inflammation) immediately after application of a topical product. Smarting, burning, and stinging sensations are examples of such reactions among users of dermatologicals. Some humectants, such as lactic acid,[79] urea,[80,81] and PCA[82] are substances well known for causing subjective sensations. Fragrances and preservatives are potential allergens in moisturizers. Virtually all moisturizers in the supermarket contain fragrances and preservatives. Fragrances are not medically necessary, but their inclusion may enhance the important patient compliance. Furthermore, not all fragrance ingredients are able to cause sensitization.

The cosmetic properties and the ease of use of the products are important parameters to achieve the desired treatment effects. Sticky and viscous formulations are less attractive than easily applied and rapidly absorbed emulsions. The patients only allow a short time for the absorption of applied products. The use of moisturizers may also create some practical problems, which can discourage the subjects from complying with a skin care program. A product with an immediate effect is more likely to be perceived as beneficial by the user, and its continued use encouraged.

Differences in dosing have been noted among self-application versus operator-assisted application of creams.[83] Self-application resulted in larger amount applied per unit area.[83,84] Moreover, jars promoted use of larger quantities than the same cream in a tube (1.7 vs. 0.7 mg/cm^2, respectively).[84] The distribution has also been found to vary over the treated region and the real surface. For example, the genitoanal area received more than sevenfold amount of cream in comparison with its proportional surface area.[83] In addition, distribution within the treated area differs depending on the type of vehicle.[85] A thick ointment with only a small percentage of water was equally distributed in the center and periphery of the treated area, whereas formulations with lower viscosity and more volatile ingredients (e.g., creams) were less evenly spread on the skin.[85] However, it may be expected that moisturizer treatment seldom requires the same application rate on different sites, because of the various severities of the dryness in the treated areas. Applied products may also be absorbed, slough off, or be transferred to other surfaces. After 8 hours, about 50% of applied cream remained on the surface.[86] Creams and ointments seem to allow higher transfer of the actives than lotions and tinctures.[87]

V. EVALUATION TECHNIQUES

The efficacy of moisturizers can be studied by different techniques. The experimental findings have to be significant and valid relative to the clinical and patient appreciations. So far, the only universally accepted technique for proving an effect on dry skin is clinical testing with visual and tactile evaluation of the effect on dryness.

Many different systems have been proposed for the measurement of treatment effects and patient acceptability. Structured assessment of the skin symptoms and

TABLE 24.3
Example of a Category Scale for Visual Scoring of the Degree of Scaling[88]

Intensity of Scaling
0 = Absent
1 = Slight (small scales only, surface lightly dull in color)
2 = Moderate (small scales in combination with larger scales (>0.05 mm), surface opaque or whitish)
3 = Severe (larger and large scales (flakes >1 mm) are prominent, surface whitish)
4 = Extreme (larger flakes covering almost the entire skin surface in the examination field)

their severity by a trained expert can be combined with information given by the patient. A number of scoring systems is available. For example, one system for an overall score of dry skin includes evaluation of the degree of scaling, roughness, redness, and cracks on a categorical scale from 0 to 4; see Table 24.3 for evaluation of degree of scaling.[88] The size of the involved area can also be estimated and the total index can be calculated.[76,88] The scale sensitivity can be augmented by increasing the number of categories or by using a variation of the interval scale as the "line scale" or visual analogue scale. Most instrumental techniques for assessing skin hydration are based on the electrical properties of the skin, such as measurement of resistance, conductance, capacitance and impedance (see other chapters in this book).

One limitation with electrical measurements is that they only give qualitative information on differences in water content at less well-defined locations within the skin. Other potential difficulties for correct interpretation of the results are that skin surface irregularities and agents other than water (e.g., urea and salts) may influence the readings of the skin hydration.[89,90] In addition, residues on the surface may give misleading results.[91,92] Application of petrolatum and dewaxed lanolin to the skin results in values indicating decreased skin hydration for some hours.[92] Petrolatum consists of a complex combination of hydrocarbons, which have high electrical resistivity. Hence, petrolatum remaining on the skin surface might depress the electrical response, so that the effect of the lipids is underestimated. Moreover, occlusion and contact with wet surfaces easily change skin hydration, and hence it is important that the probe does not alter the degree of hydration during the measurement.

VI. CONCLUSION

Moisturizers and emollients differ in composition of active ingredients and excipients. Obvious differences also exist between the cosmetic properties among products. However, it is less well known that certain humectants and lipid mixtures are more efficient than others in improving dry skin conditions. An increased understanding of the chemistry and function of dry skin and moisturizers will facilitate "tailoring" of specific moisturizers for different purposes.

REFERENCES

1. Middleton, J.D., The mechanism of water binding in stratum corneum, *Br. J. Dermatol.*, 80, 437, 1968.
2. Blank, I.H., Further observations on factors which influence the water content of the stratum corneum, *J. Invest. Dermatol.*, 21, 259, 1953.
3. Thune, P., Evaluation of the hydration and the water-holding capacity in atopic skin and so-called dry skin, *Acta Derm. Venereol. (Suppl.)*, 144, 133, 1989.
4. Lodén, M., Olsson, H., Axell, T., and Linde, Y.W., Friction, capacitance and transepidermal water loss (TEWL) in dry atopic and normal skin, *Br. J. Dermatol.*, 126, 137, 1992.
5. Denda, M., Koyama, J., Namba, R., and Horii, I., Stratum corneum lipid morphology and transepidermal water loss in normal skin and surfactant-induced scaly skin, *Arch. Dermatol. Res.*, 286, 41, 1994.
6. Elias, P.M., Wood, L.C., and Feingold, K.R., Epidermal pathogenesis of inflammatory dermatoses, *Am. J. Contact Dermatol.*, 10, 119, 1999.
7. Lucky, A.W., Leach, A.D., Laskarzewski, P., and Wenck, H., Use of an emollient as a steroid-sparing agent in the treatment of mild to moderate atopic dermatitis in children, *Pediatr. Dermatol.*, 14, 321, 1997.
8. Linde, Y.W., "Dry" skin in atopic dermatitis. A clinical study, *Acta Derm. Venereol. (Stockh.)*, 69, 311, 1989.
9. Jemec, G.B.E. and Serup, J., Scaling, dry skin and gender, *Acta Derm. Venereol.*, 177(Suppl), 26, 1992.
10. de Rigal, J., Losch, M.J., Bazin, R., Camus, C., Sturelle, C., Descamps, V., and Lévêque, J.L., Near-infrared spectroscopy: a new approach to the characterization of dry skin, *J. Soc. Cosmetic Chem.*, 44, 197, 1993.
11. Leveque, J.L., G.L. Grove G.L., de Rigal, J., Corcuff, P., and Kligman, A.M., Biophysical characterization of dry facial skin, *J. Soc. Cosmetic Chem.*, 82, 171, 1987.
12. Long, C.C. and Marks, R., Stratum corneum changes in patients with senile pruritus, *J. Am. Acad. Dermatol.*, 27, 560, 1992.
13. Horii, I., Nakayama, Y., Obata, M., and Tagami, H., Stratum corneum hydration and amino acid content in xerotic skin, *Br. J. Dermatol.*, 121, 587, 1989.
14. Berardesca, E., Fideli, D., Borroni, G., Rabbiosi, G., and Maibach, H.I., In vivo hydration and water-retention capacity of stratum corneum in clinically uninvolved skin in atopic and psoriatic patients, *Acta Derm. Venereol.*, 70, 400, 1990.
15. Tagami, H., Kanamaru, Y., Inoue, K., Suehisa, S., Inoue, F., Iwatsuki, K., Yoshikuni, K., and Yamada, M., Water sorption-desorption test of the skin *in vivo* for functional assessment of the stratum corneum, *J. Invest. Dermatol.*, 78, 425, 1982.
16. Takenouchi, M., Suzuki, H., and Tagami, H., Hydration characteristics of pathologic stratum corneum-evaluation of bound water, *J. Invest. Dermatol.*, 87, 574, 1986.
17. Werner, Y., Lindberg, M., and Forslind, B., The water-binding capacity of stratum corneum in dry non-eczematous skin of atopic eczema, *Acta Derm. Venereol.*, 62, 334, 1982.
18. Rawlings, A.V., Harding, C., Watkinson, A., Banks, J., Ackerman, C., and Sabin, R., The effect of glycerol and humidity on desmosome degradation in stratum corneum, *Arch. Dermatol. Res.*, 287, 457, 1995.

19. Finlay, A.Y., Nicholls, S., King, C.S., and Marks, R., The 'dry' non-eczematous skin associated with atopic eczema, *Br. J. Dermatol.*, 103, 249, 1980.
20. Warner, R.R., Boissy, Y.L., Lilly, N.A., Spears, M.J., McKillop, K., Marshall, J.L., and Stone, K.J., Water disrupts stratum corneum lipid lamellae: damage is similar to surfactants, *J. Invest. Dermatol.*, 113, 960, 1999.
21. Anderson, R.L., Cassidy, J.M., Hansen, J.R., and Yellin, W., Hydration of stratum corneum, *Biopolymers*, 12, 2789, 1973.
22. Jacobi, O.K., Moisture regulation in the skin, *Drug Cosmetic Ind.*, 84, 732, 1959.
23. Caspers, P.J., Lucassen, G.W., Carter, E.A., Bruining, H.A., and Puppels, G.J., *In vivo* confocal Raman microspectroscopy of the skin: noninvasive determination of molecular concentration profiles, *J. Invest. Dermatol.*, 116, 434, 2001.
24. Wellner, K., Fiedler, G., and Wohlrab, W., Investigations in urea content of the horny layer in atopic dermatitis, *Z. Hautkr.*, 67, 648, 1992.
25. Denda, M., Hori, J., Koyama, J., Yoshida, S., Nanba, R., Takahashi, M., Horii, I., and Yamamoto, A., Stratum corneum sphingolipids and free amino acids in experimentally-induced scaly skin, *Arch. Dermatol. Res.*, 284, 363, 1992.
26. Watanabe, M., Tagami, H., Horii, I., Takahashi, M., and Kligman, A.M., Functional analyses of the superficial stratum corneum in atopic dermatitis, *Arch. Dermatol.*, 127, 1689, 1991.
27. Jacobson, T.M., Yukse, I.U., Greesin, J.C., Gordon, J.S., Lane, A.T., and Gracy, R.W., Effects of aging and xerosis on the amino acid composition of human skin, *J. Invest. Dermatol.*, 95, 296, 1990.
28. Sybert, V.P., Dale, B.A., and Holbrook, K.A., Ichthyosis vulgaris: identification of a defect in filaggrin synthesis correlated with an absence of keratohyaline granules, *J. Invest. Dermatol.*, 84, 191, 1985.
29. Marstein, S., Jellum, E., and Eldjarn, L., The concentration of pyroglutamic acid (2-pyrrolidone-5-carboxylic acid) in normal and psoriatic epidermis, determined on a microgram scale by gas chromatography, *Clin. Chim. Acta*, 43, 389, 1973.
30. Fulmer, A.W. and Kramer, G.J., Stratum corneum lipid abnormalities in surfactant-induced dry scaly skin, *J. Invest. Dermatol.*, 86, 598, 1986.
31. Saint-Leger, D., Francois, A.M., Leveque, J.L., Stoudemayer, T.J., Kligman, A.M., and Grove, G., Stratum corneum lipids in skin xerosis, *Dermatologica*, 178, 151, 1989.
32. Melnik, B., Hollmann, J., Hofmann, U., Yuh, M.S., and Plewig, G., Lipid composition of outer stratum corneum and nails in atopic and control subjects, *Arch. Dermatol. Res.*, 282, 549, 1990.
33. Imokawa, G., Abe, A., Jin, K., Higaki, Y., Kawashima, M., and Hidano, A., Decreased level of ceramides in stratum corneum of atopic dermatitis: an etiologic factor in atopic dry skin? *J. Invest. Dermatol.*, 96, 523, 1991.
34. Yamamoto, A., Serizawa, S., Ito, M., and Sato, Y., Stratum corneum lipid abnormalities in atopic dermatitis, *Arch. Dermatol. Res.*, 283, 219, 1991.
35. Motta, S., Sesana, S., Ghidoni, R., and Monti, M., Content of the different lipid classes in psoriatic scale, *Arch. Dermatol. Res.*, 287, 691, 1995.
36. Paige, D.G., Morse-Fisher, N., and Harper, J.I., Quantification of stratum corneum ceramides and lipid envelope ceramides in the hereditary ichthyosis, *Br. J. Dermatol.*, 131, 23, 1994.
37. Denda, M., Koyama, J., Hori, J., Horii, I., Takahashi, M., Hara, M., and Tagami, H., Age- and sex-dependent change in stratum corneum sphingolipids, *Arch. Dermatol. Res.*, 285, 415, 1993.

38. Rawlings, A., Hope, J., Rogers, J., Mayo, A., Watkinson, A., and Scott, I., Skin dryness — what is it? *J. Invest. Dermatol.*, 100, 510, 1993.
39. Man, M.Q., Feingold, K.R., and Elias, P.M., Exogenous lipids influence permeability barrier recovery in acetone-treated murine skin, *Arch. Dermatol.*, 129, 728, 1993.
40. Barany, E., Lindberg, M., and Lodén, M., Unexpected skin barrier influence from nonionic emulsifiers, *Int. J. Pharm.*, 195, 189, 2000.
41. Takahashi, M., Yamada, M., and Machida, Y., A new method to evaluate the softening effect of cosmetic ingredients on the skin, *J. Soc. Cosmetic Chem.*, 35, 171, 1984.
42. Rieger, M.M. and Deem, D.E., Skin moisturizers. II. The effects of cosmetic ingredients on human stratum corneum, *J. Soc. Cosmetic Chem.*, 25, 253, 1974.
43. Hellgren, L. and Larsson, K., On the effect of urea on human epidermis, *Dermatologica*, 149, 89, 1974.
44. Miettinen, H., Johansson, G., Gobom, S., and Swanbeck, G., Studies on constituents of moisturizers: water-binding properties of urea and NaCl in aqueous solutions, *Skin Pharmacol. Appl. Skin Physiol.*, 12, 344, 1999.
45. Wellner, K. and Wohlrab, W., Quantitative evaluation of urea in stratum corneum of human skin, *Arch. Dermatol. Res.*, 285, 239, 1993.
46. Lodén, M., Bostrom, P., and Kneczke, M., Distribution and keratolytic effect of salicylic acid and urea in human skin, *Skin Pharmacol.*, 8, 173, 1995.
47. Batt, M.D. and Fairhurst, E., Hydration of the stratum corneum, *Int. J. Cosmetic Sci.*, 8, 253, 1986.
48. Batt, M.D., Davis, W.B., Fairhurst, E., Gerreard, W.A., and Ridge, B.D., Changes in the physical properties of the stratum corneum following treatment with glycerol, *J. Soc. Cosmetic Chem.*, 39, 367, 1988.
49. Swanbeck, G., A new treatment of ichthyosis and other hyperkeratotic conditions, *Acta Derm. Venereol. (Stockh)*, 48, 123, 1968.
50. Grice, K., Sattar, H., and Baker, H., Urea and retinoic acid in ichthyosis and their effect on transepidermal water loss and water holding capacity of stratum corneum, *Acta Derm. Venereol. (Stockh.)*, 54, 114, 1973.
51. Tagami, H., Electrical measurement of the water content of the skin surface. Functional analysis of the hygroscopic property and water-holding capacity of the stratum corneum *in vivo* and technique for assessing moisturizing efficacy, *Cosmetic Toiletries*, 97, 39, 1982.
52. Middleton, J.D. and Roberts, M.E., Effect of a skin cream containing the sodium salt of pyrrolidone carboxylic acid on dry and flaky skin, *J. Soc. Cosmetic Chem.*, 29, 201, 1978.
53. Nicholls, S., King, C.S., and Marks, R., Short term effects of emollients and a bath oil on the stratum corneum, *J. Soc. Cosmetic Chem.*, 29, 617, 1978.
54. Lodén, M., Olsson, H., Skare, L., and Axéll, T., Instrumental and sensory evaluation of the frictional response of the skin following a single application of five moisturizing creams, *J. Soc. Cosmetic Chem.*, 43, 13, 1992.
55. Mattai, J., Froebe, C.L., Rhein, L.D., Simion, F.A., Ohlmeyer, H., Su, D.T., and Friberg, S.E., Prevention of model stratum corneum lipid phase transitions *in vitro* by cosmetic additives — differential scanning calorimetry, optical microscopy, and water evaporation studies, *J. Soc. Cosmetic Chem.*, 44, 89, 1993.

56. Froebe, C.L., Simion, F.A., Ohlmeyer, H., Rhein, L.D., Mattai, J., Cagan, R.H., and Friberg, S.E., Prevention of stratum corneum lipid phase transitions *in vitro* by glycerol — an alternative mechanism for skin moisturization, *J. Soc. Cosmetic Chem.*, 41, 51, 1990.
57. Kolbe, L., Kligman, A.M., and Stoudemayer, T., Objective bioengineering methods to assess the effects of moisturizers on xerotic leg of elderly people, *J. Dermatol. Treat.*, 11, 241, 2000.
58. Alderson, S.G., Barratt, M.G., and Black, J.G., Effect of 2-hydroxyacids on guinea-pig footpad stratum corneum: mechanical properties and binding studies, *Int. J. Cosmetic Sci.*, 6, 91, 1984.
59. Takahashi, M., Machida, Y., and Tsuda, Y., The influence of hydroxy acids on the rheological properties of stratum corneum, *J. Soc. Cosmetic Chem.*, 36, 177, 1985.
60. Hall, K.J. and Hill, J.C., The skin plasticisation effect of 2-hydroxyoctanoic acid. 1: The use of potentiators, *J. Soc. Cosmetic Chem.*, 37, 397, 1986.
61. Frithz, A., Investigation of Cortesal®, a hydrocortisone cream and its water-retaining cream base in the treatment of xerotic skin and dry eczemas, *Curr. Ther. Res.*, 33, 930, 1983.
62. Pope, F.M., Rees, J.K., Wells, R.S., and Lewis, K.G.S., Out-patient treatment of ichthyosis: a double-blind trial of ointments, *Br. J. Dermatol.*, 86, 291, 1972.
63. Dahl, M.V. and Dahl, A.C., 12% lactate lotion for the treatment of xerosis, *Arch. Dermatol.*, 119, 27, 1983.
64. Ghadially, R., Halkier-Sorensen, L., and Elias, P.M., Effects of petrolatum on stratum corneum structure and function, *J. Am. Acad. Dermatol.*, 26, 387, 1992.
65. Imokawa, G., Kuno, H., and Kawai, M., Stratum corneum lipids serve as a bound-water modulator, *J. Invest. Dermatol.*, 96, 845, 1991.
66. Wehr, R., Krochmal, L., Bagatell, F., and W., R., A controlled two-center study of lactate 12% lotion and a petrolatum-based creme in patients with xerosis, *Cutis*, 37, 205, 1986.
67. Rogers, R.S., Callen, J., Wehr, R., and Krochmal, L., Comparative efficacy of 12% ammonium lactate lotion and 5% lactic acid lotion in the treatment of moderate to severe xerosis, *J. Am. Acad. Dermatol.*, 21, 714, 1989.
68. Vilaplana, J., Coll, J., Trullás, C., Axón, A., and Pelejero, C., Clinical and non-invasive evaluation of 12% ammonium lactate emulsion for the treatment of dry skin in atopic and non-atopic subjects, *Acta Derm. Venereol. (Stockh.)*, 72, 28, 1992.
69. Siskin, S.B., Quinlan, P.J., Finkelstein, M.S., Marlucci, M., Maglietta, T.G., and Gibson, J.R., The effect of ammonium lactate 12% lotion versus no therapy in the treatment of dry skin of the heels, *Int. J. Dermatol.*, 32, 905, 1983.
70. Green, L. and Cole, G.W., A comparison study of 7.5% lactic acid cream and 12% lactic acid lotion in psoriatic patients with xerosis cutis, *Cosmetic Derm.*, 7, 44, 1994.
71. Wehr, R.F., Kantor, I., Jones, E.L., and McPhee, M.E., A controlled comparative efficacy study of 5% ammonium lactate lotion versus an emollient control lotion in the treatment of moderate xerosis, *J. Am. Acad. Dermatol.*, 25, 849, 1991.
72. Serup, J., A double-blind comparison of two creams containing urea as the active ingredient. Assessment of efficacy and side-effects by non-invasive techniques and a clinical scoring scheme, *Acta Derm. Venereol.*, 177, 34, 1992.
73. Pigatto, P.D., Bigardi, A.S., Cannistraci, C., and Picardo, M., 10% urea cream (Laceran) for atopic dermatitis: a clinical and laboratory evaluation, *J. Dermatol. Treat.*, 7, 171, 1996.

74. Kuster, W., Bohnsack, K., Rippke, F., Upmeyer, H.J., Groll, S., and Traupe, H., Efficacy of urea therapy in children with ichthyosis. A multicenter randomized, placebo-controlled, double-blind, semilateral study, *Dermatology*, 196, 217, 1998.
75. Schölermann, A., Banké-Bochita, J., Bohnsack, K., Rippke, F., and Herrmann, W.M., Efficacy and safety of Eucerin 10% urea lotion in the treatment of symptoms of aged skin, *J. Dermatol. Treatm.*, 9, 175, 1998.
76. Lodén, M., Andersson, A.C., Anderson, C., Bergbrant, I.M., Frodin, T., Ohman, H., Sandstrom, M.H., Sarnhult, T., Voog, E., Stenberg, B., Pawlik, E., Preisler-Haggqvist, A., Svensson, A., and Lindberg, M., A double-blind study comparing the effect of glycerin and urea on dry, eczematous skin in atopic patients, *Acta Derm. Venereol.*, 82, 45, 2002.
77. De Groot, A.C., Nater, J.P., Lende, R., and Rijcken, B., Adverse effects of cosmetics and toiletries: a retrospective study in the general population, *Int. J. Dermatol. Sci.*, 9, 255, 1988.
78. Ya-Xian, Z., Suetake, T., and Tagami, H., Number of cell layers of the stratum corneum in normal skin — relationship to the anatomical location on the body, age, sex and physical parameters, *Arch Dermatol. Res.*, 291, 555, 1999.
79. Frosch, P.J. and Kligman, A.M., A method for appraising the stinging capacity of topically applied substances, *J. Soc. Cosmetic Chem.*, 28, 197, 1977.
80. Gabard, B., Nook, T., and H.., M.K., Tolerance of the lesioned skin to dermatological formulations, *J. Appl. Cosmetol.*, 9, 25, 1991.
81. Rietschel, R.L. and Fowler, J.F., *Fisher's Contact Dermatitis*, 4th edition, Williams & Wilkins, Baltimore, Maryland, 1995.
82. Larmi, E., Lahti, A., and Hannuksela, M., Immediate contact reactions to benzoic acid and the sodium salt of pyrrolidone carboxylic acid, *Contact Dermatitis*, 20, 38, 1989.
83. Schlagel, C.A. and Sanborn, E.C., The weights of topical preparations required for total and partial body inunction, *J. Invest. Dermatol.*, 42, 253, 1964.
84. Lynfield, Y.L. and Schechter, B.A., Choosing and using a vehicle, *J. Am. Acad. Dermatol.*, 10, 56, 1984.
85. Ivens, U.I., Steinkjer, B., Serup, J., and Tetens, V., Ointment is evenly spread on the skin, in contrast to creams and solutions, *Br. J. Dermatol.*, 145, 264, 2001.
86. Rhodes, L.E. and Diffey, B.L., Fluorescence spectroscopy: a rapid, noninvasive method for measurement of skin surface thickness of topical agents, *Br. J. Dermatol.*, 136, 12, 1997.
87. Johnson, R., Nusbaum, B.P., Horwitz, S.N., and Frost, P., Transfer of topically applied tetracycline in various vehicles, *Arch. Dermatol.*, 119, 660, 1983.
88. Serup, J., EEMCO guidance for the clinical assessment of dry skin (xerosis) and ichthyosis: clinical scoring systems, *Skin Res. Technol.*, 1, 109, 1995.
89. Potts, R.O., Stratum corneum hydration: experimental techniques and interpretations of results, *J. Soc. Cosmetic Chem.*, 37, 9, 1986.
90. Lodén, M., Hagforsen, E., and Lindberg, M., The presence of body hair influences the measurement of skin hydration with the Corneometer, *Acta Derm. Venereol.*, 75, 449, 1995.
91. Jemec, G.B., Na, R., and Wulf, H.C., The inherent capacitance of moisturising creams: a source of false positive results? *Skin Pharmacol. Appl. Skin Physiol.*, 13, 182, 2000.
92. Wepierre, J., Study of the hydrating effect of cosmetic preparations by measuring cutaneous impedance in the hairless rat, *Soap Perfum. Cosmetics*, 50, 506, 1977.

25 Infant Skin Hydration

Marty O. Visscher and Steven B. Hoath

CONTENTS

I. Introduction ... 307
II. The Full Term Infant ... 307
A. Birth ... 307
B. The First Month ... 310
C. Infant Skin Hydration: Environmental Influences ... 312
D. Adult Comparisons ... 313
III. The Premature Infant ... 313
IV. Role of Vernix Caseosa ... 317
References ... 318

I. INTRODUCTION

A discussion of infant skin hydration necessarily encompasses the development of fetal skin *in utero*, the adaptation of the infant's skin following full-term birth, and the development of an adequate skin barrier for the premature infant who is born before the process is complete. These topics revolve around a fundamental question: How does the infant develop a fully functional and competent epidermal barrier while totally submerged in amniotic fluid for 9 months? The mechanisms whereby a functional stratum corneum (SC) is formed under aqueous conditions are unclear insofar as prolonged water exposure causes maceration and barrier compromise.[1–5] Whereas considerable work still remains to be done, particularly about the subject of early fetal skin maturation, our knowledge of perinatal skin biology and barrier development has increased greatly in recent years. This chapter focuses on the physiology of term and preterm infant skin hydration in the perinatal period.

II. THE FULL TERM INFANT

A. Birth

The human infant moves from a heated (37°C) water-filled chamber to a cold, dry, terrestrial environment within minutes at the time of birth. The normal full-term infant (gestational age ≥37 weeks) has a competent epidermal barrier that provides water homeostasis and protects against infection.[6,7] Transepidermal water loss

0-8493-1443-7/05/$0.00+$1.50

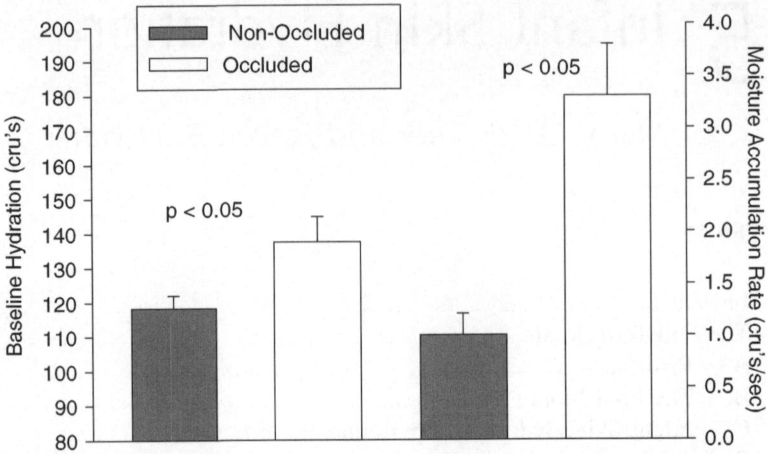

FIGURE 25.1 Differential hydration of the skin of newborn infants 4 hours after birth — effect of occlusion. The occluded region had a significantly higher baseline hydration (cru's) than the nonoccluded site. The moisture accumulation rate (cru's/sec) under the instrument probe was also significantly higher for the occluded site than for the nonoccluded region. (Data from Visscher et al.[9])

(TEWL) values are low for full-term newborns at birth, with values comparable to or lower than adults.[7,8]

Skin barrier function and hydration have been investigated during the immediate newborn period.[9,10] In a recent study, we hypothesized that infant SC undergoes programmatic changes in hydration and water binding immediately after birth and conducted a set of *in vivo* trials among a cohort of 101 healthy full-term newborns. Conventional delivery room practices dry the infants with a cotton towel and put them under a radiant warmer. Some institutions place the infant in a plastic bag from midtorso to feet to protect against evaporative heat loss. We determined the effects of both radiant warming and wrapping (i.e., occlusion, and high humidity) by placing a subset of 29 infants in a bag to midtorso. The SC properties for chest (outside bag, under warmer) and an occluded suprapubic region were evaluated within 4 hours after birth. The baseline hydration measured in capacitive reactance units (cru's) was significantly lower for the nonoccluded site (118 ± 4.1 cru's) than for the occluded region (137 ± 7.3 cru's). The rate of moisture accumulation under the instrument probe (MAT cru's/sec) was also significantly lower for the nonoccluded region (1.0 ± 0.2 cru's/sec) than for the occluded site (3.3 ± 0.5 cru's/sec), as shown in Figure 25.1. This finding is consistent with that of Hammarlund et al.,[11] who found water evaporation rates to be high and to be the major cause for heat loss for healthy infants immediately after birth. The rate of heat loss depended on environmental conditions (i.e., incubator vs. delivery room). A linear relationship between the water evaporation rate and incubator humidity was observed for 19 healthy newborns, with a calculated TEWL of 8.1 g/m^2 per hour.[12]

The effect of anatomical site on SC-water interactions was determined in a subset of 34 infants.[9] The infants were dried and placed under a radiant warmer. Moisture

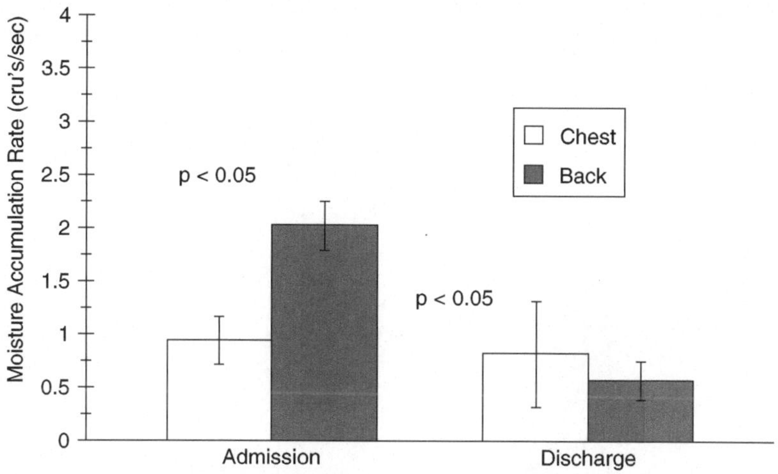

FIGURE 25.2 Differential hydration of the skin of newborn infants 4 hours after birth — effect of body site. The moisture accumulation rate was significantly lower for the chest vs. the back shortly after birth, and values were 0.9 ± 0.2 and 2.0 ± 0.5 cru's/sec, respectively. The moisture accumulation rate was lower for both sites on discharge from the nursery, with a significant decrease over the back. (Data from Visscher et al.[9])

accumulation rates were recorded for the chest and back twice within 4 hours. The first readings were taken 92 minutes after birth (range 36 to 124 minutes). The moisture accumulation rate was significantly higher for the back vs. the chest, with values of 2.0 ± 0.5 and 0.9 ± 0.2 cru's/sec, respectively (Figure 25.2). A second measurement taken 55 minutes later indicated the moisture accumulation rate to be lower for both sites, with a significant decrease over the back. High-resolution surface-illuminated images of the chest and back skin revealed that the back was significantly less dry and scaly than the chest.

Comparison of the chest and forehead sites of 37 infants within 90 minutes of birth revealed a significantly greater moisture accumulation rate for the forehead than for the chest (2.1 ± 0.6 vs. 0.8 ± 0.1 cru's/sec, respectively). The forehead skin temperature was significantly lower than for the chest, indicating that increased hydration may have produced evaporative cooling (M.O. Visscher, S.B. Hoath, unpublished data). Sedin et al.[13] found that regional differences in evaporation rate were influenced by nutritional status, gestational age, activity, and ambient environment. Yosipovitch et al.[8] studied a group of 44 full-term infants during the first 2 days of life and found significantly lower SC hydration on day 2 vs. day 1 on the forearms and palms, indicating that hydration decreases in the immediate newborn period. TEWL was significantly higher on day 2 vs. day 1 for the forearm, palms, and inguinal folds. TEWL was significantly higher overall for the forearm, palms, and inguinal area than for the soles, back, forehead, and abdomen. The SC water-handling behavior was examined for 30 full-term neonates during the first postnatal day. The moisture accumulation rate decreased dramatically during the first day (Figure 25.3),

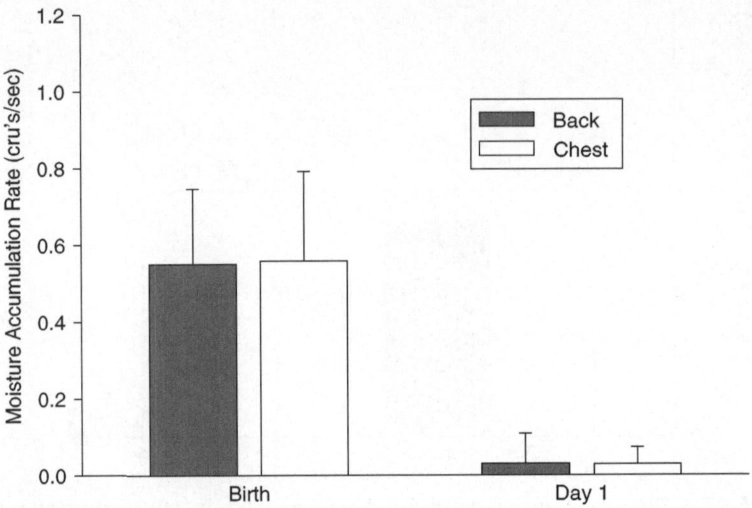

FIGURE 25.3 The moisture accumulation rate decreases significantly during the first postnatal day in a cohort of 30 full-term infants, indicating that changes occur rapidly as the infant adapts to a dry environment.

indicating that the epidermis undergoes rapid changes as the infant adapts to a dry environment.

B. The First Month

Stratum corneum hydration and barrier properties have been investigated over the first month of postnatal adaptation for 30 full-term newborns and compared to adult skin under similar environmental conditions.[14] The infants' skin showed increasing surface hydration and rate of moisture accumulation during the first 2 postnatal weeks, whereas maternal measurements did not vary significantly over the time period. Most significant was the progressive increase in the moisture accumulation rate over the first two postnatal weeks from the initial, low values at birth. The infant skin reached a plateau around day 14, and values were comparable at days 21 and 28 (Figure 25.4). Hoeger and Enzmann[10] found a similar pattern of skin adaptation with a significant increase in SC hydration during the postnatal period.[10] Saijo and Tagami[15] evaluated skin electrical conductance as a measure of skin surface hydration; these investigators found very low hydration and evaporative water loss for newborns compared with adults and with infants of 1 and 6 months of age. They attributed these findings to impaired SC water-holding capacity. Bodak and Bodemer[16] reported that the pattern of desquamation found in 60% of newborns indicates a temporary functional abnormality in the SC and recommended the use of emollients until the desquamation normalized.[16]

Such measures of surface hydration and dryness can be contrasted with TEWL values, which are low for full-term newborns.[17] Hammarlund et al.[18] reported that TEWL for full-term infants remained unchanged for the first 2 weeks of life. As a consequence, TEWL is unlikely to account for the increased rate of moisture

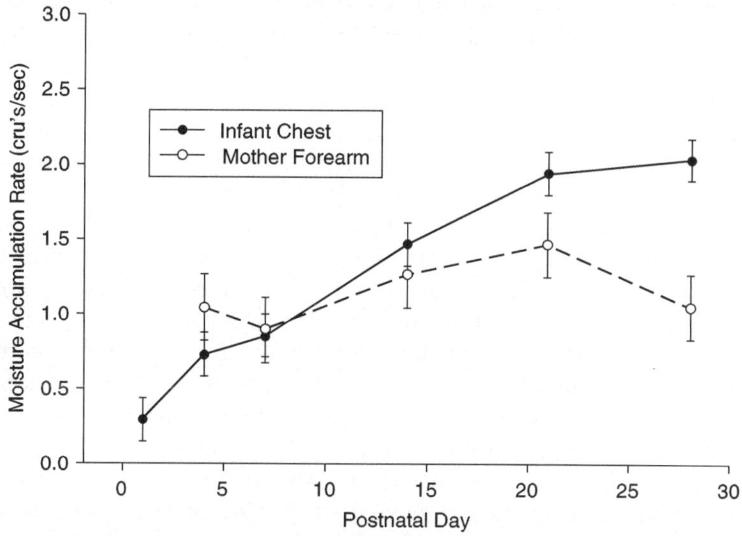

FIGURE 25.4 Differential hydration of the skin of newborn infants 4 hours after birth — effect of postnatal age. During the first 2 postnatal weeks, infant skin exhibits increasing moisture accumulation rate and surface hydration. The infant skin reached a plateau around day 14, and values are comparable at days 21 and 28. Maternal measurements do not vary significantly over the time period. (Data from Visscher et al.[14])

accumulation (Figure 25.4). In another study, the water-handling properties of the skin were evaluated by occluding the skin for 1 hour and then evaluating TEWL.[19] TEWL was found to be higher than normal, indicating an increased rate of water movement through the epidermis. The increased moisture accumulation rate could perhaps be the result of eccrine sweating. Sedin reported increases in surface water loss in full-term infants aged 4 hours to 11 days because of eccrine sweating when incubator temperatures were higher than 34°C.[20] In our investigation, infants were in open cribs or at home, and the skin sites had equilibrated to environmental conditions for 25 minutes before measurements.[14] No sweating was detected, and the low standard error of the mean indicates good experimental control of the measurements.

Water sorption–desorption evaluations also show significant changes in the upper SC during the first 14 days followed by a plateau through the next 2 weeks.[14,15] The desorption rate decreases for infants during the first 14 days and remains unchanged for mothers. The relatively rapid desorption rate at birth may indicate a greater capacity to shed exogenous water, thereby minimizing potential evaporative heat loss. The presence of a hydrophobic film on the skin surface, such as vernix caseosa, would result in increased water desorption.[21] Alternatively, low water-holding capacity coupled with low hydration is also indicative of dry skin.[15] The decrease in desorption rate over time indicates that the skin surface begins to bind exogenous water rather than to shed it. The generation of hygroscopic NMF components in the upper SC, for example, would presumably increase water holding capacity and, thereby, reduce the desorption rate.

The effects of environmental humidity on the SC have been examined with respect to hydration, TEWL, and epidermal cell proliferation.[22] In this study, groups of hairless mice were kept at high (90% relative humidity [RH]) and low (10% RH) humidity for 5 days following a period of acclimation at 40 to 50% RH. Epidermal DNA synthesis increased, beginning 12 hours after exposure to low humidity, but did not change at high humidity, even though TEWL values were similar for both groups. The SC hydration was significantly lower for the low-humidity group, indicating that the decrease in SC hydration caused an increase in DNA synthesis.[22]

Scott and Harding[23] examined the influence of decreasing humidity at birth on the epidermal protein filaggrin using a newborn rat model. Filaggrin proteolysis produces free amino acids, pyrrolidone carboxylic acid, and urocanic acid (natural moisturizing factor, NMF) that facilitate water binding of the stratum corneum at low environmental humidities.[24] In the absence of NMFs, the stratum corneum is dry, does not desquamate properly, and has poor water holding capacity. During late gestation in the animal, filaggrin is present throughout the stratum corneum. In the study cited, proteolysis occurred immediately after birth and filaggrin was found in the lower stratum corneum only.[23] Proteolysis depended on the humidity at birth and was observed at ranges of 80 to 95% RH, but not at 100% RH. The authors hypothesized that NMF generation depended on a decrease in ambient water activity. This report provides a possible explanation for the development of a dry, poorly hydrated stratum corneum during the first postnatal day in the full-term neonate reported by our laboratory.[9] In separate studies, we have determined that levels of NMF are extremely low at birth in newborn term infants but increase significantly over the first postnatal month (M.O. Visscher, S.B. Hoath, unpublished data).

To investigate the influence of varying environmental exposure following birth, we measured the SC barrier properties of skin within the diaper.[14] The diaper skin also exhibited increasing surface hydration and moisture accumulation rate during the first 14 days. However, regional differentiation was found beginning on day 7, and the diaper area skin had significantly higher baseline hydration and moisture accumulation rate compared to the chest. This result indicates that the diaper environment may influence the SC water-handling properties and the skin adaptation process. By postnatal day 21, the diaper and chest sites were not significantly different and both displayed an apparent plateau.

C. Infant Skin Hydration: Environmental Influences

Exposure of an infant's skin to water for brief periods is common during routine bathing. Here, the SC goes from a wet to a dry environment, somewhat analogous to the transition from aqueous *in utero* conditions to much drier conditions at birth. The effects of a brief (10-minute) water soak on infant skin hydration have been reported.[25] Bathing resulted in significant changes in SC water-handling properties, with decreases in the rate of moisture accumulation, erythema, visual dryness/scaling, and skin surface friction. Similar decreases in the rate of moisture accumulation and hydration were observed in an adult population in whom the volar forearms were soaked for 10 minutes.[26] The authors suggested that water soaking removes water-soluble amino acids (i.e., NMF), thereby reducing the amount of secondary bound water in the skin. The reduction in moisture accumulation after soaking may be the result of a decrease in the NMF-dependent bound water that gives rise to the higher capacitance reading before the soak.

The technical measurements from the bathing study among 3- to 6-month-old infants showed that the baseline skin hydration and the rate of moisture accumulation under occlusion were lower for the nondiaper site than for diapered skin, indicating that differentiation may occur with increasing age. Diapers provide some occlusivity at the skin surface, thereby increasing skin wetness.[27–29] Prolonged exposure to sufficiently high levels of water can result in various forms of inflammation, irritation, dermatoses, and urticaria.[3,30–32] Disruption of the stratum corneum lipid morphology occurs after 4 hours of exposure to water or urine.[4,5]

Gfatter et al.[33] examined the influences of skin cleansing formulations on skin properties.[33] They found differences in the quantity of surface lipids removed during washing and suggested that lipid reduction would influence SC hydration.

D. Adult Comparisons

Comparisons of infant vs. adult skin parameters have demonstrated that epidermal barrier properties related to water handling change markedly over the first 4 weeks of postnatal life, indicating a significant period of adaptation.[9] Infant and adult comparisons on days 1 and 2 after birth revealed lower TEWL for infants in the forehead, palms, and soles and higher TEWL than adults for forearm sites.[8] Comparison of TEWL for day 1 vs. day 2 indicated lower values on day 2 in the forearm, palms, and soles. Skin surface hydration was significantly lower in the infants on the forehead, back, and abdomen, and higher on the forearms and palms. Skin hydration and TEWL values were correlated in infants (positively) but not in adults. Skin conductance was reported for newborns aged 5 days compared with infants aged 1 month, 2 months, and 6 months and to their mothers.[15] The skin surface of the newborns was significantly drier than the older infants and the mothers. Giusti et al.[34] investigated 70 infants, aged 8 to 24 months, vs. a cohort of 30 female adults and found no significant differences between infants and adults in TEWL on either the forearms or the buttocks.[34] However, capacitance values were significantly higher for the infants, indicating higher levels of SC hydration. The authors concluded that the infant skin was more "immature" than adult skin. However, measurement of full-thickness skin from infants, children, and adults has shown no significant differences in SC thickness.[35] On the basis of these reports, differences in capacitance values for infants and adults cannot be attributed to variations in SC barrier thickness or integrity, as TEWL values were not different. Alternatively, the higher capacitance values observed for infants may indicate that infant skin is better hydrated and more plastic than adult skin. More studies are needed in this area.

III. THE PREMATURE INFANT

The premature newborn infant has an immature epidermis with a poorly formed or nonexistent stratum corneum, as evidenced by markedly elevated TEWL (Figure 25.5).[6,36,37] The very premature infant has, essentially, a wounded skin surface with

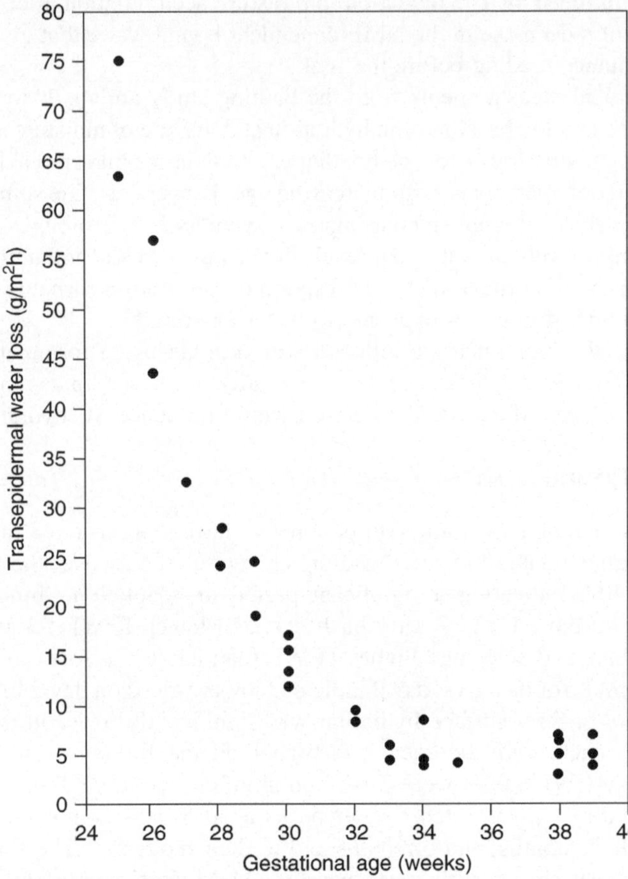

FIGURE 25.5 Transepidermal water loss is markedly increased in infants less than 26 weeks' gestational age and decreases until 33 to 34 weeks, when values are similar to those of full-term infants. The skin of the very premature infant has few cornified layers and presents as a wounded skin surface. (Data from Sedin et al.[6])

few cornified layers.[6,36–38] TEWL decreases with increasing gestational age until weeks 33 to 34, when values are comparable to those of full-term infants and adults. Wickett et al.,[39] using electrical capacitance methods, measured skin hydration during development in neonatal rodents and found a significant relationship with gestational age. Using similar methods, Okah et al.[40] studied preterm infants and found skin hydration was significantly higher for the infants less than 30 weeks gestational age (GA) than for infants born after 30 weeks (Figure 25.6). Skin hydration for infants of 32 to 34 weeks' gestation was not significantly different from that of the full-term group. Percutaneous drug absorption is greater for premature infants than for older infants, reflecting the immaturity of the SC.[17,41] As a group, premature infants can vary considerably with respect to SC barrier integrity and hydration, depending on the gestational age at birth.

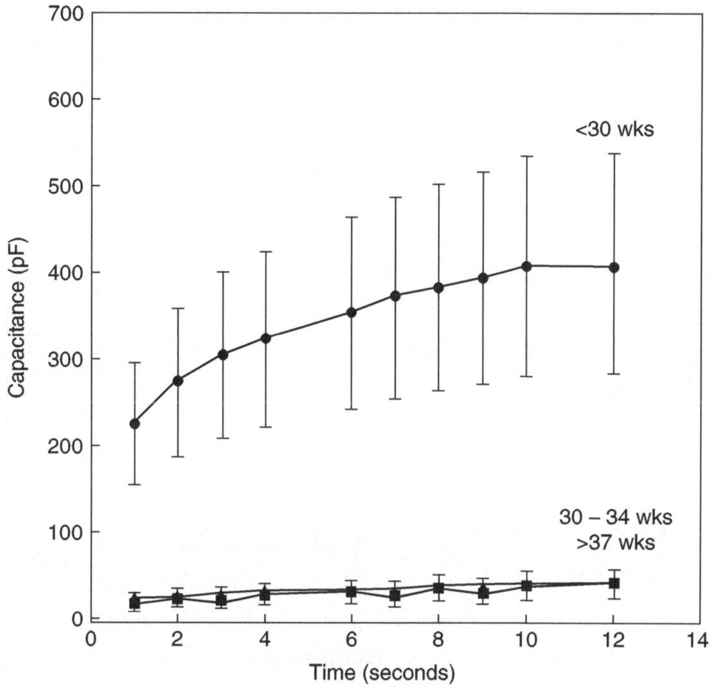

FIGURE 25.6 Surface electrical capacitance measurements in infants of varying gestational ages on the first postnatal day. Both initial readings (surface hydration) and slope (water accumulation) were significantly higher in infants born before 30 weeks gestation. (Data from Okah et al.[40])

Birth and the transition to a dry environment significantly influence SC barrier development, and TEWL gradually decreases with increasing postnatal age.[18] The epidermis of a 2-week-old postnatal preterm infant is histologically similar to that of a full-term infant.[37] The morphological changes are directly related to development of epidermal barrier properties. Okah et al.[40] examined the ontogeny of epidermal barrier maturation during the first 5 postnatal days for groups of varying gestational age.[40] The rate of transepidermal water movement under probe occlusion was significantly higher in infants less than 30 weeks GA compared with those more than 30 weeks GA. The rate of water movement changed most dramatically in the most premature infants and had decreased substantially by day 5 in infants with less than 26 weeks gestation (Figure 25.7). The rapid functional barrier maturation was demonstrated in all very premature infants and did not appear to be influenced by the severity of their illness.

SC barrier function as measured by TEWL was evaluated during the first postnatal month for a cohort of preterm infants of 24 to 25 weeks' gestational age.[42] Although TEWL decreased over time, values at 1 month were significantly higher than those observed for normal full-term newborns. Epidermal maturation occurs, but the development of a fully functional barrier requires longer than 4 weeks for infants born at 23 to 25 weeks.[43] For infants born at 23 to 24 weeks' gestation, 5 to

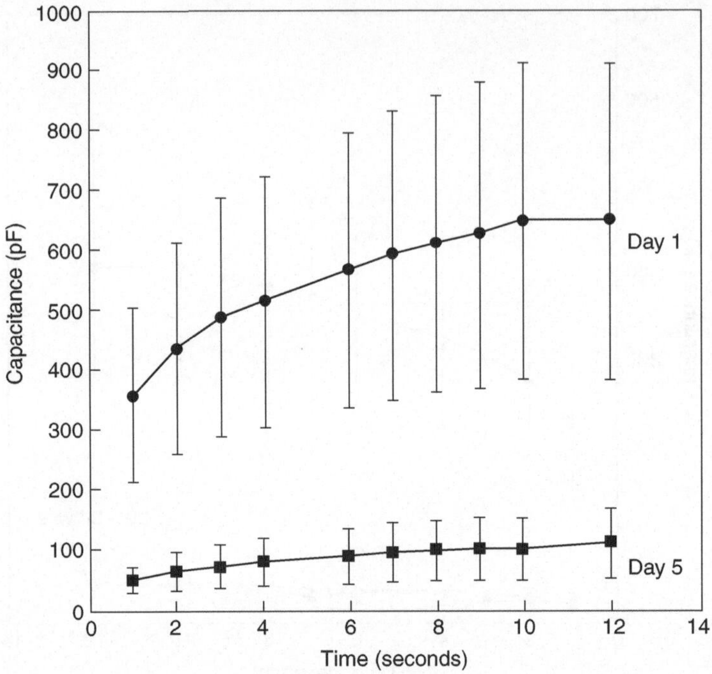

FIGURE 25.7 Surface electrical capacitance measurements in preterm infants born at less than 26 weeks' gestation on postnatal days 1 and 5. Significant decreases in both baseline surface hydration and slope are observed over this time period, indicating rapid maturation of epidermal barrier function. (Data from Okah et al.[40])

7 weeks were required for significant maturation.[44] Following exposure to relatively dry conditions for several weeks, premature infants often present with a very dry, scaly skin surface indicative of presumptive hyperproliferation and abnormal desquamation.

Semipermeable films and dressings have been used to cover the preterm neonate's skin and reduce the high TEWL.[45–47] Topical ointments (e.g., petrolatum, Aquaphor, Eucerin) have been found to significantly reduce water loss for 6 to 8 hours after application.[48–50] Nopper et al.[49] reported a decrease in nosocomial infection rate and an improvement in skin condition among preterm infants of 29 weeks' mean gestational age. Campbell et al. reported an increase in systemic candidiasis among a group of extremely low-birth-weight infants who had been treated with topical petrolatum.[51] A multicenter clinical trial of twice-daily topical application of a petrolatum-based ointment (Aquaphor) vs. no treatment among very low-birth-weight premature infants revealed a significantly higher incidence of nosocomial infection for the treatment in infants weighing 501 to 750 grams (26 weeks' mean gestational age).[52]

Human and animal studies using tape stripping to create a compromised SC barrier can be considered as models of the insufficient epidermal barrier of the preterm infant. Hanley et al.[53] found that barrier development was accelerated in

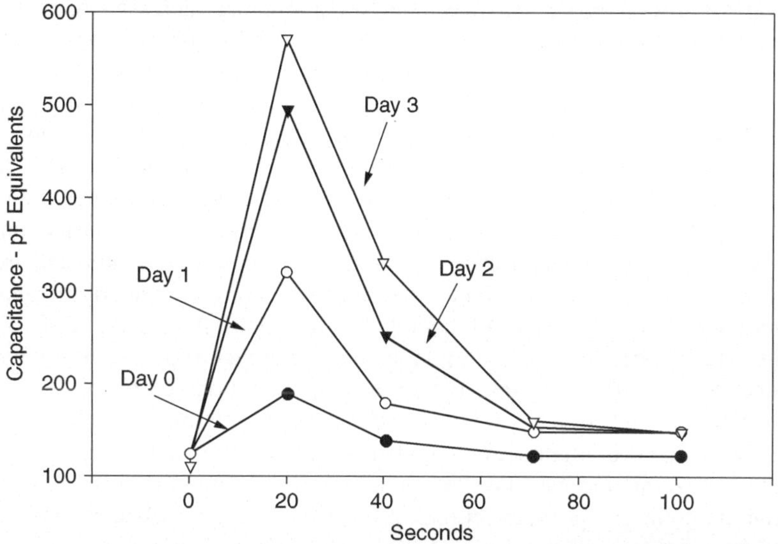

FIGURE 25.8 Ontogenetic profile of the water sorption–desorption test in the neonatal rat. As shown, the newborn (day 0) animal has minimal water sorption with a low peak at 20 seconds following exogenous water exposure. Over the next 3 postnatal days, the animal becomes progressively less hydrophobic, with higher and higher peak sorption values. This hydrophobicity is confined to the outermost peridermal layer. (Data from Wickett et al.[56])

stripped skin explants exposed to an air-medium interface relative to occluded samples and submerged controls. In an adult human model, exposure to semipermeable films accelerated barrier development following barrier compromise relative to no occlusion and complete occlusion.[54] These findings from animal and human models indicate that control of the water gradient is important for the development of a proper barrier. Confirmation of this hypothesis in the premature infant is warranted.

IV. ROLE OF VERNIX CASEOSA

Investigations of the adaptation process in both full-term and premature infants provide the contrasts necessary to gain insight regarding the development of a fully functional SC barrier. However, our initial question remains: How does the fetus develop a fully functional and competent epidermal barrier while submerged in water and amniotic fluid for 9 months? The vernix caseosa, the material that coats the fetus during the last trimester *in utero*, may facilitate this process. The full-term infant has been exposed to vernix *in utero*, whereas the preterm infant has been born without this contact.

Studies in the neonatal rat demonstrated that the skin surface is very hydrophobic at birth and sheds water effectively, presumably for thermal control and survival.[55] Measurements of water sorption–desorption behavior showed low water binding and rapid desorption. As the rat matured in the first week of life, the water sorption–desorption profile changed and water binding increased (Figure 25.8). Stripping away the outermost layer of the epidermis with adhesive tape results in a marked decrease in surface hydrophobicity. Histological assessment indicates that the surface hydrophobicity is attributable to the presence of a postnatally persistent periderm, which is removed by tape stripping. Accelerated differentiation of the periderm follows administration of antenatal steroids to the mother. This finding parallels steroid-induced maturation of other epithelial systems such as the lung.

Investigations in our laboratories support the hypothesis that vernix functions as a waterproofing agent that overlays the developing epidermis *in utero* and provides a relatively "dry" environment to facilitate cornification of the underlying nucleated keratinocytes. Vernix consists of hydrated corneocytes (terminally differentiated keratinocytes) embedded in a hydrophobic lipid matrix with specific antioxidant and antiinfective properties. *In vitro* and *in vivo* investigations of native vernix have shown important and clinically relevant differences vs. currently available treatments for use on premature infant skin, wounded or irritated skin, and compromised skin barrier.[21,56–61] Vernix protects the skin and allows for maturation using a fully natural material, as compared to conventional protectants (e.g., petrolatum). The authors hypothesize that the composition and structure of vernix facilitate the maintenance of an optimal water gradient over the surface of the body. Confirmatory studies are in progress.

REFERENCES

1. Fluhr, J.W., et al., Effects of prolonged occlusion on stratum corneum barrier function and water holding capacity, *Skin Pharmacol. Appl. Skin Physiol.*, 12(4), 193, 1999.
2. Tsai, T.F. and Maibach, H.I., How irritant is water? An overview, *Contact Dermatitis*, 41(6), 311, 1999.
3. Willis, I., The effects of prolonged water exposure on human skin, *J. Invest. Dermatol.*, 60, 166, 1973.
4. Warner, R.R., et al., Water disrupts stratum corneum lipid lamellae: damage is similar to surfactants, *J. Invest. Dermatol.*, 113(6), 960, 1999.
5. Warner, R.R., et al., Hydration disrupts human stratum corneum ultrastructure, *J. Invest. Dermatol.*, 120(2), 275, 2003.
6. Sedin, G., et al., Transepidermal water loss in full-term and pre-term infants, *Acta Paediatr. Scand. Suppl.*, 305, 27, 1983.
7. Cunico, R.L., et al., Skin barrier properties in the newborn. Transepidermal water loss and carbon dioxide emission rates, *Biol. Neonate.*, 32(3–4), 177, 1977.
8. Yosipovitch, G., et al., Skin barrier properties in different body areas in neonates, *Pediatrics*, 106(1 Pt 1), 105, 2000.
9. Visscher, M.O., Munson, K.A., Bare, D.E., and Hoath, S.B., Early adaptation of human skin following birth: a biophysical assessment, *Skin Res. Technol.*, 5, 213, 1999.
10. Hoeger, P.H. and Enzmann, C.C., Skin physiology of the neonate and young infant: a prospective study of functional skin parameters during early infancy, *Pediatr. Dermatol.*, 19(3), 256, 2002.

11. Hammarlund, K., et al., Transepidermal water loss in newborn infants. V. Evaporation from the skin and heat exchange during the first hours of life, *Acta Paediatr. Scand.*, 69(3), 385, 1980.
12. Hammarlund, K., et al., Transepidermal water loss in newborn infants. I. Relation to ambient humidity and site of measurement and estimation of total transepidermal water loss, *Acta Paediatr. Scand.*, 66(5), 553, 1977.
13. Sedin, G., et al., Water transport through the skin of newborn infants, *Ups. J. Med. Sci.*, 86(1), 27, 1981.
14. Visscher, M.O., et al., Changes in diapered and nondiapered infant skin over the first month of life, *Pediatr. Dermatol.*, 17(1), 45, 2000.
15. Saijo, S. and Tagami, H., Dry skin of newborn infants: functional analysis of the stratum corneum, *Pediatr. Dermatol.*, 8(2), 155, 1991.
16. Bodak, N. and Bodemer, C., Cutaneous hydration of the premature and new born, *Ann Dermatol. Venereol. (Stockh.)*, 129(1 Pt 2), 143, 2002.
17. Harpin, V.A. and Rutter, N., Barrier properties of the newborn infant's skin, *J. Pediatr.*, 102(3), 419, 1983.
18. Hammarlund, K., et al., Transepidermal water loss in newborn infants. VII. Relation to post-natal age in very pre-term and full-term appropriate for gestational age infants, *Acta Paediatr. Scand.*, 71(3), 369, 1982.
19. Orsmark, K., et al., *In vivo* transepidermal water loss and epidermal occlusive hydration in newborn infants: anatomical region variation, *Acta Derm. Venereol. (Stockh.)*, 60(5), 403, 1980.
20. Rutter, N. and Hull, D., Response of term babies to a warm environment, *Arch. Dis. Child*, 54(3), 178, 1979.
21. Youssef, W., et al., Surface free energy characterization of vernix caseosa. Potential role in waterproofing the newborn infant, *Skin Res. Technol.*, 7(1), 10, 2001.
22. Sato, J., et al., Loss of water from the stratum corneum induces epidermal DNA synthesis in hairless mice, *Arch. Dermatol. Res.*, 290(11), 634, 1998.
23. Scott, I.R. and Harding, C.R., Filaggrin breakdown to water binding compounds during development of the rat stratum corneum is controlled by the water activity of the environment, *Dev. Biol.*, 115(1), 84, 1986.
24. Rawlings, A.V., et al., Stratum corneum moisturization at the molecular level, *J. Invest. Dermatol.*, 103(5), 731, 1994.
25. Visscher, M.O., Chatterjee R., Ebel, J.P., LaRuffa, A.A., and Hoath, S.B., Biomedical assessment and instrumental evaluation of healthy infant skin, *Pediatr. Dermatol.*, 19(6), 473, 2002.
26. Visscher, M.O., et al., Effect of soaking and natural moisturizing factor on stratum corneum water-handling properties, *J. Cosmetic Sci.*, 54, 289, 2003.
27. Berg, R.W., Etiologic factors in diaper dermatitis: a model for development of improved diapers, *Pediatrician*, 14(Suppl 1), 27, 1987.
28. Berg, R.W., Etiology and pathophysiology of diaper dermatitis, *Adv. Dermatol.*, 3, 75, 1988.
29. Davis, J.A., et al. Comparison of disposable diapers with fluff absorbent and fluff plus absorbent polymers: effects on skin hydration, skin pH, and diaper dermatitis, *Pediatr. Dermatol.*, 6(2), 102, 1989.
30. Hurkmans, J.F., et al. Skin irritation caused by transdermal drug delivery systems during long-term (5 days) application, *Br. J. Dermatol.*, 112(4), 461, 1985.
31. Medeiros, M., Jr., Aquagenic urticaria, *J. Invest. Allergol. Clin. Immunol.*, 6(1), 63, 1996.

32. Halkier-Sorensen, L., Occupational skin diseases, *Contact Dermatitis*, 35, 1, 1996.
33. Gfatter, R., et al., Effects of soap and detergents on skin surface pH, stratum corneum hydration and fat content in infants, *Dermatology*, 195(3), 258, 1997.
34. Giusti, F., et al., Skin barrier, hydration, and pH of the skin of infants under 2 years of age, *Pediatr. Dermatol.*, 18(2), 93, 2001.
35. Fairley, J.A. and Rasmussen, J.E., Comparison of stratum corneum thickness in children and adults, *J. Am. Acad. Dermatol.*, 8(5), 652, 1983.
36. Hammarlund, K., et al., Transepidermal water loss in newborn infants. VIII. Relation to gestational age and post-natal age in appropriate and small for gestational age infants, *Acta Paediatr. Scand.*, 72(5), 721, 1983.
37. Evans, N.J. and Rutter, N., Development of the epidermis in the newborn, *Biol. Neonate*, 49(2), 74, 1986.
38. Cartlidge, P., The epidermal barrier, *Semin. Neonatol.*, 5(4), 273, 2000.
39. Wickett, R.R., et al., Use of continuous electrical capacitance and transepidermal water loss measurements for assessing barrier function in neonatal rat skin, *Skin Pharmacol.*, 8(4), 179, 1995.
40. Okah, F.A., et al., Surface electrical capacitance as a noninvasive bedside measure of epidermal barrier maturation in the newborn infant, *Pediatrics*, 96(4 Pt 1), 688, 1995.
41. West, D.P., et al., Percutaneous absorption in preterm infants, *Pediatr. Dermatol.*, 4(3), 234, 1987.
42. Agren, J., et al., Transepidermal water loss in infants born at 24 and 25 weeks of gestation, *Acta Paediatr.*, 87(11), 1185, 1998.
43. Kalia, Y.N., et al., Development of skin barrier function in premature infants, *J. Invest. Dermatol.*, 111(2), 320, 1998.
44. Nonato, L.B., et al., Transepidermal water loss in 24 and 25 weeks gestational age infants, *Acta Paediatr.*, 89(6), 747, 2000.
45. Knauth, A., et al., Semipermeable polyurethane membrane as an artificial skin for the premature neonate, *Pediatrics*, 83(6), 945, 1989.
46. Mancini, A.J., et al. Semipermeable dressings improve epidermal barrier function in premature infants, *Pediatr. Res.*, 36(3), 306, 1994.
47. Vernon, H.J., et al., Semipermeable dressing and transepidermal water loss in premature infants, *Pediatrics*, 86(3), 357, 1990.
48. Rutter, N., Percutaneous drug absorption in the newborn: hazards and uses, *Clin. Perinatol.*, 14(4), 911, 1987.
49. Nopper, A.J., et al., Topical ointment therapy benefits premature infants, *J. Pediatr.*, 128(5 Pt 1), 660, 1996.
50. Lane, A.T. and Drost, S.S., Effects of repeated application of emollient cream to premature neonates' skin, *Pediatrics*, 92(3), 415, 1993.
51. Campbell, J.R., et al., Systemic candidiasis in extremely low birth weight infants receiving topical petrolatum ointment for skin care: a case-control study, *Pediatrics*, 105(5), 1041, 2000.
52. Edwards, W.H., Conner, J.M., and Soll, R.F., The effect of Aquaphor® original emollient ointment on nosocomial sepsis rates and skin integrity in infants of birth weight 501 to 1000 grams, *Pediatr. Res.*, 49, 388A, 2001.
53. Hanley, K., et al., Acceleration of barrier ontogenesis *in vitro* through air exposure, *Pediatr. Res.*, 41(2), 293, 1997.
54. Visscher, M., et al., Effect of semipermeable membranes on skin barrier repair following tape stripping, *Arch. Dermatol. Res.*, 293(10), 491, 2001.

55. Wickett, R.R., et al., Ontogeny of water sorption-desorption in the perinatal rat, *J. Invest. Dermatol.*, 100(4), 407, 1993.
56. Hoath, S.B., et al., Role and biology of vernix, *Neonatal Infant Nursing Rev. (NINR)*, 1, 53, 2001.
57. Pickens, W.L., et al., Characterization of vernix caseosa: water content, morphology, and elemental analysis, *J. Invest. Dermatol.*, 115(5), 875, 2000.
58. Pickens, W.L., et al., Antioxidant defense mechanisms in vernix caseosa: potential role of endogenous vitamin E, *Pediatr. Res.*, 47, 425A, 2000.
59. Narendran, V., et al. Vernix caseosa contains surfactant proteins: potential role in innate immune function in the fetus, *Pediatr. Res.*, 47, 420A, 2000.
60. Narendran, V., et al., Interaction between pulmonary surfactant and vernix: a potential mechanism for induction of amniotic fluid turbidity, *Pediatr. Res.*, 48(1), 120, 2000.
61. Bautista, M.I., et al., Characterization of vernix caseosa as a natural biofilm: comparison to standard oil-based ointments, *Pediatr. Dermatol.*, 17(4), 253, 2000.

26 Hydration and Atopic Dermatitis

Chris Edwards and Ronald Marks

CONTENTS

Abstract 323
I. Introduction 324
II. Xerosis in AD 325
III. TEWL 325
IV. Hydration of Stratum Corneum 326
V. Water Binding in the Stratum Corneum 327
VI. Causes of Xerosis 328
VII. Epidermal Lipids 329
References 330

ABSTRACT

Frequently, the skin on noneczematous areas of atopic dermatitis (AD) sufferers feels rough and appears "dry." This xerosis is associated with the itch that accompanies the disorder. Reports of the incidence of xerosis in AD patients vary from 48 to 98%.

Higher levels of transepidermal water loss (TEWL) are seen in the eczematous, the nonlesional xerotic, and the apparently normal skin of AD patients. TEWL is thought to be a good marker of the epidermal diffusion barrier, so that an increase indicates an impaired barrier function in AD skin. It has also been found that measures of stratum corneum hydration and water-binding capacity are lower in AD skin, although the literature is not unanimous in this.

Ultrastructural studies have demonstrated incomplete and retarded extrusion of lamellar bodies in the upper stratum granulosum in AD skin. Other studies have shown decreased levels of ceremides and 6-fatty acids, increased levels of phospholipids and certain phosphoglycerides, and changes in arachidonic acid and its derivatives in AD skin compared to normal skin. The ceramides, and especially ceramide 1, are present in lower quantities than normal, and because the ceramides are of vital importance to the intercellular lipid bilayers, which in turn play a major role in the barrier and water-holding functions of the stratum corneum, it has been postulated that lipid abnormalities may be an etiological factor in xerosis in AD.

0-8493-1443-7/05/$0.00+$1.50

I. INTRODUCTION

Patients with AD often have asthma, hay fever, abnormal levels of IgE antibodies, and dermatitis, and there is, in about 70% of cases, a family history of predisposition to this disease.[1] There are significant immunological abnormalities found in AD; however, the ultimate cause of AD is unknown, and the clinical manifestation of the condition will depend on various constitutional and precipitating factors.

Although commonly referred to as "eczema," AD is only one of the groups of eczematous dermatoses. It is a chronic condition usually with an early age of onset characterized by itching, redness, and scaling and, subsequently, by skin thickening with exaggeration of the skin markings — lichenification. Eczematous areas and lichenification are often localized to the flexures and the facial skin. The disorder fluctuates in intensity, being most troublesome in childhood and becoming milder, more localized, and easier to control by adolescence. Although remission is common, with about 40 to 60% of cases showing clearing,[2,3] dry skin usually persists to some degree. A common finding in AD patients is a generalized xerosis on areas other than those predisposed to develop overt eczematous change. This dry skin is usually uniform, nonerythematous, finely scaling "rough" skin, and in some cases a perifollicular accentuation is superimposed.[4] Hyperlinear palms are also reported as an accompanying characteristic.[5]

The term "dry skin" is both confused and confusing. It would be better if the term were dropped altogether, as it prejudices the reader into believing that the skin (or stratum corneum) does in fact have a reduced water content. At present we do not have data to confirm or deny this point for many of the conditions to which "dry skin" is applied. The term actually signifies that the skin surface looks as though it is lacking in water, and this perception is heightened by the pharmacological effect of putting water onto the skin. The improvement seen is taken as validation of an original lack of water. This is as logical as believing that fire breaks out in, say, an old barn because of "lack of water" and then to view water putting the fire out as further evidence of this deficiency in water.

Typical histological changes are those seen with other types of eczema, and they include spongiosis, epidermal thickening, and an inflammatory cell infiltrate.[6] Sufferers from AD often exhibit hyperresponsiveness to environmental irritants or allergens.

Finlay et al.[7] found that there was epidermal thickening in the xerotic areas of patients with AD. In addition there was some hypergranulosis and patchy parakeratosis — features indicative of mild eczematous change. This was later confirmed in a study that investigated the uninvolved skin of various types of eczematous reactions.[8]

There has been some debate about the precise definition of AD, although most investigators now use the definition proposed by Hanifin and Rajka.[9] However, much of the older literature does not adhere to such an agreed and common diagnostic description, and there is still no generally accepted screening test for the disorder. This deficiency results in difficulty in comparing results from different sources, and in particular, the prevalence figures quoted in the literature may vary according to the different criteria used for diagnosis. More recently, a U.K. working group has refined the Hanifin and Rajka criteria to a list of six parameters described as "the minimum list of reliable discriminators for AD." This list includes a "history of dry skin."[10]

Despite possible differences in definition, it can be stated with some confidence that AD is a very common disorder. Surveys have put childhood prevalence at around 2 to 3%.[11,12] However, more recent studies have demonstrated a wide variation in prevalence worldwide.[13,14] It would seem that the incidence has increased over the last several decades, with a recent estimate of a 10% incidence.[15]

II. XEROSIS IN AD

In a long-term follow-up study of almost 1000 patients with AD, it was found that "dry/itchy" skin was one of the least favorable long-term prognostic features.[16] In a 50-patient study, 82% of the participants thought that their skin was "dry," and on examination, 48% had clinically obvious patches of "dry" skin.[17] Other studies have reported frequencies of subjective or objective xerosis of between 63% and 98%.[4,18,19] These studies differed in age structure, geographical location, and precise clinical definitions of dry skin.

There are three important functional properties of stratum corneum that have been studied in relation to the appearance and perception of dry skin. These are the rate of TEWL, the total water content, or hydration, and the water-binding capacity of the stratum corneum proteins.

III. TEWL

TEWL is considered a good indication of the functional status of the epidermal (i.e., stratum corneum) diffusion barrier. A damaged barrier will allow a greater rate of permeation by water vapor for the same epidermal–ambient diffusion gradient, as well as increased quantities of xenobiotics that may have profound toxic and immunological effects. However, it must be remembered that TEWL is a surrogate measure of "barrier function" and will only be relevant to the barrier to hydrophilic or aqueous substances. Whether it is relevant when considering barrier to solvents, oils, or other substances will depend on the particular substance.

Different techniques have been developed to measure the rate of TEWL,[20,21] and one well-established noninvasive technique is readily available in several commercial devices, based on the evaporimeter.[22] In this device, humidity and temperature are measured at two vertically adjacent points just above the skin surface. The test site is delineated by an open-ended small cylinder. Thus, the overlying volume of air and the humidity and temperature detectors are protected from air currents by this "chimney," which is the "working end" of the probe. From the difference in water vapor content of the air between both points, a figure for the flux of water vapor leaving the delineated test area of skin can be calculated. Recently, different designs of TEWL measurement devices have become available, using closed chambers. These offer better isolation of the measurement from changes in ambient conditions.[23]

By carefully following the recommended measurement protocol,[24] good reproducibility is obtained.[25] Extremely close agreement is found between different investigators when using the same design of evaporimeter. For undamaged, healthy skin, typical values of TEWL of up to about 8 g/m^2 per hour are found.[26,27]

Increased TEWL has been found on all skin showing eczematous change. In addition, increased TEWL is seen in both normal appearing and overtly xerotic skin of patients with AD.[28–30] In a study on patients with persistently dry skin on nonlesional sites, mean TEWL values from buttock skin was significantly higher (8.6 g/m^2 per hour) in those patients than in an age-matched control group of nonatopic subjects (5.4 g/m^2 per hour).[31] In another study,[30] AD patients with xerosis were compared with control subjects with no sign or history of dry skin or atopy. TEWL was measured on the back of the hand, on the flexor forearm, and on the lower back. On the back-of-the-hand site, the AD group mean TEWL value was 17.5 (SD 7.2) g/m^2 per hour compared to 9.8 (SD 2.2) g/m^2 per hour for controls. On the flexor forearm site, the AD group mean TEWL value was 13.3 (SD 5.8), whereas the control group mean was 4.7 (SD 1.9) g/m^2 per hour. For the lower-back site, the values were 14.4 (SD 6.6) g/m^2 per hour and 6.6 (SD 3) g/m^2 per hour for the AD and control groups, respectively.

Even in AD patients without overtly dry skin on noneczematous or dermatitic sites, TEWL was higher than the control subject on two sites out of three: nonxerotic AD group mean TEWL for back-of-hand site was 14.6 (SD 3.9) g/m^2 per hour, compared with 9.8 (SD 2.2) g/m^2 per hour for controls; flexor forearm site was 7.6 (SD 2.8) g/m^2 per hour and 4.7 (SD 1.9) g/m^2 per hour for nonxerotic AD and control groups, respectively.

IV. HYDRATION OF STRATUM CORNEUM

The presence of water is an essential prerequisite for the maintenance of normal stratum corneum structure and function. A soft and flexible skin will have an intact and optimally hydrated stratum corneum, with an approximate water content of about 10 to 30%.[32,33] However, the literature is not of a unanimous opinion that clinically or subjectively "dry skin" has a decreased level of hydration. At least two studies have suggested an increased level of hydration in AD stratum corneum.[8,34]

Unlike the measurement of TEWL, there is no single universal and direct method for the *in vivo* determination of stratum corneum water content. There are many techniques and instruments available, but even the procedures of use for individual instruments are not fully standardized. Some direct methods have been developed, but they are far from easily adaptable or available for routine use. Nuclear magnetic resonance spectroscopy and nuclear magnetic imaging do provide a direct measurement of proton density and have been used in the assessment of stratum corneum hydration,[35–37] but for *in vivo* measurement of stratum corneum hydration, the present resolution has restricted the use of nuclear magnetic imaging to stratum corneum from the heel. Other methods described include microwave,[38] thermal,[39,40] and spectroscopic[41,42] techniques, but the most common methods use measurement of electrical conductance, capacitance, or impedance (which includes contributions from both conductance and capacitance) as an indirect indication of stratum corneum hydration. The basic method relies on the behavior of the stratum corneum as a dielectric material, with enough direct current resistance to act as the material between the plates of a capacitor. Probes have two electrodes of various designs that are placed on the skin. When an alternating current or voltage is applied to the electrodes, the load presented by the probe depends in some way on the properties of the skin. This arrangement can be used to measure the capacitance, conductance, or complex (frequency-dependent) impedance. The exact behavior of the skin–electrode arrangement is extremely complex, depending on the frequency of the applied stimulus and on the electrical properties of the skin (which depend on other occult interactions of, among others, stratum corneum proteins with water). Difficulties arise in the interpretation of published results because different instruments are not directly comparable. For example, the depth to which these instruments sample varies in a complex manner with electrode arrangement and with the frequency of the alternating current or voltage applied as the measuring stimulus. Because it is well known that the hydration of the stratum corneum varies with depth, it becomes clear that the exact sample depth of any particular instrument is critical when it comes to comparing the different studies.

In studies using a capacitance measuring probe, known as the Corneometer,[43] the water content parameter of the stratum corneum of clinically dry skin of AD patients was significantly lower than that of both apparently normal skin of other AD patients and of the skin of a group of control volunteers. This device is considered to sample the lower levels of the stratum corneum.[44] This result was supported by a study using a high-frequency impedance meter, the Skicon-200,[45] which is said to sample the uppermost layers of the stratum corneum.[46] Using the flexor surface of the forearm of a group of AD patients who had dry skin but no eczematous areas or ichthyosis, skin conductance at a frequency of 3.5 MHz was measured. Control subjects ($n = 18$) had a mean conductance of 35 (SE 2) μmho, whereas the AD xerotic skin group ($n = 28$) gave a mean conductance of 21 (SE 2) μmho. There are other studies that agree with this finding. However, a study using infrared spectroscopy[34] and a study using conductance[8] concluded that there may be an increase in hydration in AD noneczematous skin.

In vitro studies of stratum corneum water content have also been reported.[43] A gravimetric determination was made of the amount of water taken up by dehydrated stratum corneum samples from normal control subjects and from AD patients with and without evidence of dry skin. Samples from the control group had a water content of 0.69 mg/mg dry stratum corneum, those from nonxerotic skin of AD patients had a water content of 0.59 mg/mg dry stratum corneum, and dry skin from AD patients had a water content of 0.32 mg/mg dry stratum corneum.

V. WATER BINDING IN THE STRATUM CORNEUM

In the stratum corneum, water exists in two states: water firmly incorporated into the protein structure, termed "bound water," with the amount thus held determined by the hygroscopicity of the stratum corneum; any water present in excess of the amount bound is more freely present in other compartments of the stratum corneum and is termed "free water."[47]

By measuring the desorption of water from fully hydrated samples of stratum corneum *in vitro*, estimates of the total amount of bound and free water can be made.[48,49] The typical desorption curve has two phases: the initial steep slope representing desorption of free water, followed by a slower desorption of bound water. A study reports the use of a microbalance to measure the weight loss of fully hydrated stratum corneum over 40 minutes in a drying chamber.[50] The stratum corneum was subsequently fully dehydrated to determine its dry weight, with the desorption then calculated as a percentage of dry weight. This procedure was carried out on samples from 12 AD patients and from a group of control subjects.

At all time points, the stratum corneum from the AD patient group had a lower water content than that from the control group, and it was concluded that stratum corneum from dry skin of AD patients binds less water than that from controls. This conclusion holds for both bound and free water.

VI. CAUSES OF XEROSIS

Many believe that there is a true association between a form of ichthyosis and AD, but this view is not shared by everyone.[51] There is also a high coincidence of nonallergic contact dermatitis, especially of the hands, in patients with AD,[52,53] as well as the increases in TEWL and lowered hydration discussed above in Section V. These factors indicate an abnormality in the stratum corneum barrier function, but whether this is primary or secondary to an underlying eczema is not completely clear. Finlay et al.[7] suggested, on the basis of the histological changes and noninvasive tests, that the alterations in the xerotic skin represented a minor degree of eczema. However, the TEWL of facial skin in normal individuals is usually higher than that of atopic xerotic skin, indicating that high TEWL, per se, is not sufficient to induce a dermatitic reaction.[54]

Further, more direct evidence of stratum corneum abnormality can be found by careful study of stratum corneum structure. One study[8] reported that stratum corneum from uninvolved skin of AD patients had smaller corneocytes than normal controls, and that they had a significantly increased tendency to clump when shed by forced desquamation. They also found an increased epidermal thickness in AD skin compared with normal control skin and an increased ratio of the length of the basal cell layer to the length of the granular cell layer, which is a parameter related to the proliferative activity of the epidermis, and indicated a degree of hyperproliferation in AD. The findings of corneocyte size and clumping were confirmed in an independent study,[45] which also found an increased number of cell layers in the stratum corneum from dry skin of AD patients compared with normal control skin. The study also confirmed the hyperproliferation noted in the earlier study by recording a reduced stratum corneum turnover time of 7 (SD 2) days for AD skin compared with 14 (SD 2) for normal skin.

VII. EPIDERMAL LIPIDS

In the "brick wall" model of SC,[55] the corneocytes ("bricks") are surrounded by intercellular lipids ("mortar"). These lipids form multilayered structures within the stratum corneum and are of central importance to the water retention[56] and permeability barrier[57,58] functions of the stratum corneum. Because there seems to be a dysfunction of both barrier and water-holding capacity in stratum corneum from AD patients, the study of epidermal lipids in AD may be expected to show some abnormalities. This is indeed the case: Early reports of quantitative differences of lipids between normal skin and skin from AD patients[20,21] have been followed up to look in more detail at the precise nature of the defective lipid metabolism.

During keratinization, phospholipids in the lower epidermis are catabolized to synthesize more nonpolar lipids higher in the epidermis. This occurs in lamellar bodies (or membrane-coating granules), which are organelles seen in the stratum granulosum that consist of parallel lipid lamellae containing glucosphingolipids, free sterols, and hydrolytic enzymes.[59] In the upper stratum granulosum, they extrude their contents into the intercellular spaces, where the lamellar disks are reorganized into the broad lipid bilayers of the intercorneal spaces. Incomplete and retarded extrusion of these membrane-coating granules has been reported in AD skin from ultrastructural electron microscopic studies.[31]

A subgroup of sphingolipids, the ceramides, are major constituents of epidermal lipids, and decreased levels of ceramides have been found in stratum corneum from AD patients.[60] More explicitly, levels of the ceramide 1 fraction are thought to be the major influence on the reduction of total ceramides.[61] It has been suggested that the role of ceramide 1 is crucial in the barrier function of skin, in that it may act as a "molecular rivet" in stabilizing the lipid lamella sheets. Similar reductions of these lipids were noted in both lesional and nonlesional skin of AD patients.[60] In another study,[45] an almost twofold elevation of phospholipids levels was found in the lesional and nonlesional epidermis from AD patients, compared with normal control epidermis. Also reported were significantly higher levels of monosaturated fatty acids in the phosphoglycerides, lower levels of n-6 fatty acids, and changes in arachidonic acid and its derivatives in lesional skin when compared with nonlesional skin of AD patients. A measure of disease severity was found to inversely correlate with the n-6 fatty acid content of nonlesional AD epidermis. The authors concluded that AD epidermis has increased activity of phospholipase A_2 and an incomplete transformation of phospholipids into other lipid classes.

Observed reduction in ceramide content of the stratum corneum of AD is seen to disappear when the xerotic change is reversed.[62,63] Irritant challenge with a standard surfactant (1% sodium lauryl sulfate in petrolatum applied for 24 hours) on atopic skin induces more severe change that also lasts longer than on normal skin.[64]

It has been postulated that the evidence of abnormalities in stratum corneum, indicating abnormal keratinization, are associated with the abnormalities in the lamellar bodies that result in the deficient synthesis of certain lipids, especially ceramides. This lipid abnormality contributes to the observed reduction in barrier function, stratum corneum hydration, and consequent "dry" skin in AD patients.

REFERENCES

1. Champion, R.H. and Parrish, W.E., Atopic dermatitis, in *Textbook of Dermatology*, 5th edition, Champion, R.H., Burton, J.L., and Ebling, F.J.G., Eds., Blackwell Scientific Publications, London, 1992.
2. Roth, H.L. and Kierland, R.R., The natural history of atopic dermatitis, *Arch. Dermatol.*, 89, 209, 1964.
3. Williams, H. and Wuthrich, B., The natural history of atopic dermatitis, in *Atopic Dermatitis*, Williams, H.C., Ed., Cambridge University Press, Cambridge, U.K., 2000, pp. 41–59.
4. Uehara, M. and Harumitsu, M., The morphological characteristics of dry skin in atopic dermatitis, *Arch. Dermatol.*, 120, 1186, 1984.
5. Mevorah, B., Marazzi, A., and Frenk, E., The prevalence of accentuated palmoplantar marking in keratosis pilaris in atopic dermatitis, autosomal dominant ichthyosis and control dermatological patients, *Br. J. Dermatol.*, 112, 679, 1984.
6. Marks, R., The pathology and pathogenesis of the eczematous reaction, in *Eczema*, Marks, R., Ed., Martin Dunitz, London, 1992.
7. Finlay, A.Y., Nicholls, S., King, C.S., and Marks, R., The "dry" non-eczematous skin associated with atopic eczema, *Br. J. Dermatol.*, 102, 249, 1980.
8. Al-Jaberi, H. and Marks, R., Studies of the clinically uninvolved skin in patients with dermatitis, *Br. J. Dermatol.*, 111, 437, 1984.
9. Hanifin, J.M. and Rajka, G., Diagnostic features of atopic dermatitis, *Acta Derm. Venereol. (Suppl.)*, 92, 44, 1980.
10. Williams, H.C., Burney, P.G., Hay, R.J., Archer, C.B., Shipley, M.J., Hunter, J.J., Bingham, E.A., Finlay, A.Y., Pembroke, A.C., and Graham-Brown, R.A., The U.K. Working Party's Diagnostic Criteria for Atopic Dermatitis. I. Derivation of a minimum set of discriminators for atopic dermatitis, *Br. J. Dermatol.*, 131(3), 383, 1994.
11. Walker, R.B. and Warin, R.P., Incidence of eczema in early childhood, *Br. J. Dermatol.*, 68, 182, 1956.
12. Hanifin, J.M., Epidemiology of atopic dermatitis, *Monogr. Allergy*, 21, 116, 1987.
13. Schultz Larsen, F., Diepgen, T., and Svensson, A., The occurrence of atopic dermatitis in north Europe: an international questionnaire study, *J. Am. Acad. Dermatol.*, 34(5 Pt 1), 760, 1996.
14. Williams, H., Robertson, C., Stewart, A., Ait-Khaled, N., Anabwani, G., Anderson, R., Asher, I., Beasley, R., Bjorksten, B., Burr, M., Clayton, T., Crane, J., Ellwood, P., Keil, U., Lai, C., Mallol, J., Martinez, F., Mitchell, E., Montefort, S., Pearce, N., Shah, J., Sibbald, B., Strachan, D., von Mutius, E., and Weiland, S., Worldwide variations in the prevalence of symptoms of atopic eczema in the international study of asthma and allergies in childhood, *J. Allergy Clin. Immunol.*, 103(1), 125, 1999.
15. Schultz-Larsen, F., Holm, N.V., and Henningsen, K., Atopic dermatitis: a genetic-epidemiological study in a population-based twin sample, *J. Am. Acad. Dermatol.*, 15, 487, 1986.
16. Rystedt, I., Prognostic factors in atopic dermatitis, *Acta Derm. Venereol.*, 65, 206, 1985.
17. Werner Linde, Y., Dry skin in atopic dermatitis, *Acta Derm. Venereol. (Suppl.)* 177, 9, 1992.
18. Kang, K. and Runmei, T., Atopic dermatitis. An evaluation of clinical and laboratory findings, *Int. J. Dermatol.*, 26, 27, 1987.

19. Svensson, A., Edman, B., and Moller, H., A diagnostic tool for atopic dermatitis based on clinical criteria, *Acta Derm. Venereol. (Suppl)*, 114, 33, 1985.
20. Rajka, G., Transepidermal water loss on the hands in atopic dermatitis, *Arch. Dermatol. Forsch.*, 251, 111, 1974.
21. Abe, T., Ohkido, M., and Yamamoto, K., Studies on skin surface barrier functions — skin surface lipids and transepidermal water loss in atopic skin during childhood, *J. Dermatol. (Tokyo)*, 5, 223, 1978.
22. Nilson, G.E., Measurement of water exchange through the skin, *Med. Biol. Eng. Comp.*, 15, 209, 1977.
23. Tagami, H., Kobayashi, H., and Kikuchi, K., A portable device using a closed chamber system for measuring transepidermal water loss: comparison with the conventional method, *Skin Res. Technol.*, 8(1), 7, 2002.
24. Pinnagoda, J., Tupker, R.A., Agner, T., and Serup, J., Guidelines for transepidermal water loss (TEWL) measurement, *Contact Dermatitis*, 22, 164, 1990.
25. Blichmann, C.W. and Serup, J., Reproducibility and variability of transepidermal water loss measurement, *Acta Derm. Venereol.*, 67, 206, 1987.
26. Nilsson, G.E., On the Measurement of Evapourative Water Loss. Methods, and Clinical Applications, Medical dissertation, Linkoping University, Sweden, 1977.
27. Scott, R.C., Oliver, G.J.A., Dugard, P.H., and Singh, H.J., A comparison of techniques for the measurement of transepidermal water loss, *Arch. Derm. Res.*, 274, 57, 1982.
28. Tupker, R.A., Pinnagoda, J., Coenraads, P.J., and Nater, J.P., Susceptibility to irritants: role of barrier function, skin dryness and history of atopic dermatitis, *Br. J. Dermatol.*, 123(2), 199, 1990.
29. Thune, P., Evaluation of the hydration and the water-holding capacity in atopic skin and so-called dry skin, *Acta Derm. Venereol. (Suppl.)*, 144, 133, 1989.
30. Werner, Y. and Lindberg, M., Transepidermal water loss in dry and clinically normal skin in patients with atopic dermatitis, *Acta Derm. Venereol.*, 65, 102, 1985.
31. Fartasch, M. and Diepgen, T.L., The barrier function in atopic dry skin, *Acta Derm. Venereol. (Suppl.)*, 176, 26, 1992.
32. Blank, H.I., Factors which influence the water content of the stratum corneum, *J. Invest. Dermatol.*, 18, 433, 1952.
33. Idson, B., Water and the skin, *J. Soc. Cosmetic Chem.*, 24, 197, 1973.
34. Gloor, M., Heymann, B., and Stuhle, T.H., Infrared-spectroscopic determination of the water content of the horny layer in healthy subjects and in patients suffering from atopic dermatitis, *Arch. Dermatol. Res.*, 271, 429, 1981.
35. Foreman, M.I., A proton magnetic resonance study of water in human stratum corneum, *Biochem. Biophys. Acta*, 437, 599, 1976.
36. Richard, S., Querleux, B., Bittoun, J., Idy-Peretti, I., Jolivet, O., Cermakova, E., and Leveque, J., In vivo proton relaxation times analysis of the skin layers by magnetic resonance imaging, *J. Invest. Dermatol.*, 97, 120, 1991.
37. Querleux, B., Richard, S., Bittoun, J., Jolivet, O., Idy-Peretti, I., Bazin, R., and Leveque, J., *In vivo* analysis of water and lipid content in human skin by high resolution magnetic resonance imaging (Abstract), *Skin Pharmacol.*, 5, 225, 1992.
38. Jacques, S.L., A linear measurement of the water content of human skin using a microwave probe, in *IEEE-EMBS*, Denver, Colorado, 1979, pp. 1–3.
39. Xiao, P. and Imhof, R.E., Optothermal measurement of water distribution within the stratum corneum, *Curr. Probl. Dermatol.*, 26, 48, 1998.

40. Dittmar, A., Martini, M.C., Planchette, L., and Lotte, J., *In vivo* and *in vitro* measurements of the skin water content by thermal conductivity, in *4th International Symposium on Bioengineering and the Skin*, Besancon, France, 1983.
41. Potts, R.O., Guzek, D.B., Harris, R.R., and McKie, J.E., A non-invasive *in vivo* technique to quantitatively measure water concentration of the stratum corneum using attenuated total reflectance infrared spectroscopy, *Arch. Derm. Res.*, 277(6), 489, 1985.
42. Petersen, E.N., The hydrating effect of a cream and white petrolatum measured by optothermal infrared spectrometry *in vivo*, *Acta Derm. Venereol.*, 71, 373, 1991.
43. Werner, Y., The water content of the stratum corneum in patients with atopic dermatitis, *Acta Derm. Venereol.*, 66, 281, 1986.
44. Triebskorn, A., Gloor, M., and Greiner, F., Comparative investigations on the water content of the stratum corneum using different methods of measurement, *Dermatologica*, 167, 64, 1983.
45. Watanabe, M., Tagami, H., Horii, I., Takahashi, M., and Kligman, A.M., Functional analyses of the superficial stratum corneum in atopic xerosis, *Arch. Dermatol.*, 127(11), 1689, 1991.
46. Obata, M. and Tagami, H., Electrical determination of water content and concentration profile in a simulation model of *in vivo* stratum corneum, *J. Invest. Dermatol.*, 92, 854, 1989.
47. Edwards, C., The electrical properties of skin, in *The Physical Nature of Skin*, Marks, R., Barton, S.P., and Edwards, C., Eds., MTP Press Ltd, Lancaster, England, 1988.
48. Scheuplein, R.J. and Morgan, L.J., "Bound water" in keratin membranes measured by a microbalance technique, *Nature*, 214, 456, 1967.
49. Tagami, H., Kanamaru, Y., Inoue, K., Suehisa, S., Inoue, F., Iwatsuki, K., Yoshikuni, K., and Yamada, M., Water sorption-desorption test of the skin *in vivo* for functional assessment of the stratum corneum, *J. Invest. Dermatol.*, 78, 425, 1982.
50. Werner, Y., Lindberg, M., and B., F., The water-binding capacity of stratum corneum in dry non-eczematous skin of atopic eczema, *Acta Derm. Venereol.*, 62, 334, 1982.
51. Fartasch, M., Diepgen, T.L., and Hornstein, O.P., Atopic dermatitis — Ichthyosis vulgaris — Hyperlinear palms — an ultrastructural study, *Dermatologica*, 178, 202, 1989.
52. Rystedt, I., Atopic background in patients with occupational hand eczema, *Contact Dermatitis*, 12, 1, 1985.
53. Nilsson, E., Mikaelsson, B., and Andersson, S., Atopy, occupation and domestic work as risk factors for hand eczema in hospital workers, *Contact Dermatitis*, 13, 216, 1985.
54. Zhen, Y.X., Suetake, T., and Tagami, H., Number of cell layers of the stratum corneum in normal skin. Relationship to the anatomical location of the body, age, sex and physical parameters, *Arch. Derm. Res.*, 291, 555, 1999.
55. Elias, P.M., Epidermal lipids, barrier function and desquamation, *J. Invest. Dermatol.*, 80(Suppl 6), 44, 1983.
56. Imokawa, G. and Hattori, M., A possible function of structural lipids in the water-holding properties of the stratum corneum, *J. Invest. Dermatol.*, 83, 282, 1985.
57. Elias, P.M. and Friend, D.S., The permeability barrier in mammalian epidermis, *J. Cell Biol.*, 65, 180, 1975.

58. Lampe, M.A., Burlingame, A.L., Whitney, J., Williams, M.I., Browne, B.E., Roitman, E., and Elias, P.M., Human stratum corneum lipids: characterisation and regional variations, *J. Lipid Res.*, 24, 120, 1983.
59. Elias, P.M., Menon, G.K., Grayson, S., and Browne, B.E., Membrane structural alterations in murine stratum corneum: relationship to the localisation of polar lipids and phospholipases, *J. Invest. Dermatol.*, 91, 3, 1988.
60. Imokawa, G., Abe, A., Jin, K., Higaki, Y., Kawashima, M., and Hidano, A., Decreased level of ceramides in stratum corneum of atopic dermatitis: an etiological factor in atopic dry skin? *J. Invest. Dermatol.*, 96, 523, 1991.
61. Yamamoto, A., Serizawa, S., and Sato, Y., Stratum corneum lipid abnormalities in atopic dermatitis, *Arch. Derm. Res.*, 283, 219, 1991.
62. Tanaka, M., Zhen, Y.X., and Tagami, H., Normal recovery of the stratum corneum barrier function following damage induced by tape stripping in patients with atopic dermatitis, *Br. J. Dermatol.*, 136(6), 966, 1997.
63. Matsumoto, M., Umemoto, N., Sugiura, H., and Uehara, M., Difference in ceramide composition between "dry" and "normal" skin in patients with atopic dermatitis, *Acta Derm. Venereol.*, 79(3), 246, 1999.
64. Tabata, N., Tagami, H., and Kligman, A.M., A twenty-four-hour occlusive exposure to 1% sodium lauryl sulfate induces a unique histopathologic inflammatory response in the xerotic skin of atopic dermatitis patients, *Acta Derm. Venereol.*, 78(4), 244, 1998.

27 Within and beyond the Skin Barrier as Seen by Electrical Impedance

Stig Ollmar and Ingrid Nicander

CONTENTS

Abstract 335
I. Background 336
II. What Is Measured? 337
III. Impedance Spectroscopy of Skin 340
IV. What Else About Barriers? 346
V. What About Differential Diagnostics? 347
References 348

ABSTRACT

Hydration is a major factor for maintaining integrity of the skin barrier. The concept of skin hydration is not well defined, however, and several factors such as cellular structure and lipid content are essential for the functionality of the skin barrier. The detection principles of different devices sense the contributing factors in different proportions, and this explains why investigators sometimes obtain contradictive results, as different "definitions" of skin hydration are tacitly used. In addition, devices based on dielectric or impedance measurements are also influenced by other factors, such as blood glucose. Part of the difficulty of defining skin hydration is to delimit the thickness of the tissue included in the measurement, which is highly dependent on frequency and electrode shape and size. Awareness of the multitude of factors involved in the response of skin hydration meters should facilitate better interpretation of the results in various contexts, and some investigators might find it interesting to use multifrequency techniques to extract information about some nonhydration parameters. It has been shown that several skin reactions and diseases can be differentiated using the information in electrical impedance spectra and the alterations residing in the skin barrier, in deeper skin layers, or in both.

0-8493-1443-7/05/$0.00+$1.50

I. BACKGROUND

Theoretically, the most direct way to assess skin hydration would be to use nuclear magnetic resonance because it would be very specific. Unfortunately, the resolution perpendicular to the skin surface of the best nuclear magnetic resonance techniques is only around 50 μm, so at best only the gross layered structure of the skin will be accessible.

The optical part of the electromagnetic spectrum contains some specific resonances to water in the near-infrared region that might be useful, but the results are modulated by the local topology of the skin. This drawback can only to some degree be compensated for by integrating over a sufficiently large area.

Going down in frequency from the optical realm of the electromagnetic spectrum, one enters into the field of dielectric spectroscopy and impedance methods, which is the scope of this chapter. Gildemeister [1] was among the first to investigate skin properties using electrical techniques. Rosendal [2] showed that hydration of the stratum corneum was a dominating factor for its conductive properties. In the meantime the Cole brothers had launched a model for the dielectric behavior of systems without DC conductivity, and K.S. Cole also created the so-called Cole equation to describe the behavior of a system in impedance space. For a comprehensive review of both the history and theory, the reader is referred to the textbook by Grimnes and Martinsen [3].

The Cole models include a so-called constant phase element, which is actually unphysical but may be justified for the sake of simplicity and efficient data reduction. The latter properties made this type of modeling, including the corresponding equivalent circuit diagrams, very popular, and it has been used without much discrimination for systems in which this type of modeling may be adequate, as well as for systems in which it certainly is not.

The Cole equation

$$Z = R_\infty + \frac{R_0 - R_\infty}{1 + (j\omega\tau)^\alpha}$$

where the complex number Z is a function of frequency f [Hz] or angular frequency ω [rad/s], denotes the tissue impedance, R_0 the resistance at zero frequency, R_∞ the resistance at infinite (extremely high) frequency, and τ a characteristic time constant (the inverse of the so-called characteristic angular frequency ω_c). For $\alpha = 1$, the equation has a one-to-one correspondence with a simple equivalent circuit made up of two pure resistors and one pure capacitor, and in some applications these components may be interpreted as, for example, extracellular conductive pathway, intracellular conductive pathway, and capacitance of the cell membrane in a limited frequency range (i.e., for one so-called dispersion). However, even in a fairly homogeneous tissue, like blood, α will be somewhat less than 1, and then there is no completely defined correspondence to pure physical elements, as the needed constant phase element is a composite approximation of a number of known or unknown properties modulating the behavior. Interpretation of the Cole-α is subject to some controversy (see pp. 223–224 in [3]), but in many applications the deviation from

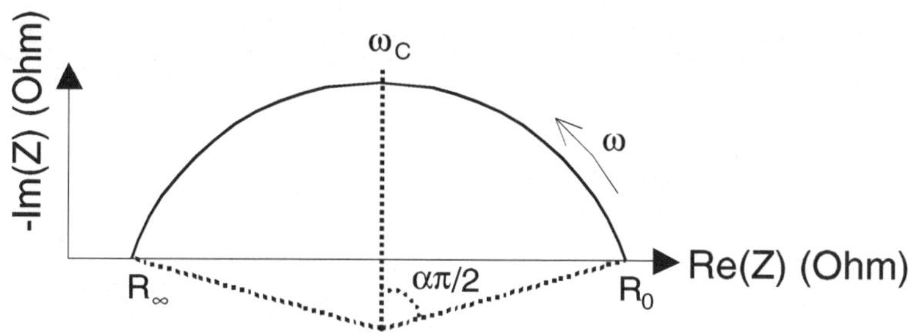

FIGURE 27.1 Nyquist plot of the Cole equation.

1 may be regarded as an index of heterogeneity. The Cole equation can be visualized in the complex number space in a so-called Nyquist plot (called Wessel diagram in [3], though some authors prefer the term Argand diagram), which is the imaginary part vs. real part. For $\alpha = 1$, this yields a semicircle with its center on the real axis; for $\alpha < 1$, the semicircle will be shifted vertically so that its center is no longer on the real axis (Figure 27.1).

There are a number of frequency-dependent conduction mechanisms at play in living systems, and this was not sufficiently sorted out until 1957, when Schwan [4] identified the three major dispersions of living tissues, the so-called α (not the Cole-α), and dispersions. Thus, it is now clear that the properties of living tissues give rise to multiple dispersions with different biological origin (Figure 27.2), although this may be overlooked if the range of investigated frequencies is too narrow. Because of heterogeneities of certain tissue types, each major dispersion also has a number of subdispersions, which may be overlooked if the density of frequency points is too low.

It has been shown by Yamamoto and Yamamoto [5] and by Martinsen et al. [8] that the frequency response of the stratum corneum cannot be modeled by the Cole equation. Still, in a chapter in the textbook *Skin Bioengineering: Techniques and Applications in Dermatology and Cosmetology*, Salter argues that it can [6]. This chapter is an honest attempt to shed some light on some electrical phenomena related to skin hydration, but is limited to a narrow frequency range and, as pointed out by Martinsen and Grimnes [7], Yamamoto is misquoted. Fitting the Cole equation to data in a limited frequency range may yield impressive χ^2 values, and because the dispersions are not comparative in magnitude, impressive χ^2 values may occasionally be obtained even in a wider frequency range while effectively filtering away relevant information. A beautiful model can be highly seductive, and some care should be taken not to apply any model to applications for which it was never intended.

II. WHAT IS MEASURED?

From a strict physical point of view, measurement of impedance will yield two numbers for each frequency, the magnitude and phase or, mathematically equivalent, the real and the imaginary parts of the impedance. If the dielectric properties of each volume element were known, it would be straightforward to compute the frequency

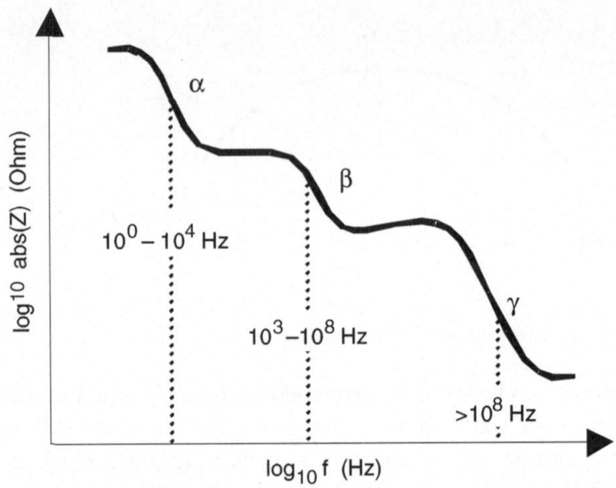

FIGURE 27.2 The major dispersions of living tissue. It is the slopes, not the plateaus, that are the dispersions. In fairly homogeneous tissues, like blood, each dispersion is clearly discernible. In very heterogeneous structures, like skin, the dispersions (in particular the α and dispersions) are overlapping in various degrees. In addition, each dispersion contains subdispersions (i.e., α_1, α_2 etc.). Note that if only a few widely separated points are taken, brute force curve-fitting to a straight line may look nice, but will effectively filter away most of the information inherent in the complete spectrum. In fact, fitting only the magnitude is totally inadequate because each impedance data point has two dimensions: the magnitude and the phase or, equivalently, the real and imaginary parts, for each frequency (cf. Figure 27.1).

response for a given electrode system, the forward problem. Given a frequency response, it is practically impossible to compute the dielectric properties of small-volume elements in the tissue under test, in particular for skin because of its high degree of heterogeneity and nonisotropy. A number of responses from different layers will be superimposed. Still, by investigating the major physical constituents of skin (i.e., a poorly conducting outer layer on top of a well-conducting deeper layer, using known dielectric properties of both), the general behavior in the frequency domain can be illuminated. It is straightforward to compute the frequency dependence of contributions from various depths of a simple two-layer model. Using a realistic size and shape of the electrode system, Martinsen et al. [8] showed that at very low frequencies around 10 Hz the stratum corneum would heavily dominate the sensed information, whereas at frequencies above around 1 MHz, most of the sensed information would originate from the viable skin. The transition between dominating layers is not linear and would be further modulated by details in addition to the rough first approximation. Variation of electrode size and shape would shift the transition along the frequency axis, whereas the fundamental principle of depth dependence on frequency is maintained.

Strictly physical, conductivity is mobility times concentration of charge carriers, where contributions from all charge carriers, mainly ions, have to be summed. This would be the dominating mechanism for aqueous solutions at low frequencies. However, any imposed structure to the solution by barriers of any sort will introduce polarization phenomena (i.e., charge will build up, discharge, and recharge with opposite polarity going through the cycles of alternating current). At higher frequencies, polar molecules will interact with the field, and this may be fairly specific, in particular at radio and microwave frequencies. The early high-frequency work was made by radar pioneers, and the reader is referred to the textbook by Hasted [9] for a comprehensive overview. Dielectric spectroscopy has been successfully used to analyze a number of biologically relevant molecules *in vitro* [10]. Attempts to do the same *in vivo*, and even noninvasively, typically fail because of the complexity of living systems, but the pharmaceutical applications imply that one has to be aware of the motley collection of factors that may influence the result.

Electrode size and shape, and in particular the distance between electrodes, is another factor to consider. In a homogeneous system, the effective depth of penetration is about half the distance of the electrodes, as a rule of thumb. In a layered system, this "truth" will be modulated (see pp. 134–135 in [3] for details). The size factor implies that the dense interdigitated electrode pattern of the Corneometer should measure more superficially than, for example, the Novameter or Skicon or DermaLab, which use coaxial patterns with larger dimensions. There are also other important differences because all use different frequencies and detect various aspects of the conductive or dielectric aspects at the used frequencies. Fluhr et al. [11,12] and Fischer el al. [13] have discussed features of various instruments. One unsolved problem is the depth penetration calibration, which will be context dependent. It will vary with the actual thickness and dielectric properties of the skin layers, in addition to the frequency of measurements and the geometrical shape and dimensions of the electrode system, and instruments based on conductive properties cannot be directly compared with instruments based on capacitive properties. The problem is not new: Impedance was used already in the 1930s to detect variations in geophysical composition [14,15].

One cannot estimate depth penetration by inserting a stack of thin isolating plastic foils, as some devices require a conductive path for proper operation, and in any case such a nonbiological insertion will distort the field from its shape in natural skin. To get a fair estimate, simulation of a simple structure using finite element methods is recommended — a two-layer structure as used by Martinsen et al. [8] should not be controversial. However, validating depth penetration in real life is difficult, and we suggest the following procedure: The measurement probe is placed on a stack of material with properties similar to stratum corneum. It could be layers of human stratum corneum trypsinated from living epidermis, shed snake skin, or maybe layers of thin paper (e.g., lens cleaning paper). The stack is placed alternately on a highly isolating material (glass, plastic) and on a highly conducting layer (metal). The stack is inundated with saline solution to be within range for instrument readings on normal stratum corneum. When the thickness of the stack is such that no difference is seen on the two different background materials, the limit depth penetration is achieved. In practice this could be defined as the thickness at which the difference in readings is within a certain distance (a percentage, e.g., 10%) of the asymptotic limit value, which is exponentially approached.

Any measurement takes time. When the electrode system (i.e., the tip of the handheld probe) is applied to the skin surface, the skin is occluded and moisture is building up in the test area as a result of continuous transepidermal water loss until equilibrium is achieved. It is the preocclusion value that is supposed to be related to the actual skin hydration. An early model of the Corneometer (type 420) delivered values repeatedly, and it became the "gold standard" to read the second displayed value in flight. The same problem befalls all devices in this category, and is usually handled by automatic triggering by a spring-loaded piston carrying the electrode system, with the spring also ensuring uniform pressure. The gradient of the moisture build-up is rapid in the beginning, so the choice of trigger point is crucial. Slow application and pressing of the piston will modulate any precise automatic triggering mechanism, and the value of the reading will further be modulated by the transepidermal water loss value, which is one determinant of the initial slope. Temperature will modulate transepidermal water loss, as well as mobility of ions.

It was mentioned above that a number of biochemical substances may influence the result. Most of these may be of little importance or may be considered slowly varying. To make things worse, however, it has been found that blood glucose level is correlated to "skin hydration" values (arbitrary units, essentially related to capacitance) [16]. Does this mean that glucose or some other metabolic agent shows up in the measurement, or is "skin hydration" a secondary effect of metabolism? In any case, because the correlation is conspicuous and blood glucose is a rapidly changing parameter, skin researchers have to be aware of this phenomenon and to also consider diabetes and other metabolic disorders, or at least time after meal (for healthy volunteers), a relevant factor when speaking about skin hydration.

It seems difficult, then, to reach consensus on a definition of skin hydration. Water will have different roles in different strata, and various aspects will be picked up by different measurement principles. The water molecule may be directly addressed by its dielectric properties at high frequencies, whereas water molecules in bulk will constitute a playing ground for charged entities, granting mobility of ions at lower frequencies. The low-frequency approach taken by Martinsen and Grimnes [7], measuring susceptibility at only one frequency, seems physically correct for very superficial hydration. Single-frequency devices may yield consistent data obviously correlated to some sort of hydration as long as all other contributing factors, including environmental factors, are held reasonably constant — which is not easy. In our view, skin hydration should not be separated from the properties of the skin barrier, which are multifactorial and thus require a multifrequency approach if assessed by electrical methods.

III. IMPEDANCE SPECTROSCOPY OF SKIN

Electrical impedance spectroscopy may assign associations to characteristics of optical spectroscopy, with very specific narrow spectral lines, but electrical impedance spectroscopy typically yields smooth spectra, partially resembling semicircles if plotted in a so-called Nyquist diagram (imaginary part of the impedance vs. the real part).

The many difficulties listed above are not automatically resolved by taking a multifrequency approach. It is not possible to design one electrode system suitable for frequencies from close to zero to up into the microwave region. At very high frequencies, the applicator will bear the attributes of an antenna, and the coupling of energy will be as wave propagation. At low to medium-high frequencies, an electromagnetic field is penetrating the tissue under test; the shape of the field depending on the size, shape, and placement of the electrodes; the applied frequency; and the distribution of electrical properties in the tissue. The electrode/tissue interface has unique properties in itself, which will be superimposed on the properties of the tissue under study. This artefact is easily removed when studying larger sections of the human body (e.g., a leg or the thorax), using a four-electrode array separating current injection and voltage detection, but because of the small dimensions of skin layers and skin surface topology, this procedure is not practical on skin.

Some of the artefacts of the electrode/tissue interface may be removed by using a wet electrode system. Great care has to be taken in choosing the wet medium, as most electrode gels contain long polymers that will interact with the applied field [17]. Gels intended for electrocardiogram electrodes typically contain polyethylene glycol, which will interfere with the measurement from around 10 kHz and may dominate the response above 100 kHz. This is the reason why our group always has used physiological saline solution. Unfortunately, our device has repeatedly been mentioned in the same context as devices designed to detect skin hydration. This is a suitable occasion to once again clarify that we never have and never will attempt to assess skin hydration with our multifrequency spectrometer. We assess multifactorial properties of the skin barrier and even deeper layers of the skin, and we understand that skin hydration in one form or another will be one of the factors involved.

Another advantage of the wet-electrode approach is that the occlusion moist gradient, affecting all devices using dry electrodes, is totally eliminated. Furthermore, artefacts resulting from sweat and environmental humidity are eliminated. The extremely high impedance of the naturally dry outermost part of the stratum corneum is dramatically reduced by saline inundation. This means that a good electrical contact with the more interesting skin barrier in the naturally wet, deeper part of the stratum corneum, as well as the deeper living strata of the skin, is achieved. The impedance of the intact skin barrier is high compared with living strata, and the impedance method is a very sensitive method for detecting any alteration of the skin barrier, as can be observed by chemically breaking down the barrier [20–26,29–31,35–38,44,46,47] or by tape stripping (Figure 27.3). The magnitude of the impedance will decrease with increasing damage, corresponding to less obstacles for the charge carriers, in particular at low frequencies and there will be characteristic changes in the shape of impedance spectra as well as of the indices described below.

It is interesting to observe how the Nyquist plots gradually display parts resembling sections of Cole semicircles while the stratum corneum is removed. This is expected, as the deeper living strata are constituted by a relatively homogeneous array of living cells. It also confirms the results by Yamamoto and Yamamoto [5] and Martinsen et al. [18], that intact stratum corneum does not obey the Cole equation.

Since the beginning of the 1990s, our group has made a number of studies of skin and oral mucosa using electrical impedance spectroscopy, starting with simple manual laboratory equipment and unsophisticated electrode systems [19–22]. In

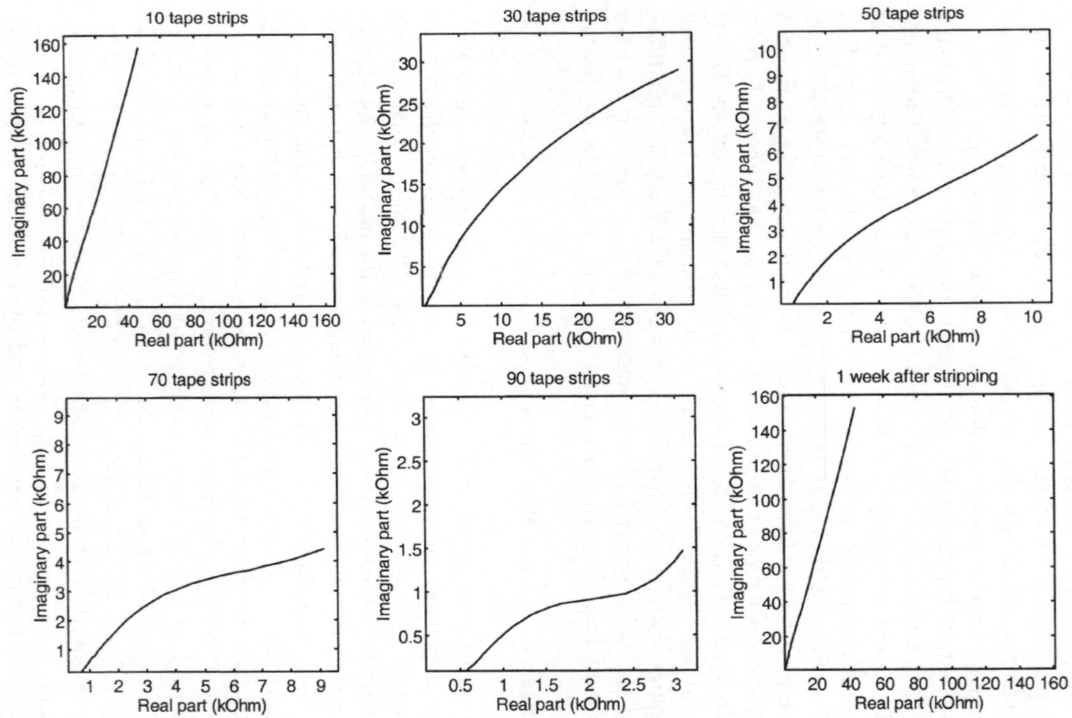

FIGURE 27.3 Nyquist plots of electrical skin impedance of a healthy volunteer in the frequency range 1 to 1000 kHz. The stratum corneum was gradually removed using common adhesive tape. Before any stripping, the shape of the graph was identical to the one after 10 strips, but magnitude was almost double. One week after this rather extensive stripping the skin barrier had recovered to the level observed after 10 strips, and after 2 weeks it was back at the level observed before any stripping at all. Note that the skin was soaked with saline solution before each measurement to obtain good contact between the skin and the electrode system. The plots thus show the effect of stripping on the function of the skin barrier and not the poor contact with the naturally dry outer layers of the stratum corneum.

most studies, sodium lauryl sulfate was used to induce experimentally controlled damage to the barrier. At this early stage, the manual operation only allowed readings at a few frequencies. Fitting data to the Cole equation did not work in the frequency range in which we operated, and data analysis was used by means of a simple index to achieve some sort of data reduction, including a normalization to reduce some of the variation between sites and individuals. Once a fully automatic impedance spectrometer was developed, full spectra of 31 frequencies logarithmically distributed in the range from 1 kHz to 1 MHz were recorded, and a mechanism for depth penetration variation was added [23,24].

Although the simple index would present an objective number related to the degree of alteration of the skin barrier or alterations in deeper layers, there was obviously more information inherent in the spectra. Because of the character of the data, we introduced four indices based on complete impedance data at two frequencies [25]. The formulation of the four indices entails normalization, which was achieved at the cost of one degree of freedom (because the indices are not entirely independent, absolute magnitude is lost in the normalization) present in the original magnitude and phase values at two frequencies. However, the set of indices contains two pairs that provide a simple description of the general behavior of the information in the spectra as long as the Nyquist plot is essentially a straight line between the two used frequencies — which is true as long as the skin barrier is intact or almost intact. If this condition is not met, the indices are not powerful enough for adequate data reduction, and piecemeal indices, or preferably multivariate techniques, using all of the information in the spectra should be used:

Magnitude index $$\mathrm{MIX} = \mathrm{abs}\left(Z_{20\mathrm{kHz}}\right)/\mathrm{abs}\left(Z_{500\mathrm{kHz}}\right);$$

Phase index $$\mathrm{PIX} = \arg\left(Z_{20\mathrm{kHz}}\right) - \arg\left(Z_{500\mathrm{kHz}}\right);$$

Real part index $$\mathrm{RIX} = \mathrm{Re}\left(Z_{20\mathrm{kHz}}\right)/\mathrm{abs}\left(Z_{500\mathrm{kHz}}\right);$$

Imaginary part index $$\mathrm{IMIX} = \mathrm{Im}\left(Z_{20\mathrm{kHz}}\right)/\mathrm{abs}\left(Z_{500\mathrm{kHz}}\right),$$

where abs(Z_i) is the magnitude (modulus) of the complex electrical impedance at the frequency *i*, arg(Z_i) is the argument (phase angle) in degrees, Re(Z_i) is the real part of the complex electrical impedance, and Im(Z_i) the imaginary part of the complex electrical impedance.

Armed with this set of indices, we investigated whether experimentally induced skin reactions elicited by different agents might differ in impedance responses. Using the anionic surfactant sodium lauryl sulfate, the almost neutral nonanoic acid, and the cationic benzalkonium chloride in concentrations producing essentially the same degree of visual response, it was found that the tissue alterations as seen by histology were clearly different, and that the impedance indices displayed specific patterns for each of these irritants [26]. The three irritants used in this study have different

TABLE 27.1
Index Patterns for Induced Skin Reactions

	Magnitude Index	Phase Index	Real Part Index	Imaginary Part Index
Sodium lauryl sulfate 0.5%	↓	↑		↓
Nonanoic acid 40%	↓			↓
Benzalkonium chloride 0.5%		↓	↑	
Nickel sulfate	↑		↓	↑

Note: Each arrow indicates a significant deviation from baseline in the direction of the arrow. Dose for the irritants was chosen to elicit similar visual appearance.

polarity, and therefore penetrate and attack the skin layers in different ways. The spatial resolution of light microscopy of biopsies or impedance spectroscopy *in vivo* is not sufficient to determine all details of the damage, but clearly the different types of alterations had a correspondence to specific alterations of the impedance spectra. Soon it was found that allergic reactions, at least in the early state, seem to be distinguishable from irritant reactions using the impedance technique and indices [27–30] (Table 27.1). The main reason is probably the difference in involvement of the skin barrier, which comes later after immunological reactions.

Obviously, the indices do not extract all the information inherent in the recorded impedance spectra. In an attempt to estimate the power of the indices as well as the differential diagnostic potential of the method, data from a few early studies were pooled and divided into a training set and a test set for a neural network experiment [31]. Data were introduced as complete spectra and as reduced data sets of indices, and the sensitivity and specificity were computed. There was a difference, but not a dramatic one, probably because the included lesions still retained a fairly functional skin barrier, so that the spectra in the Nyquist plot still had reasonable semblance of a straight line. Some of the included data were recorded by an early prototype to the spectrometer, and the varying data quality of the recordings in this study may also have obscured the fine structure of the spectra.

It is a known fact that diabetics too often develop severe ulcers that are difficult or impossible to heal. Would it be possible to get an early warning of such ulceration, or possibly screen out diabetics even before the diagnosis has been established? Using the full spectra and multivariate analysis, it was possible to separate diabetics from nondiabetics, but only after the age and sex variation had been sorted out [32]. It is known that age and sex do modulate the properties of skin in various aspects, and from the impedance point of view, this has been documented by Nicander et al. [33], using indices. One can argue about the pros and cons of different methods of extracting information from impedance spectra [34]: Cole or other lumped parameter equivalent circuit modeling, various types of indices, artificial neural networks, multivariate statistical methods, and so forth. This is a context depending common sense issue, and the principle of Occam's Razor should be applied, or as Einstein put it: Everything should be made as simple as possible, but not simpler.

Going into the essential chemistry of the skin barrier, studying the effect of removing lipids, the indices were still sufficient to observe alterations in the electrical properties [35–37]. Would we have seen more using the full spectra and more sophisticated mathematics? Yes, of course, but probably not very much because the lipid-deprived sites still retained a somewhat functional skin barrier, so that the spectra in the Nyquist plot still had reasonable semblance of a straight line. Full spectra were, however, recorded, so a more advanced analysis of the raw data can be made later if indicated. To separate diabetics from nondiabetics, as mentioned above [32], it was necessary to use the full spectra and all depth settings (i.e., a data set of 310 elements: 31 frequencies × 2 impedance parameters (magnitude and phase) ×5 depths).

The sensitivity of the impedance method in general is very high, and alterations far below the visual threshold are easily picked up using indices [38]. However, in studies involving very minute differences, the power of discrimination increases if more information in the spectra is used. It was recently demonstrated by Åberg et al. [39] that systematic variations within the often-used test area of the volar forearm exist. In this case, a new impedance spectrometer (Figures 27.4, 27.5) (SciBase II, SciBase AB, Huddinge, Sweden) was used for data collection, providing higher-data quality than earlier instruments. In particular, the new instrument is virtually immune to electromagnetic interference from the environment. The improved data quality makes it worthwhile to look for even finer details in the spectra, using more

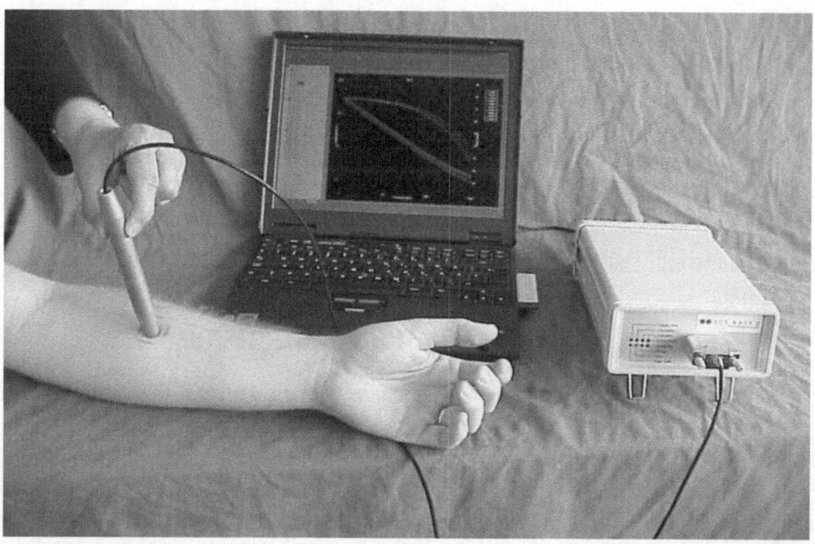

FIGURE 27.4 The electrical impedance spectrometer SciBase II equipped with a skin probe. The spectrometer is battery operated and is under computer control. The connection between the computer and the spectrometer is via IR-link, to ensure that the test person is galvanically isolated from the computer.

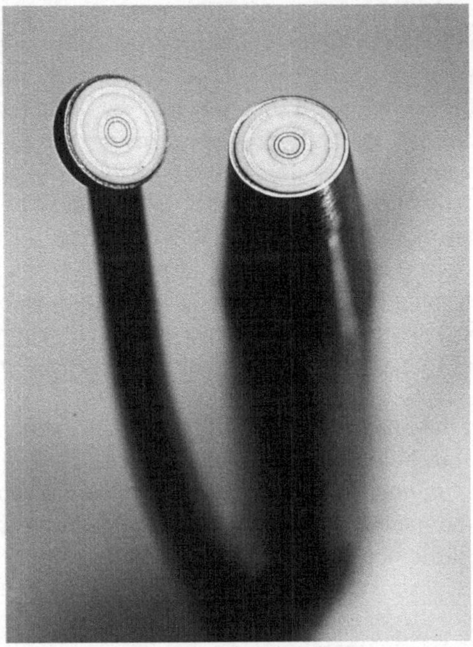

FIGURE 27.5 Handheld probes for the SciBase impedance spectrometer. The left one is intended for measurement in the oral cavity (and possibly vagina); the right one is for measurement on skin. The multiple electrode system is used to achieve selectable depth penetration.

sophisticated mathematics. The result of this study strongly emphasizes the need for randomized site locations — even within a relatively small and supposedly homogenous section of the volar forearm. Otherwise, a "significant" difference between, for example, two cosmetic preparations, might be entirely caused by a systematic error in the study design. In studies involving small alterations, it is equally important to conclude the experiments within a well-defined time period because of seasonal variations, in particular avoiding the summer season [40].

IV. WHAT ELSE ABOUT BARRIERS?

Skin is not the only bodily surface with a barrier. For example, the oral mucosa has many structural and functional similarities with skin, although residing in a wet environment [41]. Different locations vary in baseline characteristics in the same style as for skin, although different locations differ more within the oral cavity relative to each other than between various types of skin because of the dramatic differences in keratinizations in the oral mucosa [33,42]. Chemical attacks on the mucosal barrier produce similar reactions, although the magnitude of the impedance is typically lower in mucosa than in skin, and the time scale for mucosal reactions and recovery is faster than for skin [43–47]. Interestingly, some systemic diseases (e.g., Sjögren's syndrome) manifest symptoms in the oral cavity before symptoms are manifest in the skin or other organs. The oral probe used with our impedance spectrometer is also appropriate for vaginal measurements, but so far we have not conducted any regular study of this region. Still, it is worth mentioning that Bartos [48] and Bartos and Sedlacek [49] reported in 1977 that the impedance of the vaginal mucous membrane (i.e., the barrier function) changes dramatically during the menstrual cycle.

V. WHAT ABOUT DIFFERENTIAL DIAGNOSTICS?

By now, the reader should be convinced that there is more to the skin barrier than skin hydration. Skin diseases will produce alterations in the skin barrier, in living strata, or both. Barrier changes could be related to altered state of hydration, altered structure, altered chemical composition, or all of these modifications. Changes in living strata could be related to altered compartmentation between extracellular and intracellular water, altered structure, altered chemical composition, or all of these. All these factors can be picked up by impedance spectroscopy, but it may be difficult to distinguish too many of them at the same time because of the complexity of biological systems. However, starting with the encouraging result by Nicander et al. [26], further investigations indicate that whenever a histological difference can be seen, there will also be a measurable difference in the impedance spectra. Thus, using the concept of "electronic biopsy," meaning that diagnostic information obtainable from microscopic inspection of biopsies could be obtained noninvasively by analysis of impedance spectra [50], may be justified. It seems fairly simple to detect that something is not "normal" because the sensitivity of the method is high for almost any alteration, while specificity (correct classification) has to be validated in each application, and the latter may be more difficult and require elaborate multivariate analysis of comprehensive data sets. In the clinical context, however, the problem is not to replace the medical doctor with an automatic machine but, rather, to add additional information needed to decide which of a limited set of alternatives is the most likely. The method should be regarded as a diagnostic decision support tool. Many skin disorders, including cancer, may look the same on the surface but exhibit dramatically different impedance spectra. Thus, in a reasonable subset of possible disorders, specificity could be quite high.

The first paper using our impedance spectrometer on skin cancer was published in 1998 [51]. Recently, Åberg et al. [52] published a study including a number of different disorders, using the above-mentioned four indices for analysis [52]. A group in Rolla, Missouri, are using the same type of equipment, but employing other methods of data analysis [53]. A somewhat different approach has been taken by Glickman et al. [54], using an array of pointed electrodes designed to penetrate the stratum corneum. Thus, looking at disorders mainly manifested below the stratum corneum, it is only logical to minimize the contribution of the skin barrier by removing it or penetrating it, at the expense of not being invasive. One way of achieving this in a minimally invasive manner would employ microspikes [55], but this would be another chapter.

We gratefully acknowledge Peter Åberg for preparing the illustrations.

REFERENCES

1. Gildemeister, M., Über elektrischen Widerstand, Kapazität und Polarisation der Haut, *Pflügers Arch. Physiol.*, 176, 84, 1919.
2. Rosendal, T., Concluding studies on the conducting properties of human skin to alternating current, *Acta Physiol. Scand.*, 9, 39, 1945.
3. Grimnes, S. and Martinsen, Ø.G., *Bioimpedance and Bioelectricity Basics*, Academic Press, London, 2000.
4. Schwan, H.P., Electrical properties of tissue and cell suspensions, in *Advances in Biological & Medical Physics*, Vol. 5, Academic Press, New York, 1957, pp. 147–224.
5. Yamamoto, T. and Yamamoto, Y., Electrical properties of the epidermal stratum corneum, *Med. Biol. Eng.* 14, 151, 1976.
6. Salter, D.C., Examination of stratum corneum hydration state by electrical methods, in: *Skin Bioengineering, Techniques and Applications in Dermatology and Cosmetology*, Elsner, P., Barel, A.O., Berardesca, E., Gabard, B., Setup, J. (Eds.), *Curr. Probl. Dermatol.*, Basel, Karger, 1998, pp. 38–47.
7. Martinsen, Ø.G. and Grimnes, S., Facts and myths about electrical measurement of stratum corneum hydration state, *Dermatology*, 202, 87, 2001.
8. Martinsen, Ø.G., Grimnes, S., and Haug, E., Measuring depth depends on frequency in electrical skin impedance measurements, *Skin Res. Technol.*, 5, 179, 1999.
9. Hasted, J.B., *Aqueous Dielectrics*, Chapman and Hall, London, 1973.
10. Craig, D.Q.M., *Dielectric Analysis of Pharmaceutical Systems*, Taylor & Francis, London, 1995.
11. Fluhr, J.W., Gloor, M., Lazzerini, S., Kleesz, P., Grieshaber, R., and Berardesca, E., Comparative study of five instruments measuring stratum corneum hydration (Corneometer CM 820 and CM 825, Skicon 200, Nova DPM 9003, DermaLab). Part I. *In vitro*, *Skin Res. Technol.*, 5, 161, 1999.
12. Fluhr, J.W., Gloor, M., Lazzerini, S., Kleesz, P., Grieshaber, R., and Berardesca, E. Comparative study of five instruments measuring stratum corneum hydration (Corneometer CM 820 and CM 825, Skicon 200, Nova DPM 9003, DermaLab). Part II. *In vivo*, *Skin Res. Technol.*, 5, 171, 1999.
13. Fischer, T.W., Wigger-Alberti, W., and Elsner, P. Assessment of 'dry skin': current bioengineering methods and test designs, *Skin Pharmacol. Appl. Skin Physiol.*, 14, 183, 2001.
14. Stefanesco, S., Schlumberger, C., and Schlumberger, M., Sur la distribution électrique potentielle autour d'une prise de terre ponctuelle dans un terrain a couches horizontals, homogènes et isotropes, *J. Phys.*, 1, 132, 1930.
15. Roy, K.K. and Rao, K.P., Limiting depth of detection in line electrode systems, *Geophys. Prospecting*, 25, 758, 1977.
16. Elden, H.R., United States Patent No. 5,890,489.
17. Clar, E.J., Her, C.P., and Sturelle, C.G., Skin impedance and moisturization, *J. Soc. Cosmetic Chem.*, 26, 337, 1975.
18. Martinsen, Ø.G., Grimnes, S., and Sveen, O., Dielectric properties of the epidermal stratum corneum, *Med. Biol. Eng. Comput.*, 35, 172, 1997.
19. Nilsson, R., Fallan, J.O., Larsson, K.S., Ollmar, S., and Sundström, F., Electrical impedance — a new parameter for oral mucosal irritation tests, *J. Mater. Sci. Mater. Med.*, 3, 278, 1992.
20. Ollmar, S. and Emtestam, L., Electrical impedance applied to non-invasive detection of irritation in skin, *Contact Dermatitis*, 27, 37, 1992.

21. Emtestam, L. and Ollmar, S., Electrical impedance index in human skin: measurement after occlusion, in 5 anatomical regions and in mild irritant dermatitis, *Contact Dermatitis*, 28, 104, 1993.
22. Ollmar, S., Nyrén, M., Nicander, I., and Emtestam, L., Electrical impedance compared with other non-invasive bioengineering techniques and visual scoring for detection of irritation in human skin, *Br. J. Dermatol.*, 130, 29, 1994.
23. Ollmar, S., Eek, A, Sundström, F., and Emtestam, L., Electrical impedance for estimation of irritation in oral mucosa and skin, *Med. Prog. Technol.*, 21, 29, 1995.
24. Nicander, I., Ollmar, S., Lundh Rozell, B., Eek, A., and Emtestam, L., Electrical impedance measured to five skin depths in mild irritant dermatitis induced by sodium lauryl sulphate, *Br. J. Dermatol.*, 132, 718, 1995.
25. Ollmar, S. and Nicander, I., Information in multi-frequency measurement on intact skin, *Innov. Technol. Biol. Med.*, 16, 745, 1995.
26. Nicander, I., Ollmar, S., Eek, A., Lundh Rozell, B., and Emtestam, L., Correlation of impedance response patterns to histological findings in irritant skin reactions induced by various surfactants, *Br. J. Dermatol.*, 134, 221, 1996.
27. Nyrén, M., Ollmar, S., Nicander, I., and Emtestam, L., An electrical impedance technique for assessment of wheals, *Allergy*, 51, 923, 1996.
28. Nicander, I., Ollmar, S., Lundh Rozell, B., and Emtestam, L., Allergic contact reactions in the skin assessed by electrical impedance — a pilot study, *Skin Res. Technol.*, 3, 121, 1997.
29. Nyrén, M., Hagströmer, L., and Emtestam, L., Instrumental measurement of the Mantoux test: differential effects of tuberculin and sodium lauryl sulphate on impedance response patterns in human skin, *Dermatology*, 201, 212, 2000.
30. Nyrén, M., Kuzmina, N., and Emtestam, L., Electrical impedance as a potential tool to distinguish between allergic and irritant contact dermatitis, *J. Am. Acad. Dermatol.*, 48, 394, 2003.
31. Ollmar, S., Nicander, I., Ollmar, J., and Emtestam, L., Information in full and reduced data sets of electrical impedance spectra from various skin conditions, compared using a holographic neural network, *Med. Biol. Eng. Comput.*, 35, 415, 1997.
32. Lindholm-Sethson, B., Han, S., Ollmar, S., Nicander, I., Jonsson, G., Lithner, F., Bertheim, U., and Geladi, P., Multivariate analysis of skin impedance data in long-term type 1 diabetic patients, *Chemometrics Intelligent Lab. Syst.*, 44, 381, 1998.
33. Nicander, I., Nyrén, M., Emtestam, L., and Ollmar, S., Baseline electrical impedance measured at various skin sites — related to age and sex, *Skin Res. Technol.*, 3, 252, 1997.
34. Ollmar, S., Methods of information extraction from impedance spectra of biological tissue, in particular skin and oral mucosa — a critical review and suggestions for the future, *Bioelectrochem. Bioenergetics*, 45, 157, 1998.
35. Nicander, I., Norlén, L., Brockstedt, U., Lundh Rozell, B., Forslind, B., and Ollmar, S., Electrical impedance and other physical parameters as related to lipid content of human stratum corneum. *Skin Res. Technol.*, 4, 213, 1998.
36. Norlén, L., Nicander, I., Lundh Rozell, B., Ollmar, S., and Forslind, B., Differences in human stratum corneum lipid content related to physical parameters of skin barrier function *in vivo*, *J. Invest. Dermatol.*, 112, 72, 1999.
37. Nicander, I., Norlén, L., Forslind, B., and Ollmar, S., Lipid content and electrical impedance, *Curr. Probl. Dermatol.*, Vol. 26, Basel, Karger, 1998, pp. 165–176.

38. Nicander, I. and Ollmar, S., Mild and below threshold skin responses to sodium lauryl sulphate assessed by depth controlled electrical impedance, *Skin Res. Technol.*, 3, 259, 1997.
39. Åberg, P., Geladi, P., Nicander, I., and Ollmar, S. Variation of skin properties within human forearms demonstrated by non-invasive detection and multi-way analysis, *Skin Res. Technol.*, 8, 194, 2002.
40. Nicander, I. and Ollmar, S., Electrical impedance measurements at different skin sites related to seasonal variations, *Skin Res. Technol.*, 6, 81, 2000.
41. Ten Cate, A.R., Ed., *Oral Histology — Development, Structure, and Function*, 4th edition, Mosby, Toronto, 1994.
42. Nicander, I., Rundquist, L., and Ollmar, S., Electric impedance measurements at six different anatomic locations of macroscopically normal human oral mucosa, *Acta Odontol. Scand.*, 55, 88, 1997.
43. Nicander, I., Lundh Rozell, B., Rundquist, L., and Ollmar, S., Electrical impedance. A method to evaluate subtle changes of the human oral mucosa, *Eur. J. Oral Sci.*, 105, 576, 1997.
44. Nicander, I. and Ollmar, S., Electrical bio-impedance related to structural differences and reactions in skin and oral mucosa, *Ann. N.Y. Acad. Sci.*, 873, 221, 1999.
45. Rantanen, I., Nicander, I., Jutila, K., Ollmar, S., Tenovuo, J., and Söderling, E., Betaine reduces the irritating effect of sodium lauryl sulphate on human oral mucosa *in vivo*, *Acta Odontol. Scand.*, 60, 306, 2002.
46. Nicander, I., Åberg, P., and Ollmar, S., The use of different concentrations of betaine as a reducing irritation agent in soaps monitored visually and non-invasively, *Skin Res. Technol.*, 9, 43, 2003.
47. Nicander, I., Rantanen, I., Lundh Rozell, B., Söderling, E., and Ollmar, S., The ability of betaine to reduce the irritating effects of detergents assessed visually, histologically and by bioengineering methods, *Skin Res. Technol.*, 9, 40, 2003.
48. Bartos, L., Vaginal impedance measurement used for mating in the rat, *Lab. Anim. (England)*, 11, 53, 1977.
49. Bartos, L. and Sedlacek, J., Vaginal impedance measurement used for mating in the guinea-pig, *Lab. Anim. (England)*, 11, 57, 1977.
50. Ollmar, S., Making electronic biopsies into a viable future for non-invasive diagnostics with electrical impedance, *Med. Biol. Eng. Comput.*, 37(Suppl 2), 116, 1999.
51. Emtestam, L., Nicander, I., Stenström, M., and Ollmar, S., Electrical impedance of nodular basal cell carcinoma: a pilot study, *Dermatology*, 197, 313, 1998.
52. Åberg, P., Nicander, I., Holmgren, U., Geladi, P., and Ollmar S., Assessment of skin lesions and skin cancer using simple electrical impedance indices, *Skin Res. Technol.*, 9, 257, 2003.
53. Beetner, D.G., Kapoor, S., Manjunath, S., Zhou, X., and Stoecker, W.V., Differentiation among basal cell carcinoma, benign lesions, and normal skin using electric impedance, *IEEE Trans. Biomed. Eng.*, 50(8), 1020, 2003.
54. Glickman, Y.A., Filo, O., David, M., Yayon, A., Topaz, M., Zamir, B., Ginzburg, A., Rozenman, D., and Kenan, G., Electrical impedance scanning: a new approach to skin cancer diagnosis, *Skin Res. Technol.*, 9, 262, 2003.
55. Griss, P., Enoksson, P., Tolvanen-Laakso, H.K., Meriläinen, P., Ollmar, S., and Stemme, G., Micromachined electrodes for biopotential measurements. *IEEE J. Microelectromech. Syst.*, 10, 10, 2001.

28 Hydration in Psoriasis and Eczema: The Dry Surface–High Evaporative Water Loss Paradox

Jørgen Serup

CONTENTS

I. Introduction 351
II. The Water Barrier and the Water Vapor Mantel of Ambient Air 352
III. The Scaling Process in Psoriasis and Eczema 352
IV. Methodological Aspects — Measurement of Skin Surface Hydration and TEWL 353
V. The Dynamic of the Water Barrier Function of the Skin 355
VI. The Dry Surface — High Evaporative Water Loss Paradox of Psoriasis and Eczema 355
VII. Measurement of Hydration Characteristics during Healing and Evaluation of Efficacy of Therapeutics 356
References 357

I. INTRODUCTION

Psoriasis and eczema are the most prominent epidermal skin diseases. The epidermis is a highly specialized tissue of approximately 20 cell layers. There is a constant movement of keratinocytes from the basal cell layer to the surface where invisible desquamation finally takes place. The very important steep gradient in water content between the humid internal milieu and the dry ambient air is found within the epidermis. On this background it is no wonder that alterations of the skin surface hydration and the evaporative water loss become prominent features of both psoriasis and eczema.

During the last decade our knowledge of transepidermal water loss (TEWL) both in normal and pathological conditions expanded substantially. Recently the European Society of Contact Dermatitis published a review and gave guidelines for

0-8493-1443-7/05/$0.00+$1.50

the measurement of TEWL.[1] During the same period, H. Tagami of Tohoku University (Sendai, Japan) performed a series of experiments on skin surface hydration under normal and pathological conditions using an electrical method for measurement of skin surface hydration.[2]

This chapter will address the paradox of skin hydration in psoriasis and eczema, i.e., the dryness of the skin surface in contrast to the increased evaporative water loss. The review will mainly be based on my own experiments, and the extensive literature on skin hydration will not be revisited.

II. THE WATER BARRIER AND THE WATER VAPOR MANTEL OF AMBIENT AIR

The water content of the dermis and the squamous cell layer of the epidermis is approximately 75% while in the granular and horny layers it may be 10 to 30%. It is well known and often expressed that the water barrier is simply located within the epidermis. However, the constant flux of water out of the skin also known as perspiratio insensibilis also depends on the water vapor pressure of the ambient air immediately outside the skin. In the ServoMed evaporimeter the sensors for registration of humidity are mounted 3 and 6 mm above the skin surface and the gradient in water vapor pressure in the ambient air reflecting the transepidermal water can be measured. The measuring principle even works if the probe with the two sensors is removed 5 mm from the skin surface when the special protection shield with the cylindrical chamber is mounted.

Thus, we carry on our skin surface an invisible mantel of water vapor approximately 10 mm in thickness (Figure 28.1). This mantel of water vapor is, for obvious reasons, easily influenced by convection of air, use of textiles, and of course, the humidity and temperature conditions of the environment.

The water barrier function of the skin and the dryness of the skin surface is not simply dependent on the epidermis, the water-holding capacity of the stratum corneum, the efficacy of the biological barrier, etc. but ambient conditions and in particular the boundary layer of water, *the water vapor mantel,* are major influences.

This has consequences when measurements of electrical conductivity of the skin surface to indicate skin surface hydration and when measurement of TEWL are performed. The condition of the water vapor mantel also influences the epidermis itself. The skin surface is drier in the winter, and eczema is more frequent and worse in winter.[2,3]

III. THE SCALING PROCESS IN PSORIASIS AND ECZEMA

The transit time of epidermal cells from the basal cell layer to the skin surface is sped up in psoriasis and eczema to about 1 week in comparison with 3 to 4 weeks under healthy conditions.[4] The normal process of invisible desquamation, where ideally single corneocytes are shed, is a result of a gradual maturation of the

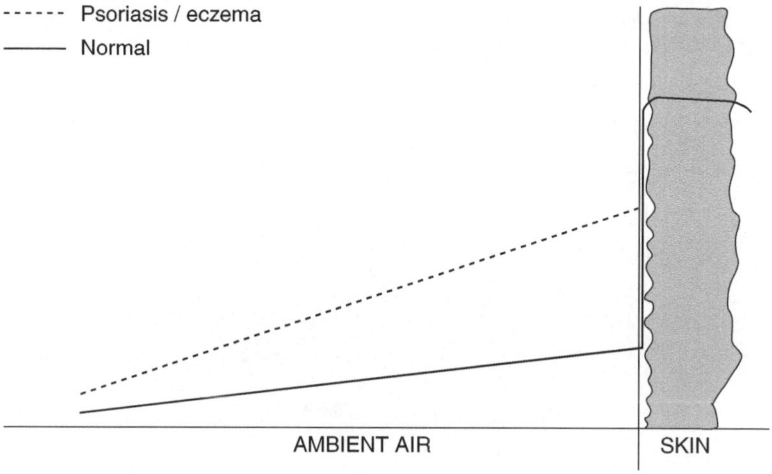

FIGURE 28.1 Water barrier function of the skin (schematic). The *water vapor mantel* of ambient air and the intraepidermal water gradient. Diseased (psoriasis, eczema) and normal skin.

keratinocytes with maturation of keratin and loosening of cell cohesion on the way out of the epidermis.[5]

In psoriasis and eczema this delicate process of keratinization and loss of cohesion is disturbed, and larger scales are discharged from the surface.

D-Squame® tape is an easy method used to demonstrate both vehicle effects and the formation of larger scales or flakes (see Figure 28.2).[6] This tape is a specialized sticky slide technique. However, psoriasis is not simply a disease of increased scaling since there is a reduced epidermal discharge in between the scales as compared to normal skin with undisturbed desquamation (Figure 28.2). Thus, the skin surface of psoriasis and eczema is not homogeneous. There is an asynchronous process with areas with visible scaling surrounded by areas with no visible scaling changing from each day.

IV. METHODOLOGICAL ASPECTS — MEASUREMENT OF SKIN SURFACE HYDRATION AND TEWL

Electrical methods for the measurement of skin surface hydration such as the Corneometer®, the Skicon®, and the Novapen® are convenient pieces of equipment that have become commonplace.[2,7,8] The principles behind electrical measurements were reviewed by Leveque and de Rigal.[9] From my experience the Corneometer® and the Skicon® correlate; however, the Skicon® is better suited when the skin surface is hydrated experimentally and absorption–desorption dynamics recorded, while the Corneometer® with the larger contact area may have advantages when psoriasis, eczema, and clinically dry skin are treated and the healing course or improvement

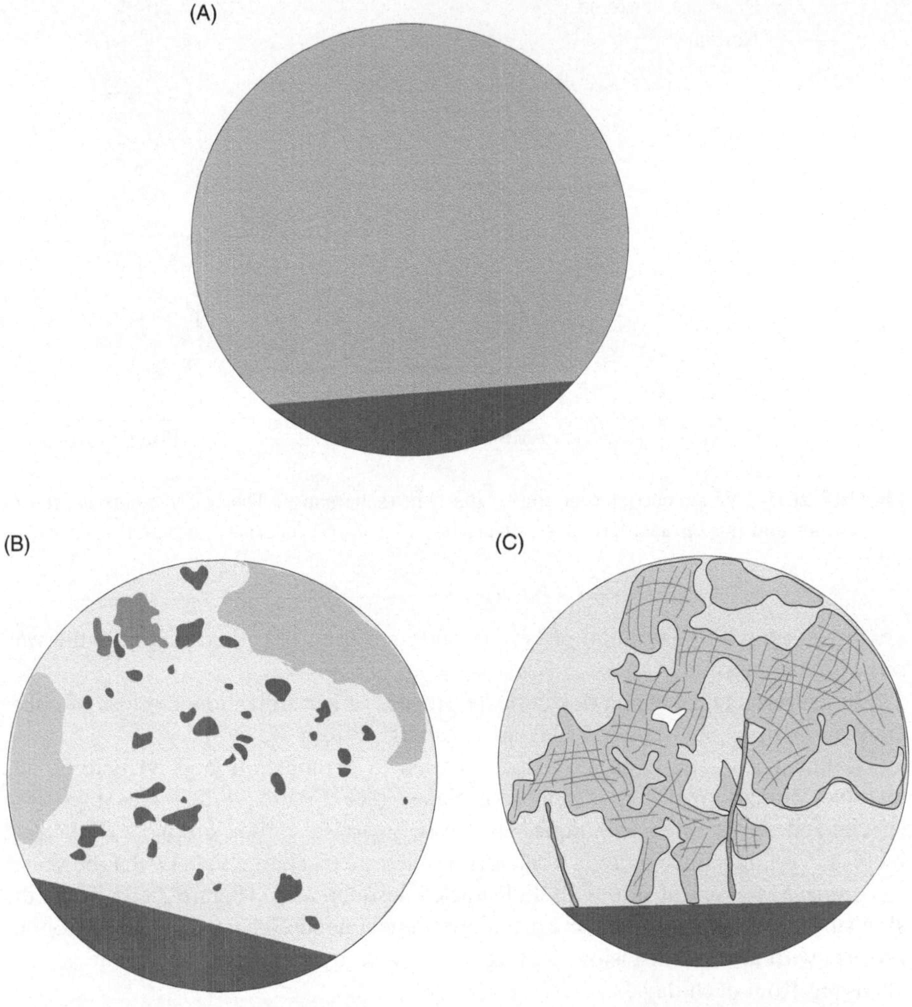

FIGURE 28.2 D-Squame® tapes obtained from normal skin (A), a psoriasis plaque (B), and solar eczema (C). Note that desquamation is *reduced* in between the scales (B and C).

is monitored.[7,8] The Novapen® is delivered with a number of probes with different specifications.

The measuring area and the physical contact with the probe depending on the geometry of the skin surface is important. If the skin surface is rough and the physical and electrical contact is poor, low values not representing an average of the hydration of the skin surface will be recorded. If the probe is small the method may be too site dependent, and it may be necessary to take a number of readings and average. The electrical methods are not water specific, and any polar material may in principle influence recordings. However, the methods have in practice proven useful for arbitrary recording of the skin surface hydration state in a broad range of situations.

Recently, infrared optothermal and magnetic resonance techniques became available, and these complicated and expensive methods are obviously more specific.

TEWL is normally measured either with the ServoMed or the Tevameter® evaporimeters. As mentioned, the flux of water out of the skin is calculated based on recordings within the 10 mm water vapor mantel of ambient air. The many variables and sources of error, which need be taken into account, were described in the guidelines mentioned above.[1]

In the evaluation of skin hydration, previous physical activity of the test subject and sweat gland function need special attention. Sweat glands are rich in the face, upper trunk, palms, and soles. Individual sweat glands have a discontinuous output. With a fiberoptic microscope it is easily demonstrated on the palm of the hand on a warm day how sweat bubbles suddenly appear and evaporate within a few seconds.

V. THE DYNAMIC OF THE WATER BARRIER FUNCTION OF THE SKIN

Equilibrium between skin surface hydration and ambient air normally takes place within seconds or a few minutes. This was clearly demonstrated by Tagami in his classical absorption–desorption experiments.[2] In such experiments the water-holding capacity of the stratum corneum can be evaluated. When the forearm is immersed in water for 5 or 20 min the conductance, capacitance, and water evaporation are increased immediately thereafter; however, after 15 to 20 min pretreatment values are obtained again.[10] During immersion for 5 and 20 min the stratum corneum undergoes swelling, and the skin surface friction increases. These experiments demonstrate that the compartment of the skin that can be loaded with and hold water is small, probably no more than five to seven cell layers, thus evaporation and adaptation to the ambient air conditions only need a little time to take place.

VI. THE DRY SURFACE — HIGH EVAPORATIVE WATER LOSS PARADOX OF PSORIASIS AND ECZEMA

Psoriasis and eczema show, as a very consistent finding, a reduction in electrical parameters such as conductance and capacitance, indicating dryness in contrast to increase of TEWL (see Figure 28.3).[11,12] This means that even if scales are exposed to a higher water vapor pressure they remain drier. Experiments under different conditions have demonstrated that pathological scales actually have a decreased ability to hold water.[2,13] As mentioned previously (Figure 28.2), dryness and scaling of lesional skin is, however, in the horizontal level not simply a continuous manifestation, but there may easily be areas of increased moisture in between scales, areas which might or might not be detected by the electrical methods due to the roughness and stiffness of the skin surface and the incomplete physical contact with the measuring probe. In subacute phase eczema the conductance may be increased.[8]

Scales are white due to their content of air, and due to the air content they are expected to have a prominent influence (low conductance) on electrical recording. If scales or epidermal discharge harvested with a D-Squame® tape are treated with

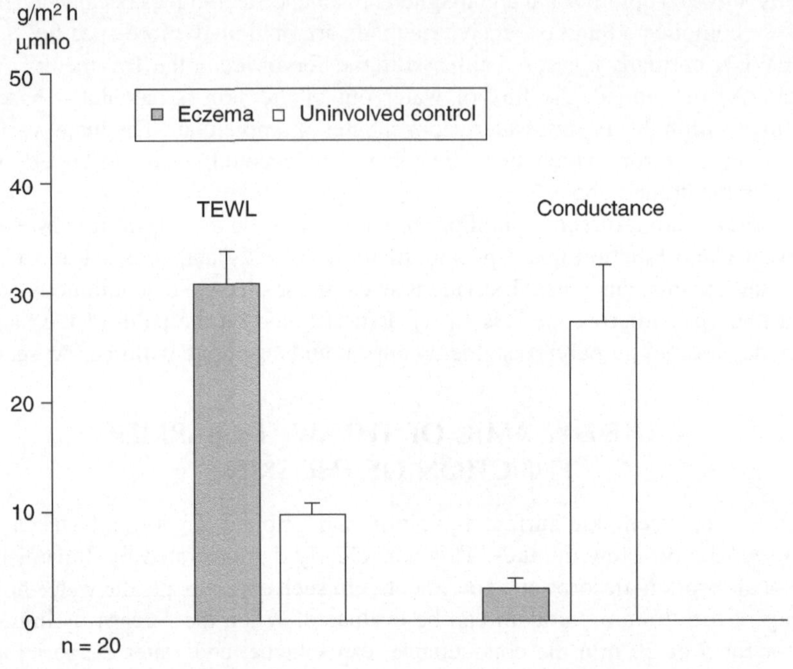

FIGURE 28.3 Transepidermal water loss ($g/m^2/h$) and electrical conductance (µmho) of eczema skin and uninvolved control of the same region. Mean and SE.

immersion oil or paraffin oil the tape instantly clears up and become translucent like glass. The oil replaces the air bubbles contained within the scales.

It has been suggested that abnormality of the intercellular phospholipid bilayer between keratinocytes plays a role in atopic eczema and possibly in the scaling process in general as well.[14] It is documented that changes of the intercellular cement are important in different types of ichthyosis.[15] In psoriasis, where rapid passage of keratinocytes through the epidermal layers and incomplete differentiation are the prominent features, disturbance of the intercellular phospholipid bilayer is probably not a critical factor.

VII. MEASUREMENT OF HYDRATION CHARACTERISTICS DURING HEALING AND EVALUATION OF EFFICACY OF THERAPEUTICS

During PUVA treatment TEWL becomes normal parallel to normalization of the skinfold thickness.[16] During treatment with dithranol, which is a local irritant, TEWL remains elevated even after the thickness is normalized. TEWL is a very sensitive parameter for the study of primary irritancy.[17] TEWL is probably a very useful parameter for objective and overall documentation of healing of psoriatic plaques with the reservation that the therapeutic will be without irritant properties.

Therapeutics for the treatment of psoriasis or eczema are often formulated in creams or ointments, which might interfere with the electrical parameters. Hydration measurements are generally useful but not the most successful bioengineering method for monitoring healing during active treatment.[8,17] However, for the documentation of ultimate healing after active treatment was stopped the electrical methods certainly have a potential not yet sufficiently explored.

The water of cream formulations evaporates from the skin surface within 10 min while skin surface lipids are increased for six or more hours, even following application of a cosmetic lotion.[18,19] On repeated application the electrical parameters improve, and the desquamation process becomes strongly influenced according to evaluations by D-Squame® tape.[6,20,21] Thus, vehicles influence both electrical parameters and desquamation. TEWL, however, is not influenced directly by vehicles, but may change slightly and improve as a result of biological changes within the epidermis resulting from such vehicles.[21] It is a myth that vehicles such as ointments and petrolatum efficiently occlude skin. A realistic layer of petrolatum or ointment applied to the skin will only have occlusive effects for 10 to 30 min.[22,23]

Hydration of psoriasis and eczema scaling and the effect of keratolytic treatment with salicylic acid was not really studied objectively in the past, despite the widespread clinical use of salicylic acid formulated 5 or 10 percent in petrolatum. In contrast, the effects of urea on electrical parameters, TEWL, and the scaling process were well documented.[19,21,24,25] It is a general misunderstanding that stinging due to urea, a small molecule with strong osmotic influence, is identical to skin irritation and cytotoxic damage.[25]

Measurement of TEWL is useful to monitor healing of psoriasis and eczema and is generally overlooked. It is widely vehicle independent, but the therapeutic must not be a local irritant. Electrical measurement of skin surface hydration is, on the other hand, vehicle dependent and dependent on roughness and surface geometry. However, the electrical methods have a not yet explored potential for documentation of ultimate healing or relapse after treatment was stopped.

REFERENCES

1. Pinnagoda, J., Tupker, R. A., Agner, T., and Serup, J., Guidelines for transepidermal water loss (TEWL) measurement, *Contact Dermatitis,* 22, 164–178, 1990.
2. Tagami, H., Impedance measurement for evaluation of the hydration state of the skin surface, in *Cutaneous Investigation in Health and Disease,* Leveque, J.-L., Ed., Marcel Dekker, New York, 79–111.
3. Agner, T. and Serup, J, Seasonal variation of skin resistance to irritants, *Br. J. Dermatol.,* 121, 323–328, 1989.
4. Dover, R. and Wright, N. A., The cell proliferation kinetics of the epidermis, in *Physiology, Biochemistry, and Molecular Biology of the Skin,* 2nd ed., Goldsmith, L. A., Ed., Oxford University Press, New York, 1991, 239–265.
5. Odland, G. F., Structure of the skin, in *Physiology, Biochemistry, and Molecular Biology of the Skin,* 2nd ed. Goldsmith, L. A., Ed., Oxford University Press, New York, 1991, 3–62.

6. Serup, J., Winther, A., and Blichmann, C., A simple method for the study of scale pattern and effects of a moisturizer: qualitative and quantitative evaluation by D-Squame® tape compared with parameters of epidermal hydration, *Clin. Exp. Dermatol.,* 14, 277–282, 1989.
7. Blichmann, C. W. and Serup, J., Assessment of skin moisture: measurement of electrical conductance, capacitance and transepidermal water loss, *Acta Derm. Venereol. (Stockholm),* 68, 284–290, 1988.
8. Agner, T. and Serup, J., Comparison of two electrical methods for measurement of skin hydration. An experimental study on irritant patch test reactions, *Bioeng. Skin* 4, 263–269, 1988.
9. Lévêque, J.-L. and de Rigal, J., Impedance methods for studying skin moisturisation, *J. Soc. Cosmet. Chem.,* 34, 419–428, 1983.
10. Stender, I. M., Blichmann, C., and Serup, J., Effects of oil and water baths on the hydration state of the epidermis, *Clin. Exp. Dermatol.,* 15, 206–209, 1990.
11. Serup, J. and Blichmann, C., Epidermal hydration of psoriasis plaques and the relation to scaling. Measurement of electrical conductance and transepidermal water loss, *Acta Derm. Venereol. (Stockholm),* 67, 357–366, 1987.
12. Blichmann, C. and Serup, J., Hydration studies on scaly hand eczema, *Contact Dermatitis,* 16, 155–159, 1987.
13. Tagami, H. and Yoshikuni, K., Interrelationship between water barrier and reservoir functions of pathologic stratum corneum, *Arch Dermatol.,* 121, 642–645, 1985.
14. Linde, Y. W., Dry skin in atopic dermatitis, *Acta Derm. Venereol. (Stockholm),* Suppl. 177, 9–13, 1992.
15. Elias, P. M., Epidermal lipids, barrier function, and desquamation, *J. Invest. Dermatol.,* 80 (Suppl.), 44s–49s, 1983.
16. Rogers, S., Measurement of plaque thickness and evaporative water loss in psoriasis with PUVA and dithranol treatment, *Clin. Exp. Dermatol.,* 18, 21–24, 1993.
17. Agner, T., Noninvasive measuring methods for the investigation of irritant patch test reactions, *Acta Derm. Venereol. (Stockholm),* Suppl. 173, 1–26, 1992.
18. Blichmann, C. W., Serup, J., and Winther, A., Effects of single application of a moisturizer: evaporation of emulsion water, skin surface temperature, electrical conductance, electrical capacitance, and skin surface (emulsion) lipids, *Acta Derm. Venereol. (Stockholm),* 69, 327–330, 1989.
19. Serup, J., A three-hour test for rapid comparison of effects of moisturizers and active constituents (urea). Measurements of hydration, scaling and skin surface lipidization by noninvasive techniques, *Acta Derm. Venereol. (Stockholm),* Suppl. 177, 29–33, 1992.
20. Serup, J., Winther, A., Blichmann, C. W., Effects of repeated application of a moisturizer, *Acta Derm. Venereol. (Stockholm),* 69, 457–459, 1989.
21. Serup, J., A double-blind comparison of two creams containing urea as the active ingredient. Assessment of efficacy and side-effects by non-invasive techniques and a clinical scoring scheme, *Acta Derm. Venereol. (Stockholm),* Suppl. 177, 34–38, 1992.
22. Blichmann, C. W. and Serup, J., The effects of moisturizers determined by measurement of transepidermal water loss, *Ugeskr. Laeg.,* 149, 1389–90, 1987.
23. Ghadially, R., Halkier-Sórensen, L., and Elias, P. M., Effects of petrolatum on stratum corneum structure and functions, *J. Am. Acad. Dermatol.,* 26, 387–396, 1992.
24. Raab, W., Harnstoff in der Dermatologie, *Hautarzt,* Suppl. 9, 1–86, 1989.
25. Serup, J., Urea revisited: including clinical uses and evaluation by bioengineering techniques, *Acta Derm. Venereol.,* Suppl. 177, 1–52, 1992.

29 Hydration and Lipids

Philip W. Wertz and Bozena B. Michniak

CONTENTS

I. Introduction 359
II. Lipids and Hydration in Normal Stratum Corneum 360
  A. Water in the Intercellular Space 360
  B. Water within the Corneocytes 360
  C. Water at the Skin Surface 361
III. Lipids and Hydration in Physically/Chemically Damaged Stratum Corneum 362
  A. Solvent-Treated Skin 362
  B. Detergent-Damaged Skin 362
  C. Ultraviolet-Exposed Skin 362
IV. Lipids and Hydration in Stratum Corneum from Diseased Skin 363
  A. Ichthyoses 363
    1. Ichthyosis Vulgaris 363
    2. Recessive X-Linked Ichthyosis 364
    3. Lamellar Ichthyosis 364
  B. Atopic Dermatitis 364
  C. Essential Fatty Acid Deficiency 365
V. Potential Use of Lipids for Restoration of Normal Hydration 365
References 366

I. INTRODUCTION

The primary function of the epidermis is to produce a stratum corneum capable of providing a watertight barrier [1]. Although the stratum corneum is an excellent water barrier, it is not completely impermeable, and under normal conditions water from the fully hydrated living layers diffuses through the stratum corneum and evaporates at the surface. Water is an essential component of the stratum corneum, and in its absence the stratum corneum would be stiff and brittle.

A water gradient exists across the stratum corneum from about 70% water in the underlying viable tissue to ambient relative humidity at the surface [2]. Molecules diffuse along the gradient, move through the stratum corneum, and evaporate at the skin surface. This is measured by various means as transepidermal water loss

0-8493-1443-7/05/$0.00+$1.50

(TEWL), or insensible perspiration. The water gradient across the stratum corneum depends on the relative humidity. The actual water content actually varies relatively little over the range of relative humidity from 20% through about 80%. Within this range the stratum corneum contains approximately 0.1 g of water per cubic centimeter. When fully hydrated, the water content of the stratum corneum approaches ten times this concentration. A small fraction of the water in the stratum corneum is tightly bound and not easily removed [3].

The goals of the present chapter are to discuss the disposition and movement of water in normal, physically damaged, and diseased stratum corneum. A mechanism will be put forth to explain the permeabilizing effect of water.

II. LIPIDS AND HYDRATION IN NORMAL STRATUM CORNEUM

A. Water in the Intercellular Space

Water provides an essential orienting force in the formation of lamellar phases by bipolar lipid molecules [4]. It is expected that there are water molecules hydrogen-bonded to the polar head groups of the lipids in the intercellular spaces of the stratum corneum. It could also be anticipated that water could move through the stratum corneum primarily along the polar head group regions and, in fact, water has been visualized within the intercellular spaces by an autoradiographic technique [5]. In addition, this pathway leads to a tortuous diffusion pathway, thereby increasing the distance that a water molecule would have to travel to completely traverse the stratum corneum. This "tortuosity" is thought to be an important aspect of the stratum corneum barrier, which would be lost if there were a transcellular pathway [6].

With many lamellar lipid systems in excess water, the interlamellar spacing increases with the water content, indicating that the excess water is between the lamellae [4]. This is not the case with the stratum corneum lipids. A 13-nm periodicity is the predominant lamellar unit revealed by small angle x-ray diffraction, and this does not alter with increasing degree of hydration [7]. At very high water content, and possibly requiring significant time to develop, a separate water phase appears within the intercellular space of the stratum corneum [8]. This may not represent any normally encountered physiological state, but rather resembles a pathological condition in which the integrity of the tissue is questionable.

B. Water within the Corneocytes

As keratinocytes differentiate, they accumulate a histidine-rich protein called profilaggrin as a component of the keratohyalin granules [9]. This is a high-molecular weight, multiply phosphorylated protein. Near the end of the differentiation program, this protein undergoes dephosphorylation and proteolysis to produce filaggrin [10].

As the differentiating cells pass into the stratum corneum, filaggrin induces the aggregation of the keratin filaments into dense bundles that fill most of the interior of the cell. The filaggrin itself then undergoes proteolytic degradation to its component amino acids [11,12]. The resulting histidine and glutamine are further converted to urocanic acid and pyrrolidone carboxylic acid, respectively. The total free amino acid content of the corneocytes is on the order of 2 M, making them very hygroscopic [12]. In addition to the free amino acids and derivatives, there are also appreciable concentrations of other low-molecular-weight molecules that contribute to the osmotic strength of the corneocyte interior [13]. These include urea and lactate produced as an end product of anaerobic glycolysis. Collectively, these low-molecular-weight molecules have been referred to as "natural moisturizing factors." The high osmotic strength within the corneocytes allows them to soak up excess water, which prevents the accumulation of water in the intercellular spaces.

As the hydration of the stratum corneum increases, the hydrophobic corneocytes take up water and swell [14]. They increase about 25% in thickness but only a few percentage points in length. Although at higher degrees of hydration there may be somewhat more water molecules associated with the polar head group regions of the intercellular lamellae, the x-ray diffraction results would indicate that this is not a major effect. However, it is possible that swelling of the corneocytes with increased hydration introduces increased curvature, thereby increasing their surface area. This would result in a situation in which a greater area within the intercellular space would have to be covered by the same amount of lipid. This could be accomplished by increasing the distance between adjacent polar head groups, perhaps by the hydrogen bonding of additional water molecules. In this view, motion along the aliphatic chains beneath each polar head group would have to increase so that they would occupy the cylindrical volume beneath the head group. The result would be more fluid, and thereby more permeable, intercellular lamellae without any change in lamellar spacing.

C. Water at the Skin Surface

Water evaporating from the skin surface in the absence of perspiration can be measured as TEWL [15,16]. For most regions of normal human skin this will be approximately 0.5 mg/cm^2 per hour at low relative humidity. Of course, this value depends somewhat on the relative humidity, but the variation is not great between approximately 20 and 80% [2]. TEWL is often taken as an indicator of barrier function, although it may not be predictive of the barrier function for hydrophobic molecules [17]. The relationship between hydration of the skin and TEWL is complex. Two topics that are closely associated are hydration for the purpose of enhancing transdermal drug delivery and the alleviation of dry skin associated with various kinds of skin damage or disease. There is no universally accepted definition of dry skin, but it is generally agreed that the skin surface is rough, or in some variants, even scaly. In general, lipid composition of the stratum corneum is altered, TEWL is elevated, and hydration is decreased. Specifics for different dry skin conditions are discussed below.

III. LIPIDS AND HYDRATION IN PHYSICALLY/CHEMICALLY DAMAGED STRATUM CORNEUM

A. Solvent-Treated Skin

When Imokawa and Hattori [18] removed stratum corneum lipids from normal human skin by surface extraction *in vivo*, they observed a decrease in skin conductance, if the solvent contact time was at least 5 minutes, that persisted for at least 4 days. This indicates a decrease in the hydration level of the stratum corneum, although the hydration level of cells beneath the most superficial layer is uncertain. In addition, the extracted skin was said to have a chapped and scaly appearance, indicating that a reduction of lipid content could contribute to dry skin conditions. More recently, Norlen et al. [14] demonstrated that extraction of lipids from stratum corneum *in vitro* reduces the capacity of the tissue to take up water and to swell.

It is proposed that the influence of lipids on degree of hydration of the stratum corneum is not the result of any direct interaction between lipids and water. Rather, it is suggested that the removal of lipids with a concomitant decrease in barrier function permits more rapid loss of water from the corneocytes, thereby reducing the level of hydration.

A number of studies with human skin, both *in vivo* and *in vitro*, have demonstrated that removal of lipids from stratum corneum results in decreased barrier function [19–22].

B. Detergent-Damaged Skin

It is well known that repeated exposure to soaps or detergents can induce a dry skin condition. In 1986, Fulmer and Kramer [23] induced dry skin on the legs of human subjects by daily treatment with 4% aqueous sodium dodecyl sulfate over a 2-week period. The treated skin had a dry and scaly appearance and reduced level of hydration as judged by measurement of electrical impedance. The total lipid content of the stratum as well as the total ceramide content were unaffected by treatment; however, the proportions of individual ceramides were altered, the free cholesterol level was reduced, cholesterol esters were increased, and the proportion of long chain fatty acids was reduced. Barrier function was not assessed. These results indicate that the detergent effect is not mediated by gross lipid removal from the stratum corneum, but may result from alteration of lipid metabolism. This may require surfactant partitioning into the skin. In a subsequent study it was shown that, in fact, very little lipid was extracted from stratum corneum by several surfactants.

C. Ultraviolet-Exposed Skin

A single acute exposure to a relatively high dose of ultraviolet light (UV), or multiple smaller daily doses for 2 weeks, has been shown to result in disruption of the permeability barrier [24–26]. In one study, several effects of UVA, UVB, and UVC were compared [27]. All three ranges of UV induced erythema with a maximum response at 24 hours. Only UVB resulted in increased TEWL and a decrease in water content of the stratum corneum. In one study, a single dose of UVB was shown to result in an increase in TEWL, but no change in water content of the stratum corneum [26]. One possible explanation for this difference in the apparent effect of UV on water content of the stratum corneum could be differences in the sensitivity of the methods used. Capacitance was used to assess water content in the study where no change was seen, whereas optothermal infrared spectroscopy was used in the other study.

A study using murine skin demonstrated that the initial effect of a single high dose of UVB is damage to the uppermost granular cells, including depletion of lamellar granules [24]. This initial damage is followed by abnormal membrane arrangement in the lower stratum corneum, associated with increased TEWL. Subsequently, there is a hyperplastic response characterized by increased synthesis of lipids and accelerated lamellar granule production. The stimulated extrusion of lamellar granule contents leads to restoration of barrier function.

IV. LIPIDS AND HYDRATION IN STRATUM CORNEUM FROM DISEASED SKIN

A. Ichthyoses

The ichthyoses are a group of disorders generally characterized by dry skin [28]. The stratum corneum water content is lower than normal at a given relative humidity, and the skin surface is usually rough or scaly. This can be associated with pruritus, leading to scratching and subsequent infection. The following is not all inclusive but includes the most common or better understood ichthyoses.

1. Ichthyosis Vulgaris

Ichthyosis vulgaris, or common dry skin, is an autosomal-dominant, genetically determined condition affecting approximately one person out of every 300 [28]. This is a retention ichthyosis with a normal transit time for cell migration from the basal layer to the stratum corneum. It is characterized by both reduced water-holding capacity within the stratum corneum [29] and elevated TEWL [30].

The fundamental defect appears to be the underproduction of profilaggrin [31,32], which, in turn, results in underproduction of free amino acids within the cornified cells, causing the reduced ability of the stratum corneum to hold water [29]. This reduced level of amino acids apparently reflects an inherently unstable profilaggrin m-RNA.

Reduced levels of serine protease activities in the stratum corneum may also contribute to skin surface roughness. Although there is no published evidence for decreased levels or altered composition of stratum corneum in ichthyosis vulgaris, the application of a lipid mixture containing, among other things, ceramides, sterol, and free fatty acid has been shown to significantly improve the water-holding capacity of the stratum corneum, as judged by capacitance measurements [33].

2. Recessive X-Linked Ichthyosis

Recessive X-linked ichthyosis (RXLI) is a relatively rare condition [28]. A retrospective study performed between 1983 and 1998 in Hiroshima, Japan found a frequency of approximately 1 per 10,000 male births [34]. Another estimated frequency of 1 per 6000 has been reported [28]. Onset of the disease is usually within the first several months after birth and always within the first year. It is a lifelong disease. It is characterized by the presence of thick, dark scales, which are most prominent on the limbs, although other regions are also involved. Palmar and plantar surfaces appear normal. Like ichthyosis vulgaris, RXLI is a retention ichthyosis.

The fundamental defect underlying RXLI is the absence or deficiency of the sterol sulfatase that normally hydrolyzes cholesterol sulfate [35]. Cholesterol sulfate accumulates to high levels in the stratum corneum and elsewhere [36]. It is now known that cholesterol sulfate is a serine protease inhibitor and may prevent normal desquamation in RXLI by inhibiting the proteases that would normally degrade desmosomal proteins [37].

One study found no difference in TEWL between RXLI skin and normal skin [38], which was consistent with a separate report of normal intercellular lamellae as judged by transmission electron microscopy [39]. However, a second report indicated significantly elevated TEWL in RXLI [30]. Furthermore, defective barrier function was also indicated in RXLI compared to normal control skin by a more rapid vascular response to topical hexyl nicotinate [30]. Electrical capacitance, reflecting stratum corneum hydration, was reported to be significantly reduced in RXLI [30,38].

3. Lamellar Ichthyosis

Lamellar ichthyosis is a rare and serious hyperproliferative autosomal recessive condition found in about one in 100,000 newborns [28]. The fundamental underlying defect is mutation of the gene coding for transglutaminase 1 [40]. At birth, the infant is covered with a fissured membrane, and such infants are referred to as collodion babies. Ultimately, large, thick plaques with raised edges cover the entire skin surface. Water loss through the cracks in the stratum corneum at the edges of the scales and infection present major problems in the management of this condition, particularly in infants. TEWL is abnormally high in lamellar ichthyosis, and the water content is very low, as judged by capacitance measurements [30]. Oral retinoids in combination with topical emollients provide effective therapy [41].

B. Atopic Dermatitis

Atopic dermatitis is characterized by dry sensitive skin, often in conjunction with allergies or asthma [15]. As in other dry skin conditions, the water-holding capacity of the stratum corneum is diminished [42], and TEWL is elevated [43]. Its etiology is uncertain, but the frequency of atopic dermatitis in the United States is more than 10% and growing [44]. Atopic dermatitis is fundamentally a disease of the immune system [45]; however, it is exacerbated by impaired skin barrier function, which may allow penetration of antigens or irritants. The altered barrier function is thought to reflect an alteration in lipid composition [46]. In addition, the amino acid content of the stratum corneum is lower than normal in people with atopic dermatitis, resulting in reduced water-holding capacity and dryness.

C. Essential Fatty Acid Deficiency

Essential fatty acid deficiency is rarely seen in humans in the absence of general malnutrition; however, it has been much studied in experimental animals as a model for hyperproliferative skin disease [47–51]. In the absence of essential fatty acids, the epidermis becomes hyperplastic, skin barrier function becomes compromised, and the skin surface becomes rough and scaly. The hyperplasia and scaling can be corrected by topical prostaglandin E2 [47]. The barrier defect reflects the requirement for linoleic acid in two specific sphingolipids — acylglucosylceramide and acylceramide [49–50]. The former consists of 30- through 34-carbon long ω-hydroxyacids amide-linked to sphingosine with glucose β-glycosidically attached to the primary hydroxyl group of the base and linoleate ester-linked to the ω-hydroxyl group [50]. The acylceramide is produced by deglycosylation of the acylglucosylceramide. Acylglucosylceramide is thought to be involved in assembly of the lamellar granules in the viable portion of the epidermis, whereas acylceramide appears to be essential for the proper organization of the lipids in the intercellular spaces of the stratum corneum. In essential fatty acid deficiency, the lamellar granules appear to form normally and to extrude their contents into the intercellular space, indicating that linoleate is not essential for these processes. However, as oleate replaces linoleate in acylceramide molecules, the barrier function of the skin declines as judged by increasing TEWL [50]. It appears that a normal linoleate-containing acylceramide is required for organization of the stratum corneum lipids into proper functional units.

V. POTENTIAL USE OF LIPIDS FOR RESTORATION OF NORMAL HYDRATION

Traditionally, moisturizer creams or lotions have been used in the treatment of dry skin conditions [15]. Some of the more effective moisturizers contain simple lipids, such as triglyceride oils. More recently, the concept of using a lipid mixture with a composition resembling that of the stratum corneum for restoration of barrier function has been introduced [52]. It has been demonstrated that mixtures of ceramide, cholesterol, and fatty acid in appropriate proportions accelerate the recovery of barrier function in mice following damage to the barrier. In a recent clinical study, a ceramide-dominant lipid-based emollient was shown to improve the condition of the skin in a group of childhood atopic dermatitis patients. TEWL decreased, and the severity scoring of atopic dermatitis values improved. It was suggested that TEWL measurements could provide an objective measure of disease progression. The results of this study lead to a suggestion that the atopic dermatitis symptoms may be driven by the defective permeability barrier, as opposed to immunologic pathogenesis.

REFERENCES

1. Attenborough, D., *Life on Earth*, Little Brown & Company, Boston, 1980.
2. Blank, I.H., et al., The diffusion of water across the stratum corneum as a function of its water content, *J. Invest. Dermatol.*, 82, 188, 1984.
3. Takenouchi, M., Suzuki, H., and Tagami, H., Hydration characteristics of pathologic stratum corneum – evaluation of bound water, *J. Invest. Dermatol.*, 87, 574, 1986.
4. Larsson, K., *Lipids — Molecular Organization, Physical Functions and Technical Applications*, The Oily Press, Dundee, 1994, chap. 3.
5. Squier, C.A. and Lesch, C.A., Penetration pathways of different compounds through epidermis and oral epithelia, *J. Oral Pathol.*, 17, 512, 1988.
6. Talreja, P., et al., Visualization of the lipid barrier and measurement of lipid pathlength in human stratum corneum, *A.A.P.S. Pharmsci.*, 3, E13, 2001.
7. Bouwstra, J.A., et al., Structure of human stratum corneum as a function of temperature and hydration: A wide-angle X-ray diffraction study, *Int. J. Pharm.*, 84, 205, 1992.
8. Van Hal, D.A., et al., Structure of fully hydrated human stratum corneum: a freeze-fracture electron microscopy study, *J. Invest. Dermatol.*, 106, 89, 1996.
9. Presland, R.B., et al., Characterization of the human epidermal profilaggrin gene. Genomic organization and identification of an S-100-like calcium binding domain at the amino terminus, *J. Biol. Chem.*, 267, 23772, 1992.
10. Lynley, A.M. and Dale, B.A., The characterization of human epidermal filaggrin. A histidine-rich, keratin filament-aggregating protein, *Biochim. Biophys. Acta*, 744, 28, 1983.
11. Scott, I.R., Harding, C.R., and Barrett, J.G., Histidine-rich protein of the keratohyalin granules. Source of the free amino acids, urocanic acid and pyrrolidone carboxylic acid in the stratum corneum, *Biochim. Biophys. Acta*, 719, 110, 1982.
12. Scott, I.R. and Harding, C.R., Filaggrin breakdown to water binding compounds during development of the rat stratum corneum is controlled by the water activity of the environment, *Dev. Biol.*, 115, 84, 1986.
13. Caspers, P.J., *In vivo* confocal Raman microspectroscopy of the skin: noninvasive determination of molecular concentration profiles, *J. Invest. Dermatol.*, 116, 434, 2001.
14. Norlen, L., Emilson, A., and Forslind, B., Stratum corneum swelling. Biophysical and computer assisted quantitative assessments, *Arch. Dermatol. Res.*, 289, 506, 1997.
15. Loden, M., Biophysical properties of dry atopic and normal skin with special reference to effects of skin care products, *Acta Derm. Venereol.*, Suppl. 192, 1, 1995.
16. Berardesca, E. and Maibach, H.I., Transepidermal water loss and skin surface hydration in the noninvasive assessment of stratum corneum function, *Dermatosen Beruf Umwelt. Occupat. Environ. Dermatoses*, 38, 50, 1990.
17. Chilcott, R.P., et al., Transepidermal water loss does not correlate with skin barrier function *in vitro*, *J. Invest. Dermatol.*, 118, 871, 2002.
18. Imokawa, G. and Hattori, M., A possible function of structural lipids in the water-holding properties of the stratum corneum, *J. Invest. Dermatol.*, 84, 282, 1985.
19. Sweeney, T.M. and Downing, D.T., The role of lipids in the epidermal barrier to water diffusion, *J. Invest. Dermatol.*, 55, 135, 1970.
20. Montiero-Riviere, N.A., et al., Effect of selective lipid extraction from different body regions on epidermal barrier function, *Pharm. Res.*, 18, 992, 2001.

21. Abrams, K., et al., Effect of organic solvents on *in vitro* human skin water barrier function, *J. Invest. Dermatol.*, 101, 609, 1993.
22. Man, M.Q., et al., Cutaneous barrier repair and pathophysiology following barrier disruption in IL-1 and TNF type I receptor deficient mice, *Exp. Dermatol.*, 8, 261, 1999.
23. Fulmer, A.W. and Kramer, G.J., Stratum corneum lipid abnormalities in surfactant-induced dry scaly skin, *J. Invest. Dermatol.*, 86, 598, 1986.
24. Holleran, W.M., et al., Structural and biochemical basis for the UVB-induced alterations in epidermal barrier function, *Photoderm. Photoimmun. Photomed.*, 13, 117, 1887.
25. Meguro, S., et al., Relationship between covalently bound ceramides and trans-epidermal water loss (TEWL), *Arch. Dermatol. Res.*, 292, 463, 2000.
26. Fluhr, J.W., et al., Testing for irritation with a multifactorial approach: comparison of eight non-invasive measuring techniques on five different irritation types, *Br. J. Dermatol.*, 145, 696, 2001.
27. Frodin, T., Molin, L., and Skogh, M., Effects of single doses of UVA, UVB and UVC on skin blood flow, water content, and barrier function measured by laser-Doppler flowmetry, optothermal infrared spectrometry, and evaporimetry, *Photo-Dermatol.*, 5, 187, 1988.
28. Williams, M.L., The ichthyoses — pathogenesis and prenatal diagnosis: a review of recent advances, *Pediatr. Dermatol.*, 1, 1, 1983.
29. Mukherjee, S. and Gupta, A.B., A statistical study on *in vivo* sorption and desorption of water in ichthyosis vulgaris, *J. Dermatol.*, 21, 78, 1994.
30. Lavrijsen, A.P., et al., Barrier function parameters in various keratinization disorders: transepidermal water loss and vascular response to hexyl nicotinate, *Br. J. Dermatol.*, 129, 547, 1993.
31. Nirunsuksiri, W., et al., Decreased profilaggrin expression in ichthyosis vulgaris is a result of selectively impaired posttranscriptional control, *J. Biol. Chem.*, 270, 871, 1995.
32. Sybert, V.P., Dale, B.A., and Holbrook, K.A., Ichthyosis vulgaris: identification of a defect in synthesis of filaggrin correlated with an absence of keratohyaline granules, *J. Invest. Dermatol.*, 84, 191, 1985.
33. Wang, F., Man, M.Q., and Elias, P.M., A lipid mixture improves skin hydration in ichthyosis vulgaris, *Int. J. Dermatol.*, 36, 876, 1997.
34. Sakura, N., et al., Frequency of steroid sulfatase deficiency in Hiroshima, *Acta Paediatrica Japonica*, 40, 63, 1998.
35. Shapiro, L.J.R., et al., X-linked ichthyosis due to steroid sulfatase deficiency, *Lancet*, 1, 70, 1978.
36. Williams, M.L. and Elias, P.M., Stratum corneum lipids in disorders of cornification, *J. Clin. Invest.*, 68, 1404, 1981.
37. Sato, J., et al., Cholesterol sulfate inhibits proteases that are involved in desquamation of stratum corneum, *J. Invest. Dermatol.*, 111, 189, 1998.
38. Johansen, J.D., et al., Skin barrier properties in patients with recessive X-linked ichthyosis, *Acta Dermato-Venereol.*, 75, 202, 1995.
39. Fartasch, M., Epidermal barrier in disorders of the skin, *Micros. Res. Technol.*, 38, 361, 1997.
40. Candi, E., et al., Transglutaminase 1 mutations in lamellar ichthyosis. Loss of activity due to failure of activation by proteolytic processing, *J. Biol. Chem.*, 273, 13693, 1998.

41. Hashimoto, K., Gee, S. and Tanaka, K., Lamellar ichthyosis: response to etretinate with transglutaminase 1 recovery, *Am. J. Dermatopathol.*, 22, 277, 2000.
42. Thune, P., Evaluation of the hydration and water-holding capacity in atopic skin and so-called dry skin, *Acta Derm. Venereol.*, Suppl. 144, 133, 1989.
43. Werner, Y. and Lindberg, M., Transepidermal water loss in dry and clinically normal skin in patients with atopic dermatitis, *Acta Derm. Venereol.*, 65, 102, 1985.
44. Oranje, A.P. and de Waard-van der Spek, F.B., Atopic dermatitis: review 2000 to January 2001, *Curr. Opin. Pediatr.*, 14, 410, 2002.
45. Tamaki, K. and Nakamura, K., The role of lymphocytes in healthy and eczematous skin, *Curr. Opin. Allergy Clin. Immunol.*, 1, 455, 2001.
46. Di Nardo, A., et al., Ceramide and cholesterol composition of the skin of patients with atopic dermatitis, *Acta Derm. Venereol.*, 78, 27, 1998.
47. Prottey, C., Investigation of functions of essential fatty acids in the skin, *Br. J. Dermatol.*, 97, 29, 1977.
48. Elias, P.M. and Brown, B.E., (1978) The mammalian cutaneous permeability barrier: defective barrier function is essential fatty acid deficiency correlates with abnormal intercellular lipid deposition. *Lab. Invest.*, 39, 574, 1978.
49. Hansen, H.S. and Jensen, B., Essential function of linoleic acid esterified in acylglucosylceramide and acylceramide in maintaining the epidermal water permeability barrier: evidence from feeding studies with oleate, linoleate, arachidonate, columbinate and α-linolenate, *Biochim. Biophys. Acta*, 834, 357, 1985.
50. Melton, J.L., et al., Effects of essential fatty acid deficiency on epidermal O-acylsphingolipids and transepidermal water loss in young pigs, *Biochim. Biophys. Acta*, 921, 191, 1987.
51. Hou, S.Y.E., et al., Membrane structures in normal and essential fatty acid-deficient stratum corneum: characterization by ruthenium tetroxide staining and X-ray diffraction, *J. Invest. Dermatol.*, 96, 215, 1991.
52. Chamlin, S.L. et al., Ceramide-dominant barrier repair lipids alleviate childhood atopic dermatitis: changes in barrier function provide a sensitive indicator of disease activity, *J. Am. Acad. Dermatol.*, 47, 198, 2002.

30 Xerotic Skin and Impairment of Barrier Function Associated with Itch

Gil Yosipovitch

CONTENTS

I. Stratum Corneum Hydration and Barrier Function and Itch 370
A. Atopic Dermatitis 370
B. Senile Xerosis 370
C. Xerosis and Uremic Itch 371
II. HIV and Xerosis 371
III. Pathophysiology of Itch in Xerotic Skin 372
References 373

Dry xerotic skin, which is characterized by a rough, scaly, and flaky skin surface, has been associated for decades with the somatosensory sensation of itch. The list of clinical conditions of dry skin and itch as shown in Table 30.1 includes common skin diseases such as atopic dermatitis and senile pruritus, as well as systemic diseases such as uremia and human immunodeficiency virus (HIV). It is important to note that not every xerotic skin is associated with itch; for example, patients with ichthyosis vulgaris usually do not complain of itch.

Although the clinical associations between itch and dry skin is well established, data of studies examining the skin hydration and transepidermal water loss, which are objective measures for skin hydration and barrier impairment, provide conflicting results on such an association. This chapter will provide a review of the existing data and possible mechanisms of itch associated with xerotic skin.

0-8493-1443-7/05/$0.00+$1.50

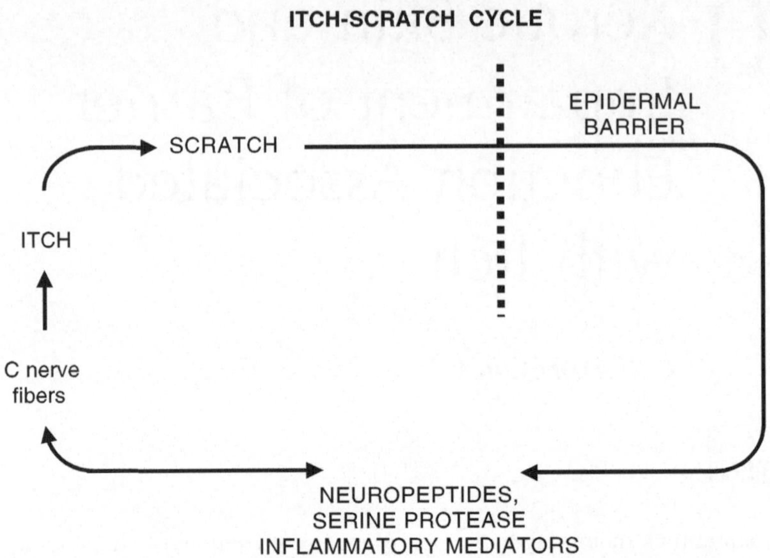

FIGURE 30.1 Mechanisms inducing itch following barrier perturbation.

I. STRATUM CORNEUM HYDRATION AND BARRIER FUNCTION AND ITCH

A. Atopic Dermatitis

Several studies have shown a significant increase in basal transepidermal water loss in atopic dermatitis, which is the most common itchy dermatosis. The higher transepidermal water loss was noted also in clinically uninvolved skin.[1–3] The degree of skin barrier impairment has been correlated to the degree of inflammation in atopic eczema, but none of the studies correlated the barrier function to itch severity.

B. Senile Xerosis

Senile xerosis is a common clinical entity associated with itch.[4] A study by Long and Marks[5] demonstrated that elderly patients (n = 13) with generalized pruritus had a clinically drier skin than age- and sex-matched control subjects (n = 13). The authors examined skin surface conductance, which is indicated to be a marker of stratum corneum water content, and found that skin conductance decreased in patients with generalized pruritus. In addition, the pruritic patients had increased intracorneal cohesion compared with controls. These results were consistent with an acquired abnormality of keratinization. A possible explanation to why xerosis causes itch in senile pruritus was suggested by Guillet et al.,[6] who found an increase in histamine release and skin hypersensitivity in 60 patients with senile pruritus. However, these results do not explain why patients with senile pruritus do not respond to oral antihistamines.

C. Xerosis and Uremic Itch

Dry skin is the most common skin manifestation of hemodialysis patients with end-stage renal disease.[7] Itch also exists in 60 to 90% of hemodialysis patients. The association between the degree of itch and the degree of skin dryness has been investigated in several studies, but most of them did not find any association. Young et al.[8] were the first to publish an association between intensity of itch and degree of skin dryness. However, other studies did not confirm any significant relationship between itch and skin dryness, using techniques for objective evaluation of skin dryness. Stahle-Backdahl[9] measured the stratum corneum hydration with capacitance and found no difference in values between hemodialysis patients with and without itch. We measured the stratum corneum hydration in 40 patients with end-stage renal failure on dialysis vs. 40 healthy controls and did not find any relationship to itch, although the skin hydration of the ill patients was significantly lower than that of controls.[10] In another study, we examined the relationship between stratum corneum hydration and skin surface pH to itch in hemodialysis patients. No relationship was found between the parameters and itch.[11] The skin surface pH was significantly elevated in hemodialysis patients in all body sites. Because skin surface pH is known to be elevated when barrier function is impaired,[12] these results provide evidence for an abnormal stratum corneum barrier function in hemodialysis patients. Kato et al.[13] studied the water content in the stratum corneum of hemodialysis patients and its relationship to the severity of itch. They used a sensitive technique of high-frequency conductance to measure the stratum corneum dryness with a surface hygrometer and, in addition, they measured resorption and desorption pre- and postdialysis treatments.[14] The results, again, did not find any correlation between skin xerosis and itch. A recent clinical study in hemodialysis patients demonstrated a significant relationship between intensity of xerosis and itch intensity. Significantly more patients with very rough skin had itch than those with slightly dry skin.[15] However, the authors did not assess skin dryness with any objective measures.

II. HIV and Xerosis

Xerosis with itch, especially in the lower legs, is common in HIV disease, occurring in over 20% of the patients.[16] In addition, treatment with protease inhibitors such as indinavir for HIV has shown to induce dry skin and itch in 40% of the patients using it.[17]

A recent study has shown that patients with HIV-associated xerosis had a significant reduction in cutaneous innervation density in the epidermis and papillary dermis in comparison to healthy skin.[18] In addition, the authors found that neuropeptide distribution of Calcitonin Gene Related Peptide was significantly reduced in the epidermis in both upper arms and upper legs, whereas substance P was only reduced in the upper arm. These results were similar to those in familial amyloid, where xerosis and itch occur and where a widespread depletion of cutaneous innervation was documented.[19]

III. PATHOPHYSIOLOGY OF ITCH IN XEROTIC SKIN

The pathophysiology involved with xerosis begins with the alteration of the structure of the stratum corneum and evolves into abnormalities in proliferation, keratinization, surface lipid, water metabolism, pH, and cytokine levels.[20] These may contribute to the sensation of itch. Recent studies in mice have provided new insights on the pathophysiology of itch in xerotic dry skin. Miyamoto et al.[21] demonstrated that dry skin induced by disruption of the cutaneous barrier function in the rostral back of mice caused a spontaneous increase in scratching. The authors used several techniques to disrupt the barrier, including sodium lauryl sulfate 1%, tape stripping, and cutaneous application of acetone/ether mixture (1:1 mixture [AE] and water following AE [AEW]). Only pretreatment with an acetone ether mixture with water (AEW) significantly increased the transepidermal water loss as well as the spontaneous scratching. The spontaneous scratching after the AEW treatment was significantly suppressed by two opioid antagonists: naloxone and naltrexone. Because these two drugs are known to inhibit itch in different systemic diseases such as cholestatic and uremic itch as well as in chronic dermatitis, it indicates that there is cross-talk between the stratum corneum and the peripheral nervous system. Degranulation of histamine occurs 48 hours after acetone treatment during exposure to a dry environment[22]; therefore, histamine could be a possible mediator for the itch in dry skin. To examine whether the AEW treatment increased the mast cell number or degranulation, Miyamoto et al.[21] also examined the number and degranulation of mast cells in AEW-treated skin and did not find any difference. In addition, they performed the same study in mast cell-deficient mice and were able to induce similar spontaneous scratching. Therefore, the mechanism of itch associated with dry skin does not seem to be associated with histamine secretion.

Possible mechanisms for itch in dry skin include induction of nerve fibers in the involved area by other mediators such as neuropeptides. Recently Takamori et al.[23] showed that the skin of patients with xerosis had a high density of intraepidermal fibers. Rat skin that was treated with AE had a significant increase in the density of these nerve fibers, with significant expression of nerve growth factor (NGF) in the epidermis. The findings indicate that dry skin induces elongation and penetration of C fibers to the epidermis by NGF. These results differ from the results in HIV xerosis as described above, where xerosis and itch were associated with reduction in the number of nerve fibers in the epidermis. A possible explanation for these discrepancies is that there are different mechanisms for itch-associated xerosis and that different neuromediators may be involved. However, all these studies indicate that xerotic itchy skin is associated with abnormal function of nerve fibers in the epidermis and papillary dermis.

Recently, a transgenic model expressing human stratum corneum of a serine protease named stratum corneum chymotryptic enzyme was developed and induced pathologic changes in the stratum corneum with hyperkeratosis and severe itch.[24] These results suggest that increased activity of this enzyme may play a role in itch associated with impaired epidermal barrier function. Other mediators that may be involved in eliciting itch in dry skin could be tryptase, cytokines like TNF alpha,

TABLE 30.1
Xerosis of the Skin Associated with Itch

1. Senile xerosis
2. Seasonal xerosis
3. Atopic dermatitis
4. Uremic xerosis
5. Human immunodeficiency virus xerosis
6. Familial amyloidosis
7. Congenital ichthyosis: lamellar ichthyosis, congenital ichthyosiform erythroderma, epidermyloytic hyperkeratosis

and interleukin 4.[25] Figure 30.1 summarizes possible mechanisms inducing itch following barrier perturbation.

In conclusion, xerotic skin can induce itch both in humans and in a murine model. The exact mechanisms are as yet unclear. A large body of evidence indicates that there is cross-talk between the stratum corneum and the nerve fibers in the epidermis and dermis that transfer the itch sensation to the central nervous system.

REFERENCES

1. Agner, T., Non-invasive measuring methods for the investigation of irritant patch test reactions. A study of patients with hand eczema, atopic dermatitis and controls, *Acta Derm. Venereol. (Suppl.)*, 173, 1, 1992.
2. Seidenari, S. and Giusti, G., Objective assessment of the skin of children affected by atopic dermatitis: a study of pH, capacitance and TEWL in eczematous and clinically uninvolved skin, *Acta Derm. Venereol.*, 75, 429, 1995.
3. Werner, Y. and Lindberg, M., Transepidermal water loss in dry and clinically normal skin in patients with atopic dermatitis, *Acta Derm. Venereol.*, 65, 102, 1985.
4. Fleischer, A.B., Pruritus in the elderly; management by senior dermatologists, *J. Am. Acad. Dermatol.*, 28, 603, 1993.
5. Long, C.C. and Marks, R., Stratum corneum changes in patients with senile pruritus, *J. Am. Acad. Dermatol.*, 27, 560, 1992.
6. Guillet, G., Zampetti, A., Czarlewski, W., and Guillet, M.H., Increased histamine release and skin hypersensitivity to histamine in senile pruritus: study of 60 patients, *J. Eur. Derm. Venereol.*, 14, 65, 2000.
7. Bencini, P.L., Montagnino, G., Citterio, A., et al., Cutaneous abnormalities in uremic patients, *Nephron*, 40, 316, 1985.
8. Young, A.W., Sweeney, E.W., Davis, D.S., et al., Dermatologic evaluation of pruritus in patients on haemodialysis, *N.Y. State J. Med.*, 73, 2670, 1973.
9. Stahle-Backdahl, M., Uremic pruritus. Clinical and experimental studies, *Acta Derm. Venereol. (Suppl.) (Stockh.)*, 145, 1, 1989.
10. Yosipovitch, G., Reis, J., Tur, E., et al., Sweat secretion, stratum corneum hydration, small nerve function and pruritus in patients with advanced chronic renal failure, *Br. J. Dermatol.*, 133, 561, 1995.

11. Yosipovitch, G., Tur, E., Morduchowicz, G., and Boner, G., Skin surface pH, moisture, and pruritus in haemodialysis patients, *Nephrol. Dial. Transplant*, 8, 1129, 1993.
12. Fluhr, J.W., Kao, J., Jain, M., et al., Generation of free fatty acids from phospholipids regulates stratum corneum acidification and integrity, *J. Invest. Dermatol.*, 117, 44, 2001.
13. Kato, A., Hamada, M., Maruyama, T., et al., Pruritus and hydration state of stratum corneum in hemodialysis patients, *Am. J. Nephrol.*, 20, 437, 2000.
14. Tagami, H., Kanamaru, Y., Inoue, K., et al., Water sorption-desorption test of the skin *in vivo* for functional assessment of the stratum corneum, *J. Invest. Dermatol.*, 78, 425, 1982.
15. Szepietowski, J.C., Sikora, M., Kusztal, M., Salomon, J., Magott, M., and Szepietowski, T., Uremic pruritus: a clinical study of maintenance hemodialysis patients, *J. Dermatol.*, 20, 621, 2002.
16. Uthayakumar, S., Nadwani, R., Drinkwater, T., et al., The prevalence of skin disease in HIV infection and its relationship to the degree of immunosuppression, *Br. J. Dermatol.*, 137, 595, 1997.
17. Calista, D. and Boschini, A., Cutaneous side effects induced by indinavir, *Eur. J. Dermatol.*, 10(4), 292, 2000.
18. Rowe, A., Mallon, E., Rosenberger, P., et al., Depletion of cutaneous peptidergic innervation in HIV associated xerosis, *J. Invest. Dermatol.*, 112, 284, 1999.
19. Ramsey, B., Terenghi, G., Polak, J.M., et al., Depleted cutaneous innervation in familial amyloid, *Clin. Exp. Dermatol.*, 21, 449, 1996.
20. Elias, P.M. and Ghadially, R., The aged epidermal permeability barrier: basis for functional abnormalities, *Clin. Geriatr. Med.*, 18, 103, 2002.
21. Miyamoto, T., Nojima, H., Shinkado, T., et al., Itch-associated response induced by experimental dry skin in mice, *Jpn. J. Pharmacol.*, 88, 285, 2002.
22. Denda, M., Sato, J., Tsuchiya, T., et al., Low humidity stimulates epidermal DNA synthesis and amplifies the hyperproliferative response to barrier disruption; implication for seasonal exacerbations of inflammatory dermatoses, *J. Invest. Dermatol.*, 111, 873, 1998.
23. Takamori, K., Takimoto, R., and Hase, T., *Mechanisms of Itch in Dry Skin — NGF Induces the Elongation/Penetration of Nerve Fibers into the Epidermis*, International Workshop for the Study of Itch, October 5–7, 2001, Singapore, pp. 36.
24. Hansson, L., Backman, A., Ny, A., et al., Epidermal overexpression of stratum corneum chymotryptic enzyme in mice: a model for chronic itchy dermatitis, *J. Invest. Dermatol.*, 118, 444, 2002.
25. Chan, L.S., Robinson, N., and Xu, L., Expression of interleukin-4 in the epidermis of transgenic mice results in a pruritic inflammatory skin disease: an experimental animal model to study atopic dermatitis, *J. Invest. Dermatol.*, 117, 977, 2001.

31 Testing and Efficacy of Barrier Creams

Hongbo Zhai and Howard I. Maibach

CONTENTS

I. Introduction ... 375
II. Testing Methodology ... 375
A. *In Vitro* Methods ... 375
B. *In Vivo* Methods ... 376
III. Conclusions ... 383
References ... 383

I. INTRODUCTION

Efficacy of barrier creams (BCs) in the prevention of contact dermatitis (CD) including allergic contact dermatitis (ACD) and irritant contact dermatitis (ICD) has been documented.[1–9] However, some reports indicate that inappropriate BC application exacerbates irritation rather than providing benefit effects.[1,2,10–16] This might be a design defect of the tested BC, but another possibility relates to the testing methodology. The methodology is important, and hence the accuracy of results depends on the choice of proper models.

To test efficacy of BC, several *in vitro* and *in vivo* methods have been developed and the subject has been extensively reviewed.[1–9,17,18] This chapter summarizes recent updated information.

II. TESTING METHODOLOGY

A. *In Vitro* Methods

Treffel et al.[15] measured the effectiveness of BC on human skin against dyes (eosin, methylviolet, and oil red O) with varying n-octanol/water partition coefficients (0.19, 29.8, and 165, respectively). BC effects were assayed by measurements of the dyes in the epidermis of protected skin samples after 30 minutes. Some BCs showed efficacy, but several revealed data contrary to manufacturer's claims. No correlation existed between the galenic (pharmaceutic) parameters of the assayed products and

0-8493-1443-7/05/$0.00+$1.50

the protection level, indicating that neither the water content nor the consistency of the formulations influenced effectiveness.

Fullerton and Menne[19] tested the protective effect of ethylenediaminetetraacetate barrier gels against nickel contact allergy. Thirty milligrams of barrier gel was applied on the epidermal side of the skin *in vitro*, and a nickel disc was applied above the gel. After 24 hours, the nickel disc was removed and the epidermis separated from the dermis. Nickel content in epidermis and dermis was quantified by adsorption differential pulse voltammetry. The amount of nickel in the epidermal skin layer on barrier gel–treated skin was significantly reduced compared with the untreated control.

Shah and Kirchner[20] evaluated moisture penetration through a thin film of skin protectants both *in vivo* and *in vitro*, using Fourier transform infrared spectroscopy. Petrolatum offered some protection against water penetration; a hydroactive polymer system (protectant) prevented moisture penetration.

Goffin et al.[21] assessed the efficacy of BC to surfactants and organic solvents with shielded variants of the corneosurfametry and corneoxenometry methods. Petrolatum exhibited the best protection being a blocker for sodium lauryl sulfate (SLS) and also provided protection against hexane–methanol.

Zhai et al.[22] used an *in vitro* diffusion system to measure the protective effective of three quaternium-18 bentonite (Q18B) gels to prevent 1% ^{35}S-SLS penetration by human cadaver skin. The accumulated amount of ^{35}S-SLS in receptor cell fluid was counted to evaluate the efficacy of the model Q-18B gels over 24 hours. These test gels significantly decreased SLS absorption when compared with the unprotected skin control samples. Protection effect percentage of the three Q18B gels was 88%, 81%, and 65%, respectively.

van Der Bijl et al.[23] conducted an *in vitro* study of the permeability of tritiated water through fresh and frozen human skin in the presence and absence of two different BCs in a flow-through diffusion system. Buffer/tritiated water was collected from the acceptor chambers at 2-hour intervals for a total of 20 hours and counted in a liquid scintillation counter. Both BCs lowered the average flux rates of tritiated water through fresh and frozen skin, but no significant differences could be detected between the two.

B. *In Vivo* Methods

Lachapelle et al.[24,25] developed a guinea pig model to evaluate the protective value of BC or gels by laser Doppler flowmetry (blood flow) and histological assessment. The histopathological damage after 10 minutes of contact to toluene was mainly confined to the epidermis. Dermal blood flow changes were relatively high on the control site compared with the gel-pretreated sites.

Fullerton and Menne[19] performed an *in vivo* patch testing with nickel-sensitive patients by using nickel discs with and without barrier gels. Test preparations and nickel discs were removed 1 day postapplication, and the test sites were evaluated. Barrier gel–treated sites significantly reduced the positive test reactions.

Frosch et al.[12–14,26] established the repetitive irritation test (RIT) in the guinea pig and in humans to evaluate the efficacy of BC using series bioengineering techniques. The cream-pretreated and untreated test skin (guinea pig or humans) was exposed daily to the irritants for 2 weeks. The resulting irritation was scored on a visual scale and assessed by biophysical (bioengineering) technique parameters. Some test creams suppressed irritation with all test parameters, though some even increased irritation.

Grunewald et al.[27] used a SLS repetitive washing model to evaluate the protective effects of BCs by measuring with bioengineering techniques on 15 human volunteers. All BCs reduced the deterioration of skin functions following 1 week repetitive washing. Subsequently, they also found that urea- and glycerol oil-in-water emulsions provided a greater protection against a lipophilic irritant (toluene) after 7 days of repetitive irritation.[28]

Marks et al.[29] investigated a topical lotion containing 5% Q18B in the prevention of experimentally induced poison ivy and poison oak ACD in susceptible volunteers. One hour before both forearms were patch tested with urushiol, 5% Q18B lotion was applied on one forearm. The test patches were removed after 4 hours and the sites interpreted for reaction 2, 5, and 8 days later. The test sites pretreated with Q18B lotion had absent or significantly reduced reactions to the urushiol compared with untreated control sites ($P < .0001$) on all test days.

Zhai and Maibach[16] measured the effectiveness of BC in an *in vivo* human model against dye indicator solutions: methylene blue in water and oil red O in ethanol, representative of model hydrophilic and lipophilic compounds. Solutions of 5% methylene blue and 5% oil red O were applied to untreated and BC-pretreated skin, with the aid of aluminum occlusive chambers, for 0 hours and 4 hours. At the end of the application time, materials were removed and consecutive skin surface biopsies obtained. The amount of dye penetrating into each strip was determined by colorimetry. Two model creams exhibited effectiveness, but one enhanced the cumulated amount of dye.

Schlüter-Wigger and Elsner[30] assessed four commercially available BCs against four standard irritants: 10% SLS, 1% sodium hydroxide (NaOH), 30% lactic acid (LA), and undiluted toluene (TOL) in the RIT in humans for 12 days. Irritation was assessed by visual scoring, transepidermal water loss (TEWL), and colorimetry. All products were very effective against SLS irritation. No BC provided significant protection against TOL. Three products showed a partially protective effect against all ionic irritants, whereas the fourth showed less protection against SLS and NaOH, and even amplification of inflammation by TOL. Wigger-Alberti and Elsner[31] evaluated petrolatum using the above model: Petrolatum was effective against SLS, NaOH, and LA irritation and provided a moderate protection against TOL. Wigger-Alberti et al.[32] also examined three other BCs and petrolatum against 10% SLS, 0.5% NaOH, 15% LA, and undiluted TOL in the RIT in humans for 9 days. All BCs exhibited a significant protective effect against irritation by SLS, NaOH, and LA. Lower efficacy was observed against TOL. In another 12-day RIT study,[33] white petrolatum provided a significant protective effect against SLS, NaOH, and TOL, but with less protective effect against LA irritation.

de Fine Olivarius et al.[34] determined BC efficacy in protecting against water, based on evaluation of color intensities when an aqueous solution of crystal violet is applied to the skin, after pretreatment with different creams. The BCs with particles gave the best immediate protection (dorsal 76%, volar 69%). The moisturizer was intermediately protective (dorsal 57%, volar 34%), whereas little protection was found for the silicone-containing cream (dorsal 16%, volar 10%).

Fartasch et al.[35] investigated protective capacity of a lipophilic BC on acute ICD by TEWL measurement. Application of the BC before and during irritation showed a decrease of TEWL by 58% (back) and 49% (arm).

Elsner et al.[36] evaluated perfluoropolyethers (PFPE) containing BC against a set of four irritants: 10% SLS, 0.5% NaOH, 15% LA, and undiluted TOL in the RIT on the person's back. Irritation was assessed by visual scoring, TEWL, and colorimetry. All PFPE preparations significantly suppressed irritation by SLS and NaOH. However, only the 4% PFPE preparation was significant against LA and TOL.

Zhai et al.[37] introduced a facile approach to screening protectants *in vivo* in human subjects. Two acute irritants and one allergen were selected: SLS, the combination of ammonium hydroxide (NH_4OH) and urea, and Rhus. The model irritants and allergen were applied with an occlusive patch for 24 hours. Inflammation was scored with an expanded 10-point scale at 72 hours postapplication. Most test protectants statistically suppressed SLS irritation and Rhus allergic reaction but not NH_4OH and urea-induced irritation. The authors further used this model to evaluate the putative skin-protective formulations.[38] All formulations failed to inhibit NH_4OH and urea irritation. Only paraffin wax in cetyl alcohol statistically ($P < .01$) reduced Rhus-ACD. Three commercial formulations markedly ($P < .001$) suppressed SLS-ICD.

Shimizu[39] used a squamometry method to evaluate a barrier protectant (tannic acid). Five percent tannic acid and distilled water (as a control) were applied to forearms for 30 minutes; these pretreated sites were dosed with different concentrations of SLS for 24 hours. Squamometric evaluation indicated that the skin damage increased with SLS concentration in a dose-dependent manner and that tannic acid significantly reduced the damage ($P < .01$).

Vidmar and Iwane[40] assessed the ability of the topical skin protectant (TSP) to protect against urushiol (Rhus)-ACD. Open urushiol patch testing was conducted on 50 Rhus-sensitive subjects. After 96 hours, dermatitis severity scores were compared between TSP protected and TSP unprotected sites by using a 9-point dermatitis scale. Results showed that TSP-protected sites had mean dermatitis scores about 2 points lower than TSP-unprotected sites ($P < .001$).

Patterson et al.[41] determined the preventive effect of a skin protectant containing dimethicone and glycerin with various inactive ingredients in an aerosol foam against SLS-ICD and poison ivy and poison oak (urushiol) ACD. Skin reaction was assessed periodically for 10 days by using a 0- to 7-point dermatitis scale. The formulation was significantly effective in reducing the irritation from SLS but did not prevent urushiol-ACD.

Zhai et al.[42] evaluated the efficacy of a dimethicone skin protectant lotion against SLS-ICD by clinical visual grading and bioengineering techniques in humans. Both forearms were pretreated either with the testing protectant lotion or with its vehicle control before contact with SLS. Thirty minutes later, 0.5% SLS was applied to the each pretreated site for 24 hours. One additional site received SLS only. The efficacy of protective effect was determined by visual scoring (VS), TEWL, skin color (a* value), and cutaneous blood flow volume (BFV). VS and TEWL data showed a significant decrease on the protectant lotion pretreated site in comparison with SLS-only treated sites, as well as to the vehicle control site. However, BFV and a* values did not show a statistical difference between either treated site.

Berndt et al.[43] investigated the efficacy of a BC and its vehicle in a field setting: two panels of 25 hospital nurses with mild signs of skin irritation were asked to use one of the test products (BC or its vehicle), and especially to use the product before contact with skin irritants over 4 weeks. Effects of both preparations were studied weekly by clinical examination and bioengineering measurements. Results showed no significant differences between BC and its vehicle. In both groups, clinical skin status improved and stratum corneum hydration increased significantly during the study. The authors concluded that the vehicle alone is capable of positively influencing skin status.

Schnetz et al.[44] introduced a standardized test procedure for the evaluation of skin protective products. A repeated short-time occlusive irritation test with a standardized protocol has been evaluated in two phases (protocols of 12 days and 5 days) in several clinical centers. Skin was treated by two irritants (0.5% SLS and toluene, two times per day for 30 minutes). Inflammation was measured by bioengineering methods (TEWL and colorimetry) and clinical scoring. The 5-day protocol was sufficient to achieve significant results. Furthermore, in spite of the expected intercenter variations resulting from heterogeneity of the individual threshold of irritation, interpretation of clinical score, and interinstrumental variability, the ranking of the vehicles regarding reduction of the irritant reaction was consistent in all centers.

McCormick et al.[45] measured the efficacy of a BC and an oil-containing lotion for protecting the hands of health care workers with severe hand irritation. Objective and subjective parameters for scaling, cracking, weeping, bleeding, and pain were scored by two blinded investigators weekly for 4 weeks. Subjects in both groups experienced marked improvement in overall hand condition (each, $P < .02$), particularly in scaling, cracking, and pain. Volunteers randomly assigned to use of the oil-containing lotion showed the greatest improvement.

Sun et al.[46] used laser-induced breakdown spectroscopy to evaluate the effect of BC on human skin; three representatives of commercial BCs advertised as being effective against lipophilic and hydrophilic substances were evaluated by measuring zinc absorbed through the stratum corneum. Four consecutive SSB were taken from biceps of the forearms of six volunteers at time periods of 0.5 and 3 hours after BC application. The BC provided appreciable protection against the penetration of both $ZnCl_2$ and ZnO into the skin when compared with control skin (not BC treated).

Allmers[47] tested two types of latex gloves with and without the use of a BC on subjects who had type I hypersensitivity reactions to natural rubber latex gloves. One hand received BC for 10 minutes before both hands used gloves for 30 minutes. BC decreased the latex glove–induced contact urticaria syndrome. Table 31.1 summarizes the recent experimental models and BC efficacy.

TABLE 31.1
BC Efficacy and Testing Models from Recent Experiments

Models		Irritants or Allergens	Barrier Cream	Efficacy	Refs.
In vitro	*In vivo*				
Human skin		Dyes (eosin, methylviolet, oil red O)	16 BC	Protection varied	Treffel et al.[15]
Human skin	Nickel-sensitive patients	Nickel disc	(EDTA) gels	Significantly reduced the amount of nickel in the epidermis *in vitro*, and significantly reduced positive reactions *in vivo*	Fullerton and Menne[19]
Thin protective product film	Humans	Moisture penetration	Petrolatum and a protectant	Petrolatum offered some protection against water penetration and a hydroactive polymer system (protectant) prevented moisture penetration	Shah and Kirchner[20]
Human skin		SLS and a mixture of hexane and methanol	Six products	Petrolatum exhibited the best protective effect to against SLS and also provided a good protection rates to organic solvents	Goffin et al.[21]
Human skin		[^{35}S]-SLS	Three Q18B gels	Protection effect (%) was 88%, 81% and 65%, respectively	Zhai et al.[22]
Fresh and frozen human skin		Tritiated water	Two BC	Both BC lowered the average flux rates of tritiated water through fresh and frozen skin	van Der Bijl et al.[23]
	Guinea pig	TOL, n-hexane, and trichlorethylene	Antisolvent gel and other 3 BC	Dermal blood flow changes were relatively high on the control site compared to the gel pretreated sites. They found that BC can against TOL, n-hexane but not to trichlorethylene	Lachapelle et al.[24,25]
	Guinea pigs and humans	SLS, sodium hydroxide, TOL, and LA	Several BC	Some suppressed irritation, some failed, and some exacerbated	Frosch et al.[12-14,26]

Humans	TOL	Several BC	All tested BC markedly reduced the irritating effect of repetitive toluene contact	Grunewald et al.[27]
Humans with a history of allergy to poison ivy/oak	Urushiol	Q18B lotion	Q-18B lotion significantly reduced reactions to the urushiol	Marks et al.[29]
Humans	Dyes (methylene blue and oil red O)	Three BC	Two exhibited effectiveness; one enhanced cumulative amount of dye	Zhai and Maibach[16]
Humans	10% SLS, 1% NaOH, 30% LA, and TOL	Four BC and white petrolatum	Different protective effects were detectable. All products were effective against SLS irritation	Schlüter-Wigger and Elsner[30]
Humans	Water	Two BC and a moisturizer	BC with particles provided the greatest immediate protection (dorsal 76%, volar 69%)	de Fine Olivarius et al.[34]
Humans	SLS	A lipophilic BC	A decrease of TEWL by 58% (back) and 49% (arm)	Fartasch et al.[35]
Humans	SLS, NaOH, LA, and TOL	BC containing PFPE	All BC significantly suppressed irritation by SLS and NaOH. Only the 4% PFPE-containing preparation was significant against LA and TOL	Elsner et al.[36]
Humans	SLS, NH_4OH and urea, Rhus	Several protectants	Most suppressed the SLS irritation and Rhus ACD, but failed to NH_4OH and urea irritation	Zhai et al.[37]
Humans	SLS	A barrier protectant (tannic acid)	Tannic acid significantly reduced SLS induced damage	Shimizu and Maibach[39]
Rhus-sensitive subjects	Urushiol	TSP	TSP protected sites had lower dermatitis scores than TSP unprotected sites	Vidmar and Iwane[40]
Humans and Rhus-sensitive subjects	SLS-ICD and poison ivy and poison oak (urushiol)-ACD	A formulation containing dimethicone and glycerin	Test formulation significantly effective in reducing the irritation from SLS but did not prevent urushiol-ACD	Patterson et al.[41]

(continued)

TABLE 31.1 (CONTINUED)
BC Efficacy and Testing Models from Recent Experiments

Models		Irritants or Allergens	Barrier Cream	Efficacy	Refs.
In vitro	*In vivo*				
	Humans	SLS	A dimethicone-containing skin protectant lotion	VS and TEWL data showed a significant decrease on the protectant lotion pretreated site in comparison to SLS only treated site as well as to vehicle control site	Zhai et al.[42]
	Nurses with mild signs of skin irritation	Occupational risk exposures	A test BC or its vehicle	No significant differences between BC and its vehicle. In both groups, clinical skin status improved and stratum corneum hydration increased significantly during the study	Berndt et al.[43]
	Humans	SLS and TOL	Three BCs	Various protective effects were detectable with 5-day study protocol to achieve significant results	Schnetz et al.[44]
	Health care workers with severe hand irritation	Working environments	One BC and an oil-containing lotion	Subjects in both groups experienced marked improvement in overall hand condition, but use of the oil-containing lotion showed greater improvement	McCormick et al.[45]
	Humans	Zinc	Three representative commercial BCs	BC provided appreciable protection against the penetration of both $ZnCl_2$ and ZnO into skin	Sun et al.[46]
	Humans	Two types of latex gloves	A test BC	BC decreased latex glove-induced contact urticaria syndrome	Allmers[47]

III. CONCLUSIONS

The *in vitro* and *in vivo* methods described above with common-use irritants and allergens in occupational use and in the home have been well developed; these provide insight into mechanisms as well as greater discriminatory potential.

In vitro models, which mimic the reaction of *in vivo* skin, are widely used to test the effects of BC because they are simple, rapid, and safe. In particular, they are recommended in screening procedures for BC candidates. Radiolabeled methods may determine the accurate protective and penetration results even with the lower levels of chemicals because of the sensitivity of radiolabeled counting. Animal experiments may be used to generate kinetic data because of a closer similarity between humans and some animals (pigs and monkeys) in percutaneous absorption and penetration for some compounds. However, no one animal, with its complex anatomy and biology, will simulate penetration in humans for all compounds. Therefore, the best estimate of human percutaneous absorption is determined by *in vivo* studies in humans. Histological assessments may define what layers of skin are damaged or protected and may provide insight into BC mechanisms. Noninvasive bioengineering techniques provide accurate, highly reproducible, and objective observations in quantifying the inflammation response to various irritants and allergens; they can assess subtle differences to supplement traditional clinical studies.

To date, however, no standardized procedure for the evaluation of BCs exists. The testing quality and the reproducibility of results still remain incompletely developed — in fact, this is one reason for the lively debate concerning the prevention of CD by BC application. Therefore, a sensible, standardized, and widely accepted method is warranted. In addition, it should be taken into consideration that, no matter whether the data were generated from *in vitro* or *in vivo* models, they were probably generated in laboratory conditions. Because many factors may influence the effectiveness of the BC,[48–50] we suggest that BC actual benefits should be evaluated in the workplace as a supplement to model experiments.

REFERENCES

1. Frosch, P.J., Kurte, A., and Pilz, B., Biophysical techniques for the evaluation of skin protective creams, in *Noninvasive Methods for the Quantification of Skin Functions*, Frosch, P.J. and Kligman, A.M., Eds., Springer, Berlin, 1993, 214.
2. Lachapelle, J.M., Efficacy of protective creams and/or gels, in *Prevention of Contact Dermatitis, Curr Probl Dermatol* Elsner, P., Lachapelle, J.M., Wahlberg, J.E., and Maibach, H.I., Eds., Karger, Basel, 1996, 182.
3. Zhai, H. and Maibach, H.I., Percutaneous penetration (Dermatopharmacokinetics) in evaluating barrier creams, in *Prevention of Contact Dermatitis*, Elsner, P., Lachapelle, J.M., Wahlberg, J.E., and Maibach, H.I., Eds., Karger, Basel, 1996, 193.
4. Wigger-Alberti, W. and Elsner, P., Do barrier creams and gloves prevent or provoke contact dermatitis? *Am. J. Contact Dermatitis*, 9, 100, 1998.

5. Zhai, H. and Maibach, H.I., Efficacy of barrier creams (skin protective creams), in *Cosmetics. Controlled Efficacy Studies and Regulation*, Elsner, P., Merk, H.F., and Maibach, H.I., Eds., Springer, Berlin, 1999, 156.
6. Wigger-Alberti, W. and Elsner, P., Protective creams, in *Cosmeceuticals. Drugs vs. Cosmetics*, Elsner, P. and Maibach, H. I., Eds., Marcel Dekker, New York, 2000, 189.
7. Maibach, H.I. and Zhai, H., Evaluations of barrier creams, in *Strategies to Protect the Health of Deployed U.S. Forces. Force Protection and Decontamination*, Wartell, M.A., Kleinman, M.T., Huey, B.M., and Duffy, L.M., Eds., National Academy Press, Washington, D.C., 2000, 217.
8. Wigger-Alberti, W. and Elsner, P., Barrier creams and emollients, in *Handbook of Occupational Dermatology*, Kanerva, L., Elsner, P., Wahlberg, J.E., and Maibach, H.I., Eds., Springer, Berlin, 2000, 490.
9. Zhai, H. and Maibach, H.I., Barrier creams — skin protectants: can you protect skin? *J. Cosmetic Dermatol.*, 1, 20, 2002.
10. Goh, C.L., Cutting oil dermatitis on guinea pig skin. (I). Cutting oil dermatitis and barrier cream, *Contact Dermatitis*, 24, 16, 1991.
11. Goh, C.L., Cutting oil dermatitis on guinea pig skin. (II). Emollient creams and cutting oil dermatitis, *Contact Dermatitis*, 24, 81, 1991.
12. Frosch, P.J., Schulze-Dirks, A., Hoffmann, M., Axthelm, I., and Kurte, A., Efficacy of skin barrier creams. (I). The repetitive irritation test (RIT) in the guinea pig, *Contact Dermatitis*, 28, 94, 1993.
13. Frosch, P.J., Schulze-Dirks, A., Hoffmann, M., and Axthelm, I., Efficacy of skin barrier creams. (II). Ineffectiveness of a popular "skin protector" against various irritants in the repetitive irritation test in the guinea pig, *Contact Dermatitis*, 29, 74, 1993.
14. Frosch, P.J., Kurte, A., and Pilz, B., Efficacy of skin barrier creams. (III). The repetitive irritation test (RIT) in humans, *Contact Dermatitis*, 29, 113, 1993.
15. Treffel, P., Gabard, B., and Juch, R., Evaluation of barrier creams: an *in vitro* technique on human skin, *Acta. Derm. Venereol.*, 74, 7, 1994.
16. Zhai, H. and Maibach, H.I., Effect of barrier creams: human skin *in vivo*, *Contact Dermatitis*, 35, 92, 1996.
17. Zhai, H. and Maibach, H.I., Models assay for evaluation of barrier formulations, in *Hand Eczema*, 2nd edition, Menné, T. and Maibach, H.I., Eds., CRC Press, Boca Raton, Florida 2000, 333.
18. Zhai, H. and Maibach, H.I., Tests for Skin Protection: Barrier Effect, in *Handbook of Cosmetic Science and Technology*, Barel, A.O., Maibach, H.I., and Paye, M., Eds., Marcel Dekker, New York, 2001, 823.
19. Fullerton, A. and Menne, T., *In vitro* and *in vivo* evaluation of the effect of barrier gels in nickel contact allergy, *Contact Dermatitis*, 32, 100, 1995.
20. Shah, S. and Kirchner, F., *In vitro* and *in vivo* evaluation of water penetration through skin protectant barriers, *Skin Res. Technol.*, 3, 114, 1997.
21. Goffin, V., Piérard-Franchimont, C., and Piérard, G.E., Shielded corneosurfametry and corneoxenometry: novel bioassays for the assessment of skin barrier products, *Dermatology*, 196, 434, 1998.
22. Zhai, H., Buddrus, D.J., Schulz, A.A., Wester, R.C., Hartway, T., Serranzana, S., and Maibach, H.I., *In vitro* percutaneous absorption of sodium lauryl sulfate (SLS) in human skin decreased by Quaternium-18 bentonite gels, *In Vitro Mol. Toxicol.*, 12, 11, 1999.

23. van Der Bijl, P., van Eyk, A.D., Cilliers, J., and Stander, I.A., Diffusion of water across human skin in the presence of two barrier creams, *Skin Pharmacol. Appl. Skin Physiol.*, 13, 104, 2000.
24. Mahmoud, G. and Lachapelle, J.M., Evaluation of the protective value of an antisolvent gel by laser Doppler flowmetry and histology, *Contact Dermatitis*, 13, 14, 1985.
25. Lachapelle, J.M., Nouaigui, H., and Marot, L., Experimental study of the effects of a new protective cream against skin irritation provoked by the organic solvents n-hexane, trichlorethylene and toluene, *Dermatosen*, 38, 19, 1990.
26. Frosch, P.J. and Kurte, A., Efficacy of skin barrier creams. (IV). The repetitive irritation test (RIT) with a set of 4 standard irritants, *Contact Dermatitis*, 31, 161, 1994.
27. Grunewald, A.M., Gloor, M., Gehring, W., and Kleesz, P., Barrier creams. Commercially available barrier creams versus urea- and glycerol-containing oil-in-water emulsions, *Dermatosen*, 43, 69, 1995.
28. Grunewald, A.M., Lorenz, J., Gloor, M., Gehring, W., and Kleesz, P., Lipophilic irritants: protective value of urea- and of glycerol-containing oil-in-water emulsions, *Dermatosen*, 44, 81, 1996.
29. Marks, J.G. Jr., Fowler, J.F. Jr., Sheretz, E.F., and Rietschel, R.L., Prevention of poison ivy and poison oak allergic contact dermatitis by quaternium-18 bentonite, *J. Am. Acad. Dermatol.*, 33, 212, 1995.
30. Schlüter-Wigger, W. and Elsner, P., Efficacy of 4 commercially available protective creams in the repetitive irritation test (RIT), *Contact Dermatitis*, 34, 278, 1996.
31. Wigger-Alberti, W. and Elsner, P., Petrolatum prevents irritation in a human cumulative exposure model *in vivo*, *Dermatology*, 194, 247, 1997.
32. Wigger-Alberti, W., Rougier, A., Richard, A., and Elsner, P., Efficacy of protective creams in a modified repeated irritation test. Methodological aspects, *Acta Derm.Venereol.*, 78, 270, 1998.
33. Wigger-Alberti, W., Caduff, L., Burg, G., and Elsner, P., Experimentally induced chronic irritant contact dermatitis to evaluate the efficacy of protective creams *in vivo*, *J. Am. Acad. Dermatol.*, 40, 590, 1999.
34. de Fine Olivarius, F., Hansen, A.B., Karlsmark, T., and Wulf, H.C., Water protective effect of barrier creams and moisturizing creams: a new *in vivo* test method, *Contact Dermatitis*, 35, 219, 1996.
35. Fartasch, M., Schnetz, E., and Diepgen, T.L., Characterization of detergent-induced barrier alterations — effect of barrier cream on irritation, *J. Invest. Dermatol.*, 3, 121, 1998.
36. Elsner, P., Wigger-Alberti, W., and Pantini, G., Perfluoropolyethers in the prevention of irritant contact dermatitis, *Dermatology*, 197, 141, 1998.
37. Zhai, H., Willard, P., and Maibach, H.I., Evaluating skin-protective materials against contact irritants and allergens. An *in vivo* screening human model, *Contact Dermatitis*, 38, 155, 1998.
38. Zhai, H., Willard, P., and Maibach, H.I., Putative skin-protective formulations in preventing and/or inhibiting experimentally-produced irritant and allergic contact dermatitis, *Contact Dermatitis*, 41, 190, 1999.
39. Shimizu, T. and Maibach, H.I., Squamometry: an evaluation method for a barrier protectant (tannic acid), *Contact Dermatitis*, 40, 189, 1999.
40. Vidmar, D.A. and Iwane, M.K., Assessment of the ability of the topical skin protectant (TSP) to protect against contact dermatitis to urushiol (Rhus) antigen, *Am. J. Contact Dermatitis*, 10, 190, 1999.

41. Patterson, S.E., Williams, J.V., and Marks, J.G., Jr., Prevention of sodium lauryl sulfate irritant contact dermatitis by Pro-Q aerosol foam skin protectant, *J. Am. Acad. Dermatol.*, 40, 783, 1999.
42. Zhai, H., Brachman, F., Pelosi, A., Anigbogu, A., Ramos, M.B., Torralba, M.C., and Maibach, H.I., A bioengineering study on the efficacy of a skin protectant lotion in preventing SLS-induced dermatitis, *Skin Res. Technol.*, 6, 77, 2000.
43. Berndt, U., Wigger-Alberti, W., Gabard, B., and Elsner, P., Efficacy of a barrier cream and its vehicle as protective measures against occupational irritant contact dermatitis, *Contact Dermatitis*, 42, 77, 2000.
44. Schnetz, E., Diepgen, T.L., Elsner, P., Frosch, P.J., Klotz, A.J., Kresken, J., Kuss, O., Merk, H., Schwanitz, H.J., Wigger-Alberti, W., and Fartasch, M., Multicentre study for the development of an *in vivo* model to evaluate the influence of topical formulations on irritation, *Contact Dermatitis*, 42, 336, 2000.
45. McCormick, R.D., Buchman, T.L., and Maki, D.G., Double-blind, randomized trial of scheduled use of a novel barrier cream and an oil-containing lotion for protecting the hands of health care workers, *Am. J. Infection Control*, 28, 302, 2000.
46. Sun, Q., Tran, M., Smith, B., and Winefordner, J.D., In-situ evaluation of barrier-cream performance on human skin using laser-induced breakdown spectroscopy, *Contact Dermatitis*, 43, 259, 2000.
47. Allmers, H., Wearing test with 2 different types of latex gloves with and without the use of a skin protection cream, *Contact Dermatitis*, 44, 30, 2001.
48. Packham, C.L., Evaluation of barrier creams: an *in vitro* technique on human skin (letter), *Acta. Derm. Venereol.*, 74, 405, 1994.
49. Wigger-Alberti, W., Maraffio, B., Wernli, M., and Elsner, P., Self-application of a protective cream. Pitfalls of occupational skin protection, *Arch. Dermatol.*, 133, 861, 1997.
50. Wigger-Alberti, W., Maraffio, B., Wernli, M., and Elsner, P., Training workers at risk for occupational contact dermatitis in the application of protective creams: efficacy of a fluorescence technique, *Dermatology*, 195, 129, 1997.

32 Testing of Moisturizers

Marie Lodén and Magnus Lindberg

CONTENTS

I. Background ... 387
  A. Introduction ... 387
  B. The Concept of Dry Skin ... 388
  C. Water-Retaining Treatments ... 389
    1. Lipids ... 390
    2. Humectants ... 390
II. Product-Testing Strategies Using the Bioengineering Device ... 390
  A. Transepidermal Water Loss ... 391
    1. Short-Term Effects ... 391
    2. Long-Term Effects ... 393
    3. Things to Remember ... 394
  B. Surface Topography ... 394
    1. Short-Term Effects ... 395
    2. Long-Term Effects ... 395
    3. Things to Remember ... 396
  C. Electrical Measurements ... 396
    1. Short-Term Effects ... 396
    2. Long-Term Effects ... 398
    3. Things to Remember ... 399
  D. Magnetic Resonance Imaging ... 399
  E. Infrared Spectroscopy ... 399
    1. Short-Term Effects ... 399
    2. Long-Term Effects ... 399
    3. Things to Remember ... 400
III. Ethics, Design, and Statistics ... 400
IV. Final Comments ... 402
V. Acknowledgments ... 402
References ... 402

I. BACKGROUND

A. Introduction

Moisturizers are used to restore or to maintain a normal function of the stratum corneum (SC). Mostly they are used for the indication of so-called dry skin. Product

0-8493-1443-7/05/$0.00+$1.50

testing of moisturizers is facilitated by the use of bioengineering devices for objective evaluation of treatment effects.

Noninvasive measurements have the advantage of not interfering with the studied area, in contrast to biopsies. The use of multiple instruments also makes it possible to measure several features at one time, which will increase the confidence of the observation. Furthermore, the treated area can be studied repeatedly to monitor changes with time. Several instruments also detect subclinical (nonvisible) changes, which may facilitate understanding of the mechanism and distinguish between treatment effects. The benefit for the consumer may also be enhanced if the data correlate to a perceptible and clinically interesting change of the skin.

When multiple variables are used, the critical ones should be identified and the procedure for how they were selected should be discussed (e.g., by reference to publications, guidelines, or recommendations by regulatory authorities). Consistency among multiple measures of the same attribute increases the confidence of the observation. This chapter is an effort to provide some information into how the performance of moisturizers can be objectively measured.

B. The Concept of Dry Skin

The condition of dry skin, which afflicts everyone at some time, is visually characterized by roughness and scaling. The skin may also feel less flexible. Among dermatologists the condition is called xerosis. In more severe cases, cracking, bleeding, and erythema may be present.

The term "dry skin" is not generally accepted. Some relate it to the lack of water in the SC, whereas others consider dry skin to belong to a group of disorders with a rough surface.[1] Dry skin is not related to oily skin.[2,3] The lack of an accepted universal definition of dry skin allows for confusion in the evaluation of product effects. Many studies are performed on normal skin and conclusions drawn to the so-called dry skin.

In the 1950s, the treatment of dry skin conditions underwent a change of direction as a result of the new classic experiments by Blank.[4] In these studies he observed that the dehydrated SC is very hard and brittle in spite of any natural oils it may contain. Since then, the main goal in restoring skin elasticity has been to add and hold water by using humectants instead of replacing lost oils. However, during the last decades, it has been demonstrated that the intercellular lipids of the SC also play an important role in its water-holding properties.[5–7] Thus, the key objective of today's moisturizers may be to restore lost lipids in addition to hydrating the SC via hygroscopic substances.

From the consumer point of view, beneficial therapy for dry skin may not always be to increase its hydration. For instance, skin elasticity may be increased by the use of α-hydroxyacids.[8–10] In addition, the presence of natural moisturizing factors (NMFs) in the SC seems to be important for elasticity.[11] Water itself cannot restore elasticity if the NMFs are removed.[11] Table 32.1 lists the various materials that have been reported to be found in the NMF.[12] A glossy and appealing skin under dry conditions has also been suggested as being possible to obtain by the treatment.[13]

TABLE 32.1
Composition of Natural Moisturizing Factor[12]

Ingredients	Percentage
Amino acids	40.0
Pyrrolidone carboxylic acid	12.0
Lactate	12.0
Urea	7.0
Na, Ca, K, Mg, phosphate, chloride	18.5
NH_3, uric acid, glucosamine, creatinine	1.5
Unidentified	9.0

Furthermore, glycerin may prevent crystal formation of SC lipids and preserve their normal structure even when the skin is underhydrated.[14]

C. Water-Retaining Treatments

The term "moisturizer" is often used synonymously with emollient, as in this chapter. Moisturizers, however, often contain humectants to hydrate the SC, whereas emollients lubricate and occlude the skin (see below). Both types are complex formulations composed of numerous ingredients, but all are based on water or lipids. Most improve smoothness of the skin and its overall condition, but we are not in the position to exactly describe the mechanism.

Single applications of emollients to the skin induce physiological changes, which can be different from cumulative effects following repeated applications for a period of time. The frictional properties of the surface may change.[15] Ingredients may disappear from the surface by evaporation, diffusion, or metabolism depending on their nature. Studies show that after 8 hours, about 50% of applied cream remained on the surface.[16] Creams and ointments seem to allow higher transfer of the active ingredients than lotions and tinctures.[17]

One of the main changes seen in the SC after topical application of emollients is smoothing of the surface. The immediate smoothing effect is mainly the result of the filling of spaces between partially desquamated skin flakes.[18] The hydration of SC also increases. Cumulative effects (i.e., long-term improvement or deterioration of the skin condition or adaptation phenomena) can be noticed over a treatment period. The "regression method" can be used to identify products that have long-lasting effects beyond mere cosmetic camouflaging of scales.[19] Formulations can be considered to have no real effect if scaling returns to the pretreatment level within a few days after stopping applications.

Numerous techniques have been used to describe skin condition and to determine efficacy of emollients.[20,21] Hydration of the SC produces changes in its physical properties, such as electrical properties and elasticity. The physical parameters may reflect not only the hydration of the skin but also the properties and distribution of product components both in and on the skin. Because many measurements are made at poorly defined depths within the skin, and readings of the hydration are influenced by other substances than water, it is important to be aware of the composition that is to be evaluated.

There are two major substance classes used to increase skin hydration: lipids and humectants. They work according to two different principles: occlusion and attraction of water.

1. Lipids

Occlusion from lipids implies a simple reduction of the loss of water from the outside of the skin. This causes a buildup of water in the SC. Common occlusive substances are petrolatum, beeswax, lanolin, and various oils. However, when combined with other ingredients in skin care products, their occlusiveness may be reduced.[22,23]

A new way of looking at the occlusiveness of topically applied lipids is their ability to restore the lipid layer around the corneocytes,[24–26] thereby retaining the moisture content in the corneocytes. Disruption of the lipid membranes in the SC makes the skin more susceptible to drying-out effects.[5] The NMF substances in the SC became equally accessible to water extraction, as if the skin was treated with ether. It has also been demonstrated that the application of solvent-extracted skin lipids to experimentally induced scaly skin causes an increase in conductance values, which is related to the amount of structural lipids (ceramides) in the used preparation.[6,26] With increasing age, and in atopic dermatitis, there is a decreased level of ceramides in the skin.[27]

2. Humectants

Attraction of water by humectants in the SC is believed to increase the degree of hydration. Low-molecular-weight humectants like urea, lactic acid, pyrrolidone carboxylic acid, and salts are easily absorbed by the SC, whereas those with higher molecular weight, such as cellulose-derived polymers, mucopolysaccharides (e.g., hyaluronic acid) and proteins (e.g., collagen), cannot be absorbed. It can be assumed that the latter will stay on the skin surface and bind water to the top surface.[18] Whether topical application of high-molecular-weight substances can increase the water content in the SC is not conclusively shown; however, they have found cosmetic usage as moisturizers.

II. PRODUCT-TESTING STRATEGIES USING THE BIOENGINEERING DEVICE

In recent years, increased attention has been focused on the measurement of product efficacy to develop evidence in support of claims of product superiority. However, several factors need to be considered before measurements can take place (Table 32.2). Substitute or surrogate parameters may be used, although they are not a direct measure of the benefit of the product. Carefully validated surrogate parameters often allow less expensive studies to provide answers to questions that would typically require much larger trials if the targeted endpoint were used. However, special precautions are recommended in recognizing such parameters. The existence

TABLE 32.2
Factors to Consider During Measurements of Skin Properties

Correctly calibrated? Standards included?
Measured area representative for the area of interest?
Interference from other surface material, e.g., hair product residues?
Correctly calibrated? Standards included?
Proper measuring climate, e.g., temperature, humidity, air convection?

of a measurable physical effect does not guarantee the existence of a benefit to the user of the product.

In the following text we will discuss some of the available biophysical methods and the results obtained with them in studies on moisturizers and their effects on the skin.

A. Transepidermal Water Loss

Water in the SC is in dynamic equilibrium with that in the underlying tissues and that in the atmosphere. At room humidities below 100%, the net transport is from the skin to the surrounding air. The rate of evaporation can be measured conveniently, using an evaporimeter.[28] There is a complex relationship between water content and transepidermal water loss (TEWL) in the SC (Figure 32.1). An increased TEWL has been observed both in dry skin of atopic patients[29–31] and in hyperhydrated skin (Figure 32.1). Cracks, defects in the intercellular bilayer lipids, and decreased projected corneocyte area may explain the impaired barrier function in dry skin. In normal skin, the diffusion of water is facilitated by hyperhydration.[32] Increased degree of hydration reduces the diffusional resistance of SC,[33,34] and occlusion is often used to increase the bioavailability of topical drugs. Hence, efficient moisturizing agents may in theory increase TEWL through a hydration-induced increase in the diffusion coefficient of water.[35] TEWL has also been suggested to correlate to the penetration of other topical substances.[36]

Volatile agents other than water may interfere with the TEWL readings, and erroneously elevated values were obtained for some hours after application of a propylene glycol–containing lotion to the skin.[37] This was the result of interactions between propylene glycol and the sensor in the instrument. Guidelines for measurement of TEWL have been published by the standardization group of the European Contact Dermatitis Society[38] and the European Group for Efficacy Measurements of Cosmetics and Other Topical Products[39] and can be found elsewhere in this book.

1. Short-Term Effects

Measurements of TEWL have been used to describe the immediate effects of emollients on skin hydration.[40–42] Application of water-containing products to the skin is characterized by evaporation of product water during the following 10 to 30 minutes.[40,41,43] Emollients could increase the degree of hydration by delivery of their

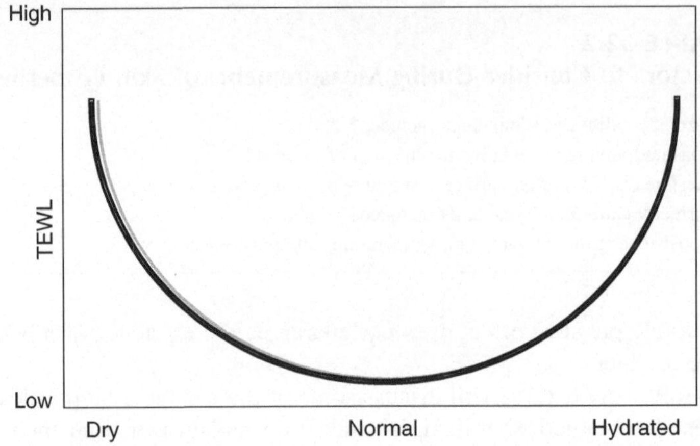

FIGURE 32.1 Transepidermal water loss may be high in both dry skin and in hyperhydrated skin.

water to the skin and by occlusion.[35,40] These two mechanisms were studied using emollients with different amounts of water.[40] One of the tested products was an ordinary cream containing about 70% water and 30% nonvolatiles, and another was pure petrolatum with 100% nonvolatiles. The skin covered with the ordinary cream for 5 minutes became hydrated as a result of absorption of water from the cream. This could be detected after removal of cream residue from the surface using soft tissue immersed in anhydrous ether and subsequently measuring TEWL. The absorbed amount of water was observed as an increased TEWL over the basal level for some minutes after cleaning the skin. In contrast, the area covered with petrolatum showed decreased TEWL because of occlusion[40] (Figure 32.2). Removal of petrolatum after 5 minutes did not increase the water loss above the normal value. However, if petrolatum was applied for 40 minutes, the trapping of water in the skin because of occlusion could be measured.[40] Release of trapped water because of occlusion has also been reported by others.[41,42] Thus, this technique proved useful in illustrating the mechanisms by which emollients can increase the hydration of the skin; namely, immediate hydration resulting from absorption of water from the products and trapping water resulting from occlusive mechanisms.

The immediate decrease in TEWL after application of petrolatum is followed by an increase after about 3 hours.[35] This increase in water loss can be explained by an increased diffusion coefficient of water as the SC become more hydrated.[32,35]

Similar experiments with occlusion have been used to investigate the water-retention capacity of SC.[44] The study was based on differences in the rate of desorption from normal control skin and clinically uninvolved skin in atopic patients. The skin was occluded for 24 hours using plastic chambers. After removing the chambers, visible excess water was wiped away and the TEWL was measured for 25 minutes. The water loss curves showed different patterns of decay in the three groups. Atopic patients had lower values throughout the study, which was considered to be to the result of a reduced amount of water in the atopic skin. One possible

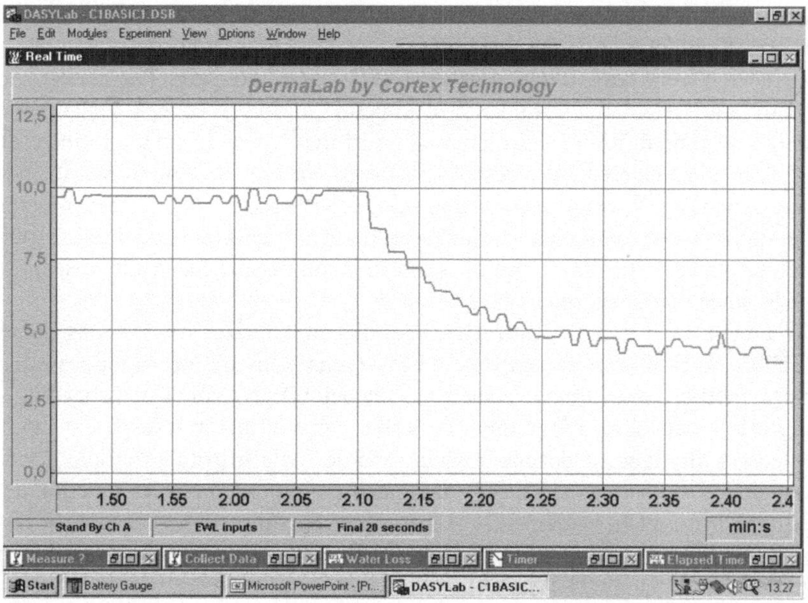

FIGURE 32.2 Application of petrolatum to the skin reduces transepidermal water loss by occlusion. In this subject the value decreases from about 10 g/m^2 per hour to less than 5 g/m^2 per hour.

pitfall in this kind of study is the wiping of the skin. If all surface water is not removed, the results will be strongly influenced by the evaporation of nonabsorbed water.

The water concentration profile in the SC before and after application of a moisturizer has been calculated from the rate of water loss from the surface.[45] The moisturizer decreased the TEWL, and the percentage increase of water in various layers could be estimated using Fick's equation and knowledge about the diffusivity in the SC. One factor to consider in these types of experiments is the possible presence of humectants in the products. All of the absorbed/trapped water will not be easily released if the humectants penetrate into the skin.

2. Long-Term Effects

Little is known about the effect on the barrier function or the water content of SC following repeated application of moisturizers. Changes in TEWL seem to be the signal for recovery of barrier structure and function following the injury.[46] Normalization of barrier function and SC lipid replenishment is clearly linked with epidermal lipid synthesis.[46] Application of water-impermeable membrane hampers the increase in epidermal sterol and fatty acid synthesis.[47–49] Occlusive emollients (e.g., petrolatum) might therefore be suspected to delay barrier recovery by forming an inert, epicutaneous, and occlusive membrane. This was demonstrated not to be true in a recent study in humans and in hairless mice.[24] The skin barrier was disrupted by acetone swabbing until elevated rates of TEWL occurred. A thin layer of petrolatum was then applied to the skin several times a day for four days, and the effect on barrier recovery was followed using TEWL as an indicator. The results showed that petrolatum accelerated, rather than impeded, barrier recovery. Visualization of the depth of penetration of petrolatum showed that it was localized to intercellular domains in the SC.[24] Other types of emollients have also been shown to increase the rate of healing of surfactant-irritated skin, as assessed by TEWL measurements.[50] Furthermore, it was found that regular use of the emollients reduced the susceptibility to the surfactant.[50] In other studies, some moisturizers reduced the susceptibility, whereas other increased the skin reaction.[51,52]

Treatment of dry skin with urea-containing moisturizer for 3 weeks decreased the TEWL.[53] This was accompanied by increased hydration, as assessed using electrical methods and clinical evaluation (see below). Dry skin was defined as either having a low capacitance or high flakiness (evaluated using transparent tape). No volunteers with signs of dermatological disorder or a history of atopic dermatitis were included. A decrease in TEWL indicated an improved barrier function against water, which might be expected when dry skin becomes more normal. In another study it was found that treatment of normal skin for 4 weeks with urea-free moisturizers did not change TEWL.[54]

3. Things to Remember

- Measurement with an open chamber is strongly influenced by air convection in the surroundings.
- Ambient humidity, skin temperature, and sweating influence the TEWL.
- Volatile agents, other than water, in the products could influence the readings of TEWL if measurements are made immediately after application of the products.
- Because both dry and hydrated skin show increased TEWL compared to normal, a change in the readings must be viewed cautiously. The experimental design and other evaluations are important parameters in supporting obtained data.

B. Surface Topography

Analysis of the skin surface topography has been used for evaluations of the effects of moisturizers on the skin structure. Visual examination of the skin surface reveals a system of furrows, and closer examination of the intermediate plateaus shows the presence of a secondary, finer furrow system.[55,56] Higher magnification reveals the arrangement of the corneocytes.

The effect of skin treatments can be assessed after taking replicas of the skin. The surface feature of a negative or positive replica of the skin is then objectively quantified in numerical terms using image analysis or profilometry.[57–59] Imaging techniques are dependent on the interaction of an illuminated beam with the surface, and the subsequent production of shadows, and they are therefore somewhat indirect.

Profilometric description of surface topography was first developed for metallic surfaces. The measurements follow the same operational principles for most investigators. A transducer is moved horizontally along the surface of the sample, or conversely, the sample is moved under the stylus. The features of the surface induce vertical movement in the stylus, which are then converted into electrical signals and subsequently digitized, yielding the profile of the cross-section of the surface.

Parameters widely used to describe skin profiles fall into three categories: amplitude parameters that refer to vertical aspects of the profile (e.g., peak heights and furrow depths), spacing parameters that refer to horizontal distance between features along the plane of the skin surface, and so-called "hybrid" parameters that combine the size of vertical features with the distance between them. Many parameters are specified standard measurements according to Deutches Institut fur Normung and International Organization of Standarization.

Values for common amplitude parameters have been reported to increase when skin hydration increases.[60] The reason might be a swelling of SC.[61] Dry atopic skin also shows an increase in amplitude parameters compared with normal skin[62] Thus, both moist and dry skin may have a rough surface. Most studies indicate that dry skin is characterized by fewer peaks in the profile line than normal skin.[62,63]

1. Short-Term Effects

Single applications of emollients are effective in "smoothing" the skin, as assessed from replicas of the surface.[18,64] Surface contour measurements demonstrated a flattening of relief for some hours after application. However, the relief on the replicas reflected not the true surface, but a surface whose irregularities were filled with residues from the emollient.

Similar results have been reported from other groups.[65] Conventional amplitude parameters showed a significant decrease during the first 2 hours after application of a hydrating product. In addition, the mean spacing between peaks decreased. During the next 6 hours, these parameters slowly recovered their initial values.

With an image analysis system, a change in shadow area, variance, skewness, and kurtosis were found after application of an emollient.[66] Wiping the product residue from the skin restored a more typical pattern to the skin.

2. Long-Term Effects

It has been shown that treatment of dry skin on the legs for 21 days influenced the topography of the surface,[67] as measured with a profilometer on a replica of the skin. The number of peaks increased significantly during the study, whereas the mean size of the peaks was not changed by the treatment.

Treatment of the cheek for 4 weeks with an oil/water cream has also been reported to increase the roughness of the skin, as measured using image analysis and Fourier transformation.[58] The furrows became clear and the texture finer after the treatment. On the day of measurement, the subjects were asked not to apply the cream and to wash their face 2 hours before taking the replicas.

3. Things to Remember

- When characterizing the skin surface, its anisotrophy must be considered. The roughness parameter values are dependent on the orientation of the tracings regarding the major grooves and ridges present in the surface pattern.
- Improperly mixed material and air bubbles in the replica are possible artifacts in replicating the topography.
- If the skin surface is covered by a product, the replica will not yield a faithful image of the skin topography; likewise, there is some question of whether the scaling type of dry skin can be replicated satisfactorily.
- The weight and tip geometry of the stylus used can influence the result. Soft replicas allow furrows to be created by the stylus in its traversal of the replica.
- The slower the rate of scanning, the less likely the stylus will skip over small furrows.

C. Electrical Measurements

Electrical measurements are frequently used to determine the water content of SC. A detailed review of instruments and their possibilities and limitations is given in other chapters in this book.

1. Short-Term Effects

A method for evaluating the water-retaining capacity of SC has been developed by Tagami et al.[68] The test procedure consists of measurements of the skin conductance to the high-frequency electric current of 3.5 MHz. The skin was hydrated for 10 seconds with a droplet of water and then measured at intervals of 30 seconds for 2 minutes to evaluate the water-holding capacity. Under usual ambient conditions, normal skin surface showed a high rise in conductance just after application of water, which was followed by a rapid falloff within 30 seconds thereafter by gradual return to the prehydration levels by 2 minutes. The study showed that the scaly skin of patients with various kinds of scaly dermatoses is always dry because it is less hygroscopic and is markedly less capable of binding water than normal skin. The superficial portion of the horny layer is also less hygroscopic and is somewhat less capable of binding water than the deeper portions.

A similar approach has been used to access the hydrating effects of emollients.[69] Application of a urea cream increased the capacitance immediately, as was noted following application of water (see above). The increased values rapidly decreased and then gradually returned to the former levels. A hydrophilic ointment with the same amount of water gave a similar pattern, although the values were lower. In contrast to these water-containing preparations, petrolatum did not induce any initial increase, but there was a slow increase in the capacitance with time. This demonstrated that petrolatum indirectly increases skin hydration as a result of its occlusive properties. Application of aqueous solutions of ordinary humectants (glycerin, sodium pyrrolidone carboxylate, urea, and sodium lactate) surprisingly did not increase the hygroscopicity of the skin.[69] The reason for this nondetectable effect is not known.

The water-holding capacity of emollient-treated skin was also studied using the above-mentioned techniques. A urea cream, a hydrophilic ointment, and petrolatum were applied in the same manner as above.[69] After 2 hours, water was applied to the surface for 10 seconds. Compared with untreated skin, a remarkable increase in conductance values was observed on the areas treated with the urea cream. The skin area treated with a hydrophilic ointment revealed only a slight increase, whereas petrolatum showed a decrease in hygroscopicity. A decrease in hygroscopy after treatment with petrolatum might be an artifact, as it is possible that a hydrophobic film was formed on the skin, hindering water from coming in contact with the skin. Furthermore, lipids are characterized by high electrical resistivity and could therefore reduce the overall skin conductivity.

Nonabsorbed cream components and water in the creams can also influence the capacitance values.[70,71] A single application of various emollients to the skin increased the capacitance (Corneometer measurements). However, after removal of cream residue from the surface (2 hours after application), an immediate decrease was noted on the areas treated with creams containing various humectants. On the area treated with pure petrolatum, a slight increase in the capacitance was noted. These results indicate that the capacitance value reflects not only the capacitance of the skin but also that of cream residues on the skin surface. The observed increase after removal of petrolatum may well be the result of the high electrical resistivity of petrolatum, as also noted by others.[42,72] Application of petrolatum and dewaxed lanolin indicated decreased skin hydration (higher values of the impedance) for some hours. This result was related to the amount of material applied to the skin and was considered erroneous.[72] A dry occlusive film takes up a significant volume under the probe and depresses the response, whereas nonabsorbed humectants attract water and increase the response.

Single applications of emollients with and without glycerin showed that glycerin greatly potentiated the hydrating properties on normal skin, as measured with an impedance meter.[73] Application of pure water showed increased values for less than 10 minutes, emulsions without glycerin for about 15 minutes, and the glycerin-containing products for at least 2 hours. Other measured parameters include TEWL, skin roughness, and friction. TEWL was markedly decreased by the application of glycerin in emulsion systems at all time points between 1 and 7 hours, in comparison with the nonglycerin control. Furthermore, the skin surface roughness was less pronounced after treatment with the glycerin emulsion. A difference, although not significant, between the emulsions was noted as long as 24 hours after application. Aqueous glycerin solution increased the skin friction for at least 3 hours, whereas pure water affected the friction for only a few minutes. In contrast to this study, others have reported glycerin to increase TEWL.[74,75] One reason for these conflicting data may be differences in ambient humidity. At high humidity, glycerin may reduce the TEWL, but at low humidity the TEWL might increase.[14]

A single application of a glycerin-containing lotion also showed increased levels of capacitance (Corneometer) and conductance (Skicon-100) for at least 6 hours.[43]

This was believed to be the result of penetration of emulsion lipids with consequences for epidermal hydration.

A similar test method has been used to assess the effect of moisturizers on skin surface hydration.[76,77] Liberal amounts of the products were applied to the skin, and after various times the conductance[76,77] and capacitance[76] were measured. In these studies, differences were found between the products, but measurements with both types of instruments resulted in the same rank order between the products[76]; however, the influence of nonabsorbed components on the results was not discussed.

2. Long-Term Effects

In a multicenter study at 11 centers, comprising 368 human volunteers, the efficacy of two emollients was evaluated using subjective and objective methods.[54] Skin capacitance (Corneometer), TEWL, skin surface lipids (Sebumeter), and skin roughness were measured, and a questionnaire was used for the subjective analysis. The creams were used for 4 successive weeks. The only objective parameter that changed significantly during the treatment period was the skin capacitance, which increased; the volunteers also perceived their skin as being better. No difference was found between the products.

The healing properties of several lotions after repeated insults with sodium lauryl sulfate has been studied using conductance measurements.[78] Skin areas treated with lotions after sodium lauryl sulfate exposure displayed less damage, as determined by conductance measurements and subjective evaluations. Before taking the readings, special efforts were made to eliminate lotions from the sites treated with sodium lauryl sulfate. Moreover, in an attempt to compensate for the eventuality of failure to remove lotions completely, the corresponding control sites were treated with the lotions for a few minutes before washing. However, one major problem with this study is the measuring probe. The electrode was filled with electrolyte paste and was held on the skin for 2 minutes. The interactions between the paste and the skin might well have influenced the readings.

The efficacy of five hand and body lotions was evaluated in a double-blind study on 250 subjects.[79] The products were used for 4 weeks. Clinical grades of the skin condition and instrumental measurements of elasticity and water content showed that one of the products tested was significantly more effective than the other. Electrical impedance measurements (capacitance and conductance) were taken using IBS Skin Surface Hydrometer and elasticity using the Dynamic Modulus Tester. The ranking of the other products differed between the methods of assessment. The reason for this and whether the difference was significant or not was not discussed.

Two creams containing urea at different concentrations were compared in a double-blind study on two parallel groups for a total of 45 individuals with dry skin.[53] Each individual used one of the creams for 3 weeks on one forearm. The final evaluation took place not less than 12 hours after the last application. The hydration status of the skin increased by the treatments, but no difference in efficacy between the creams could be detected. Parameters measured were skin capacitance (Corneometer) and conductance (Skicon-100). Furthermore, a dermatologist evaluated the skin condition in a blind fashion, and the volunteers rated the improvement of dryness. These methods showed good correlation. Apart from measurements of the electrical properties, the degree of scaling of the skin and TEWL were assessed.

3. Things to Remember

- One difficulty with electrical measurements is the uncertainty of where the detected water is located. It may well be on the outside of the skin, bound to nonabsorbed humectants in previously applied products.
- Compounds other than water could influence electrical measurements.
- The intimacy of the contact between the measuring probe and the skin poses a problem for the precise interpretation of the results.
- If the probe occludes the surface, then contact time needs to be considered as water will accumulate beneath the probe and affect the value.
- If the detector is small, then more measurements need to be taken to minimize local site variation.
- Variation of the probe pressures against the skin can alter the readings.

D. Magnetic Resonance Imaging

Magnetic resonance imaging is a technique used in high-technology medicine to look inside the living body in a noninvasive and chemically specific way.

Magnetic resonance imaging has been used to demonstrate the moisturizing process in the SC from petrolatum.[79] After application of petrolatum to the fingertip, a buildup of water in the SC was visualized. Because magnetic resonance imaging is harmless, it can be used repeatedly to visualize the process of moisturization over time. This methodology needs to be further developed with respect to use in the measurement of water.

E. Infrared Spectroscopy

Water is an intense absorber of infrared spectroscopy (IR).[21,80] In near-IR, water molecules show two clear absorption bands with amplitudes high enough for identification in the spectrum obtained from human skin *in vivo*.[81,82] This technique can thus be used to determine the amount of water in SC.

1. Short-Term Effects

Using IR, the water content of the uppermost layers (few microns) of the SC has been quantitatively measured. During 5-hour occlusion of the skin by petrolatum, the water weight fraction in the surface layers of the SC increased from 7 to 15%.[80]

2. Long-Term Effects

The ability of four emollients to maintain skin in smooth condition for some length of time was studied[81] using the dry leg regression method.[19] The skin condition was rated by trained evaluators and measured using near-IR measurements. The score and IR were highly correlated; the product ranks and magnitude were similar. Near-

IR was equally good as an expert grader in differentiating between product treatments.[81]

Another modified commercial apparatus for near-IR was used in combination with conductance measurements (DermoDiag) and clinical evaluation to study the effect of moisturizers on dry skin *in vivo*.[82] In separate *in vivo* experiments, a good correlation was obtained between the water content in isolated SC and the absorbance. It was also shown that the sum of clinical parameters (roughness, scaling, etc.) correlated well with the absorbance and correlated better than the conductance values. A closer examination showed the electrical conductance was related to the roughness of the skin, whereas above a certain roughness, no change in conductance was found. The lack of correlation between conductance and roughness was mainly the result of the poor quality of the electrical contact between the skin surface and the measurement electrode.[82] Some of the tested moisturizers reduced the skin dryness after 4 weeks of treatments, as evaluated clinically as well as using the instruments. One emollient was rated as ineffective by the conductance measurements, whereas the skin was improved according to clinical evaluation and IR measurements.

3. Things to Remember

- Several probes are occlusive and may cause a false water buildup at the skin's surface during measurements.
- The depth of measurements within the skin is not well defined.

III. ETHICS, DESIGN, AND STATISTICS

Human volunteer studies should be performed only by suitably trained, qualified, and experienced personnel (Table 32.3). Only carefully conducted experiments will be able to provide definite information when small differences between treatments are to be established.

The health and welfare of the subjects must be protected all along, and a thorough expert review has to preclude the occurrence of significant risk to the volunteers. The probability and magnitude of harm and discomfort should be ethically acceptable (i.e., the likelihood of trivial reactions may be accepted, whereas the probability of serious disabilities should be very remote).

Studies on humans should be performed in accordance with the recommendations guiding physicians in biomedical research that were adopted in 1964 by the 18th World Medical Assembly, in Helsinki, Finland, with later revisions. Ethics can be considered as a personal attitude toward what is right or wrong, but they must be subordinated to and coordinated with a public norm. In addition to international guidelines, human studies should always be performed in compliance with the national regulations of the country in which the testing is being conducted.

The first stage in planning an experiment is the precise definition of its objectives. It is not enough to say that we should compare cream A with cream B. We need to define the most relevant variable to be measured and when measurements should be done. Is a single application enough, or should the data be based on long-term

TABLE 32.3
Checklist for Planning and Evaluating an Efficacy Study

Clear objectives?
Safety assessment done?
Necessary approvals obtained?
Appropriate design?
Relevant outcome variables?
Instruments in a good condition?
Appropriate selection of participants?
Handling of dropouts?
Results properly described?
Correct statistics?
Adequate conclusions?
Applicability to a wider population?

treatment, inclusion, and exclusion criteria for the test subject, age, or sex? Anatomical site and its condition: dry or normal? How should we establish the skin condition? Are concurrent medication/skin care treatments allowed? Is a "wash-out" period before entering the study necessary? If the applications made by the test subjects, then how should the compliance be checked?

Technical details in the experimental setup and the number of subjects that would be adequate to be included in the study need to be considered. It is also important to go through the statistics in the beginning of an experiment, rather than at the end, as often happens in practice.

In drawing conclusions from a study, we must be able to say what confidence we have in our assertions and how likely it is that the conclusions are based on purely chance, accidental, or coincidental effects. The question most often asked is whether there is a difference between the products and whether the difference is statistically significant. In biological studies, it is accepted practice to consider difference as statistically significant when the value for α is less than 0.05. At that level, we are accepting the possibility of being wrong 1 in 20 times. If we were to reject a hypothesis when it is true, we will have committed an error referred to as type I error; this has a probability equal to α (which is the significance level). However, if we concluded that there was no difference, when in fact there was a difference, then a type II error was committed, and the probability of this error is equal to β. For most trials β is set at 0.9.

Another important consideration is the number of test subjects required to satisfy the study objective. The magnitude between samples will cause an increase or decrease in the number of tested subjects, as does the change in α and β. The greatest changes occur as the magnitude of difference decreases — the smaller the difference, the more subjects required before this difference will be detected. In skin-care treatments, half-sided comparisons are often possible. In general, this reduces the number of subjects, as the intraindividual variation is often less than the interindividual. Another possibility for reducing the number of subjects is to do a crossover study. In such a study, all subjects receive both treatments, one after another. Sometimes a wash-out period between the two treatments is needed.

The reader should reflect carefully on the practical implications. By increasing the number of tested subjects, one might find statistically significant differences between products that are of no practical value. However, many potential useful skin treatments may have been erroneously discarded as a result of the inadequate size of the clinical trials.[83] To start a study without having considered the experimental design and statistics is a misuse of resources that could be better employed on more valuable activities.

IV. FINAL COMMENTS

Dry, xerotic states of skin are often considered merely cosmetic nuisances without medical significance. However, moisturizers are useful not only for diminishing signs of dryness but also as treatment adjuncts for inflammatory diseases. They contain a variety of substances and transmit their signals to the skin. Carefully conducted studies on moisturizers will facilitate our understanding on the mechanism by which they improve dry skin conditions. Properly formulated they will enhance our quality of life.

V. ACKNOWLEDGMENTS

The valuable technical assistance in preparation of this manuscript by Ms. Izabela Buraczewska is highly appreciated.

REFERENCES

1. Rurangirwa, A., Pierard-Franchimont, C., Le, T., Ghazi, A., and Pierard, G.E., Corroborative evidence that "dry" skin is a misnomer, *Bioeng. Skin*, 3, 35, 1987.
2. Leveque, J.L., Grove G.L., de Rigal, J., Corcuff, P., and Kligman, A.M., Biophysical characterization of dry facial skin., *J. Soc. Cosmetic Chem.*, 82, 171, 1987.
3. Piérard, G.E., What does "dry skin" mean? *Int. J. Dermatol.*, 26, 167, 1987.
4. Blank, I.H., Factors which influence the water content of the stratum corneum, *J. Invest. Dermatol.*, 18, 433, 1952.
5. Middleton, J.D., The mechanism of water binding in stratum corneum, *Br. J. Dermatol.*, 80, 437, 1968.
6. Imokawa, G. and Hattori, M., A possible function of structural lipids in the water-holding properties of the stratum corneum, *J. Invest. Dermatol.*, 84, 282, 1985.
7. Imokawa, G., Kuno, H., and Kawai, M., Stratum corneum lipids serve as a bound-water modulator, *J. Invest. Dermatol.*, 96, 845, 1991.
8. Alderson, S.G., Barratt, M.G., and Black, J.G., Effect of 2-hydroxyacids on guinea-pig footpad stratum corneum: mechanical properties and binding studies, *Int. J. Cosmetic Sci.*, 6, 91, 1984.
9. Takahashi, M., Machida, Y., and Tsuda, Y., The influence of hydroxy acids on the rheological properties of stratum corneum, *J. Soc. Cosmetic Chem.*, 36, 177, 1985.

10. Hall, K.J. and Hill, J.C., The skin plasticisation effect of 2-hydroxyoctanoic acid. 1: The use of potentiators, *J. Soc. Cosmetic Chem.*, 37, 397, 1986.
11. Jokura, Y., Ishikawa, S., Tokuda, H., and Imokawa, G., Solid state 13C-NMR studies on elastic property of the stratum corneum, *J. Invest. Dermatol.*, 104, 806, 1995.
12. Jacobi, O.K., Moisture regulation in the skin., *Drug Cosmetic Ind.*, 84, 732, 1959.
13. Friberg, S.E., Micelles, microemulsions, liquid crystals, and the structure of stratum corneum lipids., *J. Soc. Cosmetic Chem.*, 41, 155, 1990.
14. Froebe, C.L., Simion, F.A., Ohlmeyer, H., Rhein, L.D., Mattai, J., Cagan, R.H., and Friberg, S.E., Prevention of stratum corneum lipid phase transitions *in vitro* by glycerol—an alternative mechanism for skin moisturization, *J. Soc. Cosmetic Chem.*, 41, 51, 1990.
15. Lodén, M., Olsson, H., Skare, L., and Axéll, T., Instrumental and sensory evaluation of the frictional response of the skin following a single application of five moisturizing creams, *J. Soc. Cosmetic Chem.*, 43, 13, 1992.
16. Rhodes, L.E. and Diffey, B.L., Fluorescence spectroscopy: a rapid, noninvasive method for measurement of skin surface thickness of topical agents, *Br. J. Dermatol.*, 136, 12, 1997.
17. Johnson, R., Nusbaum, B.P., Horwitz, S.N., and Frost, P., Transfer of topically applied tetracycline in various vehicles, *Arch. Dermatol.*, 119, 660, 1983.
18. Garber, C.A. and Nightingale, C.T., Characterizing cosmetic effects and skin morphology by scanning electron microscopy, *J. Soc. Cosmetic Chem.*, 27, 509, 1976.
19. Kligman, A.M., Regression method for assessing the efficacy of moisturizers, *Cosmetics Toiletries*, 93, 27, 1978.
20. Quattrone, A.J. and Laden, K., Physical techniques for assessing skin moisturization, *J. Soc. Cosmetic Chem.*, 27, 607, 1976.
21. Potts, R.O., Stratum corneum hydration: experimental techniques and interpretations of results, *J. Soc. Cosmetic Chem.*, 37, 9, 1986.
22. Wepierre, J. and Andragui, M., Factors in the occlusivity of aqueous emulsions, *J. Soc. Cosmetic Chem.*, 33, 157, 1982.
23. Choudhury, T.H., Marty, J.P., Orecchioni, A.M., Seiller, M., and Wepierre, J., Factors in the occlusivity of aqueous emulsions. Influence of humectants, *J. Soc. Cosmetic Chem.*, 36, 255, 1985.
24. Ghadially, R., Halkier-Sorensen, L., and Elias, P.M., Effects of petrolatum on stratum corneum structure and function, *J. Am. Acad. Dermatol.*, 26, 387, 1992.
25. Imokawa, G., Akasaki, S., Hattori, M., and Yoshizuka, N., Selective recovery of deranged water-holding properties by stratum corneum lipids, *J. Invest. Dermatol.*, 87, 758, 1986.
26. Imokawa, G., Akasaki, S., Minematsu, Y., and Kawai, M., Importance of intercellular lipids in water-retention properties of the stratum corneum: induction and recovery study of surfactant dry skin, *Arch. Dermatol. Res.*, 281, 45, 1989.
27. Imokawa, G., Abe, A., Jin, K., Higaki, Y., Kawashima, M., and Hidano, A., Decreased level of ceramides in stratum corneum of atopic dermatitis: an etiologic factor in atopic dry skin? *J. Invest. Dermatol.*, 96, 523, 1991.
28. Nilsson, G.E., Measurement of water exchange through skin, *Med. Biol. Eng. Comput.*, 15, 209, 1977.
29. Lodén, M., Olsson, H., Axell, T., and Linde, Y.W., Friction, capacitance and transepidermal water loss (TEWL) in dry atopic and normal skin, *Br. J. Dermatol.*, 126, 137, 1992.

30. Werner, Y. and Lindberg, M., Transepidermal water loss in dry and clinically normal skin in patients with atopic dermatitis, *Acta Derm. Venereol.*, 65, 102, 1985.
31. Thune, P., Evaluation of the hydration and the water-holding capacity in atopic skin and so-called dry skin, *Acta Derm. Venereol. Suppl.* 144, 133, 1989.
32. Blank, I.H., Moloney, J., Emslie, A.G., Simon, I., and Apt, C., The diffusion of water across the stratum corneum as a function of its water content, *J. Invest. Dermatol.*, 82, 188, 1984.
33. Ryatt, K.S., Mobayen, M., Stevenson, J.M., Maibach, H.I., and Guy, R.H., Methodology to measure the transient effect of occlusion on skin penetration and stratum corneum hydration *in vivo*, *Br. J. Dermatol.*, 119, 307, 1988.
34. Tiemessen, H.L.G.M., Bodde, H.M., and Junginger, H.E., A silicone membrane sandwich method to measure drug transport through isolated human stratum corneum having a fixed water content, *Int. J. Pharmacol.*, 56, 87, 1989.
35. Cooper, E.R. and van Duzee, B.F., Diffusion theory analysis of transepidermal water loss through occlusive films, *J. Soc. Cosmetic Chem.*, 27, 555, 1976.
36. Dupuis, D., Rougier, A., Lotte, C., Wilson, D.R., and Maibach, H.I., *In vivo* relationship between percutaneous absorption and transepidermal water loss according to anatomic site in man, *J. Soc. Cosmetic Chem.*, 37, 351, 1986.
37. Morrison, B.M., ServoMed evaporimeter: precautions when evaluating the effect of skin care products on barrier function, *J. Soc. Cosmetic Chem.*, 43, 161, 1992.
38. Pinnagoda, J., Tupker, R.A., Agner, T., and Serup, J., Guidelines for transepidermal water loss (TEWL) measurement. A report from the Standardization Group of the European Society of Contact Dermatitis, *Contact Dermatitis*, 22, 164, 1990.
39. Rogiers, V., EEMCO guidance for the assessment of transepidermal water loss in cosmetic science, *Skin Pharmacol. Appl. Skin Physiol.*, 14, 117, 2001.
40. Lodén, M., The increase in skin hydration after application of emollients with different amounts of lipids., *Acta Derm. Venereol.*, 72, 327, 1992.
41. Rietschel, R.L., A method to evaluate skin moisturizers *in vivo*, *J. Invest. Dermatol.*, 70, 152, 1978.
42. Wilson, D.R., Berardesca, E., and Maibach, H., *In vivo* transepidermal water loss and skin surface hydration in assessment of moisturization and soap effects, *Int. J. Cosmetic Sci.*, 10, 201, 1988.
43. Blichmann, C.W., Serup, J., and Winther, A., Effects of single application of a moisturizer: evaporation of emulsion water, skin surface temperature, electrical conductance, electrical capacitance, and skin surface (emulsion) lipids, *Acta Derm. Venereol.*, 69, 327, 1989.
44. Berardesca, E., Fideli, D., Borroni, G., Rabbiosi, G., and Maibach, H.I., *In vivo* hydration and water-retention capacity of stratum corneum in clinically uninvolved skin in atopic and psoriatic patients, *Acta Derm. Venereol.*, 70, 400, 1990.
45. Wu, M.S., Yee, D.J., and Sullivan, M.E., Effect of a skin moisturizer on the water distribution in human stratum corneum, *J. Invest. Dermatol.*, 81, 446, 1983.
46. Grubauer, G., Elias, P.M., and Feingold, K.R., Transepidermal water loss: the signal for recovery of barrier structure and function, *J. Lipid Res.*, 30, 323, 1989.
47. Grubauer, G., Feingold, K.R., and Elias, P.M., Relationship of epidermal lipogenesis to cutaneous barrier function, *J. Lipid Res.*, 28, 746, 1987.
48. Feingold, K.R., Brown, B.E., Lear, S.R., Moser, A.H., and Elias, P.M., Effect of essential fatty acid deficiency on cutaneous sterol synthesis, *J. Invest. Dermatol.*, 87, 588, 1986.

49. Menon, G.K., Feingold, K.R., Moser, A.H., Brown, B.E., and Elias, P.M., De novo sterologenesis in the skin. II. Regulation by cutaneous barrier requirements, *J. Lipid Res.* 26, 418, 1985.
50. Hannuksela, A. and Kinnunen, T., Moisturizers prevent irritant dermatitis, *Acta Derm. Venereol.*, 72, 42, 1992.
51. Held, E., Sveinsdottir, S., and Agner, T., Effect of long-term use of moisturizers on skin hydration, barrier function and susceptibility to irritants, *Acta Derm. Venereol. (Stockh.)*, 79, 49, 1999.
52. Lodén, M., Urea-containing moisturizers influence barrier properties of normal skin, *Arch. Dermatol. Res.*, 288, 103, 1996.
53. Serup, J., A double-blind comparison of two creams containing urea as the active ingredient: assessment of efficacy and side-effects by non-invasive techniques and a clinical scoring scheme, *Acta Derm. Venereol.* (Stockh.), 177, 34, 1992.
54. Bimczok, R., Ansmann, A., Biefieldt, S., Billek, D., Driller, H., Feistkorn, G., Heinze, F., Huttinger, R., Komp, B., Lautenschläger, H., Leneveu, M.-C., Motitschke, L., Pohl, L., Reng, A., Schulze, H.-J., Thomaskamp, B., Tolkiehn, K., Tronnier, H., Wekel, H.-U., and Wittern, K.-P., A multicenter comparison of different test methods for the assessment of the efficacy of skin care products with 368 human volunteers, *J. Soc. Cosmet. Chem.*, 45, 1, 1994.
55. Lavker, R.M., Kwong, F., and Kligman, A.M., Changes in skin surface patterns with age, *J. Gerontol.*, 35, 348, 1980.
56. Wagner, G. and Goltz, R.W., Human cutaneous topography, a new photographic technique: observations on normal skin, *Cutis*, 23, 830, 1979.
57. Grove, G.L., Grove, M.J., Leyden, J.J., Lufrano, L., Schwab, B., Perry, B.H., and Thorne, E.G., Skin replica analysis of photodamaged skin after therapy with tretinoin emollient cream, *J. Am. Acad. Dermatol.*, 25, 231, 1991.
58. Hayashi, S., Mimura, K., and Nishijima, Y., Changes in the surface configuration of the skin caused by ageing and application of cosmetics: three-dimensional analysis according to a new system based on image analysis and Fourier transformation, *Int. J. Cosmetic Sci.*, 11, 67, 1989.
59. Cook, T.H., Profilometry of skin—a useful tool for the substantiation of cosmetic efficacy, *J. Soc. Cosmetic Chem.*, 31, 339, 1980.
60. Murahata, R.I., Crowe, D.M., and Roheim, J.R., Evaluation of hydration state and surface defects in the stratum corneum: comparison of computer analysis and visual appraisal of positive replicas of human skin, *J. Soc. Cosmetic Chem.*, 35, 327, 1984.
61. Wolfram, M.A., Wolejsza, N.F., and Laden, K., Biomechanical properties of delipidized stratum corneum, *J. Invest. Dermatol.*, 59, 421, 1972.
62. Linde, Y.W., Bengtsson, A., and Lodén, M., "Dry" skin in atopic dermatitis. II. A surface profilometry study, *Acta Derm. Venereol.*, 69, 315, 1989.
63. Cook, T.H., Craft, T.J., Brunelle, R.L., Norris, F., and Griffin, W.A., Quantification of the skin's topography by skin profilometry, *Int. J. Cosmetic Sci.* 4, 195, 1982.
64. Nicholls, S., King, C.S., and Marks, R., Short term effects of emollients and a bath oil on the stratum corneum, *J. Soc. Cosmetic Chem.*, 29, 617, 1978.
65. Mignot, J., Zahouani, H., Rondot, D., and Nardin, P., Morphological study of human skin relief, *Bioeng. Skin*, 3, 177, 1987.
66. Miller, D.L., A simple, low cost image analysis system for evaluation of the skin surface, *Bioeng. Skin*, 3, 243s, 1987.

67. Cook, T.H. and Craft, T.J., Topographics of dry skin, non-dry skin, and cosmetically treated dry skin as quantified by skin profilometry, *J. Soc. Cosmetic Chem.*, 36, 143, 1985.
68. Tagami, H., Kanamaru, Y., Inoue, K., Suehisa, S., Inoue, F., Iwatsuki, K., Yoshikuni, K., and Yamada, M., Water sorption-desorption test of the skin *in vivo* for functional assessment of the stratum corneum, *J. Invest. Dermatol.*, 78, 425, 1982.
69. Tagami, H., Electrical measurement of the water content of the skin surface. Functional analysis of the hygroskopic property and water-holding capacity of the stratum corneum *in vivo* and technique for assessing moisturizing efficacy, *Cosmetics Toiletries*, 97, 39, 1982.
70. Lodén, M. and Lindberg, M., The influence of a single application of different moisturizers on the skin capacitance, *Acta Derm. Venereol.*, 71, 79, 1991.
71. Jemec, G.B., Na, R., and Wulf, H.C., The inherent capacitance of moisurusing creams: a source of false positive results? *Skin Pharmacol. Appl. Skin Physiol.*, 13, 182, 2000.
72. Wepierre, J., Study of the hydrating effect of cosmetic preparations by measuring cutaneous impedance in the hairless rat, *Soap Perfum. Cosmetics*, 50, 506, 1977.
73. Batt, M.D., Davis, W.B., Fairhurst, E., Gerreard, W.A., and Ridge, B.D., Changes in the physical properties of the stratum corneum following treatment with glycerol, *J. Soc. Cosmetic Chem.*, 39, 367, 1988.
74. Reiger, M.M. and Deem, D.E., Skin moisturizers. II. The effect of cosmetic ingredients on human stratum corneum, *J. Soc. Cosmetic Chem.*, 25, 253, 1974.
75. Lieb, L.M., Nash, R.A., Matias, J.R., and Orentreich, N., A new *in vitro* method for transepidermal water loss: a possible method for moisturizer evaluation, *J. Soc. Cosmetic Chem.*, 39, 107, 1988.
76. Serup, J., A three-hour test for rapid comparison of effects of moisturizers and active constituents (urea). Measurement of hydration, scaling and skin surface lipidization by noninvasive techniques, *Acta Derm. Venereol.* (Suppl.) 177, 29, 1992.
77. Korstanje, C., Ligtvoet, F.P.T., and Van Hemert, K.H.F., Differential effects of dermatological cream bases with respect to skin surface capacity: a study design in volunteers, *J. Dermatol. Treatm.*, 2, 137s, 1992.
78. Serban, G.P., Henry, S.M., Cotty, V.F., and Marcus, A.D., *In vivo* evaluation of skin lotions by electrical capacitance. I. The effect of several lotions on the progression of damage and healing after repeated insult with sodium lauryl sulfate, *J. Soc. Cosmetic Chem.*, 32, 419, 1981.
79. Dunlap, F.E., Clinical evaluation of a highly effective hand and body lotion, *Curr. Ther. Res.*, 35, 72, 1984.
80. Potts, R.O., Guzek, D.B., Harris, R.R., and McKie, J.E., A noninvasive, *in vivo* technique to quantitatively measure water concentration of the stratum corneum using attenuated total-reflectance infrared spectroscopy, *Arch. Dermatol. Res.*, 277, 489, 1985.
81. Walling, P.L. and Dabney, J.M., Moisture in skin by near-infrared reflectance spectroscopy, *J. Soc. Cosmetic Chem.*, 40, 151, 1989.
82. de Rigal, J., Losch, M.J., Bazin, R., Camus, C., Sturelle, C., Descamps, V., and Leveque, J.L., Near infra-red spectroscopy: a new approach to the characterization of dry skin, *J. Soc. Cosmet. Chem.*, 14, 197, 1993.
83. Williams, H.C. and Seed, P., Inadequate size of 'negative' clinical trials in dermatology, *Br. J. Dermatol.*, 128, 317, 1993.

Index

Page references followed by f indicate a figure.
Page references followed by t indicate a table.

A

Abdomen, 148f, 149f, 151f, 290
Abnormal barrier development, 270
Absorption, 53
Accuracy of instruments, 87
ACD. *see* Allergic contact dermatitis (ACD)
Acetone/ether (AE), 270, 372
Acetone/ether with water (AEW), 372
Acidification, 28f, 29
Acid mantle, 29
Activation, occlusion and, 56f
Acute radiation dermatitis, 71
Acylceramide, 5, 6
Acylglucosylceramides, 5–6
ADI. *see* Autosomal dominant ichthyosis (ADI)
Adults/adulthood
  changes in pH in, 32f
  infant skin *vs.* skin of, 313
  transepidermal water loss (TEWL) during, 145
Adverse skin reactions, 298, 300
Age
  barrier recovery and, 153f
  percutaneous absorption and, 150f
  permeability barrier and, 154t
  pH and, 32f
  skin irritation and, 152f
  transepidermal water loss (TEWL) and, 66–67, 89
Aging. *see also* Elderly
  of human skin, 143–155
  percutaneous penetration and skin, 147–148
  transepidermal water loss (TEWL) and, 145–147, 146t, 147f, 151
Air circulation, transepidermal water loss (TEWL) measurements and, 66
Air convections, measuring transepidermal water loss (TEWL) and, 87–88
Allergic contact dermatitis (ACD)
  Corticosteroids and, 110
  and irritation, 107–108
  patch test for, 108
  skin hypersensitivity and, 106
  transepidermal water loss (TEWL) and topical agents for, 109–110
  transepidermal water loss (TEWL) values in patients with, 105–108
Alpha-hydroxy acids, 298t
Ambient air
  humidity and temperature of, 88, 289–290
  water barrier function of, 352
  water vapor mantel of, 352
Amino acid cystine, 15
Ammonium lactate, 188
Amplitude parameters, skin profiles and, 395
Anatomical sites
  electrical response to measurements, 290
  transepidermal water loss (TEWL) and, 67, 89
Ankle, 148f, 149f, 151f
Antiinflammatory drugs, 109–110
Antioxidants, 199t
Appendage-free skin, 29
Arm, upper, 148f, 151f
Asians
  stratum corneum in, 129
  transepidermal water loss (TEWL) in, 130
Atopic dermatitis (AD)
  on another body region, 118–119
  basal transepidermal water loss (TEWL) and, 370
  characteristics of, 364–365
  eczematous skin in, 160
  noneczematous skin in, 161–163
  overview of, 324–325
  skin hydration and, 323–329
  skin hyperactivity and, 106
  sodium lauryl sulfate (SLS) and, 163
  xerosis in, 325
Atopic eczema, 174
Atopics, transepidermal water loss (TEWL) and, 162t, 163
Autoradiography, 55
Autosomal dominant ichthyosis (ADI), 161
Autosomal dominant ichthyosis vulgaris, 175

B

Back (lower, upper), 148f, 149f, 151f
Bacteria, occlusion and, 56
Balsam of Peru, 188
Barrier creams (BCs)
  efficacy and testing models of, 380–382t
  in vitro methods for testing, 375–376
  in vivo methods for testing, 376–379
Barrier disruption, calcium and, 37–38
Barrier function. *see also* Transepidermal water loss (TEWL)
  baseline, at healthy skin sites, 106
  chemical factors and, 198
  disruption of, 198–199
  dry skin and defects in, 174–176
  in humans, 129–130
  influences on repair of, 199–205
  modulation of, 70
  moisturizers and, 393–394
  pathological factors and, 198
  physical factors and, 198
  preexposure, 117–118
  repair of, 198–199
  of stingers, 138
  of stratum corneum, 98–99, 315
  techniques for assessment of, 176
  transepidermal water loss (TEWL) as a measure for, 100
Barrier homeostasis, ions and, 36–41
Barrier recovery
  age and, 153f
  disruption and repair, 198–199
  influences on barrier repair, 199–205
  overview of, 197–198
  protective mechanism of the epidermis, 198
Basal transepidermal water loss (TEWL), atopic dermatitis and, 370
Baseline barrier function, at healthy skin sites, 106
Bath oils, 296
BCs. *see* Barrier creams (BCs)
Behavioral factors, dry skin and, 174t
BET. *see* Statistical thermodynamic methods (BET)
BFV. *see* Blood flow volume (BFV)
Bioengineering
  devices, product testing strategies using, 390–400
  future directions for investigations, 195t
  studies about measurements, 192–193t
Blacks
  exposure to sodium lauryl sulfate (SLS), 132
  skin irritation in, 132
  stratum corneum in, 129
  transepidermal water loss (TEWL) in, 130
  in vitro transepidermal water loss (TEWL) values in, 131
Blood flow velocity, in subclinical cutaneous irritation, 137
Blood flow volume (BFV), 379
Bound water, 21, 288
B-scanning methods, irritant skin and, 79–80
Burning
  for assessing skin permeability, 176
  sensory irritation and, 135, 136

C

Calcium
  barrier disruption and, 37–38
  electric potential and, 38
  in the epidermis, 36–37
  gradient, ontogenesis of, 36
  gradient, origin of, 38–39
  importance of, 39–40
  ion, in epidermis and cellular compartments, 36
  pathologic conditions and, 40
  role in barrier homeostasis, 36–41
  role in barrier recovery, 37f
  as second messenger, 39
Calibration
  of the Corneometer probe, 255
  of instruments, 87
  procedures, 280–281
Calorimetric measurement, 20
Capacitance, 289, 290, 316f, 326–327
Capsaicin, for burning, 136
Carboxyl groups (COO–), 20
Caveolins, 9
Ceramides
  in the epidermis, 6
  human epidermal, 8f
  long-chained, 98
  in stratum corneum, 130
Cerebrosides, 5
Chamber scarification technique, 116
Chemical factors, barrier function and, 198
Chemical irritation, 221–222
Chemicals, exogenous, 191–192
Chemistry, 7–8
Childhood, changes in pH during, 32f
Cholesterol, epidermis and, 4–5
Cholesterol sulfate, 6–7
Cholesteryl ester, 7
Chromometry
  for assessing the irritant potential of a chemical, 101

use with stingers and nonstingers, 190
Chronic inflammatory skin disease, atopic eczema as, 174
Chronic irritation model, 226t, 228t
Chronic SLS/toluene irritation model, 226t, 227f
Circadian rhythm
of skin surface pH, 33
transepidermal water loss (TEWL) measurement and, 68, 90
Cleansing products
pH values and, 34
skin irritation and, 69
Clinical studies, about the efficacy of moisturizers, 231
Clinical symptoms, dry skin and, 172
Closed chamber system, 65, 84
Closed-patch testing, 115
Coagulase-negative staphylococci, pH and, 35
Cole equation, 237f, 336–337
Collodion babies, 165, 364
Colorimeter, 108
Compound
concentration of, 115
duration of exposure in one-time occlusive patch test, 115–116
predictive irritancy and type of, 114–115
Computerized DermaLab TEWL probe
overview of, 276f, 278–279
schematic drawing of, 278f
steady-state assessments using, 282f
test results from performance test of, 279f
using dual probe, 283–284
Conductance. *see* Electrical conductance
Confocal microscopy, for stratum corneum, 23
Congenital keratinization disorders, 165–166
Contact time, instruments and, 86
Corneocytes, 98, 360–361
Corneometer
calibration of the probe, 255
capacitor scheme of, 253f
CM 820, 251f
CM 825
advantages of the development of, 256
classification of skin type with, 259, 259t
continuous measurement with, 260f
linearity of the probe, 257
measurement of ambient conditions with, 260f
penetration depth of the measurement with, 257–259, 258f
probe, 252f
single measurements with, 259f
CM 420 probe with plastic foil, 251f
compared to other bioengineering instruments, 256
conductor track of, 253f, 254f
history of the development of, 250, 250f
for measuring SC water content, 288
readings of, 289
for skin surface hydration measurement, 353–354
software, 259–260
structure of the sensor, 253–255
technical data, 256
Cortex Technology, 276
Cortical cells, in hair, 17
Corticosteroids, allergic contact dermatitis and, 110
Cosmetics
barrier repair and, 203–204, 203t
claim support for, 69–70
development of ingredients and finished products, 71
dry skin and, 175
sensitive skin and, 135
Creams, 296, 297, 300, 357
Cumulative irritation, 223–228, 224f
Cutometer, 270
Cyber-DERM, Inc., 279
Cystine content, proteins and, 18t

D

Damaged skin, 68, 90, 214–215, 215f
Damaged stratum corneum, 362–363
DASYLab, 279
Data, analysis of, 284
DDE. *see* Dynamic data exchange (DDE)
Delipidized skin, 243–244
Depolarizing thermocurrents, 20
DermaLab® system, 64, 85, 276, 279. *see also* Computerized DermaLab TEWL probe
Dermal Phase Meter (DPM)
readings of, 289
vs. time curves for neonatal rats, 269f
Dermal Torque Meter, 270
Dermatitis. *see also* Seborrheic dermatitis
acute radiation, 71
allergic contact. *see* allergic contact dermatitis (ACD)
atopic. *see* Atopic dermatitis (AD)
contact, 68
irritant contact, 71, 113–114
occupational, predictive testing in, 120–121
Desquamation, 175
Detergent-damaged skin, 362
Detergents
and disruption of the barrier function, 198

pH values and, 34
putative action spectrum of, 114
Developmental changes, in pH, 32–33
Diagnostic programs, DermaLab TEWL Probe, 279–280
Dia-Stron DTM Torque Meter, 288
Dielectric characteristics, in subclinical cutaneous irritation, 137
Dielectric spectroscopy, 339
Differential scanning calorimetry, 17
Dihydrosphingosines, 6
Dimethicone skin protectant lotion, 378
Dioleoylphosphatidylethanolamine, 8
Diphenylcyclopropenone patch test, 108
Diseased skin
chemical irritation and, 221–222
cumulative irritation and, 223–228, 224f
hydration and transepidermal water loss (TEWL) in, 214–215, 215f
hydration in stratum corneum from, 363–365
lipids in stratum corneum from, 363–365
physical irritation and, 222–223
Diseases. *see also* Dermatitis
eczematous, skin pH values and, 33
ichthyosis, skin pH values and, 33–34
non-insulin-dependent diabetes mellitus, 34
transepidermal water loss (TEWL) and, 101
Disruption, of barrier function, 198–199
Domain mosaic model, 9, 198
Dressings
hydrocolloidal, 52
occlusive, 52, 53
Dry skin (DS). *see also* Ichthyosis vulgaris; Xerosis
atopic dermatitis and, 161–163
causes of, 174t
characteristics of, 172–173, 172t
chemical analysis of, 172–173
chemistry of, 296–297
clinical conditions of, 370f
clinical symptoms of, 172
concept of, 388–389
cosmetics and, 175
and defects in the barrier function, 174–176
environmental factors and, 174t
and exposure to solvents, 174t, 175
function of, 296–297
genetic factors and, 174–175, 174t
lipids for treatment of, 176–177
moisturizers and, 176, 177, 178t, 299t
morphology and function of, 173
Plastic Occlusion Stress Test (POST) in, 241–242
sensory irritation and, 135
tactile characteristics and, 172t
as a term, 324
transepidermal water loss (TEWL) and, 171–180
treatment effects of, 176–179
D-Squame® tapes, 353, 354f
Dual-probe instruments, measurement of TEWL using, 283–284
Dynamic data exchange (DDE), 279

E

Eczema. *see also* Atopic dermatitis (AD); Noneczematous skin
atopic, 174
as disease, 33
high evaporative water loss and, 355–356
hydration in, 351–357
measuring, 105–106
scaling process in, 352–353
skin in atopic dermatitis, 160
Edwina software, 265
Efficacy study, checklist for planning and evaluating, 401t
Elderly. *see also* Aging
skin irritation of, 149–150
transepidermal water loss (TEWL) and, 145–147, 146t, 147f, 151
Electrical conductance
of eczema, 356f
hydration of stratum corneum and, 20, 326–327
Electrical measurements, 301, 396–399
Electric potential, barrier repair and, 201–202, 202t
Electrodes, 339, 341
Electrode/tissue interface, 341
Electron spin resonance imaging, for assessing pH, 29
Emollients
definition of, 212–213
evaluation of the efficacy of, 398–399
hydrating effects of, 396–397
infrared spectroscopy and, 399–400
long-term effects on skin hydration, 393–394
short-term effects on skin hydration, 391–393
as a term, 295. *see also* Moisturizers
Endogenous factors, human surface pH and, 31–34
Enthalpimetric measurements, 23f
Entry echo, 77
Environment. *see also* Humidity; Temperature
dry skin and, 174t
electrical assessment of skin hydration and, 289–290

transepidermal water loss (TEWL) measurement and, 66, 87–89
Epidermal calcium gradient, 36
Epidermal echogenicity, hydration values and, 79f
Epidermal effects
on intraepidermal processes, 50–51
role of water and, 49–50
Epidermal lipids, 329
Epidermis
calcium distribution in, 36–37
in the ceramides, 6
fatty acids in, 4, 7
function of, 359
irritant substances and, 79–80
lipid classes in, 4–7
protective functions of, 199t
protective mechanisms of, 198–199
Erosion, for assessing skin permeability, 176
Errors in measurement, instruments and, 85–87, 288–289
Erythema, as sign of irritation, 135
Essential fatty acid deficiency, 365
Ethics, 400–401
Ethnic differences, transepidermal water loss (TEWL) values and, 67
European Center for the Validation of Alternative Methods Workshop, 71
European Contact Dermatitis Society, 391
European Expert Group Measurement of Cosmetics, 91–92
European Group for Efficacy Measurements of Cosmetics and Other Topical Products, 391
European Society of Contact Dermatitis, 276, 351
European Union Cosmetic Legislation, 69
Evaporation phase, in testing moisturizers, 216–217
Evaporimeter®, 64, 85, 108, 325
Experiment, planning, 400–402
Experimental studies, on diseased skin, 221
Exposure methods, predictive irritancy and, 115–117

F

Facial skin, stinging and, 189
Fatty acids
in epidermis, 4, 7
essential, deficiency of, 365
in the stratum corneum, 8
Fick's law of diffusion, 152–154
Filaggrin, 360–361
Filaggrin-to-histidine proteolysis, 30
Fluorescence lifetime imaging microscopy, for measuring pH, 29
Forearms
dorsal, 131, 148f, 149f, 151f
sonographic images of the skin of, 78f
volar, 31, 131, 148f, 149f, 151f
Forehead, 148f, 149f, 151f, 290
4-hour patch test, 115, 124
Fourier-transform infrared spectroscopy (FTIR), 288
Fragrances, moisturizers and, 300
Free water, 21–22, 288, 296
FTIR. *see* Fourier-transform infrared spectroscopy (FTIR)

G

GA. *see* Gestational age (GA)
Galactosylceramides, 5
Gels, 296
Gender
human surface pH and, 31–33
topographic variation and differences in, 31
transepidermal water loss (TEWL) values and, 66–67
Genetic factors
dry skin and, 174–175, 174t
irritancy and, 119
Germ-free skin, 29
Gestational age (GA), 314
Glycerin-containing lotion, 397
Glycerol, barrier repair and, 204
Glycolipids, 5–6
Gradient Estimation Method of Nilsson, 276–278

H

Hair
amount of lipids in, 22
cystine content of proteins in, 18t
free water and, 21
sorption isotherms for, 19f
stratum corneum (SC) and, 17–18
swelling, stratum corneum (SC) and, 23–24
Hardware
Edwina software, 265
NOVA Dermal Phase Meter (DPM) 9003 (*see* NOVA Dermal Phase Meter (DPM) 9003)
NOVA Petite, 265, 266f, 271, 271f
Healing, measurement of hydration during, 356–357
Healthy skin

effect of hydration on, 214
experiments on, 216–221
Hemodialysis, pH values and, 34
Hispanics
exposure to sodium lauryl sulfate (SLS) and, 132
skin irritation in, 132
Histidine-to-urocanic acid pathway, 29–30
Human immunodeficiency virus (HIV), 371
Human volunteer studies, 400–402
Humectants
moisturizers and, 297
skin hydration and, 390
testing, 219–220
Humidity
ambient air, 88, 289–290
dry skin and, 174t, 175
environmental temperature and, 66
of environment and barrier repair, 199–200, 200t
instruments and, 86
Hydration. *see* Skin hydration
Hydrocolloidal dressings, 52
Hydrocortisone, absorption in woman, 148
Hydrogen-bounded water, 296
Hydrophilic cells, 98
Hyperactivity, of skin, 106–107
Hyperhydration, 296
Hypernatremic dehydration, 165
Hypothermia, 165
Hyprid parameters, skin profiles and, 395

I

ICD. *see* Irritant contact dermatitis (ICD)
Ichthyosis/ichthyoses, 33–34, 174–175, 363
Ichthyosis vulgaris, 33, 172, 175, 363–364
Imaginary part index, 343, 344t
Immersion test, 117
Impedance
measurement of, 337–340
sensitivity of, 345
for stratum corneum hydration measurement, 326–327
Impedance spectrometer
electrical, SciBase II, 345f
handheld probes for, 346f
Impedance spectroscopy
changes in living strata and, 347
of skin, 340–346
use on skin cancer, 347
Indinavir, for xerosis, 371
Individuals, transepidermal water loss (TEWL) and, 89–91, 290
Infants. *see also* Infant skin
full term
adult comparisons, 313
birth of, 307–310
differential hydration of the skin of, 308f, 311f
first month of, 310–312
moisture accumulation rate for, 310f
skin hydration, environmental influences and, 312–313
newborn. *see* Newborns
premature, 313–317, 316f
role of vernix caseosa, 317–318
Infant skin
adult skin *vs.*, 313
environment and hydration of, 308f, 311f, 312–313, 314, 314f
Sorption–Desorption Test (SDT) in, 243
Inflammatory disorders, psoriasis as, 174
Inflammatory mediators, 199t
Infrared absorption, 20
Infrared measurements, 19
Infrared spectroscopy (IR), 399–400
Instruments
accuracy of, 87
error in measurements and, 85–87, 288–289
transepidermal water loss (TEWL) measurement and, 85
Intercellular lipid domains, 197–198
Intercellular space, water in, 360
Interindividual variations, skin and, 90–91
Intermediate water, 22
Intracellular compartments, calcium in, 36
Intraepidermal processes, epidermal effects on, 50–51
Intraindividual variations, skin and, 90–91
In vitro irritancy testing, 114, 121
In vitro transepidermal water loss values, in different races, 131
In vivo irritancy testing, 114
In vivo transepidermal water loss, in different races, 131–132
Ionizing radiation, transepidermal water loss (TEWL) values and, 66
Ions, role in barrier homeostasis, 36–41
Iontophoresis
negative and positive, 38
vs. patch testing, 108
IR. *see* Infrared spectroscopy (IR)
Irritant contact dermatitis (ICD)
aging and, 149–150
dimethicone skin products and, 71
overview of, 113–114
transepidermal water loss (TEWL) and development of, 163

Irritant substances
epidermis and, 79–80
modes of action of, 114–115
Irritation. *see also* Sensory irritation; Skin irritation
Irritative skin reactions, reduction of, 70
Itch
clinical conditions of, 370f
hemodialysis patients and, 371
pathophysiology of, in xerotic skin, 372–373
senile xerosis and, 370
sensory irritation and, 136
xerosis of skin and, 373t

K

Keratinocytes
calcium ion and, 39–40
occlusion and, 53, 55, 56f
Keratins
amino acid cystine and, 15
environmental factors and, 16
interaction of water and, 15–24
sorption isotherm of, 18–19
water and physical properties of, 16, 16f
Keratohyalin granules, 360

L

Lactic acid sting test (LAST), 136, 188–189
Lamellar bodies, 98
Lamellar body secretion, calcium ion and, 39
Lamellar ichthyosis, 68, 175, 364
Lamellar lipid systems, 360
Langerhans' cells, 107
Laser Doppler flowmetry, 101, 108, 109, 190
Laser Doppler velocimetry, 54
LAST. *see* Lactic acid sting test (LAST)
Legs, hydration values of, 290
Lesions, 7
Light sources, 66, 88
Lipid classes, in epidermis, 4–7
Lipid extraction, 23
Lipidization phase, in testing moisturizers, 217–219
Lipids
epidermal, 329
in hair, 22
nonpolar, 3
in normal stratum corneum, 360–361
skin hydration and, 390
of stratum corneum, 7
in stratum corneum from diseases skin, 363–365
for treating dry skin, 176–177
use for restoration of normal hydration, 365
Liposomes, from stratum corneum lipids, 10–11
Liquid crystalline phase, 8
Liquid ordered phase, 9
Living tissue, major dispersions of, 338f
Long-chained ceramides, 98
Lotion alba aquosa (LAA), 227
Lotions, 296

M

Magnesium, role in barrier homeostasis, 40
Magnetic resonance imaging (MRI)
for measuring SC water content, 288
moisturizers and, 399
Magnetic resonance spectroscopy, 19
Magnitude index, 343, 344t
MAT. *see* Moisture Accumulation Test (MAT)
Mathematical analysis of isotherms, 20
Measurement techniques, transepidermal water loss (TEWL), 64–65
Membrane antiporters, 30
Men, sensitive skin in, 135
Methyl salicylate, for burning, 136
Microwave, 326
Mitochondria, 5
Model
chronic irritation, 227t, 228t
chronic SLS/toluene irritation, 226t, 227f
short-term SLS irritation, 225f
Moisture accumulation rate, 310f
Moisture Accumulation Test (MAT)
correlation with electrical instruments, 271
correlation with other methods, 270
description of, 238–239
effect of extraction and soaking on slopes of, 270f
NOVA Dermal Phase Meter and, 268–270
psoriatic skin and, 240
and skin irritation, 239
typical curve, 239f
Moisturizers
alpha-hydroxy acids in, 298t
barrier function and, 393–394
chemistry of, 297–298
definition of, 212–213
dry skin and, 176, 177, 178t, 299t
effect on transepidermal water loss (TEWL) in dry skin diseases, 178t
effects of, 120
evaluation techniques of, 300–301

function of, 297–298
hydration of the stratum corneum and, 213–214
mode of action of, 215
occlusive, 212
skin hydration and, 295–301
as a term, 295–296
testing efficacy of, 212–232
transepidermal water loss (TEWL) and, 214, 217f, 218f, 219f, 391–394
treatment with, 298–299
urea-containing, 179, 394
use of, 387–388
Mucosal atopy, skin susceptibility and, 119

N

Nail, cystine content of proteins in, 18t
Natural moisturizing factor (NMF)
composition of, 172, 295, 296, 389t
corneocytes and, 295
dry skin and, 172
elasticity and, 388–389
keratinization and, 238
in stratum corneum, 296
Negative iontophoresis, 38
Nerve growth factor (NGF), 372
Neste, Van, 239
Netherton's syndrome, 165
Newborns. *see also* Infants
changes in pH in, 32f
cutaneous permeability barrier function of, 145
NGF. *see* Nerve growth factor (NGF)
Nickel contact allergy, 376
Nickel contact dermatitis, antiinflammatory drug for, 109
Nickel patch test, 107
Nickel sulfate patch tests, transepidermal water loss (TEWL) values and, 109t
NMF. *see* Natural moisturizing factor (NMF)
NMR. *see* Nuclear magnetic resonance spectroscopy (NMR); nuclear magnetic resonance spectroscopy (NMR)
Noneczematous skin
in atopic dermatitis, 161–163
of atopics, transepidermal water loss (TEWL) of, 162t
Noninflammatory scaling disorder, ichthyoses as, 174
Noninvasive bioengineering techniques, 176
Noninvasive methodology
for improving topical therapeutic treatment, 71
use in safety testing on human skin, 71
Nonlamellar inverted hexagonal phase, 8
Nonpolar lipids, definition of, 3
Nonstingers
sebum content of the skin of, 194
stingers *vs.*, 190–191
Normal barrier, moisture accumulation test use for, 270
NOVA Dermal Phase Meter (DPM) 9003
change in reading, 267f
correlation with Skicon, 271
description of, 264–265, 264f
for measuring SC water content, 288
NOVA Petite and, 271, 271f
precautions in using, 271
static measurements, 266
values with glycerin treatment, 267f
Novapen®, for measuring skin surface hydration, 353–354
NOVA Petite, 265, 266f, 271, 271f
Nuclear magnetic imaging, 326
Nuclear magnetic resonance spectroscopy (NMR), 288, 326
Nyquist plot
of cole equation, 237f
of electrical skin impedance, 342f

O

Occlusion
absorption and, 53
definition of, 49
epidermal effects, 49–51
factors influenced by, 56f
influence of time, 51–52
measurement parameters and techniques of, 54–550
occlusive materials and wound healing, 52
therapeutic use of, 53–54
Occlusive dressings, 52, 53
Occlusive emollients, 393
Occlusive moisturizers, 212
Occupational dermatitis
predictive testing in, 120–121
studies on factors influencing development of, 122–123t
Occupational settings, protection of the skin in, 70
Oil-in-water (O/W) emulsion, 217, 297
Ointments, 296, 300, 316, 357
One-time occlusive patch test, 115–117, 121
Open-chamber method, 64, 84, 276
Open-exposure technique, 116

Open-irritation test, 116
Oral mucosa, 346

P

Palms, 144, 148f, 149f, 151f, 290
Patch tests
  and transepidermal water loss (TEWL), 108–109
  in vivo, 376
Pathological factors, barrier function and, 198
Pathomechanism, sensory irritation and, 136
PCA. *see* Pyrrolidone carboxylic acid (PCA)
Percutaneous absorption, age dependence of, 150f
Percutaneous penetration, skin aging and, . 147–148, 148f
Perfluoropolyethers (PFPE), 378
Permeability barrier, age and, 154t
Person-linked variables, transepidermal water loss (TEWL) and, 66–69, 89–91
Perspiratio insensibilis, 352
Petroleum bath oils, 215
PFPE. *see* Perfluoropolyethers (PFPE)
Phase behavior, 8–9
Phase index, 343, 344t
PH of stratum corneum
  antimicrobial function of, 35–36
  changes during life, 32f
  definition of, 28
  factors affecting human surface pH, 31–33
  importance for the barrier, 34–35
  mechanisms leading to acidification, 29–30
  methods of measurements of, 28–29
Phospholipids, epidermis and, 4–5
Phospholipid-to-free fatty acid pathway, 30
Physical factors, barrier function and, 198
Physical irritation, 222–223
Physiological aspects, barrier repair and, 202–203, 204t
Phytosphingosines, 6
Plastic Occlusion Stress Test (POST)
  for assessing psoriasis, 240–241
  dehydration curve obtained with, 241f
  in dry and irritated skin, 241–242
  for measuring water in stratum corneum, 238
  psoriatic skin and, 242
Polar compounds, 117
Positive iontophroeisis, 38
POST. *see* Plastic Occlusion Stress Test (POST)
Postauricular, 148f, 149f, 151f
Potassium, role in barrier homeostasis, 40
Potentiometric assessment, for measuring pH, 28
Predictive irritancy testing
  extrinsic factors of, 114–117
  intrinsic factors of, 117–120
  as a method, 113–114, 124
  in the occupational setting, 120–121
Preexposure barrier function, transepidermal water loss (TEWL) and, 117–118
Preservatives, moisturizers and, 300
Probes
  handheld, 346f
  protection covers, instruments and, 86–87
  temperature, transepidermal water loss (TEWL) values and, 66
Product-testing strategies, bioengineering device and, 390–400
Profilaggrin, 360
Proliferation, occlusion and, 55, 56f
Propionibacteria, pH and, 35
Protease inhibitors, for xerosis, 371
Protein-rich nonviable cells, 197–198
Proteins, 15
Protofibrils, 17
Proton density, 326
Provocation tests, hand dermatitis and, 121
Psoriasis
  high evaporative water loss and, 355–356
  hydration in, 351–357
  as inflammatory disorder, 174
  Moisture Accumulation Test (MAT) and, 240
  natural moisturizing factor (NMF) and, 172
  occlusion and, 51, 54
  Plastic Occlusion Stress Test (POST) and, 242
  scaling process in, 352–353
  Sorption–Desorption Test (SDT) in, 242–243
  transepidermal water loss (TEWL) in, 68, 164–165
Psychological aspects, barrier repair and, 202–203, 204t
Pulse-echo principle, 77
Pyrrolidone carboxylic acid (PCA), 296–297

Q

Quaternium-18 bentonite (Q18B), 376

R

Race, transepidermal water loss (TEWL) measurement and, 89
Racial differences. *see also* Asians; Blacks; Whites
  barrier repair and, 205

biophysical parameters and, 130–133
human surface pH and, 31
Raman spectroscopy, 17
Real part index, 343, 344t
Recessive X-linked ichthyosis (RXLI), 364
Repair, of barrier function, 198–199
Repetitive irritation test (RIT), 376–377
Reticular distances, 17
Rigid lamellar phase, 8
RIT. *see* Repetitive irritation test (RIT)
RXLI. *see* Recessive X-linked ichthyosis (RXLI)

S

SC. *see* Stratum corneum (SC)
Scaling, 301
Scoring systems, 301, 301t
SDS. *see* Sodium dodecyl sulfate (SDS)
SDT. *see* Sorption–Desorption Test (SDT)
Seborrheic dermatitis, clinical study about, 231
Sebum, 98
Secretory phospholipase A_2 pathway (sPLA$_2$), 30
Self-assessed sensation, 136–137. *see also* Burning; Dry skin (DS); Stinging
Senile xerosis, 370
Sensitive skin
cosmetics and, 135
definition of, 187–188
epidemiology of, 135
in men, 135
sensory irritation component of, 188–189
in women, 135
Sensory irritation
sensitive skin syndrome and, 188–189
studies correlating skin irritation and, 137t
subjective, 135–136
transepidermal water loss (TEWL) and, 138–139, 139t
Sensory irritation syndrome, 136, 188–189
Servo Med Evaporimeter, 276, 280
Sex
electrical response to measurements and, 290
transepidermal water loss (TEWL) measurement and, 89
Shaving process, positive effect on, 70
Short-term SLS irritation model, 225t
Single gel phase model, 9
6-hydroxysphinogosines, 6
Skicon
for measuring SC water content, 288
for measuring skin surface hydration, 353–354
NOVA DPM and, 271
readings of, 289
Skin. *see also* Skin irritation
baseline barrier function and, 106
biophysical properties of, 130–132
damaged, 68, 214–215, 215f
delipidized, 243–244
detergent-damaged, 362
in different races, 129–130
diseased. *see* Diseased skin
diseases. *see* Diseases
dry. *see* Dry skin (DS)
eczematous. *see* Eczema
ethnic differences and, 67
factors in measuring properties of, 391t
hyperactivity of, 106–107
impedance spectroscopy of, 340–346
increase in hydration, 70
index patterns for reactions, 344t
noninvasive methodology in safety testing of cosmetics on, 71
oil-in-water cream and, 220
protection in occupational settings, 70
reduction of irritative reactions in, 70
regional variations in, 67
sensitive. *see* Sensitive skin
sites, barrier repair and, 203
solvent-treated, 362–363
sonographic images of, 78f
susceptibility to irritation, 179f
transepidermal water loss (TEWL) measurement and diseases of, 90
ultraviolet-exposed, 362–363
water barrier function of, 353f
Skin, diseases, experimental studies on, 221
Skin, dry. *see* Dry skin (DS)
Skin, facial. *see* Facial skin
Skin, infant. *see* Infant skin
Skin, sensitive. *see* Sensitive skin
Skin barrier
disruption of, 198
domain mosaic model of, 198
skin diseases and, 347
Skin barrier properties, transepidermal water loss (TEWL) measurements and, 144–145
Skin Bioengineering: Techniques and Applications in Dermatology and Cosmetology, 337
Skin cancer, impedance spectrometer on, 347
Skin care products, improvement of moisturizing properties of, 70
Skin cleansing, transepidermal water loss (TEWL) measurement and, 88–89
Skin hydration. *see also* Transepidermal water loss (TEWL); Water

atopic dermatitis and, 323–329
creams for, 296, 297
definition of, 213
effect on healthy skin, 214
electrical assessments of, 287–291
of infants. *see* Infant skin
lipids and, 390
measurement of, 101, 216
moisturizers and, 295–301
of newborn infant. *see* Newborns
in normal stratum corneum, 360–361
in psoriasis, 351–357
short-term effects of emollients on, 391–393
in stratum corneum, 326–327
in stratum corneum from diseased skin, 363–365
values, epidermal echogenicity and, 79f
Skin irritation. *see also* Irritant contact dermatitis (ICD)
age and, 152f
and allergic contact dermatitis (ACD), 107–108
assessment of, 102
chemicals and, 221–222
cleansing products and, 69
concomitant atopic dermatitis and, 118–119
cumulative, 223–228, 224f
dryness as sign of, 135
elderly and, 149–150
erythema as sign of, 135
genetic factors and, 119
irritants and personal habits and, 120
Moisture Accumulation Test (MAT) and, 239
mucosal atopy and, 119
physical, 222–223
Plastic Occlusion Stress Test (POST) and, 241–242
preexposure barrier function and, 117–118
racial differences and, 132
studies correlating sensory irritation and, 137t
subclinical cutaneous irritation, 136–138
subjective/sensory, 135–136
Skin mildness, transepidermal water loss (TEWL) and, 69–70
Skin pH, in subclinical cutaneous irritation, 137
Skin profiles parameters, 395
Skin surface
measurement of hydration, 353–355
sweating, 68
temperature, 68, 90
topography, 394–396
water at, 361
Skin surface water loss (SSWL), 240
Skin thickness, in subclinical cutaneous irritation, 137
SLS. *see* Sodium lauryl sulfate (SLS)
Soap effect, 114
Soaps, 296, 362
Sodium, role in barrier homeostasis, 40–41
Sodium dodecyl sulfate (SDS), 10
Sodium lauryl sulfate (SLS)
atopic dermatitis and, 163
exposure to, 114
irritation and, 132
reactions of, 115
repetitive washing model, 377
skin hypersensitivity and, 106, 107f
Soles, thickness of, 144
Solid or gel phase, 8
Solvents
and disruption of the barrier, 198
dry skin and exposure to, 174t, 175
Solvent-treated skin, 362
Sorption–Desorption Test (SDT)
in delipidized skin, 243–244
description of, 242
in infant skin, 243
for measuring water retention in the stratum corneum, 238
in psoriatic skin, 242–243
of stratum corneum function, 267–268
curves, 268f
Sorption energy value, 20–21
Sorption isotherm
binding energy with the protein, 20f
influence of lipids and hydrosoluble on, 22
of keratins, 18–19
Spacing parameters, skin profiles and, 395
Spectroscopic technique, 326
Sphingosines, 6
SPLA$_2$. *see* Secretory phospholipase A$_2$ pathway (sPLA$_2$)
SSWL. *see* Skin surface water loss (SSWL)
Stable cubic phase membranes, 9
Staphylococcus aureus
occlusion and, 54
pH and, 35
Static measurements, NOVA Dermal Phase Meter, 266
Statistical thermodynamic methods (BET), 21, 22
Stingers
attributes of, 138
barrier function of, 138
definition of, 136
sebum content of the skin of, 194
skin hydration of, 194
vs. nonstingers, 190–191
Stinging
for assessing skin permeability, 176
chemically induced, factors determining, 189

facial skin and, 189
sensory irritation and, 135, 136
Stratum corneum lipid liposomes
commercial applications of, 10–11
experimental applications of, 10
Stratum corneum (SC)
acidification, consequences of, 28f
anatomical variability of hydration of, 149f
barrier function of, 98–99, 315
in blacks, 129–130
components of, 197–198
constituents of, 98–99
Corneometer for hydration measurement of, 250
cystine content of proteins in, 18t
damaged, hydration and lipids in, 362–363
evaluating water content of, 216
evaluating water-retaining capacity of, 396
and hair, 17–18, 23–24
hydration from diseased skin, 363–365
hydration of, 34, 326–327
lipids, 7–9, 17
lipids from diseased skin, 363–365
mechanisms leading to acidification, 29–30
normal, hydration and lipids in, 360–361
pH of, 28–41
as a physical barrier, 144
rate of transepidermal water loss (TEWL), 325–326
sorption isotherms for, 19f
techniques for measuring water content in, 288
ultrasound and water in, 77–80
water and, 49–50, 213–214, 237–238
water-binding capacity in, 327–328
Stratum granulosum, 98
Stress
barrier repair and, 202–203
transepidermal water loss (TEWL) values and, 68–69
Studies
clinical, about transepidermal water loss (TEWL) and SC hydration, 231
experimental, on diseased skin, 221
human volunteer, 400–402
Subclinical cutaneous irritation, 136–138
Subjective/sensory irritation, 135–136
Sulfur fractions, amino acid analyses of, 18t
Surface plane, instruments and measurement of, 85–86
Sweat glands, occlusion and, 51
Sweating
electrical response to measurements and, 291
skin surface temperature and, 68
transepidermal water loss (TEWL) and, 90, 99

T

Tape-stripped skin, 222–223, 223f
Tape stripping
for compromised SC barrier, 316–317
and disruption of the barrier function, 198
effects of occlusion and, 50, 51
Temperature
ambient air, 88, 289–290
instruments and, 86
low, dry skin and, 174t, 175
probe, 66
skin surface, 68, 90, 291
transepidermal water loss (TEWL) values and, 66
TEWA-meter, 85
Tewameter®, 64, 65, 66, 85
TEWL. *see* Transepidermal water loss (TEWL)
Therapeutics, evaluation of efficacy of, 356–357
Thermal analysis, 21
Thermal techniques, 326
Thighs, 148f, 149f, 151f, 290
Tightly bound water, 20–21, 288, 296
Tightness, sensory irritation and, 135
Tissue, major dispersions of, 338f
TNF. *see* Tumor necrosis factor (TNF)
Toluene, 227
Topical agents, and transepidermal water loss (TEWL), 109–110
Topical skin protectant (TSP), 378
Transepidermal water loss (TEWL). *see also* Skin hydration
aging and, 145–147, 146t, 147f, 151
applications of measurements, 69–71
in atopic dermatitis, 160–163
basal atopic dermatitis and, 370
as biophysical method, 391–394
in congenital keratinization disorders, 165–166
definition of, 97–98
in different races, 130
in dry skin, 392f
environment and, 87–89
and exogenous chemicals, 191–192
hydration and, 214
in hyperhydrated skin, 392f
individuals and, 89–91
in infants less than 26 weeks, 314f
instruments and, 85
of irritated sites treated by LAA, 230f

as a measure for barrier function, 100
measurement of
experimental, 216
overview of, 99–100
skin surface hydration and, 353–355
techniques in, 64–65
under-state conditions, 281–282
using dual-probe instruments, 283–284
variables affecting, 66–69
moisturizers and, 214, 217f, 218f, 219f, 391–394
patch tests and, 108–109
and percutaneous penetration, 145f
as predictor for the development of irritant dermatitis in atopics, 163
in psoriasis, 164–165
racial differences and, 132
sensory irritation and, 138–139, 139t
skin barrier properties and measurements of, 144–145
stratum corneum and rate of, 325–326
in subclinical cutaneous irritation, 137
as a tool to determine disease activity, 101
and topical agents, 109–110
use in determining the irritant potential of chemicals, 100–101
values at patch sites with nickel sulfate, 109t
Transient Thermal Transfer (TTT), 288
Trans-urocanic acid (tUCA), 29–30
TSP. *see* Topical skin protectant (TSP)
TTT. *see* Transient Thermal Transfer (TTT)
TUCA. *see* Trans-urocanic acid (tUCA)
Tumor necrosis factor (TNF), 114
24-hour occlusive patch test, 115
20 MHz B-scanning, 77

U

Ultrasound
irritant skin and, 79–80
and water in stratum corneum, 77–80
Ultraviolet absorbing molecules, 199t
Ultraviolet damage, protective effects against, 70
Ultraviolet-exposed skin, 362–363
Ultraviolet light (UV), 362
Unventilated chambers, for measuring transepidermal water loss (TEWL), 99–100
Urea, 220
Urea-containing moisturizers, 179, 394
Uremic itch, xerosis and, 371
UV. *see* Ultraviolet light (UV)

V

Vapometer device, 65, 85
Vapor Pressure Gradient Estimation Method, 276–278
Vasodilatation, for assessing skin permeability, 176
Ventilated chambers method, 84, 99–100
Vernix caseosa, role of, 317–318
Visual scoring (VS), 115, 379
Volar forearms. *see* Forearms
VS. *see* Visual scoring (VS)

W

Wash test, 116
Water. *see also* Skin hydration
bound, 21
within the corneocytes, 360–361
free, 21–22
interaction of keratin and, 15–24
in the intercellular space, 360
physical properties of keratin and, 16, 16f
role of, epidermal effects and, 49–50
in stratum corneum (SC), 213–214, 237–238
tightly bound, 20–21
types of, 288
and ultrasound in stratum corneum, 77–80
Water barrier function
of ambient air, 352
dynamics of, 355
of the skin, 253f
Water-binding capacity, in the stratum corneum, 327–328
Water content
Corneometer and measuring, 288
Dia-Stron DTM Torque Meter and measuring, 288
Fourier-transform infrared spectroscopy (FTIR) and measuring, 288
magnetic resonance imaging (MRI) and measuring, 288
Skicon 200 and measuring, 288
in stratum corneum, 216, 288, 391–394, 396–399
Water evaporation (WE), 86
Water gradient, 359–360
Water-holding capacity (WHC), 238
Water-in-oil (W/O) emulsion, 217, 297
Water-retaining treatments, 389–390
Water retention, occlusion and, 56f
Water-soluble material (WSM), 22
Water sorption/desorption test, 267–268
Water vapor mantel, of ambient air, 352

WE. *see* Water evaporation (WE)
WHC. *see* Water-holding capacity (WHC)
Whealing and flare, for assessing skin permeability, 176
Whites
  exposure to sodium lauryl sulfate (SLS), 132
  skin irritation in, 132
  stratum corneum in, 129
  transepidermal water loss (TEWL) in, 130
  in vitro transepidermal water loss (TEWL) values in, 131
Women
  percutaneous absorption of hydrocortisone in, 148
  sensitive skin in, 135
Wool, 16, 20
Wound healing
  occlusion and, 53–54
  occlusive materials and, 52
WSM. *see* Water-soluble material (WSM)

X

Xenobiotic-metabolizing enzymes, 199t
Xerosis. *see also* Dry skin (DS)
  in atopic dermatitis, 325
  causes of, 328
  characteristics of skin, 369
  human immunodeficiency virus (HIV) and, 371
  moisturizers and, 176, 177, 178t, 299t
  pathophysiology of itch and, 372–373
  senile, 370
  of skin and itch, 373t
  and uremic itch, 371
Xerotic skin. *see* Xerosis
X-linked recessive ichthyosis, 175
X-ray diffraction, 19

Y

Y-aminobutyric acid, 204

Z

Zero drift, instruments and, 86
Zeroing, instruments and, 85